Lecture Notes in Computer Science   4598

Commenced Publication in 1973
Founding and Former Series Editors:
Gerhard Goos, Juris Hartmanis, and Jan van Leeuv

Guohui Lin (Ed.)

Computing and Combinatorics

13th Annual International Conference, COCOON 2007
Banff, Canada, July 16-19, 2007
Proceedings

 Springer

Volume Editor

Guohui Lin
Algorithmic Research Group and Bioinformatics Research Group
Department of Computing Science
University of Alberta
Edmonton, Alberta T6G 2E8, Canada
E-mail: ghlin@cs.ualberta.ca

Library of Congress Control Number: Applied for

CR Subject Classification (1998): F.2, G.2, I.3.5, C.2.3-4, E.1, E.5, E.4

LNCS Sublibrary: SL 1 – Theoretical Computer Science and General Issues

ISSN   0302-9743
ISBN-10   3-540-73544-5 Springer Berlin Heidelberg New York
ISBN-13   978-3-540-73544-1 Springer Berlin Heidelberg New York

Springer is a part of Springer Science+Business Media

springer.com

© Springer-Verlag Berlin Heidelberg 2007
Printed in Germany

Typesetting: Camera-ready by author, data conversion by Scientific Publishing Services, Chennai, India
Printed on acid-free paper   SPIN: 12088638   06/3180   5 4 3 2 1 0

Preface

The Annual International Computing and Combinatorics Conference is an annual forum for exploring research, development, and novel applications of computing and combinatorics. It brings together researchers, professionals and industrial practitioners to interact and exchange knowledge, ideas and progress. The topics cover most aspects of theoretical computer science and combinatorics related to computing. The 13th Annual International Computing and Combinatorics Conference (COCOON 2007) was held in Banff, Alberta during July 16–19, 2007. This was the first time that COCOON was held in Canada.

We received 165 submissions, among which 11 were withdrawn for various reasons. The remaining 154 submissions under full consideration came from 33 countries and regions: Australia, Brazil, Canada, China, the Czech Republic, Denmark, Finland, France, Germany, Greece, Hong Kong, India, Iran, Ireland, Israel, Italy, Japan, the Netherlands, Norway, Pakistan, Poland, Romania, Russia, Slovakia, South Korea, Spain, Sweden, Switzerland, Taiwan, Turkey, the UK, the USA, and the US minor outlying islands.

After a six week period of careful reviewing and discussions, the program committee accepted 51 submissions for oral presentation at the conference. Based on the affiliations, 1.08 of the accepted papers were from Australia, 7.67 from Canada, 3.08 from China, 1 from the Czech Republic, 2 from Denmark, 1 from France, 5.42 from Germany, 0.08 from Greece, 2.18 from Hong Kong, 0.33 from India, 0.17 from Ireland, 1.83 from Israel, 1.5 from Italy, 2.9 from Japan, 0.17 from the Netherlands, 2.67 from Norway, 0.5 from Poland, 1 from Switzerland, 1 from Taiwan, 0.08 from Turkey, 1.33 from the UK, 12.33 from the USA, and 0.33 from the US minor outlying islands. The program of COCOON 2007 also included three keynote talks by Srinivas Aluru, Francis Y. L. Chin, and Ming Li.

Finally, we would like to express our gratitude to the authors of all submissions, the members of the program committee and the external reviewers, the members of the organizing committee, the keynote speakers, our generous sponsors, and the supporting organizations for making COCOON 2007 possible and enjoyable.

July 2007   Guohui Lin

Organization

COCOON 2007 was sponsored by the Department of Computing Science, the University of Alberta, and the Informatics Circle of Research Excellence (iCORE).

Program Committee Chair

Guohui Lin (University of Alberta, Canada)

Program Committee Members

Therese Biedl (University of Waterloo, Canada)
Andreas Brandstädt (Universität Rostock, Germany)
Zhi-Zhong Chen (Tokyo Denki University, Japan)
Ovidiu Daescu (University of Texas at Dallas, USA)
Donglei Du (University of New Brunswick, Canada)
Patricia Evans (University of New Brunswick, Canada)
Yong Gao (University of British Columbia - Okanagan, Canada)
Raffaele Giancarlo (University of Palermo, Italy)
Wen-Lian Hsu (Academia Sinica, Taiwan)
Xiao-Dong Hu (Chinese Academy of Sciences, China)
Ming-Yang Kao (Northwestern University, USA)
Naoki Katoh (Kyoto University, Japan)
Valerie King (University of Victoria, Canada)
Michael A. Langston (University of Tennessee, USA)
Kim Skak Larsen (University of Southern Denmark, Denmark)
Bin Ma (University of Western Ontario, Canada)
Rajeev Motwani (Stanford University, USA)
Mohammad R. Salavatipour (University of Alberta, Canada)
David Sankoff (University of Ottawa, Canada)
Yaoyun Shi (University of Michigan, USA)
Zhiyi Tan (Zhejiang University, China)
Caoan Wang (Memorial University of Newfoundland, Canada)
Lusheng Wang (City University of Hong Kong, China)
Osamu Watanabe (Tokyo Institute of Technology, Japan)
Dong Xu (University of Missouri - Columbia, USA)
Jinhui Xu (State University of New York at Buffalo, USA)
Alexander Zelikovsky (Georgia State University, USA)
Huaming Zhang (University of Alabama in Huntsville, USA)
Kaizhong Zhang (University of Western Ontario, Canada)
Xizhong Zheng (BTU Cottbus, Germany)
Yunhong Zhou (HP, USA)
Binhai Zhu (Montana State University, USA)

Organizing Committee

Guohui Lin (Chair; University of Alberta, Canada)
Zhipeng Cai (University of Alberta, Canada)
Yi Shi (University of Alberta, Canada)
Meng Song (University of Alberta, Canada)
Jianjun Zhou (University of Alberta, Canada)

External Reviewers

Irina Astrovskaya
Dumitru Brinza
Zhipeng Cai
Shihyen Chen
Qiong Cheng
Arthur Chou
Ye Du
Martin R. Ehmsen
Konrad Engel
Tomas Feder
Fedor Fomin
Lance Fortnow
Loukas Georgiadis
Stephan Gremalschi
MohammadTaghi Hajiaghayi
Qiaoming Han
Jan Johannsen
Valentine Kabanets
George Karakostas
Ker-I Ko
Ekkehard Koehler
Guy Kortsarz
Lap Chi Lau
Hanno Lefmann
Weiming Li
Xueping Li
Xiaowen Liu
Jingping Liu
Shuang Luan
Elvira Mayordomo
Daniele Micciancio

Lopamudra Mukherjee
Shubha Nabar
Javier Pena
Xiaotong Qi
Tim Roughgarden
Adrian Rusu
Amin Saberi
Yi Shi
Amir Shpilka
Vikas Singh
Ram Swaminathan
Dilys Thomas
Iannis Tourlakis
Sergei Vassilvitskii
Jacques Verstraete
Peter Wagner
Duncan Wang
Kelly Westbrooks
Avi Wigderson
Gang Wu
Xiaodong Wu
Lei Xin
Dachuan Xu
Ying Xu
Jin Yan
Yang Yang
Guochuan Zhang
Li Zhang
Jianjun Zhou
Luis Zuluaga

Table of Contents

The Combinatorics of Sequencing the Corn Genome

Srinivas Aluru

Department of Electrical and Computer Engineering, Iowa State University
aluru@iastate.edu

Abstract. The scientific community is engaged in an ongoing, concerted effort to sequence the corn (also known as maize) genome. This genome is approximately 2.5 billion nucleotides long with an estimated 65-80team of university and private laboratory researchers under the auspices of NSF/USDA/DOE is working towards deciphering the majority of the sequence information including all genes, determining their order and orientation, and anchoring them to genetic/physical maps. In this talk, I will present some of the combinatorial problems that arise in this context and outline the role of graph, string and parallel algorithms in solving them.

G. Lin (Ed.): COCOON 2007, LNCS 4598, p. 1, 2007.
© Springer-Verlag Berlin Heidelberg 2007

Online Frequency Assignment in Wireless Communication Networks

Francis Y.L. Chin

Department of Computer Science,
The University of Hong Kong, Hong Kong
chin@cs.hku.hk

Abstract. Wireless communication has many applications since its invention more than a century ago. The frequency spectrum used for communication is a scarce resource and the Frequency Assignment Problem (FAP), aiming for better utilization of the frequencies, has been extensively studied in the past 20-30 years. Because of the rapid development of new wireless applications such as digital cellular network, cellular phone, the FAP problem has become more important.

In Frequency Division Multiplexing (FDM) networks, a geographic area is divided into small cellular regions or cells, usually regular hexagons in shape. Each cell contains one base station that communicates with other base stations via a high-speed wired network. Calls between any two clients (even within the same cell) must be established through base stations. When a call arrives, the nearest base station must assign a frequency from the available spectrum to the call without causing any interference with other calls. Interference may occur, which distorts the radio signals, when the same frequency is assigned to two different calls emanating from cells that are geographically close to each other. Thus the FAP problem can be viewed as a problem of multi-coloring a hexagon graph with the minimum number of colors when each vertex of the graph is associated with an integer that represents the number of calls in a cell.

FAP has attracted more attention recently because of the following:

a) Online analysis techniques: FAP problem is known to be NP-complete and many approximation algorithms have been proposed in the past. As frequency assignments have to be done without knowledge of future call requests and releases, online algorithms have been proposed and competitive analysis has been used to measure their performance.
b) New technology and application: Wideband Code-Division Multiple-Access (W-CDMA) technology is a new technology used for the implementation of 3G cellular system. Orthogonal Variable Spreading Factor (OVSF) codes are used to satisfy requests with different data rate requirements. FAP with OVSF code trees representing the frequency spectrum becomes an important problem.

G. Lin (Ed.): COCOON 2007, LNCS 4598, p. 2, 2007.
© Springer-Verlag Berlin Heidelberg 2007

Information Distance from a Question to an Answer

Ming Li

School of Computer Science, University of Waterloo, Waterloo,
Ontario N2L 3G1 Canada
mli@uwaterloo.ca

Abstract. We know how to measure distance from Beijing to Toronto.
However, do you know how to measure the distance between two information carrying entities? For example: two genomes, two music scores,
two programs, two articles, two emails, or from a question to an answer?
Furthermore, such a distance measure must be application-independent,
must be universal in the sense it is provably better than all other distances, and must be applicable.

From a simple and accepted assumption in thermodynamics, we have
developed such a theory. I will present this theory and will present one
of the new applications of this theory: a question answering system.

A New Field Splitting Algorithm for Intensity-Modulated Radiation Therapy*

Danny Z. Chen[1], Mark A. Healy[1], Chao Wang[1,**], and Xiaodong Wu[2,***]

[1] Department of Computer Science and Engineering
University of Notre Dame
Notre Dame, IN 46556, USA
{chen,mhealy4,cwang1}@cse.nd.edu
[2] Department of Electrical and Computer Engineering
Department of Radiation Oncology
University of Iowa
Iowa City, Iowa 52242, USA
xiaodong-wu@uiowa.edu

Abstract. In this paper, we present an almost linear time algorithm for the problem of splitting an intensity map of radiation (represented as an integer matrix) into multiple subfields (submatrices), subject to a given maximum allowable subfield width, to minimize the total delivery error caused by the splitting. This problem arises in intensity-modulated radiation therapy (IMRT) for cancer treatments. This is the first field splitting result on minimizing the total delivery error of the splitting. Our solution models the problem as a shortest path problem on a directed layered graph, which satisfies the staircase Monge property. Consequently, the resulting algorithm runs in almost linear time and generates an optimal quality field splitting.

1 Introduction

In this paper, we study a geometric partition problem, called *field splitting*, which arises in intensity-modulated radiation therapy (IMRT). IMRT is a modern cancer treatment technique that aims to deliver highly conformal prescribed radiation dose distributions, called *intensity maps* (IMs), to target tumors while sparing the surrounding normal tissues and critical structures. The effectiveness of IMRT hinges on its ability to accurately and efficiently deliver the prescribed

* This research was supported in part by the National Science Foundation under Grant CCF-0515203 and NIH NIBIB Grant R01-EB004640-01A2.
** Corresponding author. The research of this author was supported in part by two Fellowships in 2004-2006 from the Center for Applied Mathematics of the University of Notre Dame.
*** The research of this author was supported in part by a faculty start-up fund from the University of Iowa and in part by a seed grant award from the American Cancer Society through an Institutional Research Grant to the Holden Comprehensive Cancer Center, the University of Iowa, Iowa City, Iowa, USA.

IMs. An IM is a dose prescription specified by a set of nonnegative integers on a uniform 2-D grid (see Figure 1(b)) with respect to an orientation in the 3-D space. The value in each grid cell indicates the intensity level of prescribed radiation at the body region corresponding to that IM cell. The delivery of an IM is carried out by a set of cylindrical radiation beams orthogonal to the IM grid.

One of the current most advanced control tools for IM delivery is the *multileaf collimator* (MLC) [15]. An MLC consists of a fixed number of pairs of tungsten alloy leaves of the same rectangular shape and size (see Figure 1(a)). The opposite leaves of each pair are aligned to each other, and can move (say) up or down to form an *x*-monotone rectilinear polygonal beam-shaping region. The cross-section of a cylindrical radiation beam is shaped by such a region. In delivering a radiation beam for an IM, all IM cells exposed under the beam receive a uniform radiation dose proportional to the exposure time of the beam. The mechanical design of the MLCs restricts what kinds of beam-shaping regions are allowed [15]. A common constraint is called the **maximum field size**: Due to the limitation on the fixed number of MLC leave pairs and the overtravel distance of the leaves, an MLC cannot enclose an IM of a too large size (called the *field size*).

Two key criteria are used to measure the quality of an IMRT treatment. (1) The **treatment time** (efficiency): Minimizing the treatment time is crucial since it not only lowers the treatment cost for each patient but also increases the patient throughput of the hospitals; in addition, it reduces the risk associated with the uncertainties in the treatment. (2) The **delivery error** (accuracy): Due to the special geometric shapes of the MLC leaves [14,15] (i.e., the "tongue-and-groove" interlock feature), an MLC-aperture cannot be delivered perfectly. Instead, there is a *delivery error* between the planned dose and actual delivered dose [14] (called the *"tongue-and-groove"* error in medical literature [15]). Minimizing the delivery error is important because according to a recent study [4], the maximum delivery error can be up to 10%, creating underdose/overdose spots in the target region.

The limited size of the MLC necessitates that a large-size IM field be split into two or more adjacent subfields, each of which can be delivered separately by the MLC subject to the maximum field size constraint [5,9,16]. But, such IM splitting may result in prolonged treatment time and increased delivery error, and thus affect the treatment quality. The **field splitting problem**, roughly speaking, is to split an IM of a large size into multiple subfields whose sizes are all no bigger than a threshold size, such that the treatment quality is optimized.

A few field splitting algorithms are known in the literature [3,10,12,17], which address various versions of the field splitting problem. However, all these field splitting solutions focused on minimizing the *beam-on time* while ignoring the issue of reducing the delivery error. The beam-on time of a treatment is the time while a patient is exposed to actual irradiation [15], which is closely related to the total treatment time. In fact, when splitting an IM into multiple subfields to minimize the total beam-on time, the splitting is along the direction that is perpendicular to the direction of field splitting that aims to minimize the total

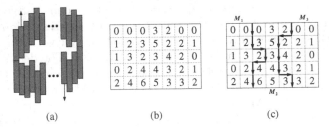

0	0	0	3	2	0	0
1	2	3	5	2	2	1
1	3	2	3	4	2	0
0	2	4	4	3	2	1
2	4	6	5	3	3	2

(a)   (b)   (c)

Fig. 1. (a) An MLC. (b) An IM. (c) An example of splitting an IM into three subfields, M_1, M_2, and M_3, using y-monotone paths.

delivery error. For example, in Figure 1(c), the splitting of the IM is along the x-axis (using y-monotone paths) and aims to minimize the total delivery error; a corresponding splitting for minimizing the total beam-on time would be along the y-axis (say, using x-monotone paths). Thus, these two optimization criteria for field splitting, i.e., minimizing the total beam-on time and minimizing the total delivery error, are geometrically complementary to each other, and together provide a somewhat complete solution to the field splitting problem (for example, we may first split an IM along the y-axis, minimizing the total beam-on time, and then split each resulting subfield of the IM along the x-axis, minimizing its delivery error). Although it is useful to consider field splitting to minimize the delivery error, to our best knowledge, no field splitting algorithms are known aiming to minimize the total delivery error.

Several papers (e.g., [2,11,14]) have discussed how to minimize the delivery error of IMRT treatments (assuming no field splitting is needed). Chen *et al.* [2] showed that for an IM M of size $m \times n$ (no larger than the maximum allowable field size $l \times w$), the minimum amount of error for delivering M is captured by the following formula (note that M contains only nonnegative integers):

$$Err(M) = \sum_{i=1}^{m}(M_{i,1} + \sum_{j=1}^{n-1}|M_{i,j} - M_{i,j+1}| + M_{i,n}). \qquad (1)$$

They also gave an algorithm for achieving this minimum error [2]. Geometrically, if we view a row of an IM as representing an x-monotone rectilinear curve f, called the *dose profile curve* (see Figure 2(a)), then the (minimum) delivery error associated with this IM row is actually the total sum of the lengths of all vertical edges on f.

In this paper, we consider the following **field splitting using y-monotone paths (FSMP) problem**: Given an IM M of size $m \times n$ and a maximum allowable field size $l \times w$, with $m \leq l$ and $n > w$, split M using y-monotone paths into $d = \lceil \frac{n}{w} \rceil$ (≥ 2) subfields $M_1, M_2, \ldots, M_d$, each with a size no larger than $l \times w$, such that the total delivery error of these d subfields is minimized (e.g., see Figures 2(b)-2(c)). Here, d is the minimum number of subfields required to deliver M subject to the maximum allowable field size $l \times w$.

We present the first efficient algorithm for the above FSMP problem. Our algorithm runs in almost linear time. In our approach, we model the FSMP

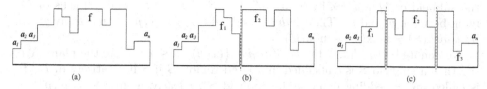

Fig. 2. (a) The dose profile curve f of one row of an IM. The (minimum) delivery error, $Err(f)$, of the row is equal to the sum of the lengths of all vertical edges on the curve f. (b) Splitting the one-row IM in (a) into two subfields. Note that $Err(f_1) + Err(f_2) \geq Err(f)$ always holds. (c) Splitting the one-row IM in (a) into three subfields. $Err(f_1) + Err(f_2) + Err(f_3) \geq Err(f)$ always holds.

problem as a shortest path problem in a directed acyclic graph (DAG) of $O(n)$ vertices and $O(nw)$ edges. Interestingly, the edge weights of this DAG satisfy the staircase Monge property [1,13] and the edges of the graph can be represented implicitly. Thus, we can solve this shortest path problem by examining only a very small portion of the edges of the graph. One frequent operation in this shortest path algorithm is to compute the weights of the examined edges, which takes $O(mw)$ time if directly using a formula (see Section 3.3.1). We improve the time complexity of this operation to $O(m)$ by using a range-minima data structure [8]. Our final FSMP algorithm runs in $O(mn\alpha(w))$ time, where $\alpha(\cdot)$ is the inverse Ackermann's function.

Our techniques have other applications and extensions. Our FSMP algorithm can be easily adapted to the field splitting scenarios in which we seek to minimize the total horizontal or vertical complexity of the subfields [6] (all we need to do is to redefine the weights of the edges of the DAG). These criteria are believed to be closely related to the beam-on time of the treatment [6].

The rest of the paper is organized as follows. Section 2 gives some observations and notation. Section 3 presents our FSMP algorithm. Section 4 extends our solution to a related intensity map partition problem to minimize the total horizontal or vertical complexity of the resulting subfields. Section 5 shows some implementation and experimental results.

2   Preliminaries

For an IM M of size $m \times n$, we can encode any y-monotone path p on M as $(p(1), p(2), \ldots, p(m)) \in [0, n]^m$, where $p(i)$ is the position at which p crosses the i-th row of M (i.e., p crosses the i-th row between the $(p(i))$-th and $(p(i) + 1)$-th elements). There are two special y-monotone paths of M: $\mathbf{0} = (0, 0, \ldots, 0)$ and $\mathbf{n} = (n, n, \ldots, n)$, which are the leftmost and rightmost paths, respectively. We assume that all paths in this paper are y-monotone on M. Define $\min(p) = \min_{i=1}^{m}\{p(i)\}$ and $\max(p) = \max_{i=1}^{m}\{p(i)\}$, i.e., $\min(p)$ and $\max(p)$ can be viewed as the smallest and largest column indices on a path p, respectively.

For two paths p' and p'', we say that $p' \leq p''$ if $p'(k) \leq p''(k)$ holds for every k, i.e., p' lies entirely on or to the left of p''. If $p' \leq p''$, we denote by $S(p', p'')$

the subfield of M *induced* by p' and p'', i.e., $S(p',p'')$ consists of all cells of M lying between p' and p''. The *width* of a subfield $S = S(p',p'')$ is naturally given by $width(S) = \max(p'') - \min(p')$.

We denote by $band(a,b)$ the *band region* $\{(x,y) \mid a \le x \le b\}$ in the plane. We say that a subfield S is embedded in a band region B if S lies entirely in B (B is called an embedding band of the subfield S). Clearly, a subfield $S(p',p'')$ is embedded in $band(a,b)$ if and only if $a \le \min(p') \le \max(p'') \le b$.

Given an IM M of size $m \times n$ and the maximum allowable field size $l \times w$ (with $m \le l$ and $n > w$), a y-monotone path set $\mathcal{T} = \{p_1, p_2, \ldots, p_{d-1}\}$ is said to form a *feasible splitting* of M if (1) $\mathbf{0} \le p_1 \le p_2 \le \cdots \le p_{d-1} \le \mathbf{n}$, and (2) $width(S(p_{k-1}, p_k)) \le w$ for every $k = 1, 2, \ldots, d$ (with $p_0 = \mathbf{0}$ and $p_d = \mathbf{n}$).

It is clear that for any feasible field splitting $\{p_1, p_2, \ldots, p_{d-1}\}$, the following properties hold:

(i) For every $k = 1, 2, \ldots, d-1$, $\min(p_k) \in [kw - \Delta, kw]$, where $\Delta = w\lceil n/w\rceil - n$ ($0 \le \Delta \le w - 1$).

(ii) $S(\mathbf{0}, p_1)$ (resp., $S(p_{d-1}, \mathbf{n})$) is embedded in $band(0, w)$ (resp., $band(n-w, n)$); for every $k = 2, 3, \ldots, d-2$, $S(p_k, p_{k+1})$ is embedded in at least one of the following bands: $band(g, g+w)$, with $g = kw - \Delta, kw - \Delta + 1, \ldots, kw$.

(iii) If $S(p_{k-1}, p_k)$ and $S(p_k, p_{k+1})$ are embedded in bands B' and B'', respectively, then the path p_k must lie in the common intersection of the two bands, i.e., $p_k \subseteq B' \cap B''$.

3   Our Algorithm for the FSMP Problem

3.1   An Overview of the Algorithm

Our FSMP algorithm starts with the observation that whenever we use a y-monotone path p to split the given IM M, the total delivery error of the resulting subfields will *increase* by a value of $2 \cdot \sum_{i=1}^{m} \min\{M_{i,p(i)}, M_{i,p(i)+1}\}$, with $M_{i,0} = M_{i,n+1} = 0$ (see Figures 2(b)-2(c)). The reason is that based on the formula $Err(M)$ for summing up the delivery error of M (i.e., Formula (1) in Section 1), when without a splitting between the two adjacent cells $M_{i,p(i)}$ and $M_{i,p(i)+1}$ of M, the delivery error incurred due to these two cells is $|M_{i,p(i)} - M_{i,p(i)+1}|$, but when with a splitting between these two adjacent cells, the delivery error incurred at these two cells is $M_{i,p(i)} + M_{i,p(i)+1}$ (since after the splitting, $M_{i,p(i)}$ becomes the last cell of a row in one subfield and $M_{i,p(i)+1}$ becomes the first cell of a row in another subfield). The net increase in the delivery error from without to with a splitting between $M_{i,p(i)}$ and $M_{i,p(i)+1}$ is $2 \cdot \min\{M_{i,p(i)}, M_{i,p(i)+1}\}$ (see Figures 2(b)-2(c)). Without affecting the correctness of our FSMP algorithm, we will use $\sum_{i=1}^{m} \min\{M_{i,p(i)}, M_{i,p(i)+1}\}$ as the (increased) cost of a y-monotone path p for splitting M (by dropping the coefficient 2). Note that this cost depends only on p and M, which we call the *cost* of p and denote by $cost(p)$. Then, the FSMP problem is equivalent to finding a feasible splitting $\mathcal{T}$ of cardinality $d - 1$ that minimizes $\sum_{p \in \mathcal{T}} cost(p)$.

The above observation helps us to model the FSMP problem as a shortest path problem in a directed graph $G = (V, E)$, which is defined as follows. (1)

G consists of d layers of vertices, with each vertex corresponding to a possible band in M. Precisely, the first layer (i.e., layer 1) contains one vertex $v_{1,0}$, which corresponds to $band(0, w)$ ($\stackrel{\Delta}{=} B_{1,0}$); the last layer (i.e., layer d) contains one vertex $v_{d,0}$, which corresponds to $band(n - w, n)$ ($\stackrel{\Delta}{=} B_{d,0}$); the k-th layer ($2 \leq k \leq d - 1$) contains $\Delta + 1$ vertices, $v_{k,0}, v_{k,1}, \ldots, v_{k,\Delta}$, with $v_{k,j}$ corresponding to $band(kw - j, kw - j + w)$ ($\stackrel{\Delta}{=} B_{k,j}$), where $\Delta = w\lceil n/w \rceil - n$ ($= O(w)$). (2) For each vertex v_{k,j_k} in G, there is a directed edge from v_{k,j_k} to every vertex $v_{k+1,j_{k+1}}$ ($0 \leq j_{k+1} \leq \Delta$) in the next layer as long as $B_{k,j_k} \cap B_{k+1,j_{k+1}} \neq \emptyset$. Clearly, for each vertex v_{k,j_k}, the vertices on the $(k + 1)$-th layer to which an edge from v_{k,j_k} goes form a contiguous subsequence (called an *interval of vertices*), and the first and last vertices of this interval for v_{k,j_k} can be easily determined. (3) For any edge $e = (v_{k,j_k}, v_{k+1,j_{k+1}})$ in G, we define its *induced y-monotone path*, denoted by $p^*_{k,j_k,j_{k+1}}$, as the minimum cost path lying in $B_{k,j_k} \cap B_{k+1,j_{k+1}}$; the weight of the edge $e = (v_{k,j_k}, v_{k+1,j_{k+1}})$ is the cost of its induced y-monotone path $p^*_{k,j_k,j_{k+1}}$.

Our algorithm then computes a shortest path from $v_{1,0}$ to $v_{d,0}$ in G, which has exactly $d - 1$ edges.

Two key questions for our FSMP algorithm remain to be answered.

(1) How does a shortest $v_{1,0}$-to-$v_{d,0}$ path in G relate to an optimal field splitting of the given IM M? As we will show in Section 3.2, it turns out that the set of $d - 1$ y-monotone paths induced by the $d - 1$ edges on such a shortest path yields an optimal splitting of M.

(2) How can we efficiently compute a shortest $v_{1,0}$-to-$v_{d,0}$ path in G? It is easy to see that G is a DAG with $O(d(\Delta + 1)) = O(n)$ vertices and $O(d(\Delta + 1)^2) = O(n^2/d)$ weighted edges. Thus, in a topological sort fashion, a shortest path in G can be computed in $O(n^2\tau/d)$ time, where τ is the time for computing the induced y-monotone path for each edge of G. In Section 3.3, we will show that by exploiting the staircase Monge property of the graph G [1,13] and using a range-minima data structure [8], we can dramatically speed up the shortest path computation. Our final FSMP algorithm runs in almost linear time on M.

3.2   Correctness of the Algorithm

In this section, we show the correctness of our FSMP algorithm, i.e., a shortest $v_{1,0}$-to-$v_{d,0}$ path in the graph G defined in Section 3.1 indeed corresponds to an optimal splitting of the given IM M.

The following lemma states the fact that any $v_{1,0}$-to-$v_{d,0}$ path in G induces a feasible splitting of M (we leave the proof to the full paper).

Lemma 1. *Let $\pi = v_{1,j_1(=0)} \rightarrow v_{2,j_2} \rightarrow \cdots \rightarrow v_{d-1,j_{d-1}} \rightarrow v_{d,j_d(=0)}$ be a $v_{1,0}$-to-$v_{d,0}$ path in G. Then the set of $d - 1$ y-monotone paths induced by all edges of π, i.e., $\{p^*_{1,j_1,j_2}, p^*_{2,j_2,j_3}, \ldots, p^*_{d-1,j_{d-1},j_d}\}$, gives a feasible splitting of the IM M.*

We now show that a shortest $v_{1,0}$-to-$v_{d,0}$ path in G induces an optimal splitting of M.

Lemma 2. *Let $\pi' = v_{1,j_1'(=0)} \to v_{2,j_2'} \to \cdots \to v_{d-1,j_{d-1}'} \to v_{d,j_d'(=0)}$ be a shortest $v_{1,0}$-to-$v_{d,0}$ path in G. Then the set of $d-1$ y-monotone paths induced by all edges of π', i.e., $\{p_{1,j_1',j_2'}^*, p_{2,j_2',j_3'}^*, \ldots, p_{d-1,j_{d-1}',j_d'}^*\}$, gives an optimal splitting of the IM M.*

Proof. Recall that the FSMP problem seeks a feasible splitting $\mathcal{T}$ of cardinality $d-1$ that minimizes $\sum_{p \in \mathcal{T}} cost(p)$. To prove the lemma, it suffices to show that for any feasible splitting $\{p_1, p_2, \ldots, p_{d-1}\}$ of M, $\sum_{k=1}^{d-1} cost(p_k) \geq \sum_{k=1}^{d-1} cost(p_{k,j_k',j_{k+1}'}^*)$ holds.

Since $\{p_1, p_2, \ldots, p_{d-1}\}$ is a feasible splitting of M, by Property (ii) in Section 2, there exist $j_1, j_2, \ldots, j_{d-1} \in [0, \Delta]$, such that $S(p_k, p_{k+1})$ is embedded in the band B_{k,j_k}. For every $k = 1, 2, \ldots, d-1$, by Property (iii) in Section 2, we have $p_k \subseteq B_{k,j_k} \cap B_{k+1,j_{k+1}}$ (for convenience, we define $j_0 = j_d = 0$). Hence there is an edge connecting v_{k,j_k} and $v_{k+1,j_{k+1}}$ in G. It follows that $\pi = v_{1,j_1(=0)} \to v_{2,j_2} \to \cdots \to v_{d-1,j_{d-1}} \to v_{d,j_d(=0)}$ is a $v_{1,0}$-to-$v_{d,0}$ path in G.

Recall that $p_{k,j_k,j_{k+1}}^*$ is a minimum cost y-monotone path that lies in $B_{k,j_k} \cap B_{k+1,j_{k+1}}$. Thus we have $cost(p_k) \geq cost(p_{k,j_k,j_{k+1}}^*)$. It follows $\sum_{k=1}^{d-1} cost(p_k) \geq \sum_{k=1}^{d-1} cost(p_{k,j_k,j_{k+1}}^*)$. Observe that the right-hand side of the above inequality equals the total weight of the $v_{1,0}$-to-$v_{d,0}$ path π in G, which cannot be smaller than $\sum_{k=1}^{d-1} cost(p_{k,j_k',j_{k+1}'}^*)$, the total weight of the shortest path π'. It thus follows that $\sum_{k=1}^{d-1} cost(p_k) \geq \sum_{k=1}^{d-1} cost(p_{k,j_k',j_{k+1}'}^*)$. $\square$

3.3   Improving the Time Complexity of the Algorithm

In this section, we will focus on the more detailed aspects of our algorithm. More specifically, we will address how to efficiently compute the weights of the edges of G and a shortest $v_{1,0}$-to-$v_{d,0}$ path in G.

3.3.1 Computing the Weights of Edges of G

One frequent operation in our FSMP algorithm is to compute the weight of a given edge of G. Recall that for an edge $e = (v_{k,j_k}, v_{k+1,j_{k+1}})$ in G, its weight $w(e)$ is defined as the cost of $p_{k,j_k,j_{k+1}}^*$, which is a minimum cost y-monotone path that lies in $B_{k,j_k} \cap B_{k+1,j_{k+1}}$.

Observe that $B_{k,j_k} \cap B_{k+1,j_{k+1}} = band(kw - j_k, kw - j_k + w) \cap band((k+1)w - j_{k+1}, (k+1)w - j_{k+1} + w) = band((k+1)w - j_{k+1}, (k+1)w - j_k)$. The existence of the edge $e = (v_{k,j_k}, v_{k+1,j_{k+1}})$ in G implies $B_{k,j_k} \cap B_{k+1,j_{k+1}} \neq \emptyset$, which in turn implies $j_k \leq j_{k+1}$.

It is not difficult to show that

$$
\begin{aligned}
w(e) &= \min\{cost(p) \mid p \subseteq B_{k,j_k} \cap B_{k+1,j_{k+1}}\} \\
&= \min\{\textstyle\sum_{i=1}^m \min\{M_{i,p(i)}, M_{i,p(i)+1}\} \mid p(i) \in [(k+1)w - j_{k+1}, \\
&\quad (k+1)w - j_k], 1 \leq i \leq m\} \\
&= \textstyle\sum_{i=1}^m \min\{M_{i,j} \mid j \in [(k+1)w - j_{k+1}, (k+1)w - j_k + 1]\}
\end{aligned}
\tag{2}
$$

Since $j_k, j_{k+1} \in [0, \Delta]$, it takes $O(\Delta)$ time to compute $\min\{M_{i,j} \mid j \in [(k + 1)w - j_{k+1}, (k + 1)w - j_k + 1]\}$, and consequently $O(m\Delta) = O(mw)$ time to compute $w(e)$ if we directly use Formula (2) above.

However, since all entries of M are static (i.e., do not change their values during the computation), we can speed up the computation of an edge weight by using a range-minima data structure [8] for each row $R_i(M) = \{M_{i,j}\}_{j=1}^{n}$ of M, $i = 1, 2, \ldots, m$. Note that for an array of values, a range-minima data structure can be built in linear time, which allows each query of finding the minimum value in any interval of the array to be answered in $O(1)$ time [8]. In our algorithm, we build a range-minima data structure for each row $R_i(M)$ of M, in $O(n)$ time per row. This enables us to compute $\min\{M_{i,j} \mid j \in [(k+1)w - j_{k+1}, (k+1)w - j_k + 1]\}$ as a range-minima query, in $O(1)$ time. In this way, we can compute the weight of any edge of G in only $O(m)$ time, after an $O(mn)$ time preprocess (for building the range-minima data structures for the m rows of M).

3.3.2 Computing a Shortest $v_{1,0}$-to-$v_{d,0}$ Path in G

In this section, we will show that the graph G defined in Section 3.1 satisfies the staircase Monge property [1,13]. This property enables us to compute a shortest $v_{1,0}$-to-$v_{d,0}$ path in G in almost linear time.

Lemma 3. *Let $v_{k,j}, v_{k,j'}, v_{k+1,r}$, and $v_{k+1,r'}$ be four vertices of G, with $j < j'$, $r < r'$, and $2 \leq k < d - 1$. If $e_1 = (v_{k,j'}, v_{k+1,r})$ is an edge of G, then $e_2 = (v_{k,j}, v_{k+1,r'})$, $e_3 = (v_{k,j}, v_{k+1,r})$, and $e_4 = (v_{k,j'}, v_{k+1,r'})$ are also edges of G. Moreover, $w(e_3) + w(e_4) \leq w(e_1) + w(e_2)$.*

Proof. Since $j', r \in [0, \Delta]$, it is clear that $e_1 = (v_{k,j'}, v_{k+1,r}) \in E(G) \Leftrightarrow B_{k,j'} \cap B_{k+1,r} \neq \emptyset \Leftrightarrow band(kw - j', kw - j' + w) \cap band((k + 1)w - r, (k + 1)w - r + w) \neq \emptyset \Leftrightarrow r \geq j'$. Similarly, $e_2 \in E(G) \Leftrightarrow r' \geq j$, $e_3 \in E(G) \Leftrightarrow r \geq j$, and $e_4 \in E(G) \Leftrightarrow r' \geq j'$. Since $r < r'$ and $j < j'$, we have $e_1 \in E(G) \Rightarrow j < j' \leq r < r' \Rightarrow e_2, e_3, e_4 \in E(G)$.

To show that $w(e_3) + w(e_4) \leq w(e_1) + w(e_2)$, it suffices to show that for any $i \in [1, m]$, $\min_{j \in I_3}\{M_{i,j}\} + \min_{j \in I_4}\{M_{i,j}\} \leq \min_{j \in I_1}\{M_{i,j}\} + \min_{j \in I_2}\{M_{i,j}\}$, where $I_1 = [(k + 1)w - r, (k + 1)w - j' + 1]$, $I_2 = [(k + 1)w - r', (k + 1)w - j + 1]$, $I_3 = [(k + 1)w - r, (k + 1)w - j + 1]$, and $I_4 = [(k + 1)w - r', (k + 1)w - j' + 1]$. Since $j < j' \leq r < r'$, clearly $I_1 = I_3 \cap I_4$ and $I_2 = I_3 \cup I_4$. It follows that $\min_{j \in I_1}\{M_{i,j}\} \geq \max\{\min_{j \in I_3}\{M_{i,j}\}, \min_{j \in I_4}\{M_{i,j}\}\}$. and $\min_{j \in I_2}\{M_{i,j}\} = \min\{\min_{j \in I_3}\{M_{i,j}\}, \min_{j \in I_4}\{M_{i,j}\}\}$, Hence, we have $\min_{j \in I_3}\{M_{i,j}\} + \min_{j \in I_4}\{M_{i,j}\} = \max\{\min_{j \in I_3}\{M_{i,j}\}, \min_{j \in I_4}\{M_{i,j}\}\} + \min\{\min_{j \in I_3}\{M_{i,j}\}, \min_{j \in I_4}\{M_{i,j}\}\} \leq \min_{j \in I_1}\{M_{i,j}\} + \min_{j \in I_2}\{M_{i,j}\}$. □

Lemma 3 actually shows that the $(\Delta + 1) \times (\Delta + 1)$ matrix $W^{(k)} = (w_{j,r}^{(k)})$ ($2 \leq k < d - 1$), with $w_{j,r}^{(k)} = w((v_{k,j}, v_{k+1,r}))$ for $0 \leq j, r \leq \Delta$, is a staircase Monge matrix [1,13]. (For convenience, an entry $w_{j,r}^{(k)} = +\infty$ if $(v_{k,j}, v_{k+1,r}) \notin E(G)$.) Recall that an edge $(v_{k,j}, v_{k+1,r}) \in E(G) \Leftrightarrow r \geq j$. Hence, the matrix $W^{(k)}$ can be represented implicitly, such that each of its entry can be obtained, whenever needed, in $O(m)$ time as shown in Section 3.3.1. It is thus easy to see that, by using the staircase Monge matrix searching algorithms [1,13], given the shortest paths from $v_{1,0}$ to all vertices on the k-th layer, only $O((\Delta + 1)\alpha(\Delta + 1)) = O(w\alpha(w))$ edges need to be examined in order to find the shortest paths from $v_{1,0}$ to all vertices on the $(k + 1)$-th layer, where $\alpha(\cdot)$ is the inverse Ackermann's function. Hence, only $O(dw\alpha(w)) = O(n\alpha(w))$ edges in total are examined for computing a shortest $v_{1,0}$-to-$v_{d,0}$ path in G. Thus, we have the following result.

Theorem 1. *Given an IM M of size $m \times n$, and an maximum allowable field width w, the FSMP problem on M can be solved in $O(mn\alpha(w))$ time.*

4   An Extension

In this section, we show that our FSMP technique can be adapted to the field splitting scenarios in which we seek to minimize the total horizontal or vertical complexity of the resulting subfields [6]. It is sufficient for us to focus on solving the horizontal complexity case.

The *horizontal complexity $HC(M')$* of a subfield M' of size $l \times w$ is defined as follows $HC(M') = \sum_{i=1}^{l}(M'_{i,1} + \sum_{j=1}^{w-1} \max\{0, M'_{i,j+1} - M'_{i,j}\})$. The horizontal complexity is closely related to the minimum beam-on time for delivering the subfield M', which can be computed as $\max_{i=1}^{l}\{M'_{i,1} + \sum_{j=1}^{w-1} \max\{0, M'_{i,j+1} - M'_{i,j}\}\}$ [7]. An intensity map with a smaller horizontal complexity is likely to be delivered more efficiently (i.e., with a smaller treatment time). Thus, it is desirable to split an IM into multiple subfields while minimizing the total horizontal complexity of the resulting subfields. We consider in this section the following **field splitting with minimized total horizontal complexity (FSHC) problem**: Given an IM M of size $m \times n$ and a maximum allowable field size $l \times w$, with $m \leq l$ and $n > w$, split M using y-monotone paths into $d = \lceil \frac{n}{w} \rceil$ (≥ 2) subfields $M_1, M_2, \ldots, M_d$, each with a size no larger than $l \times w$, such that the total horizontal complexity of these d subfields is minimized.

A key observation for solving the FSHC problem is that whenever we use a y-monotone path p to split the given IM M, the total horizontal complexity of the resulting subfields will increase by a value of $\sum_{i=1}^{m} \min\{M_{i,p(i)}, M_{i,p(i)+1}\}$, where $M_{i,0} = M_{i,n+1} = 0$ and $p(i)$ denotes the position at which p crosses the i-th row of M. The reason is that, similar to that for the delivery error, the *net increase* in the horizontal complexity from without any split between $M_{i,p(i)}$ and $M_{i,p(i)+1}$ to with such a split is $\min\{M_{i,p(i)}, M_{i,p(i)+1}\}$ [17]. We use $cost(p)$ to denote the increased horizontal complexity induced by the y-monotone path p for splitting M, that is, $cost(p) = \sum_{i=1}^{m} \min\{M_{i,p(i)}, M_{i,p(i)+1}\}$. Then, the FSHC problem is equivalent to finding a feasible splitting $\mathcal{T}$ of cardinality $d - 1$ that minimizes $\sum_{p \in \mathcal{T}} cost(p)$. As in Section 3, we model the FSHC problem as a shortest path problem in a directed acyclic graph G with d layers, which satisfies the staircase Monge property. By using a range-minima data structure [8], the weight of any edge of G can be computed in $O(m)$ time, after an $O(mn)$ time preprocessing. The staircase Monge property enables us to compute the shortest path in G by examining only $O(n\alpha(w))$ edges in total. Hence, we have the following result.

Theorem 2. *Given an IM M of size $m \times n$, and a maximum allowable field width w, the FSHC problem on M can be solved in $O(mn\alpha(w))$ time.*

5   Implementation and Experiments

To study the performance of our new FSMP algorithm with respect to clinical applications, we implemented the algorithm using the C programming language on Linux and Unix systems, and experimented with the resulting software on over 60 sets of clinical intensity maps obtained from the Dept. of Radiation Oncology, Univ. of Maryland Medical School. We conducted a comparison with a common simple splitting scheme that splits intensity maps along vertical lines. The algorithm for this simple splitting scheme is based on computing a d-link shortest path in a weighted directed acyclic graph, which we designate as FSVL (field-splitting using vertical lines). Both implementations are designed to take as input the intensity maps generated by CORVUS, one of the current most popular commercial planning systems.

The widths of the tested intensity maps range from 21 to 27. The maximum intensity level of each field is normalized to 100. The minimum net error increase of each intensity map is calculated for the given maximum subfield width. Table 1(a) compares the sums of the total net delivery error increases for splitting each of 63 intensity map fields with the maximum subfield widths w ranging from 7 to 15. For each of the tested fields, the total net delivery error increase using our FSMP algorithm is always less than or equal to the net delivery error increase using FSVL. For all 63 tested fields, the sums of the total net delivery error increases of FSVL and FSMP are 488338 and 351014, respectively. In terms of the net delivery error increase, our FSMP algorithm showed an improvement of about 28.1% over FSVL on the medical data we used. (It should be pointed out that theoretically, it can be shown that the output of FSVL can be arbitrarily worse than FSMP's in the worst case.)

Table 1. (a) Comparisons of the total sums of net delivery error increase for 63 intensity-modulated fields with the maximum allowable subfield widths w ranging from 7 to 15. (b) Net error increase using the FSMP algorithm with respect to the maximum allowable subfield width w, for three intensity fields from a single clinical case.

w	7	8	9	10	11	12	13	14	15	Total
FSVL	144590	84088	52390	35248	51434	33268	33204	29386	24730	488338
FSMP	142840	65854	34958	14846	41564	18808	18550	9778	3816	351014

(a)

w	7	8	9	10	11	12	13	14	15
Field 1	4000	3120	560	200	680	440	320	200	200
Field 2	4600	2120	920	120	760	600	560	360	0
Field 3	520	160	40	40	1400	200	160	0	0

(b)

On individual intensity-modulated fields, the total net delivery error generally decreases as the maximum allowable subfield width w increases. However, there are some cases in which the net increase in delivery error actually increases with an increased subfield width w. In these relatively rare cases, it may actually be more advantageous to split a field into more subfields rather than less if a smaller total delivery error caused by the field splitting is desired (of course, the increased number of resulting subfields may very well cause a considerable increase in the total delivery time). Table 1(b) shows the net delivery error changes for splitting 3 intensity-modulated fields from a single clinical case using our FSMP algorithm with varying maximum allowable subfield widths w.

The FSMP program runs very fast, as predicted by its theoretical linear time bound. It completed the field splitting in less than one second.

References

1. Aggarwal, A., Park, J.: Notes on Searching in Multidimensional Monotone Arrays. In: Proc. 29th Annual IEEE Symp. on Foundations of Computer Science, pp. 497–512. IEEE Computer Society Press, Los Alamitos (1988)
2. Chen, D.Z., Hu, X.S., Luan, S., Wang, C., Wu, X.: Mountain Reduction, Block Matching, and Applications in Intensity-Modulated Radiation Therapy. In: Proc. of 21th ACM Symposium on Computational Geometry, pp. 35–44. ACM Press, New York (2005)
3. Chen, D.Z., Wang, C.: Field Splitting Problems in Intensity-Modulated Radiation Therapy. In: Proc. of 17th International Symp. on Algorithms and Computation, pp. 690–700 (2006)
4. Deng, J., Pawlicki, T., Chen, Y., Li, J., Jiang, S.B., Ma, C.-M.: The MLC Tongue-and-Groove Effect on IMRT Dose Distribution. Physics in Medicine and Biology 46, 1039–1060 (2001)
5. Dogan, N., Leybovich, L.B., Sethi, A., Emami, B.: Automatic Feathering of Split Fields for Step-and-Shoot Intensity Modulated Radiation Therapy. Phys. Med. Biol. 48, 1133–1140 (2003)
6. Dou, X., Wu, X., Bayouth, J.E., Buatti, J.M.: The Matrix Orthogonal Decomposition Problem in Intensity-Modulated Radiation Therapy. In: Proc. 12th Annual International Computing and Combinatorics Conf., pp. 156–165 (2006)
7. Engel, K.: A New Algorithm for Optimal Multileaf Collimator Field Segmentation. Discrete Applied Mathematics 152(1-3), 35–51 (2005)
8. Gabow, H.N., Bentley, J., Tarjan, R.E.: Scaling and Related Techniques for Geometric Problems. In: Proc. 16th Annual ACM Symp. Theory of Computing, pp. 135–143. ACM Press, New York (1984)
9. Hong, L., Kaled, A., Chui, C., Losasso, T., Hunt, M., Spirou, S., Yang, J., Amols, H., Ling, C., Fuks, Z., Leibel, S.: IMRT of Large Fields: Whole-Abdomen Irradiation. Int. J. Radiat. Oncol. Biol. Phys. 54, 278–289 (2002)
10. Kamath, S., Sahni, S., Li, J., Palta, J., Ranka, S.: A Generalized Field Splitting Algorithm for Optimal IMRT Delivery Efficiency. The 47th Annual Meeting and Technical Exhibition of the American Association of Physicists in Medicine (AAPM), 2005. Also Med. Phys. 32(6), 1890 (2005)
11. Kamath, S., Sahni, S., Ranka, S., Li, J., Palta, J.: A Comparison of Step-and-Shoot Leaf Sequencing Algorithms That Eliminate Tongue-and-Groove. Phys. Med. Biol. 49, 3137–3143 (2004)

12. Kamath, S., Sahni, S., Ranka, S., Li, J., Palta, J.: Optimal Field Splitting for Large Intensity-Modulated Fields. Med. Phys. 31(12), 3314–3323 (2004)
13. Klawe, M.M., Kleitman, D.J: An Almost Linear Time Algorithm for Generalized Matrix Searching. Technical Report RJ 6275, IBM Research Division, Almaden Research Center (August 1988)
14. Luan, S., Wang, C., Chen, D.Z., Hu, X.S., Naqvi, S.A., Wu, X., Yu, C.X.: An Improved MLC Segmentation Algorithm and Software for Step-and-Shoot IMRT Delivery without Tongue-and-Groove Error. Med. Phys. 33(5), 1199–1212 (2006)
15. Webb, S.: Intensity-Modulated Radiation Therapy. Institute of Cancer Research and Royal Marsden NHS Trust (2001)
16. Wu, Q., Arnfield, M., Tong, S., Wu, Y., Mohan, R.: Dynamic Splitting of Large Intensity-Modulated Fields. Phys. Med. Biol. 45, 1731–1740 (2000)
17. Wu, X.: Efficient Algorithms for Intensity Map Splitting Problems in Radiation Therapy. In: Proc. 11th Annual International Computing and Combinatorics Conf., pp. 504–513 (2005)

A New Recombination Lower Bound and the Minimum Perfect Phylogenetic Forest Problem

Yufeng Wu and Dan Gusfield

Department of Computer Science
University of California, Davis
Davis, CA 95616, U.S.A.
{wuyu,gusfield}@cs.ucdavis.edu

Abstract. Understanding recombination is a central problem in population genetics. In this paper, we address an established problem in Computational Biology: compute lower bounds on the minimum number of historical recombinations for generating a set of sequences [11,13,9,1,2,15]. In particular, we propose a new recombination lower bound: the forest bound. We show that the forest bound can be formulated as the minimum perfect phylogenetic forest problem, a natural extension to the classic binary perfect phylogeny problem, which may be of interests on its own. We then show that the forest bound is provably higher than the optimal haplotype bound [13], a very good lower bound in practice [15]. We prove that, like several other lower bounds [2], computing the forest bound is NP-hard. Finally, we describe an integer linear programming (ILP) formulation that computes the forest bound precisely for certain range of data. Simulation results show that the forest bound may be useful in computing lower bounds for low quality data.

1 Introduction

Meiotic recombination is an important biological process which has a major effect on shaping the genetic diversity of populations. Recombination takes two equal length sequences and produces a third sequence of the same length consisting of some prefix of one sequence, followed by a suffix of the other sequence. Estimating the frequency or the location of recombination is central to modern-day genetics. Recombination also plays a crucial role in the ongoing efforts of association mapping. Association mapping is widely hoped to help locate genes that influence complex genetic diseases. The increasingly available population genetic variation data provides opportunities for better understanding of recombination.

In this paper, we assume the input data consists of *single nucleotide polymorphisms (SNPs)*. A SNP is a nucleotide site where exactly two (of four) different nucleotides occur in a large percentage of the population. That is, a SNP has binary states (0 or 1). A haplotype is a binary vector, where each bit (called site) of this vector indicates the state of the SNP site for this sequence. Throughout this paper, the input to our computational problems is a set of (aligned) haplotypes (i.e. a binary matrix with n rows and m columns).

G. Lin (Ed.): COCOON 2007, LNCS 4598, pp. 16–26, 2007.
© Springer-Verlag Berlin Heidelberg 2007

An established computational problem on recombination is to determine the *minimum* number of recombinations needed to generate a set of haplotypes from an ancestral sequence, using some specified model of the permitted site *mutations*. A mutation at a SNP site is a change of state from one nucleotide to the other nucleotides at that site. Throughout this paper, we assume that any SNP site can mutate at most once in the entire history of the sequences, which is supported by the standard *infinite sites model* in population genetics.

Given a set of binary sequences M, we let $Rmin(M)$ denote the minimum number of recombinations needed to generate the sequences M from any ancestral sequence, allowing only one mutation per site over the entire history of the sequences. The problem of computing or estimating $Rmin(M)$ has been studied in a number of papers, for example [11,13,9,1,2,15]. A variation to the problem occurs when a specific ancestral sequence is known in advance. No polynomial-time algorithm for either problem is known, and the second problem is known to be NP-hard [16,3]. Therefore, the problem of computing a good lower bound on the minimum number of recombinations has attracted much attention.

In this paper, we present a new lower bound on $Rmin(M)$, which has a static and intuitive meaning. This lower bound (which we call the forest bound) is closely related to the minimum perfect phylogenetic forest problem, an extension of the classic binary perfect phylogeny problem. We then demonstrate that the forest bound is provably higher than a well-known bound: the optimal haplotype bound [13,15] [1]. We resolve the complexity of computing the forest bound in a negative way with a NP-hardness proof. Finally, we give an integer linear programming formulation whose solution computes the forest bound exactly. We show empirically that this formulation can be solved in practice for data with small number of sites.

2 Background

2.1 Recombination and ARGs

For haplotype data composed of (binary) SNPs, the simplest evolutionary history that derives these haplotypes is the classic binary perfect phylogeny (if we assume the infinite sites model of mutations). The perfect phylogeny problem is to build a (rooted) tree whose leaves are labeled by rows in M and edges labeled by columns in M, and a column can label at most one edge. For the binary perfect phylogeny problem, Gusfield [7] developed a linear time algorithm. However, often the real biological data does not have a perfect phylogeny, which is partly due to recombination. In this case, we need a richer model of evolution: The evolutionary history of a set of haplotypes H, which evolve by site mutations (assuming one mutation per site) and recombination, is displayed on a directed acyclic graph called an "Ancestral Recombination Graph (ARG)" [6] (also called

---

[1] Throughout this paper, when we say bound A is higher than bound B, we mean that bound A is guaranteed to never be lower than bound B, and that there are examples where it is strictly higher.

phylogenetic networks in some literature). An ARG N, generating n sequences of m sites each, is a directed acyclic graph containing exactly one node (the root) with no incoming edges, and exactly n leaves with one incoming edge each. Every other node has one or two incoming edges. A node with two incoming edges is called a "recombination" node (and the two incoming edges are called recombination edges). Each site (integer) from 1 to m is assigned to exactly one edge in N, and none is assigned to any edge entering a recombination node. The sequences labeling the leaves of N are the extant sequences, i.e., the input sequences. See Gusfield, et al. [8] for a more detailed explanation.

An ARG N is called a minARG if N uses exactly $Rmin(M)$ recombinations. The ARG N may derive sequences that do not appear in M. These sequences are called Steiner sequences. Sequences in M are called input sequences.

2.2   Lower Bounds on $Rmin(M)$

There are a number of papers on *lower bounds* on $Rmin(M)$ [11,13,9,2,15]. In [12,13], Myers and Griffiths introduced the *haplotype lower bound*, which, when combined with additional ideas in [12,13] significantly outperforms the previous lower bounds. The haplotype bound, $h(M)$, is simple and efficiently computable. Consider the set of sequences M arrayed in a matrix. Then $h(M)$ is the number of distinct rows of M, minus the number of distinct columns of M, minus one. It is easy to establish that this is a lower bound on $R_{min}(M)$. Simulations show that $h(M)$ by itself is a very poor bound, often a negative number. However, when used with a few more tricks, it leads to impressive lower bounds. One such trick is to compute the haplotype bound on a subset of sites from M that need not be contiguous. For a subset of sites S (not necessarily contiguous), let $M(S)$ be M restricted to the sites in S, and $h(S)$ be the haplotype bound computed on $M(S)$. It is easy to see that $h(S)$ is also a lower bound on the M. The *optimal haplotype bound* is the *highest* $h(S)$ over all choices of S. Since there are an exponential number of subsets, complete enumeration of subsets quickly becomes impractical, and the problem of computing optimal haplotype bound has also been shown to be NP-hard [2]. However, Song et al. [15] showed that integer linear programming (ILP) can be used to efficiently compute the optimal haplotype bound for the range of data of current biological interest. They also showed the optimal haplotype bound is often equal to $Rmin(M)$ in certain biological datasets.

Myers and Griffiths [13] also introduced the so-called "history bound". The history bound is provably higher than the haplotype bound [1]. However, the history bound is defined only by a computational procedure (described below), and there is no *static* and intuitive meaning provided for this bound in [13], independent of the procedure to compute it. This makes both it difficult to find alternative methods to compute the history bound, or to understand and improve it. To compute the history bound for a set of binary sequences M, we initialize $R = 0$. A site c in M is called non-informative when entries in column c have only a single 0 or a single 1. A *cleanup step* is defined as the removal an any non-informative site in M, or the merging of two duplicate rows in M

into one row. A *row removal step* arbitrarily picks one row in M for removal, provided that no cleanup step is possible. A history is defined by an execution of the following algorithm:

Repeat a) and b) until M contains only one sequence:
a) Perform cleanup steps until no more are possible.
b) Perform one row removal step; increment R by one.

The history lower bound is equal to the *minimum value* of R over *all* possible histories (i.e. the ways of choosing a row in the row removal step). The correctness of the history bound can be proved by induction [13]. Computing the history bound is NP-hard and a dynamic programming algorithm with $O(2^n m)$ running time is given in Bafna et al. [2] (which improves upon the original implementation by Myers and Griffiths [13]).

3   The Forest Bound and the Minimum Perfect Phylogenetic Forest (MPPF) Problem

3.1   Definition of the Forest Bound

The optimal haplotype bound is currently one of the strongest lower bounds. In the following, we define and describe a lower bound that can be proved to be higher than the optimal haplotype bound.

Given an arbitrary ARG N, suppose we remove *all* recombination edges. N is then decomposed into connected components, each of which is a directed perfect phylogeny (sometimes simply referred to as a directed tree). Some of the tree edges are labeled by site mutations in the original ARG, and thus, we have a forest $\mathcal{F}(N)$ of perfect phylogenies, created by removing all recombination edges. In what follows, we will consider each of these trees after ignoring the edge directions. An important property of these trees in $\mathcal{F}(N)$ is that there is no duplicate mutations at a site in two trees in $\mathcal{F}(N)$. In other words, if a site s labels an edge in tree $T_1 \in \mathcal{F}(N)$, another tree $T_2 \in \mathcal{F}(N)$ can not have a mutation at s. This implies that sequences in T_2 have a uniform value (either all-0 or all-1) at site s. Also note that $\mathcal{F}(N)$ partitions the rows in M, where each partition is a perfect phylogeny and each row in M appears as a label in one of the perfect phylogenies. We call such partitioning of M *perfect* partitioning. It is easy to see that perfect partitioning always exists: a trivial partitioning simply has n partitions, where each partition has a single row. Obviously, there exists a way of partitioning the rows of M such that the *number* of partitions is *minimized*. This motivates the following optimization problem.

The Minimum Perfect Phylogenetic Forest (MPPF) Problem. Given a binary matrix M, find a set of a *minimum* number of perfect phylogenies that derives M s.t. each row is derived by some perfect phylogeny and for any site s, mutations at s occur at most once in at most one tree. We denote the minimum number of perfect phylogenies $F_{min}(M)$.

Note that $F_{min}(M) = 1$ iff M has a perfect phylogeny. That is, there exists a single tree that derives all sequences in M iff M has a perfect phylogeny.

On the other hand, when there is no perfect phylogeny for M, we need more than one tree to derive all the sequences in M. The MPPF problem asks to find the minimum number of trees in the forest. This problem is related to the well-studied maximum parsimony problem (i.e. the Steiner tree problem in phylogeny). In maximum parsimony, we construct a *single* tree (with back or recurrent mutations) which minimizes the number of site mutations. The MPPF problem asks for constructing the minimum number of trees, each of which is a perfect phylogeny, and each site can mutate once in at most one tree.

Now we define the forest bound.

Forest bound. For a binary matrix M, the forest bound is equal to $F_{min}(M)-1$.

Lemma 1. *The forest bound is a valid lower bound on $Rmin(M)$.*

Proof. Suppose we trim a minARG N by removing all recombination edges in N and we have a forest with $k \geq F_{min}(M)$ trees. Note that N is connected and we need at least one recombination to connect a tree to the rest of trees. So $R_{min}(M) \geq k-1 \geq F_{min}(M) - 1$. The reason that we subtract 1 is because we can start from a tree and this tree does not need a recombination to link it.   □

Now we explain the reason for our interest in the forest bound. Unlike the forest bound, the history bound lacks a static definition. Below we show that the forest bound is higher than the haplotype bound, and hence, the forest bound is the highest lower bound that we know of which has a simple static definition.

Lemma 2. *For a perfect phylogenetic forest $\mathcal{F}$ with n_s Steiner nodes, the number of trees k is equal to $n + n_s - m$.*

Proof. Suppose each tree $T_i \in \mathcal{F}$ contains n_i distinct sequences (nodes) for $i = 1 \ldots k$. Here, $\sum_{i=1}^{k} n_i = n + n_s$, where n_s is the number of Steiner nodes in the ARG N. We know for each tree, there are $n_i - 1$ edges with mutations. Let m_i denote the number of mutations in tree T_i. This means $m_i = n_i - 1$. We require each column to mutate *exactly* once, and it is easy to show that this constraint does not change the forest bound. So we have $m = \sum_{i=1}^{k} m_i = \sum_{i=1}^{k}(n_i - 1) = n + n_s - k$ mutations in the forest. So, $k = n + n_s - m$.   □

Lemma 3. *This forest bound is always higher than the haplotype bound, but lower than the history bound.*

Proof. We first show that the forest bound is provably higher than the haplotype lower bound. Suppose a minimum forest has $k = F_{min}(M)$ trees. From Lemma 2, $k = n + n_s - m \geq n - m$. So $k - 1 \geq n - m - 1$, which is the haplotype bound.

Now we show that the history bound is higher than the forest bound. From the algorithm to compute the history bound, it can be seen that the method produces a phylogenetic forest. However, in contrast to the definition of a phylogenetic forest given above, the forest produced by the history bound has additional time-order constraints: the trees in the forest can be time-ordered such that if site s mutates in a tree T_i, the states at s for sequences in earlier trees must be ancestral states (i.e. not the derived states). But since the forest produced by the history bound is a valid phylogenetic forest, its number of trees in that forest cannot be smaller than the forest bound.   □

We now relate the forest bound to the optimal haplotype bound.

Theorem 1. *The forest bound is higher than the optimal haplotype bound.*

Proof. By Lemma 3 we know that the forest bound applied to any subset of site is higher than the haplotype bound applied to the same subset of sites. In particular, if S^* is the subset of sites of M (called *optimal subset*) that gives the optimal haplotype bound, then the forest bound applied to S^* is higher. Hence to prove the theorem we only need to show that the forest bound applied to all of M cannot increase by restricting to a subset of sites in M.

For a given data M, suppose we have a minimum phylogenetic forest $\mathcal{F}$ for M with $F_{min}(M)$ trees. Now we consider $\mathcal{F}(S)$ when we restrict our attention to S, a subset of sites. To derive $\mathcal{F}(S)$ from $\mathcal{F}$, we remove all mutation sites in $\mathcal{F}$ that are not in S and cleanup the forest by removing edges with no mutations, and collapsing identical sequences. It is important to note that $\mathcal{F}(S)$ has *at most* $F_{min}(M)$ trees. This is because when we remove sites not in S, we may need to link up two previously disjoint trees (and thus make the number of trees smaller), but we can never *increase* the number of trees. Thus, we know $F_{min}(M(S))$ can not be higher than $F_{min}(M)$.   □

Theorem 1 and Lemma 3 say that the forest bound is higher than the optimal haplotype bound but lower than the history bound. Hence we conclude,

Corollary 1. *The optimal haplotype bound cannot be higher than the history bound* [2].

Experiments show that the forest bound can be strictly higher than the optimal haplotype bound and improve the overall recombination lower bound. For example, consider a simple matrix containing 5 rows and 5 columns: 10001, 00010, 00100, 11011 and 01101. The optimal haplotype bound for this data is 1, while it is not hard to see that a perfect phylogenetic forest contains at least 3 components. Thus, the forest bound for this data is 2.

As mentioned earlier, it is known that the optimal haplotype bound and the history bound are both NP-hard to compute [2]. However, if the forest bound could be computed efficiently, we would not need to compute the optimal haplotype bound, but could instead use the forest bound. Unfortunately, the forest bound is also NP-hard to compute.

3.2   The Complexity of the Forest Bound

Theorem 2. *The MPPF problem is NP-hard.*

---

[2] Myers [12] asserted (with no proof) that the history bound is higher than the optimal haplotype bound. Here we furnish the proof to this claim.

Proof. The high-level construction of our proof is inspired by Foulds and Gra-ham's NP-completeness proof of Steiner tree in perfect phylogeny problem [4].

As in [4], we reduce from the known NP-complete problem of Exact Cover by 3-sets (X3C) [5]. Recall that the general form of X3C is as the following:

$\mathcal{S} = \{S_1, S_2, \ldots, S_n\}$, where each $|S_i| = 3$ and $S_i \subseteq \{1, 2, \ldots, 3m\} = I_{3m}$, for $1 \le i \le n$.
Does $\mathcal{S}$ contain (non-overlapping) m sets $S_{i_1}, \ldots, S_{i_m}$ whose union is I_{3m}?

High-Level Idea. We construct a binary matrix M for $\mathcal{S}$ (the collection of sets), such that for each set S_i, the set of corresponding sequences in M can be generated on a perfect phylogeny. Thus, if there is a solution for X3C, we have m perfect phylogenies that use up all site mutations, and a collection of isolated sequences (also trivially perfect phylogenies) and the total number of trees is $F_{min}(M)$. If there is no solution for X3C, the number of trees in any perfect phylogenetic forests is more than $F_{min}(M)$. To enforce this property, two sequences corresponding to the same set S_i will have a small Hamming distance. For two sequences corresponding to different sets, their Hamming distance will be large. So, if two far apart sequences are placed into the same tree, there will be too many Steiner sequences needed to connect them, and thus by Lemma 2 and proper manipulation of the construction, we will have more trees than $F_{min}(M)$ in such forest.

WLOG we assume there is no duplicate sets in $\mathcal{S}$. Given an instance of X3C, we construct a binary matrix M as follows. We let $K = m + 1$. Note that $2K - 3 > m$, when $m > 1$. For each S_i, we construct a set of $3K$ sequences of length $3mK$. We have K sequences corresponding to each of the three elements in S_i. Each of these sequences is composed of $3m$ blocks of K sites. Each block is arranged sequentially in the increasing order for each integer in S_i. The sequences are constructed as follows. Suppose we are constructing the j_{th} sequence ($j \in \{1 \ldots K\}$) for an element $p \in S_i$. For block (of number q) $B_{i,p,j,q}$ in this sequence, if the corresponding integer $q \notin S_i$, then block $B_{i,p,j,q}$ contains all 1. If $q \in S_i$ and $q \ne p$, then we set $B_{i,p,j,q}$ to be all 0. If $q \in S_i$ and $q = p$, we set the j_{th} bit in $B_{i,p,j,q}$ to 1 and 0 for all other bits. Note that for a given row associated to a set S_i, all bits corresponding to elements not in this set are 1. Also note that for the K sequences corresponding to an integer $q \in S_i$, the K blocks $B_{i,p,j,q}$ form a diagonal matrix with all 1 on the main diagonal. One example is shown in Table 1 for the simple case when $m = 2$.

The following facts (proof omitted) about M are important.

P1. There are *no* two identical sequences in M.
P2. The $3K$ sequences corresponding to a single set S_i have a star-shaped perfect phylogeny, with the center sequence as the only Steiner sequence.
P3. For two sequences s_1, s_2 coming from the same set S_i, the Hamming distance between s_1, s_2 is 2. For two sequences s_1, s_2 coming from different sets S_i, S_j, the Hamming distance is at least $2K - 2$.

Now we claim that *X3C* problem has a solution (i.e. union of $S_{i_1}, \ldots, S_{i_m}$ is equal to I_{3m}) if and only if there is a phylogenetic forest for M with exactly $3nK - 3mK + m$ perfect phylogenies.

Table 1. Example of the constructed matrix when $m = 2$ (i.e. there are 6 elements in I_{3m}), and thus $K = 3$. The table lists the constructed rows for a set $\{1, 2, 4\}$.

	1	1	1	2	2	2	3	3	3	4	4	4	5	5	5	6	6	6
r_1	1	0	0	0	0	0	1	1	1	0	0	0	1	1	1	1	1	1
r_2	0	1	0	0	0	0	1	1	1	0	0	0	1	1	1	1	1	1
r_3	0	0	1	0	0	0	1	1	1	0	0	0	1	1	1	1	1	1
r_4	0	0	0	1	0	0	1	1	1	0	0	0	1	1	1	1	1	1
r_5	0	0	0	0	1	0	1	1	1	0	0	0	1	1	1	1	1	1
r_6	0	0	0	0	0	1	1	1	1	0	0	0	1	1	1	1	1	1
r_7	0	0	0	0	0	0	1	1	1	1	0	0	1	1	1	1	1	1
r_8	0	0	0	0	0	0	1	1	1	0	1	0	1	1	1	1	1	1
r_9	0	0	0	0	0	0	1	1	1	0	0	1	1	1	1	1	1	1

We first show that given a solution (i.e. $S_{i_1}, \ldots, S_{i_m}$) of X3C, we can build a forest with $3nK - 3mK + m$ trees. From property P2, we construct m perfect phylogenies, each from $3K$ sequences corresponding to each of S_{i_j}. Then, we treat the remaining $3nK - 3mK$ as isolated sequences (trivially perfect phylogenies). So the total number of perfect phylogeny is $3nK - 3mK + m$.

Now we show the other direction: if there is a phylogenetic forest for M that has $3nK - 3mK + m$ perfect phylogenies, then there is a solution for problem X3C. We first argue that no two sequences from different sets S_i, S_j can appear together in a same perfect phylogeny. For contradiction, suppose s_1, s_2 coming from different sets are together in one perfect phylogeny. Consider the path from s_1 to s_2 in the perfect phylogeny. From Property P3, the Hamming distance between s_1 and s_2 is at least $2K - 2$. This means there are at least $2K - 3$ intermediate nodes on that path whose states changes from s_1 to s_2 on these $2K - 2$ sites. It is also easy to see that none of these $2K - 3$ nodes can be part of M, since each sequence in M has either exactly a single 1 or all 1s within a block and intermediate nodes between s_1, s_2 must contain from 2 to $K - 1$ 1s for the block corresponding to the element not shared by S_i, S_j. That is, we know that the phylogenetic forest contains at least $2K - 3$ Steiner nodes. But from Lemma 2, we will have at least $3nk - 3mK + (2K - 3)$ perfect phylogenies, which is larger than $3nK - 3mK + m$ since $2K - 3 > m$. That is a contradiction, and thus each phylogeny can only have sequences derived from the same set S_i.

Now it is easy to see that within the forest there can be at most m non-degenerated trees. This is because each non-degenerated tree contains at least 1 Steiner node (see Property P3). Also note that we can not have fewer than m non-degenerated trees. To prove this, suppose for contradiction, that we have at most $m - 1$ non-degenerate perfect phylogenies. Since each such tree comes from a single set S_i, if there are at most $m - 1$ trees, there are at most $3(m - 1)K$ nodes in the trees. Then there are at least $3nK - 3(m-1)K = 3nK - 3mK + 3K$ degenerated trees. Since $K = m + 1$, we know we will have more than $3nK - 3mK + m$ (isolated) trees. That is a again a contradiction.

Therefore, we know we will have exactly m non-degenerated trees. Now we need to show that these m non-degenerated trees correspond to a solution for X3C. Note that each of such trees does correspond to a set in S. What we need to show is that every element in I_{3m} is covered, and no element is covered more than once. Suppose a tree has a block whose corresponding integer is not covered by other picked sets, then we can easily enlarge the tree by adding the sequences of that block. Now we want to argue that no two sets picked by this phylogenetic forest can overlap. For contradiction, suppose there is an overlap between S_i, S_j when we select all $3K$ sequences corresponding to S_i, S_j. Then for the corresponding two trees, there must be mutations in *both* trees for the overlapped sites. This contradicts the assumption that the set of trees are perfect phylogenies with no duplicate mutations (sites). Therefore, the given phylogenetic forest leads to a valid X3C solution.   □

Corollary 2. *Computing the forest bound is NP-hard.*

A Variant of the MPPF Problem. The MPPF problem requires that if a mutation occurs at a site in one tree, then this site does not mutate in any other tree. Now, suppose we allow a site to mutate in more than one of the perfect phylogenies (but still mutate at most once in any single perfect phylogeny). This problem is NP-complete even when we just want to partition the matrix into two perfect phylogenies. We omit the proof due to the space limit.

4   Practical Computation of the Forest Bound

In this section, we focus on using integer programming to compute the *exact* forest bound for data within certain range.

4.1   Computing the Forest Bound Precisely Using Integer Linear Programming

Consider an input matrix M with n rows and m sites. Our goal is to compute the minimum forest that derives the input sequences. There are 2^m possible sequences (which form a hypercube) that could be part of the minimum forest. Of course, the n input sequences must appear in this forest. From Lemma 2, in order to compute the forest bound we need to *minimize* the number of Steiner nodes. Thus, we create a variable v_i for each sequence s_i in the set of 2^m possible sequences at Steiner nodes, where $v_i = 1$ means sequence s_i appears in the forest. Next, we create a variable $e_{i,j}$ for two sequences s_i, s_j that differ at exactly one column. We create constraints to ensure $e_{i,j} = 1$ *implies* $v_i = 1$ and $v_j = 1$. We define a set E_c as the set of $e_{i,j}$ where s_i, s_j differ exactly at the single site c. The infinite sites mutation model requires that exactly one $e \in E_c$ has value 1.

Optimization goal Minimize $(\sum_{i=1}^{2^m} v_i) - m - 1$
Subject to
$\quad v_i = 1$, for each row $s_i \in M$.
$\quad e_{i,j} \le v_i$, and $e_{i,j} \le v_j$, for each edge (s_i, s_j).
$\quad \sum_{(s_i,s_j)\in E_c} e_{i,j} = 1$, for each site c

Binary Variables

v_i for each sequence s_i with m binary characters

$e_{i,j}$ for each pair of sequences s_i, s_j such that $d(s_i, s_j) = 1$.

The formulation can also be extended easily to handle the situation where there are missing values in the input data, which is important for handling real biological data. To handle missing data in the ILP formulation, for a sequence s_i with missing values, we change the constraint $v_i = 1$ to $\sum_j v_j \geq 1$, for each sequence s_j that matches the values of s_i at all non-missing positions. Our experience shows that the formulation can be solved reasonably fast for data with up to 8 sites (by a powerful ILP package CPLEX).

4.2 Simulations of Data with Missing Data

Now we describe computations of the forest bound and how they compare to the haplotype bound on simulated data. In this simulation study, to make comparison easier, we do not use the composite method [13], which often gives higher lower bound. We show here that the forest bound can be effectively computed for certain range of data with missing values.

We generated 100 datasets with Hudson's program MS [10] for each parameter setting. We fix the number of sites in these data to 7 or 8. We want to compare the forest bound with the optimal haplotype bound when the data contains various level of missing data. Currently, the only known method computing optimal haplotype bound with missing data can only work with very small data. So instead, we compare with a weaker haplotype bound method (implemented in program HapBound [15]) that can handle missing data but not always give the optimal bound [14] when there is missing data. Missing values are added to the datasets by setting an entry to be missing with a fixed probability P_{mv}.

Table 2. Comparing the forest bound with haplotype bound. We report the percentage of datasets where the forest bound is strictly higher than the haplotype bound.

% Missing value	0%	10%	20%	30%
20 rows, 7 sites	0%	0%	0%	3%
20 rows, 8 sites	0%	1%	0%	0%
30 rows, 7 sites	0%	1%	0%	8%
30 rows, 8 sites	0%	0%	0%	7%

Table 2 shows that the forest bound can outperform HapBound in some cases. On the other hand, the haplotype bound method used by the program HapBound appears to be quite good for the range of data we tested. Our tests show that when the missing value level is low or moderate, the program HapBound performs quite well for the range of data we generated. However, on random data, the forest bound was seen to often give higher bounds than the optimal haplotype bound (results not shown). In fact, for 50 simulated random datasets with

15 rows and 7 sites (and with no missing entries), 10% of the data had a strictly higher forest bound compared to the optimal haplotype bound (and 20% of the data had a strictly higher history bound than the optimal haplotype bound).

Acknowledgments. The research reported here is supported by grants CCF-0515278 and IIS-0513910 from National Science Foundation.

References

1. Bafna, V., Bansal, V.: The number of recombination events in a sample history: conflict graph and lower bounds. IEEE/ACM Trans. on Computational Biology and Bioinformatics 1, 78–90 (2004)
2. Bafna, V., Bansal, V.: Inference about Recombination from Haplotype Data: Lower Bounds and Recombination Hotspots. J. of Comp. Bio. 13, 501–521 (2006)
3. Bordewich, M., Semple, C.: On the computational complexity of the rooted subtree prune and regraft distance. Annals of Combinatorics 8, 409–423 (2004)
4. Foulds, L.R., Graham, R.L.: The Steiner Tree in Phylogeny is NP-complete, Advances in Applied Math, v. 3 (1982)
5. Garey, M., Johnson, D.: Computers and intractability, Freeman (1979)
6. Griffiths, R.C., Marjoram, P.: Ancestral inference from samples of DNA sequences with recombination. J. of Comp. Bio. 3, 479–502 (1996)
7. Gusfield, D.: Efficient algorithms for inferring evolutionary history. Networks 21, 19–28 (1991)
8. Gusfield, D., Eddhu, S., Langley, C.: Optimal, efficient reconstruction of phylogenetic networks with constrained recombination. J. Bioinformatics and Computational Biology 2, 173–213 (2004)
9. Gusfield, D., Hickerson, D., Eddhu, S.: An Efficiently-Computed Lower Bound on the Number of Recombinations in Phylogenetic Networks: Theory and Empirical Study. Discrete Applied Math 155, 806–830 (2007)
10. Hudson, R.: Generating Samples under the Wright-Fisher neutral model of genetic variation. Bioinformatics 18(2), 337–338 (2002)
11. Hudson, R., Kaplan, N.: Statistical properties of the number of recombination events in the history of a sample of DNA sequences. Genetics 111, 147–164 (1985)
12. Myers, S.: The detection of recombination events using DNA sequence data, PhD dissertation. Dept. of Statistics, University of Oxford, Oxford, England (2003)
13. Myers, S.R., Griffiths, R.C.: Bounds on the minimum number of recombination events in a sample history. Genetics 163, 375–394 (2003)
14. Song, Y.S., Ding, Z., Gusfield, D., Langley, C., Wu, Y.: Algorithms to distinguish the role of gene-conversion from single-crossover recombination in the derivations of SNP sequences in populations. In: Apostolico, A., Guerra, C., Istrail, S., Pevzner, P., Waterman, M. (eds.) RECOMB 2006. LNCS (LNBI), vol. 3909, Springer, Heidelberg (2006)
15. Song, Y.S., Wu, Y., Gusfield, D.: Efficient computation of close lower and upper bounds on the minimum number of needed recombinations in the evolution of biological sequences. Bioinformatics 421, i413–i422 (2005) Proceedings of ISMB 2005
16. Wang, L., Zhang, K., Zhang, L.: Perfect Phylogenetic Networks with Recombination. J. of Comp. Bio. 8, 69–78 (2001)

Seed-Based Exclusion Method
for Non-coding RNA Gene Search

Jean-Eudes Duchesne[1], Mathieu Giraud[2], and Nadia El-Mabrouk[1]

[1] DIRO – Université de Montréal – H3C 3J7 – Canada
{duchesnj,mabrouk}@iro.umontreal.ca
[2] Bioinfo/Sequoia – LIFL/CNRS, Université de Lille 1 – France
giraud@lifl.fr

Abstract. Given an RNA family characterized by conserved sequences and folding constraints, the problem is to search for all the instances of the RNA family in a genomic database. As seed-based heuristics have been proved very efficient to accelerate the classical homology based search methods such as BLAST, we use a similar idea for RNA structures. We present an exclusion method for RNA search allowing for possible nucleotide insertion, deletion and substitution. It is based on a partition of the RNA stem-loops into consecutive seeds and a preprocessing of the target database. This algorithm can be used to improve time efficiency of current methods, and is guaranteed to find all occurrences that contain at least one exact seed.

1 Introduction

The last 20 years have seen an explosion in the quantity of data available for genomic analysis. Much work has been devoted to speeding up data mining of proteins or gene coding DNA, but these sequences account for only a fraction of the genome. In addition, many non-coding RNA genes (ncRNAs) are known to play key roles in cellular expression, yet few efforts have been made to facilitate their search in large scale databases. Classical homology based search methods like Blast [1] often fail when searching for non-coding genes since the input is stripped from structural information down to its bare sequence. Searching algorithms that permits inputs with structural information should yield better results.

Historically, the first computer scientists to interest themselves with ncRNAs have created tailor made algorithms for specific RNA families such as tRNAs [6,4,9]. Other more general search tools where created to give control of the biological context to the user [3,12,7]. Still these tools lacked the capacity to efficiently parse large genomic databases. Klein and Eddy provided a database specialized search tool [8] for ncRNA including structural information, but is self admittedly slow for large scale databases. More recently, Zhang and Bafna presented a method to efficiently filter databases with a set of strings matching a profile to specific parameters [2]. Their experimentation gave rise to specialized filters for specific RNA families. As such this strategy would require prior

G. Lin (Ed.): COCOON 2007, LNCS 4598, pp. 27–39, 2007.
© Springer-Verlag Berlin Heidelberg 2007

knowledge on the RNA families of interest when generating database, this can become restrictive in some experimental contexts which would benefit from an all-purpose filtering method for ncRNA. Although this can be offset by combining filters with different parameters in an attempt to maximize efficiency and accuracy.

In addition to the capacity of parsing large genomic databases, as sequence and structure constraints are established from a restricted set of an RNA family representatives, any search method should account for a certain flexibility and deviation from the original consensus, allowing for possible mismatches and insertion/deletion (indel) of nucleotides. In particular RNAMotif [12] (one of the most popular and time efficient tool for RNA search), does not explicitly allow for base pair indels. In [5], we have considered a more general representation of folding constraints and developed an approximate matching algorithm allowing for for both mismatches and indels of base pairs. The major drawback of the method was its time inefficiency.

In this paper, we develop a seed-based exclusion method allowing for mismatches and indels, able to speed up existing RNA search methods. Similar heuristics have been proved very efficient to accelerate the classical homology based methods. In particular, PatternHunter [11,10] based on multiple spaced seeds has become one of the most popular method for sequence search at a genomic scale. Recently, Zhang et al.[16] proposed a formalization of the filtering problem and a demonstration that the combination of several filters can improve the search of ncRNAs. Here, we develop a new seed-based heuristic for RNA search, using seeds with distance and folding constraints. It is based on a partition of the RNA stem-loops into consecutive seeds and a preprocessing of the target database storing the occurrences of all possible seeds in a hash table. The search phase then reports, in constant time, the position lists of all seeds of the query stem-loops, and uses extension rules to account for possible errors. The heuristic is guaranteed to find all occurrences containing at least one exact seed.

The rest of the paper is organized as follows. Section 2 presents the basic concepts and definitions, and introduces the general idea of the Sagot-Viari algorithm [15] that will be used in our algorithm's search phase. Section 3 describes our new exclusion method. In Section 4, we study the choice of seeds and anchor elements. Finally, we present our experimental results in Section 5, and show how our method can be used in conjunction with RNAMotif to improve its running time.

2   Preliminary Definitions

2.1   RNA Structures

An RNA *primary structure* is a strand of consecutive nucleotides linked by phosphodiester bonds: Adenine (A), Cytosine (C), Guanine (G) and Thymine (T). When transcribed from DNA to RNA, thymine is substituted into uracil (U). As such, U and T are considered synonymous for most purposes. We denote Σ_{DNA} the alphabet of nucleotides $\{A, C, G, T\}$.

Considering that an individual nucleotide's main biological property is to form a structural bond with other nucleotides, primary structure alone ill-defines ncRNAs. An RNA *secondary structure* is represented by a series of base pairings, the most frequent ones being the canonical Watson-Crick A-T and C-G. The secondary structure is organized in a set of nested stems and loops, where a stem is a sequence of paired and unpaired nucleotides, and a loop is a sequence of unpaired nucleotides. A stem followed by a loop is called a *stem-loop* (Figure 1.(a)).

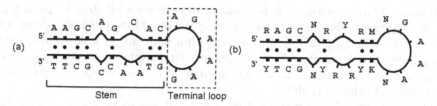

Fig. 1. (a) A stem-loop with canonical base pairings represented as dots; (b) A stem-loop descriptor. (a) is an occurrence of (b).

It is well documented that the functional properties of an RNA molecule is dependent on its final structure obtained by additional foldings over its secondary structure. Our work relies on the hypothesis that there is enough signal in the primary and secondary structure to find the overall molecule with this simplified view.

2.2 Descriptors

Descriptors are user-defined sets of conserved elements of a specific molecule's primary and secondary structure. They are often obtained from multiple alignments of different instances of the same molecule's sequence from various species, but how a good descriptor is obtained is beyond the scope of this current work.

In [5], we have introduced a rigorous and very flexible representation of folding constraints in term of "secondary expressions". In this paper, we focus on a more restrictive descriptor form, though allowing to represent most of the RNA families found in the literature. The considered constraints are:

1. Positions characterized by a possible subset of nucleotides and represented by a degenerate alphabet, the IUPAC code, over all possible substitutions (Table 1). For example, N allows for any nucleotide at the observed position.
2. Correlated constraints due to canonical base pairings. For example, in Figure 1.(b), the left-most pairing (R, Y) means that the upper nucleotide can be either A or G, but if it is A (respec. G) then the opposite nucleotide should be T (respec. C).
3. Bounded range of possible lengths for unpaired parts of the structure.

Table 1. The standard IUPAC code defines symbols for sets of nucleotides

A : A	K : G | T	B : C | G | T
C : C	M : A | C	D : A | G | T
G : G	R : A | G (purine)	V : A | C | G
T : T	S : C | G	H : A | C | T
	W : A | T	
	Y : C | T (pyrimidine)	N : A | C | G | T

2.3   The Sagot-Viari Notations

The Sagot-Viari algorithm [15] is designed to search for all stem-loops in a genomic sequence, allowing for possible mispairings. More precisely, given four
parameters s, e, d_{min} and d_{max}, the algorithm finds all possible stem-loops in
the genome G characterized by a maximum stem length s, a loop of size d with
$d_{min} \le d \le d_{max}$, and a maximum number of e mispairings and nucleotide
insertion and deletion (indels).

The interesting design feature of their method was to keep separate the two
complementary parts of the stems until the final reconstruction step. Another
way to look at their method is to consider that they filter a complete genome for
sequences that can potentially form a stem-loop structure but differ the actual
verification until the sequences have been extended to the full length of the
pattern.

We first introduce some basic notations. Given a sequence $u = u_1 u_2 \ldots u_n$,
we denote by $u_{i,j}$ the subsequence $u_{i,j} = u_i u_{i+1} \ldots u_j$. The sequence $\overline{u}$ is the
complementary inverse of u. For example, if $u = $ AATGC, then $\overline{u} = $ GCATT.
Given a sequence u of size k on Σ_{DNA}, we denote by $Oc(u)$ the list of positions
of all occurrences of u in the genomic sequence G, eventually within a threshold
of error e. The occurrences list of a stem-loop described by u is:

$$S_{(u,\overline{u})} = \{(p, q) \,|\, p \in Oc(u), q \in Oc(\overline{u}), \mathsf{good}(p, q)\}$$

The predicate $\mathsf{good}(p, q)$ checks the distance ($d_{\min} \le q - p \le d_{\max}$) and the
error constraints.

The algorithm proceeds by successive *extensions* and *filtering* steps, starting
from sets $Oc(\alpha)$ for each $\alpha \in \Sigma_{DNA}$. Each set $Oc(u_{i,j})$ could be constructed by
extending $Oc(u_{i+1,j})$ and $Oc(u_{i,j-1})$. However, a majority of the positions in
$Oc(u)$ can be eliminated before the final filtering. In fact, the algorithm never
computes any Oc list beyond the initial step. It considers only *possible occurrences* (of the stem-loop) position lists:

$$POc(u_{i,j}) = \{p \in Oc(u_{i,j}) \,|\, \exists q \in Oc(\overline{u_{i,j}}), \mathsf{good}(p, q)\}$$

The lists $POc(u_{i+1,j})$ and $POc(u_{i,j-1})$ are extended and merged into one list
$POc'(u_{i,j})$. Filtering that list for the distance and error constraints give rise to

the list $POc(u_{i,j})$. At the end, the solution set $S_{(u,\overline{u})}$ is obtained by a (quadratic) filtering between $POc(u)$ and $POc(\overline{u})$.

The POc and POc' lists are represented through stacks, and all the extension and filtering operations are done in linear time relative to the size of the stacks.

3   An Exclusion Method for RNA Search

Given an RNA descriptor D and a genomic sequence (or database) G, the goal is to find the position list S_D of the occurrences of D in G, possibly with an error threshold e. We propose to search the descriptor D starting with a set of n anchor sequences extracted from the descriptor's stem-loops. A heuristic based on an exclusion method is developed for an efficient search of anchor sequences: each anchor is partitioned into consecutive (and overlapping) seeds of a given size, and a preliminary step consists in building a seed database over the genomic sequence G. In section 4, we discuss the choice of appropriate "constraining" anchors allowing a good speed-up with a convenient sensibility.

A high level sketch of the exclusion method is given below and is schematized in Figure 2. Details are in the following subsections.

1. **Preprocessing phase:** Build a seed database over the genomic sequence G.
2. **Partition phase:** Choose a set of anchor sequences from D (with their relative distance constraints) and a set of seed-shapes, and partition the anchor sequences into consecutive seeds.
3. **Anchor search phase:** Query the database for the seeds, giving lists of occurrences Oc. Then extend occurrences and filter them while checking length, error and folding constraints.
4. **Check phase:** Check whether each RNA candidate verifies the descriptor constraints that were not used as anchors in the search phase.

3.1   The Preprocessing Phase

The genomic database G is first processed to output all elementary motifs of a given size. The preprocessing phase is designed to allow for a constant time access to the position list of all occurrences of elementary motifs, represented by seeds. Rigorous definitions follow.

A *seed of size k* or *k-seed* is a sequence of size k on the alphabet Σ_{DNA}. To allow the possibility of spaced seeds, we define two types of characters: # and -, where - denotes the don't care character. A *k-seed-shape* is a sequence of k elements from the alphabet $\{\#, -\}$.

Given a set of seed-shapes, the preprocessing phase builds a hash table containing an entry for each set of sequences with the same # positions. For example, for seed-shape ##-#, $AGAC$ et $AGTC$ are stored at the same position. We discuss the choice of appropriate seed-shapes and lengths in Section 4.

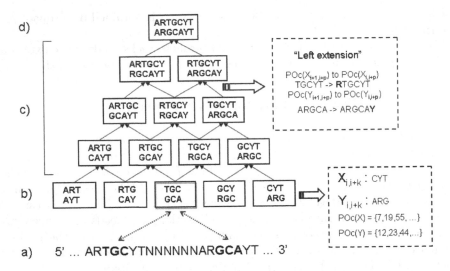

Fig. 2. (a) A specific descriptor sequence with the set of anchors $\mathcal{A} = \{ARTGCYT, ARGCAYT\}$ of common length $m = 7$. The distance constraints are $d^{1,2}_{min} = d^{1,2}_{max} = 6$. Anchors are partitioned into consecutive 3-patterns. The elements in bold represent a single pair of seeds (seed shape $\#\#\#$); (b) The initial step of the search phase is to query the database for all positions of the selected seeds. In the figure, boxes are labeled by their implicit sequences. Their actual data is the lists of positions of these sequences, as illustrated by the rightmost box; (c) Next, the algorithm iterates over a series of extensions and merges to filter the seeds that cannot possibly extend into the desired motif. Each level represents a single iteration. A single box receives incoming extensions from two sources, hence the need for merging sets of positions into a single set. One of these extension is shown in greater detail; (d) After the final iteration, the algorithm returns a list of candidate positions for the full anchor sequences. Each position needs to be validated to confirm the presence or absence of the desired motif at the given position in the genome.

3.2   The Partition Phase

The RNA descriptor is first parsed to extract a given number of *anchors* that are ordered in a priority search list (see section 4). More precisely an anchor $\mathcal{A}$ is a set of sequences $\{\mathcal{A}^1, \cdots, \mathcal{A}^l\}$ on the IUPAC alphabet, with a set of distance constraints $\{(d^{i,j}_{min}, d^{i,j}_{max})\}$. Anchor sequences can be related with complementary relations, but that is not mandatory.

A sequence of size k over the IUPAC alphabet is called a *k-pattern*. For a given length k, each anchor sequence $\mathcal{A}^i$ is partitioned into its consecutive k-patterns $\mathcal{A}^i_{1,k}, \mathcal{A}^i_{2,k+1}, \cdots \mathcal{A}^i_{m-k+1,m}$. For a given k-seed-shape sh, we then report the set of seeds corresponding to each k-pattern. A formal definition follows.

Definition 1. *Let $u = u_1 \cdots u_k$ be a k-pattern and $sh = sh_1 \cdots sh_k$ be a k-seed-shape. We say that a seed $s = s_1 \cdots s_k$ is a representative of u with respect to sh iff, for any i such that $sh_i = \#$, $s_i \in u_i$.*

Given a k-seed-shape sh and a k-pattern u, we denote by $L(u)$ the list of seed representative of u with respect to sh. For example, if $u = ARYC$ and sh=##-# we have $L(u) = \{AAAC, AACC, AAGC, AATC, AGAC, AGCC, AGGC, AGTC\}$. We also denote by $L(\mathcal{A}^i)$ the list of representative of all k-patterns of $\mathcal{A}^i$. The partition phase reports the lists $L(u)$ of each k-pattern of each anchor sequence $\mathcal{A}^i$.

A final definition is required for the following section. Given a genomic database G and a k-pattern u, the list of all occurrence positions of $L(u)$ in G is denoted by $Oc(u)$. For example, if $G = TAGACTAAAC$ and u is the k-pattern introduced above, then $Oc(u) = \{2, 7\}$.

3.3   The Anchor Search Phase

For clarity of presentation, we describe the search phase for an anchor with two anchor sequences of the same length, and a unique seed-shape of size k. Generalization to anchors of different lengths only requires a final step to extend the longest anchor sequence. Generalization to anchors with more than two sequences requires to consider one POc and POc' list per sequence. Anchors with a single sequence are usually inefficient to consider during the search phase of an RNA descriptor. Generalization to multiple seed-shapes is straightforward.

Let $\mathcal{A} = \{X, Y\}$ be the considered anchor, with the distance constraint (d_{min}, d_{max}), and m be the common length of X and Y. Let k be the size of the considered seed-shape, and the consecutive k-patterns of each anchor sequence be $X_{1,k} \cdots X_{m-k+1,m}$ (respec. $Y_{1,k} \cdots Y_{m-k+1,m}$).

The initialization step consists in computing $m - k + 1$ pairs of lists $(Oc(X_{i,i+k-1}), Oc(Y_{i,i+k-1}))$ with respect to the genomic sequence G. Following the partition phase, each seed is an entry in the hash table and accessed in constant time. Following the Sagot-Viari methodology (Section 2.3), the two lists are then traversed and filtered with respect to the distance constraints (d_{min}, d_{max}). The list's elements are of the form (pos, num_errors), where pos represents a position in the genome and num_errors is the minimum number of errors between the G subsequence at position p and the considered k-pattern with respect to the seed-shape (errors are computed on the # positions of the seed-shape).

The following $m - k$ steps extend the consecutive k-seed surviving lists to $k + 1$-seeds, then $k + 2$-seeds, until the m-seeds surviving lists representing the complete anchor. As allowed seed lengths vary from k to m, we will number the following steps from k to m.

Step **p**, *for* $\mathbf{k} \le \mathbf{p} < \mathbf{m}$:
For each i, $1 \le i < m - p + 1$ do:

1. **Extend left** $POc(X_{i+1,i+p})$ to $POc(X_{i,i+p})$ and respectively $POc(Y_{i+1,i+p})$ to $POc(Y_{i,i+p})$ iff $1 \le i$. To do so we use the Sagot-Viari rules of model construction (extension by the character X_i or Y_i) with the exception that elements of both $POc(X_{i,i+p})$ and $POc(Y_{i,i+p})$) need to satisfy one condition out of the match, mismatch, insertion and deletion. This is because we allow

for errors in both X and Y with respect to the descriptor while the original Sagot-Viari algorithm did not have that restriction.

2. **Extend right** $POc(X_{i,i+p-1})$ to $POc(X_{i,i+p})$ and respectively $POc(Y_{i,i+p-1})$ to $POc(Y_{i,i+p})$ iff $i+p \leq m$. This extension mirrors the previous step but uses equivalent symmetric extensions to add characters X_{i+p} and Y_{i+p} respectively.

3. **Merge** the two resulting lists into a new pair of lists $POc'(X_{i,i+p})$ and $POc'(Y_{i,i+p})$. If the resulting lists contain consecutive elements representing the same position but with different numbers of errors, we keep a single copy with the minimum number of errors.

4. **Filter** $POc'(X_{i,i+p})$ and $POc'(Y_{i,i+p})$ with respect to the distance, folding and error bound constraints. The resulting lists are $POc(X_{i,i+p})$ and $POc(Y_{i,i+p})$.

In contrast with the original Sagot-Viari algorithm, errors should be allowed for both anchor sequences. The filtering step should then account, not only for the distance constraint, but also for the combined error constraints. Moreover if X and Y are two strands of a given stem, then folding constraints must be checked.

At the end of the search phase, the two remaining lists $POc(X) = POc(X_{1,m})$ and $POc(Y) = POc(Y_{1,m})$ contain all possible occurrences of both anchor sequences. The last step is then to return all occurrence pairs $S_{X,Y}$ respecting the distance, error and folding constraints.

3.4   The Check Phase

The rest of the descriptor D should finally be validated against the positions of the anchor $\mathcal{A}$. For this purpose any existing RNA search method can be used, such as BioSmatch [5] or RNAMotif if indels are not allowed.

4   Choosing the Anchor Sequences and Seed Shapes

The first idea is to choose the most constraining anchor sequences (those that are likely to give rise to the minimum number of occurrences in the database), that is those with the lowest p-value. Statistical work on structured motifs of form $X\,x(\ell, \ell+\delta)\,Y$, where X and Y are correlated by secondary structure constraints, have been done in [14]. The difficulty arise from the overlapping structure of the patterns. The p-value can be computed by brute enumeration or by sampling.

However, the most constraining anchors are not necessarily the easiest to parse. Indeed, degenerated symbols (representing sets with more than one nucleotide) can give rise to a large list of seed representative in the partition phase (see section 3.2). More precisely, anchor sequences of the same size and with the same occurrence probability may give rise to different lists of seeds representatives depending on the distribution of their degenerated positions. For example, in Table 2, though both anchor sequences **ARTGCYT** and **ACTNCAT** have the same occurrence probability of $1/4^6$ under the Bernoulli model, the second

Table 2. Size of the lists involved in each stage of the extension phases, for an exact search with the seed-shape ### on a 10M test database (from E. coli. and B. subtilis)

Anchor $\mathcal{A}$	Number of seeds in $L(\mathcal{A})$	seed occurrences in the database	seeds remaining after extensions			
			step 3	step 4	step 5	step 6
ACTGCAT	5	856408	17846	871	42	4
ARTGCYT	8	1427873	35303	2380	206	10
ACTNCAT	14	2130755	60553	3599	167	6

sequence gives rise to a larger list of seed representative leading to a much larger list of occurrences in the database. As the first extension phase of our algorithm is the most time-consuming phase, a good estimation of the total time comes from the number of seed representative.

This is further illustrated in Figure 3 where all possible anchors represented by pairs of 5-patterns from the 5S RNA helix III (see Figure 3) are searched in a database of bacterial genomes. Not surprisingly, the anchors with the fewest seeds are significantly faster. Therefore, among the most constraining anchor sequences (those with the lowest p-value), we choose those that give rise to the shortest list of seed representative $L(\mathcal{A})$.

Finally, Table 3 shows the sensibility and the speed obtained with different seed-shapes. It appears that longer shapes lead to a smaller execution time, but at the cost of a lower sensibility: some sequences are missed. Here the best compromise is to use the spaced seed-shape ##-# : the parsing time is more than 40% smaller than the time needed by RNAMotif and the sensibility remains at 99%.

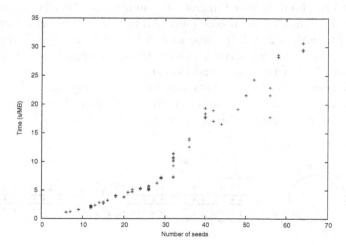

Fig. 3. Relation between the speed of the exclusion method and the number of seed representative of the anchor. We tested each possible anchor represented by pairs of 5-patterns from the 5S RNA helix III, with the seed-shape ####. The horizontal axis gives the number of seeds corresponding to each anchor pair, and the vertical axis the time taken for the search on a 130 MB bacterial database.

Table 3. Speed and sensitivity of the Exclusion method. The descriptor is an helix with 6-base stems and a loop $x(10, 50)$, searched with 1 error. It occurs 2110 times in on a 10M test database (sequences from E. coli. and B. subtilis). Sensibilities of our method are lowered by 1% due to an additional heuristic in stack transversal during the search phase. The time ratios are against the time for RNAMotif.

	RNAMotif v. 3.0.4	Exclusion			
		###	##-#	####	##-##
sensibility	100%	97%	99%	68%	67%
preprocessing time (ms)	–	684	927	1068	1262
parsing time (ms)	2709	4144 (153%)	1512 (56 %)	352 (13%)	135 (5%)
total time (ms)	2709	4828 (178%)	2439 (90 %)	1420 (51%)	1407 (45%)

5   Testing on RNA Stem-Loops

Here, we tested our new method for both quality and speed, by comparing with RNAMotif. Indeed though RNAMotif has the limitation of ignoring possible nucleotide insertions and deletions, it is an exact method thus giving a good benchmark to test our heuristic's sensitivity. Moreover it is the fastest RNA search method developed so far.

We considered three RNA families: 5S rRNAs and RNase P RNAs as in [5] as well as group II introns. In each case, the most conserved region was considered, namely helix III for 5S rRNAs, P4 region for RNase P RNAs and domain V for group II introns. The tests are performed exclusively on stem-loop signatures because of technical limitations in our current implementation. This will be extended to full structures of ncRNAs in the near future. We used a database containing 25 randomly selected microbial genomes from GenBank representing a total of more than 75 million base pairs. All tests were performed on an intel Pentium 4 PC with a 2800MHz processor, 2 GB of memory and running Fedora Core 2. The stem-loop signatures were chosen to represent various testing conditions and parameters (stem and loop size).

We considered the seed-shape ####, and two anchor sequences of size 5. Since computational time rises exponentially with the number of seed representative generated by anchor pairs, a cutoff value was selected to avoid anchors likely

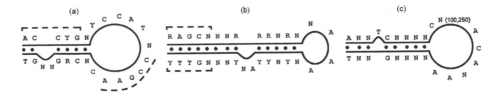

Fig. 4. The stem-loop signature used for (a) 5S rRNAs helix III, (b) group II intron domain V and (c) RNase P RNA P4 region. The dotted lines represent the specific anchor sequences selected for searching.

to generate large initial sets of occurrences from the database. It was set to 16 seed representative, based on experimental results (Figure 3). This cutoff could be raised or lowered on execution to influence speed (lower cutoff) or sensitivity (higher cutoff), but 16 has been a good compromise thus far. Both the 5s rRNA and the Intron group II consensus had several anchor pairs of size 5 falling under that cutoff. The chosen anchors are illustrated in Figure 4. The anchor sequences used for the Intron group II consensus are related by folding constraints, where as those used for the 5s rRNA are only related by distance constraints. Unfortunately, no suitable pair of anchor sequences was found for the RNAse P. This is more likely a limitation of the current implementation rather than the method since the whole consensus structure could have yielded for adequate anchors which were not present in the P4 region. However, this illustrates that certain ncRNA might not have sufficient conserved regions to select adequate anchor sequences.

We tested the ability of our exclusion method to speed up RNAMotif, in other words, the check phase was completed by using RNAMotif. Running times (Table 4, third column) are clearly improved for both 5S rRNAs and group II intron domain V. This clearly shows that the exclusion method can shave off significant amount of computational time for ncRNA searching methods. Finally, we used our method not as a filtering strategy but rather as a stand alone algorithm. In other words the exclusion method was used over the full ncRNA stem-loop signatures (Table 4, last column). As expected from the many degenerated positions in the structure consensus, the execution times are fairly slow for this setup. Here we can clearly see the relationship between conservation and execution times with the most conserved consensus structure (5S rRNAs) being significantly faster to search than the other candidates.

Table 4. Computation times obtained by running our exclusion algorithm and RNAMotif v3.0.0 on the stem-loop signature considered for each structured motif family on a database of bacterial genomes. For each method we show the times obtained when the full helix is searched and when only the most conserved subsets are searched. No suitable anchor subset was available for the RNase P RNA P4 region.

	RNAMotif	RNAMotif with Exclusion method	Exclusion on full helix
5S RNA, helix III	2.6 s/Mb	1.1 s/Mb	12.7 s/Mb
Intron group II, domain V	3.3 s/Mb	2.7 s/Mb	31.0 s/Mb
RNase P RNA P4 region	3.1 s/Mb	–	59.1 s/Mb

The database contained 71 annotated sequences of the 5S rRNA and 17 sequences of the group II intron. Of these 89 annotated ncRNA genes, only 2 weren't found by RNAMotif, both of the 5S rRNA variety. The exact same results were found by the exclusion method in combination with RNAMotif. In the case of the RNase P RNA, although we couldn't find a suitable pair of anchor sequences, using the exclusion method as a stand alone algorithm did provide the same predictions as RNAMotif, where 36 of 39 annotated sequences were found. In other words, no loss in sensitivity was observed over the tested data when compared to an exhaustive method like RNAMotif.

6   Conclusion

We have developed an exclusion method allowing for nucleotide mismatches and indels, that can be used in combination with other existing RNA search methods to speed up the search. We have shown that given sub-motifs with small degeneracy values, a hashing method built on the preprocessing of the target database can significantly improve search times. The idea is to select in the descriptor anchors which yield the least computation. That being the case, it's not given that any descriptor contains enough consecutive conservations to permit sublinear filtering. By using distance constraints we can significantly reduce the number of needed consecutive conserved positions by introducing gaps between pairs of anchors.

Furthermore, we have shown that restricting these features to the helical structures alone is not an efficient method to filter a database. This result concords with previous literature on the subject of finding signals in secondary structure alone [13]. Generalizing the problem to seeds with distance constraints without considering the secondary structures yields the best results as it takes into account signal in both secondary and primary sequence.

This filtering method is still in its early stage as we can explore many other features. It is evident from our current results that there is a bias for selecting small elements and using the largest possible seed to gain the greatest speed increase. In [16], Zhang et al. present a more robust way to select target anchors from a pattern by creative use of the pigeonhole principle. In this paper we address the same sensitivity issue through the use of "don't care" characters which gives added flexibility in choosing the anchor sequences for seeding the search. We have not yet determined if both approaches are compatible and can be defined into a single model. In any case, we plan to incorporate Zhang's filter definition into our future work to facilitate the comparison and/or addition of our parameters. Furthermore, we plan to generalize the method to an arbitrary number of anchors separated by constraint distances. This could be a viable avenue to limit the number of initial candidates to process and further lower computational times.

References

1. Altschul, S.F., Gish, W., Miller, W., Myers, E.W., Lipman, D.J.: Basic local alignment search tool. Journal of Molecular Biology 215, 403–410 (1990)
2. Bafna, V., Zhang, S.: FastR: Fast database search tool for non-coding RNA. In: Proceedings of IEEE Computational Systems Bioinformatics (CSB) Conference, pp. 52-61 (2004)

3. Eddy, S.R.: RNABOB: a program to search for RNA secondary structure motifs in sequence databases (1992) http://bioweb.pasteur.fr/docs/man/man/rnabob.1.html#toc1
4. El-Mabrouk, N., Lisacek, F.: Very fast identification of RNA motifs in genomic DNA. Application to tRNA search in the yeast genome. Journal of Molecular Biology 264, 46–55 (1996)
5. El-Mabrouk, N., Raffinot, M., Duchesne, J.E., Lajoie, M., Luc, N.: Approximate matching of structured motifs in DNA sequences. J. Bioinformatics and Computational Biology 3(2), 317–342 (2005)
6. Fichant, G.A., Burks, C.: Identifying potential tRNA genes in genomic DNA sequences. Journal of Molecular Biology 220, 659–671 (1991)
7. Gautheret, D., Major, F., Cedergren, R.: Pattern searching/alignment with RNA primary and secondary structures. Comput. Appl. Biosci. 6(4), 325–331 (1990)
8. Klein, R., Eddy, S.: RSEARCH: Finding homologs of single structured RNA sequences (2003)
9. Laslett, D., Canback, B.: ARAGORN, a program to detect tRNA genes and tmRNA genes in nucleotide sequences. Nucleic Acids Research 32, 11–16 (2004)
10. Li, M., Ma, B., Kisman, D., Tromp, J.: PatternHunter II: Highly Sensitive and Fast Homology Search. Journal of Bioinformatics and Computational Biology 2(3), 417–439 (2004) Early version in GIW 2003
11. Ma, B., Tromp, J., Li, M.: PatternHunter: faster and more sensitive homology search. Bioinformatics 18(3), 440–445 (2002)
12. Macke, T., Ecker, D., Gutell, R., Gautheret, D., Case, D.A., Sampath, R.: RNAmotif – a new RNA secondary structure definition and discovery algorithm. Nucleic Acids Research 29, 4724–4735 (2001)
13. Rivas, E., Eddy, S.R.: Secondary Structure Alone is Generally Not Statistically Significant for the Detection of Noncoding RNAs. Bioinformatics 16(7), 583–605 (2000)
14. Robin, S., Daudin, J.-J., Richard, H., Sagot, M.-F., Schbath, S.: Occurrence probability of structured motifs in random sequences. J. Comp. Biol. 9, 761–773 (2002)
15. Sagot, M.F., Viari, A.: Flexible identification of structural objects in nucleic acid sequences: palindromes, mirror repeats, pseudoknots and triple helices. In: Hein, J., Apostolico, A. (eds.) Combinatorial Pattern Matching. LNCS, vol. 1264, pp. 224–246. Springer, Heidelberg (1997)
16. Zhang, S., Borovok, I., Aharonovitz, Y., Sharan, R., Bafna, V.: A sequence-based filtering method for ncRNA identification and its application to searching for riboswitch elements. Bioinformatics 22(14), e557–e565 (2006)

A New Quartet Approach for Reconstructing Phylogenetic Trees: Quartet Joining Method

Lei Xin, Bin Ma, and Kaizhong Zhang

Computer Science Department, University of Western Ontario,
London N6A 5B7, Canada
lxin3@uwo.ca, bma@csd.uwo.ca, kzhang@csd.uwo.ca

Abstract. In this paper we introduce a new quartet-based method for phylogenetic inference. This method concentrates on reconstructing reliable phylogenetic trees while tolerating as many quartet errors as possible. This is achieved by carefully selecting two possible neighbor leaves to merge and assigning weights intelligently to the quartets that contain newly merged leaves. Theoretically we prove that this method will always reconstruct the correct tree when a completely consistent quartet set is given. Intensive computer simulations show that our approach outperforms widely used quartet-based program TREE-PUZZLE in most of cases. Under the circumstance of low quartet accuracy, our method still can outperform distance-based method such as Neighbor-joining. Experiments on the real data set also shows the potential of this method. We also propose a simple technique to improve the quality of quartet set. Using this technique we can improve the results of our method.

1 Introduction

With the accumulation of phylogenetic data in recent years, the computational biology community has shown great interest in the reconstruction of large evolutionary trees from smaller sub-trees[5,9,10,11,12]. The quartet-based method may be the simplest and most natural approach for this kind of problem. This approach usually takes two major steps to complete the reconstruction. First, a set of four-leaf subtrees i.e. quartets are built for every possible four sequences in a DNA or protein sequence set using other phylogeny methods such as maximum likelihood (ML) [2] or neighbor-joining (NJ) [8]. Then a combinatorial technique is applied to reconstruct the entire evolutionary tree according to the topology relations between the quartets built in the first step. It is well known that with the currently existing methods it is hard to build accurate quartet set[1]. So the efficiency of a quartet-based method really depends on how successful it is in tolerating quartet errors. The advantage of this approach is that it can be quite flexible. It can use the distance matrix if in the first step NJ is used. It can also be totally independent of the distance matrix if ML or some other methods are used.

In the last ten years, many efforts have been made to develop efficient quartet-based algorithms. A prominent approach is known as quartet puzzling (QP) [9].

G. Lin (Ed.): COCOON 2007, LNCS 4598, pp. 40–50, 2007.

The corresponding program package is called TREE-PUZZLE which is widely used in practice. Recently even a parallelized version was developed for this package[11]. The main dispute around quartet puzzling is that it is outperformed by faster neighbor-joining method in computer simulations. One attempt to improve quartet puzzling is weighted optimization (WO). Although WO is better than QP, it is still outperformed by neighbor-joining in most of cases. The authors stated in their paper[5]: *"Despite the fact that there were about 90% of the correct quartets in Q_{max}, we observed that the correct tree was not well inferred in comparison with other inference methods* (including ML and NJ)". A recent paper takes another approach to solve the problem of quartet errors[12]. This approach builds a series of evolutionary trees rather than a single final tree. Their results show that with multiple trees this approach can ensure high accuracy in computer simulations. However, we noticed if only one final tree was allowed, their method was still outperformed by neighbor-joining except on one type of tree model.

In this article, we concentrate on the second step of reconstructing entire phylogenetic tree from the quartet set. We introduce a new approach based on quartet: quartet joining (QJ). This method achieves error toleration by carefully selecting two possible neighbor leaves to merge and assigning weights intelligently to the quartets that contain newly merged leaves. Theoretically we can prove that this method will always reconstruct correct tree when a completely consistent quartet set is given. Intensive computer simulations also show that our approach outperforms TREE-PUZZLE package in most of cases. Under the circumstance of low quartet accuracy, our method can outperform distance-based method such as Neighbor-joining. The potential of this algorithm is also demonstrated by the experiments on the real data set. This algorithm executes in $O(n^4)$ time which is much faster than QP's $O(Mn^4)$ where M is usually greater than 1000. n is the number of sequences in the data set.

This article is organized as follows: in Section 2, we introduce some necessary notations and definitions. In Section 3, we describe quartet joining method and prove its important property. In Section 4, computer simulation results and experiment results on real data set are given. In Section 5, a simple technique is introduced to improve the results from section 4. Conclusion and future work are discussed in Section 6.

2   Notations and Definitions

In the field of molecular phylogeny, DNA or protein sequences are represented by leaf nodes on the evolutionary tree. A quartet is a set of four leaf nodes which is associated with seven possible pathway structures: three of them are fully resolved unrooted trees, three are partially resolved trees which we can not distinguish between two of three fully resolved tree, the last one is a fully unresolved tree. A figure of three fully resolved trees and one fully unresolved tree of leaf nodes $\{a, b, c, d\}$ are shown in figure 1. We will use $\{ab|cd\}$, $\{ad|cb\}$, $\{ac|bd\}$, $\{abcd\}$ to represent these four different structures. Clearly a tree with n leaf

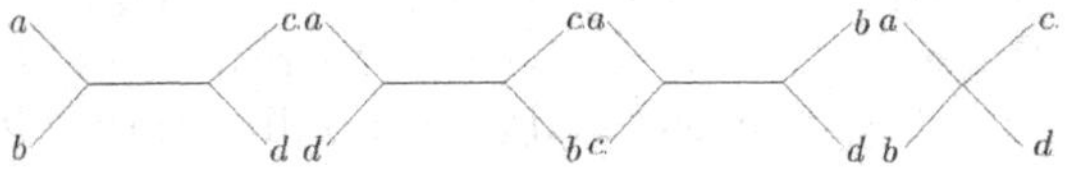

Fig. 1. Four different quartet structures

nodes will have C_n^4 quartets. This quartet set is denoted by Q_T. We say a quartet is *consistent* with an evolutionary tree T if it belongs to Q_T. A quartet set Q is *completely consistent* with evolutionary tree T if $Q = Q_T$.

Two leaf nodes are called *neighbors* if they are connected to the same internal node of the evolutionary tree. This is an important notation and will be mentioned frequently in this paper. When two leaf nodes are merged together to form a new node of the tree, this new node is called *supernode*. Any of four nodes of a quartet could be supernode. If all four nodes of a quartet are single leaf nodes, this quartet is called *single-node quartet* otherwise it is called *supernode quartet*. In this paper we will use capital letters to represent supernodes and small letters to represent single leaf nodes.

3   Quartet-Joining Algorithm

The quartet-joining algorithm follows the bottom-up scheme to build an unrooted evolutionary tree. It starts from n leaf nodes. At every step it merges two possible neighbors according to the quartet set and form one new leaf node until three nodes are left. At last, it joins these three nodes together to get a fully resolved evolutionary tree. The input and output of quartet-joining algorithm are listed below:

Input: A set of quartets.
Output: A fully resolved tree with sequences on leaf nodes.

The key to this kind of algorithms is how to decide which two nodes are most likely to be neighbors on the true evolutionary tree. We notice that for a given tree of n leaf nodes, if any two leaf nodes i, j are neighbors on the tree then the number of the quartets of the form $\{ij|kl\}$ is C_{n-2}^2. If any two leaf nodes are not neighbors then the number of the quartets of the form $\{ij|kl\}$ must be less than C_{n-2}^2 In other words, the quartets number of a pair of leaf nodes which are neighbors is maximal among any pair of leaf nodes. This fact has been used by several other methods to deduce neighbors. In this article, rather than apply the fact directly, we design a special mechanism to fully mine the information contained in the quartet set. Here we define *support* to be the weight that supports two leaf nodes to be neighbors. We use *confidence* to represent the weight of quartets including single-node quartets and supernode quartets. Now we need to choose support and confidence wisely such that the algorithm would not have bias on any type of trees. An intuition is that the confidence of supernode quartet can be defined as the percentage of consistent single node quartets it contains. The support will then be the sum of confidence of all the quartets that support two nodes to be pairs. However experiments show this definition has bias on certain type of trees. So here we use a different approach. In the quartet-joining algorithm, the confidence of the supernode quartet $\{XY|UV\}$ is defined as:

$$C(X, Y; U, V) = \sum_{x \in X, y \in Y, u \in U, v \in V} w(x, y; u, v) \tag{1}$$

where X, Y, U, V are supernodes and we use $|X|$ to denote the number of single nodes which are contained in the supernode X. $w(x, y; u, v)$ is the weight of input quartets.

For a supernode quartet $\{XY|UV\}$, a single-node quartet induced from $\{XY|UV\}$ is the quartet whose four single nodes are taken from supernodes X, Y, U, V separately. So the number of single-node quartets that can be induced from $\{XY|UV\}$ is $|X||Y||U||V|$. When every supernode only contains one single node, $C(X, Y; U, V)$ degenerates to $w(x, y; u, v)$. The support of a supernode pair (X, Y) is defined as:

$$supp(X, Y) = \frac{\overline{supp}(X, Y)}{T(X, Y)} \tag{2}$$

where

$$\overline{supp}(X, Y) = \sum_{U \neq V, \ U, V \neq X, Y} C(X, Y; U, V) \tag{3}$$

$$T(X, Y) = |X||Y| \sum_{U \neq V, \ U, V \neq X, Y} |U||V| \tag{4}$$

From the formula above, we can see the support of a pair (X, Y) is basically the averaged weights of the quartets that support (X, Y) to be neighbors. The advantage of this definition is that it balances the weights contributed by supernode quartets and single node quartets. Thus there will be no information loss during the process of merging leaf nodes. All the information from quartet set is utilized. Therefor QJ method can reconstruct the evolutionary tree more accurately than other quartet-based methods. Later on we will also prove that this definition of support will ensure reconstructing the true evolutionary tree when a completely consistent quartet set is given.

If at every step we calculate $C(X, Y; U, V)$ and $supp(X, Y)$ directly, it results in an algorithm of $O(n^5)$ time complexity. So we will use the equations below to update the values of $C(X, Y; U, V)$ and $supp(X, Y)$ instead of recalculating all the values of the two matrices at every step. This will give us an algorithm of $O(n^4)$ time complexity.

Let $C_n(X, Y; U, V)$ and $supp_n(X, Y)$ denote the matrices at step n. A_1, A_2 are the supernodes to be merged at step $n - 1$ and A_1, A_2 will be replaced by a new node A. Then we have the updating formulae:

$$C_n(A, B; C, D) = C_{n-1}(A_1, B; C, D) + C_{n-1}(A_2, B; C, D) \tag{5}$$

For the number of single nodes contained in supernode A, we have:

$$|A| = |A_1| + |A_2| \tag{6}$$

Let S_n denote the remaining nodes set at step n. For the support of the pairs that contain the new node, we have:

$$\overline{supp_n}(A, B) = \sum_{C \neq D, C, D \in S_n} C_n(A, B; C, D) \tag{7}$$

For other nodes, we have:

$$\overline{supp_n}(C, D) = \overline{supp_{n-1}}(C, D) - \sum_{V \in S_{n-1}} C_{n-1}(C, D; A_1, V) \tag{8}$$

$$- \sum_{V \in S_{n-1}} C_{n-1}(C, D; A_2, V) + C_{n-1}(C, D; A_1, A_2) \tag{9}$$

$$+ \sum_{V \in S_n} C_n(C, D; A, V) \tag{10}$$

$$= \overline{supp_{n-1}}(C, D) - C_{n-1}(C, D; A_1, A_2) \tag{11}$$

The formula above comes from fact that all the confidences don't change except confidences concerning A_1, A_2 and A. We will delete nodes A_1 or A_2 and insert node A. So for the new support matrix, we just subtract the influence of A_1 and A_2 then add back the influence of A. We also add one $C_{n-1}(C, D; A_1, A_2)$ just because we subtract it twice before. The we use equation 5, the final updating formula can be simplified as (11). Also we need to update $T(X, Y)$. For the new nodes we just recalculate using equation (4). For other nodes:

$$T_n(C, D) = T_{n-1}(C, D) - |A_1||A_2||C||D| \tag{12}$$

The quartet-joining algorithm are formally described below:

1. Use $w(x, y, u, v)$ to initialize confidence matrix $C(X, Y; U, V)$.
2. Compute support matrix $supp(X, Y)$ from confidence matrix using equation (2) (3) (4).
3. Find the pair (i, j) which has the maximal value in the support matrix
4. Connect nodes i, j to an internal node then replace them with a supernode k: delete i, j from the set of organisms and insert k.
5. Update confidence matrix , support matrix and $T(X, Y)$ with the updating equations above.
6. Repeat 3 to 5 until only three nodes remain.
7. Connect the three remaining nodes to an internal node. Output the final tree.

We have the following important property for quartet-joining algorithm:

Theorem 1. *If the input quartet set is completely consistent with evolutionary tree T, the quartet-joining algorithm will reconstruct the exact evolutionary tree T.*

To prove this theorem we need to prove a lemma first. Let $T_0 = T$. T_n is a subtree of T with some nodes merged and removed at step n. A figure of T_n is shown in figure 2. If two nodes A, B of T_{n-1} are merged at step $n-1$, T_n will be a subtree of T_{n-1} with A, B removed. Here we point out that although A, B are supernodes in T, they are treated as single nodes in T_{n-1}.

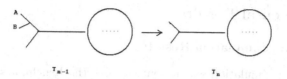

Fig. 2. Definition of T_n tree

Lemma 1. *If the input quartet set is completely consistent with evolutionary tree T, in the quartet-joining algorithm, $C_n(A, B; C, D)$ equals $|A||B||C||D|$ if and only if quartet topology $\{AB|CD\}$ is consistent with tree T_n. Otherwise $C_n(A, B; C, D) = 0$.*

We use induction to prove this lemma.

Proof. At step 0, every supernode only contains 1 single node. Because the input quartet set is completely consistent with T, it is clear that $C_0(A, B; C, D) = 1 = |A||B||C||D|$ when $\{AB|CD\}$ is consistent with tree T_0. $C_0(A, B; C, D) = 0$ when $\{AB|CD\}$ is not consistent with tree T_0.

Assume lemma holds at step $n - 1$. Assume A_1, A_2 are the neighbors of T_{n-1} to be merged at step $n - 1$. Because A_1, A_2 are neighbors, both $\{A_1B|CD\}$ and $\{A_2B|CD\}$ will be consistent with T_{n-1} or both will be inconsistent with T_{n-1} at the same time. We assume lemma holds at step $n-1$, so $C_{n-1}(A_1, B; C, D) = C_{n-1}(A_2, B; C, D) = |A||B||C||D|$ or $C_{n-1}(A_1, B; C, D) = C_{n-1}(A_2, B; C, D) = 0$. When $\{A_1B|CD\}$ and $\{A_2B|CD\}$ are consistent with T_{n-1}, from the definition of T_n, we can see at this time $\{AB|CD\}$ is consistent with T_n. From equation (5), we have $C_n(A, B; C, D) = |A_1||B||C||D| + |A_2||B||C||D|$. Take a look of equation (6), we will know $C_n(A, B; C, D) = |A||B||C||D|$. When $\{A_1B|CD\}$ and $\{A_2B|CD\}$ are inconsistent with T_{n-1}, $\{AB|CD\}$ is inconsistent with T_n and $C_{n-1}(A_1, B; C, D) = C_{n-1}(A_2, B; C, D) = 0$. Surely at this time $C_n(A, B; C, D) = 0$. For other nodes, confidence concerning them don't change, lemma also holds. $\square$

Now we prove Theorem 1.

Proof. From Lemma 1 and definition of $T_n(C, D)$ in equation (12), we know $supp_n(A, B) \leq 1$ and $supp_n(A, B) = 1$ if and only if (A, B) are neighbors in tree T_n. This means that the maximal value in the support matrix will give us a pair of neighbors. Notice the fact that for a bottom up tree-building scheme, if at every step, a pair of neighbors on the evolutionary tree are merged, this will finally give us the exact evolutionary tree T. Thus Theorem 1 holds. $\square$

Quartet-joining algorithm will take $n - 3$ steps to build a fully resolved tree, n is the number of sequences to be studied. Initialization needs $O(n^4)$ time. At every step updates of confidence and support matrix need $O(n^3)$. So the time complexity for the algorithm is $O(n^4)$. Because we need to keep the confidence matrix, it is a four-dimension matrix, the space complexity is also $O(n^4)$.

4   Experimental Results

4.1   Computer Simulation Results

In the computer simulation experiment, we use the benchmarks developed by Ranwez and Gascuel to test and compare different phylogeny methods [5]. Six model trees, each consisting of eight leaf nodes, are used to generate test sets under various situations. Among them, AA, BB, AB are molecular clock-like trees while the other three CC, DD, CD, have different substitution rate among lineages. Three evolutionary rates are considered: their maximal pair-wise divergences (MD) are 0.1, 0.5, 0.8 substitution per site ranging from very low to fast. For each group of parameters, sequences of length 300, 600 and 900 sites are generated. A total data set of 86,400 DNA sequences are generated using Seq-Gen[4]. The evolutionary model used here is F81.

In this paper, we concentrate on reconstructing evolutionary tree from a set of quartets. So we just use a very simple method to assign weights to quartets. Let D_{ij} to be the distance between generated sequences i and j. The weights of input quartets are computed as:

$$
w(x, y; u, v) = \begin{cases} 1 & D_{xy} + D_{uv} < D_{xu} + D_{yv}, D_{xy} + D_{uv} < D_{xv} + D_{yu}; \\ \frac{1}{2} & D_{xy} + D_{uv} = D_{xu} + D_{yv} < D_{xv} + D_{yu} \\ & or\, D_{xy} + D_{uv} = D_{xv} + D_{yu} < D_{xu} + D_{yv}; \\ \frac{1}{3} & D_{xy} + D_{uv} = D_{xv} + D_{yu} = D_{xu} + D_{yv}; \\ 0 & \text{else.} \end{cases} \tag{13}
$$

Three phylogeny methods are tested on this data set: QJ, NJ and QP. Here we emphasize again that QJ is not a distance based method. It can reconstruct the phylogenetic tree without the distance matrix as long as the quartet set is provided. QP has the same property. On the other hand, NJ is a pure distance based method. Its reconstruction must be based on the distance matrix. In this experiment, for QP method, the tree appears most frequently in the building process is selected as the output.

The final results are shown in Fig 3. The Y-axis in Fig 3 is the average correctly reconstructed tree percentage for six model trees. The X-axis is the expected number of substitution in DNA sequences i.e. MD times sequence length. The corresponding data table is shown in table 1. The figures in the table are the correct tree percentage for each type of trees. The figures in the parenthesis are the correct quartet percentage i.e. the percentage of input quartet set that actually belongs to the quartet set of model trees.

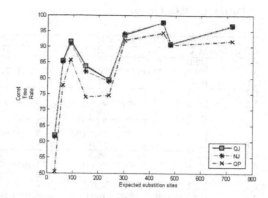

Fig. 3. Average correctly reconstructed tree percentage

From Fig 3, we can clearly see that in the average sense, the performance of QJ and NJ are quite close: their performance lines are almost overlapping except a small portion where QJ outperforms NJ slightly. But when compared to QP, the performance advantage of QJ is quite obvious: QJ outperforms QP in all the combination of evolutionary rate and sequence length. If we check the data table, we can find some other patterns. First is that within the evolutionary rates considered, faster evolutionary rates correspond to more accurate quartet set. When the correct quartet rate is relatively low, QJ performs slightly better than NJ. For example, when sequence length=300, MD=0.1, for the model tree AA, the correct quartet rate is 88%, QJ outperforms NJ about 4%. However, if the correct quartet rate is high, both QJ and NJ can do very well. Another thing should be noticed is that compared to other five tree models, QP method performs better on CD type of tree. When the evolutionary rates are high, it even can outperform QJ and NJ slightly.

4.2 Real Data Set Experiment

The real data set comes from GenBank. They are mitochondrial DNA sequences of 36 different mammals. The evolutionary tree of these 36 mammals are reconstructed by biologists using maximum likelihood method. [6].

Each of these 36 DNA sequences has about 16k sites. To estimate the pairwise evolutionary distances between these sequences, we will use the metric of strings introduced by Bin Ma and Ming Li[3]. We apply QJ, NJ and QP methods to this data set and compare the results with the tree reconstructed by maximum likelihood method. The RF distances[7] between the trees reconstructed by above three methods and the tree rebuilt by maximum likelihood method is shown in Table 2.We can see from the table that on the real data set, the results of quartet-joining method is closer to the result of maximum likelihood method than the other two methods.

Table 1. Data for each type of trees

	Algorithms	WITH CLOCK			NO CLOCK		
		AA	BB	AB	CC	DD	CD
sequence length 300	MD≈ 0.1 QJ	32.0	73.5	76.5	56.5	71.5	61.5
	NJ	28.0(88)	77.0(96)	76.5(96)	51.0 (92)	72.5 (96)	63.5 (95)
	QP	8.0	65.5	67.0	34.0	75.0	53.5
	MD≈ 0.5 QJ	59.5	97.5	96.0	73.0	93.0	83.5
	NJ	55.5 (91)	98.0 (98)	96.5 (98)	65.5 (94)	92.5 (98)	84.5(95)
	QP	24.5	90.5	86.5	67.5	89.5	85.5
	MD≈ 0.8 QJ	55.5	94.5	95.5	60.5	97.0	74.0
	NJ	51.0(91)	95.0 (97)	96.0 (98)	54.5 (92)	96.0 (98)	81.5 (97)
	QP	23.5	92.0	87.0	69.0	93.5	82.0
sequence length 600	MD≈ 0.1 QJ	63.5	97.0	96.5	78.0	92.5	85.0
	NJ	62.0 (94)	97.5 (99)	96.0(99)	77.5 (96)	92.0 (99)	86.0(98)
	QP	42.0	92.5	92.0	62.5	92.5	84.5
	MD≈ 0.5 QJ	80.5	100.0	100.0	89.0	100.0	93.0
	NJ	82.5 (96)	99.5 (99)	100.0 (99)	85.5 (96)	100.0 (99)	97.0(98)
	QP	66.5	99.0	99.5	91.5	98.0	97.5
	MD≈ 0.8 QJ	78.5	99.0	100.0	78.0	100.0	89.5
	NJ	75.5 (95)	99.5 (99)	100.0 (99)	76.0 (93)	100.0 (99)	93.0(97)
	QP	53.0	98.5	98.0	95.5	99.5	98.0
sequence length 900	MD≈ 0.1 QJ	78.0	99.5	99.0	79.5	97.5	96.0
	NJ	77.0 (95)	98.5 (99)	99.0 (99)	75.5 (95)	97.5 (99)	98.0(99)
	QP	64.0	97.5	98.5	65.5	95.5	93.0
	MD≈ 0.5 QJ	92.5	100.0	99.5	94.5	100.0	99.5
	NJ	92.5 (98)	100.0 (100)	99.5 (99)	94.0 (98)	100.0 (99)	100.0(99)
	QP	81.5	99.5	99.5	86.0	99.5	100.0
	MD≈ 0.8 QJ	93.5	100.0	100.0	93.5	100.0	92.5
	NJ	93.0 (97)	100.0 (100)	100.0 (100)	89.5 (97)	100.0 (99)	97.0(99)
	QP	75.0	100.0	100.0	76.0	100.0	99.5

Table 2. Real Data Set Results

Algorithms	RF distances
QJ	12
NJ	16
QP	20

5   Improvement of Experimental Results

No matter how successfully a quartet-based method can tolerate quartet errors, the accuracy of input quartet set will still influence final results. Here, we propose a simple technique than can improve our experimental results when the evolutionary rate is low. Notice that for a quartet 1,2,3,4, if we have

$$D_{13} + D_{24} < D_{14} + D_{23}, D_{13} + D_{24} < D_{12} + D_{34} \qquad (14)$$

but

$$D_{12} + D_{34} \neq D_{14} + D_{23} \qquad (15)$$

Then we will have unresolved case as shown in Fig 4.

It could either take the form of $\{13|24\}$ or $\{12|34\}$. The internal branch length a, b can be directly computed from pairwise distance. The intuition is if a is longer, then the quartet will have more chance to take the form of $\{13|24\}$. If

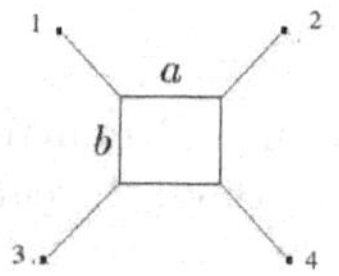

Fig. 4. Unresolved quartet

Table 3. Improved Results

			WITH CLOCK			NO CLOCK		
		Algorithms	AA	BB	AB	CC	DD	CD
sequence length 300	MD≈ 0.1	QJ Imp	36.5	75.5	76.5	66	78.5	67.5
		QJ	32.0	73.5	76.5	56.5	71.5	61.5
		NJ	28.0	77.0	76.5	51.0	72.5	63.5
		QP	8.0	65.5	67.0	34.0	75.0	53.5

b is longer, the quartet will have more chance to take the other form. So here instead of assigning a binary value to quartet weights, we will assign probability values estimated from a and b. But we find this technique works only when the evolutionary rate is low. The improved results are shown in Table 3. From the first row, we can clearly see the improvement of QJ method. By using this technique, QJ totally outperform NJ when MD equals 0.1.

6   Conclusion

In this paper, we propose a new quartet approach for reconstructing phylogenetic trees: quartet-joining method. QJ method will execute in $O(n^4)$ time. Theoretically it will build a fully resolved tree that perfectly represents the input quartet set when the quartet set is completely consistent with the evolutionary tree. In computer simulations, it totally outperforms popular QP method in the average sense. When compared to NJ, their performances are quite close except QJ performs a little bit better. However when a special technique is used, QJ can gain quite performance advantage over NJ when the evolutionary rate is low. We also test QJ method on the real data set and the results show its output is closer to maximum likelihood method than QP and NJ. Currently QJ gets its input quartet set from a very simple method. We expect when more complex methods like maximum likelihood are used to generate quartet weights, QJ can gain some performance advantage over NJ under most conditions.

Acknowledgement

This research is supported by Natural Sciences and Engineering Research Council of Canada.

References

1. Adachi, J., Hasegawa, M.: Instability of quartet analyses of molecular sequence data by the maximum likelihood method: the cetacean/artiodactyla relashionships Cladistics, Vol. 5, pp.164-166 (1999)
2. Felsenstein, J.: Evolutionary trees from DNA sequences: a maximum likelihood approach. J. Mol. Evol. 17, 368–376 (1981)
3. Li, M., Chen, X., Li, X., Ma, B., Paul, M.B.: The Similarity Metric. IEEE Transsections On Information Theory, vol. 50(12) (2004)
4. Rambaut, A., Grassly, N.C.: Seq-Gen: An application for the Monte Carlo simulation of DNA sequence evolution along phylogenetic trees. Comput. Appl. Biosci. (1996)
5. Ranwez, V., Gascuel, O.: Quartet-Based Phylogenetic Inference:Improvement and Limits. Mol. Biol. Evol. 18(6), 1103–1116 (2001)
6. Reyes, A., Gissi, C., Pesole, G., Catzeflis, F.M., Saccone, C.: Where Do Rodents Fit? Evidence from the Complete Mitochondrial Genome of Sciurus vulgaris. Molecular Biology and Evolution 17, 979–983 (2000)
7. Robinson, D.F., Foulds, L.R.: Comparison of phylogenetic trees. Math. Biosci. 53, 131–147 (1981)
8. Saitou, N., Nei, M.: The Neighbor-joining Method: A New Method for Reconstructing Phylogenetic Trees. Mol. Bio. Evol. 4(4), 406–425 (1987)
9. Strimmer, K., Goldman, N., Von Haeseler, A.: Quartet puzzling:a quartet maximum-likeihood method for reconstructing tree topologies. Mol. biol. E 13, 964–969 (1996)
10. Strimmer, K., Goldman, N., Von Haeseler, A.: Bayesian probabilities and quartet puzzling. Mol. biol. E 14, 210–211 (1997)
11. Schmidt, H.A., Strimmer, K., Vingron, M.: TREE-PUZZLE: maximum likelihood phylogenetic analysis using quartets and parallel computing. Bioinformatics 18(3), 502–504 (2002)
12. Zhou, B.B., Tarawneh, M., Wang, C.: A Novel Quartet-based Method for Phylogenetic Inference. In: Proceeding of 5th IEEE Symposium on Bioinformatics and Bioengineering 2005, IEEE Computer Society Press, Los Alamitos (2005)

Integer Programming Formulations and Computations Solving Phylogenetic and Population Genetic Problems with Missing or Genotypic Data

Dan Gusfield[1], Yelena Frid[1], and Dan Brown[2]

[1] Department of Computer Science, University of California, Davis
gusfield@cs.ucdavis.edu
[2] David R. Cheriton School of Computer Science, University of Waterloo, Canada
browndg@cs.uwaterloo.ca

Abstract. Several central and well-known combinatorial problems in phylogenetics and population genetics have efficient, elegant solutions when the input is complete or consists of haplotype data, but lack efficient solutions when input is either incomplete, consists of genotype data, or is for problems generalized from decision questions to optimization questions. Unfortunately, in biological applications, these harder problems arise very often. Previous research has shown that integer-linear programming can sometimes be used to solve hard problems in practice on a range of data that is realistic for current biological applications. Here, we describe a set of related integer linear programming (ILP) formulations for several additional problems, most of which are known to be NP-hard. These ILP formulations address either the issue of missing data, or solve *Haplotype Inference Problems* with objective functions that model more complex biological phenomena than previous formulations. These ILP formulations solve efficiently on data whose composition reflects a range of data of current biological interest. We also assess the biological quality of the ILP solutions: some of the problems, although not all, solve with excellent quality. These results give a practical way to solve instances of some central, hard biological problems, and give practical ways to assess how well certain natural objective functions reflect complex biological phenomena. Perl code to generate the ILPs (for input to CPLEX) is on the web at wwwcsif.cs.ucdavis.edu/~gusfield.

1   Introduction

Several well-studied problems in computational phylogenetics and population genetics have efficient, elegant solutions when the data is "simple" or "ideal", but lack efficient solutions when the data is more complex, due to recombination, homoplasy, missing entries, site incompatibility, or the need to use genotypic rather than haplotypic data. Similarly, optimization variants of many decision problems are often more difficult to solve. In this paper we discuss integer linear

G. Lin (Ed.): COCOON 2007, LNCS 4598, pp. 51–64, 2007.
© Springer-Verlag Berlin Heidelberg 2007

programming (ILP) formulations for several such problems and report on empirical investigations done with these formulations. For most of the problems, the concept of incompatibility is fundamental.

Definition: Given a matrix M whose entries are 0's and 1's, two sites (or columns) p and q in M are said to be *incompatible* if and only if there are four rows in M where columns p and q contain all four of the ordered pairs 0,1; 1,0; 1,1; and 0,0. The test for the existence of all four pairs is called the "four-gamete test" in population genetics.

The concept of incompatibility is central to many questions concerning phylogenetic trees and population histories [14,25,7] for the following reason. Considering each row of M as a binary sequence, the classic Perfect Phylogeny Theorem says that there is a rooted phylogenetic tree that derives the sequences in M, starting from some unspecified root sequence and using only one mutation per site, if and only if *no* pair of sites of M are incompatible. Moreover, if the root sequence is specified, then the phylogenetic tree is *unique*. For expositions of this classic result, see [9,25]. The assumption of one mutation per site is called the "infinite sites" assumption in population genetics.

2   Missing Data Problems

When there are no missing values in M, the decision question of whether there are incompatible pairs of sites can be answered in linear time [8], and of course the number of incompatible pairs can trivially be computed in polynomial time. However, the situation is more interesting and realistic when some entries of M are missing, and this leads to three natural, biologically motivated problems, which we call M1, S1, and R1.

2.1   Imputing Values to Minimize Incompatibility

Problem M1: Given a binary matrix M with some entries missing, fill in (impute) the missing values so as to *minimize* the number of incompatible pairs of sites in the resulting matrix M'.

Problem M1 generalizes the following decision question: Can the missing values be imputed so that the the sequences in M' can be generated on a perfect phylogeny? That decision question has an efficient, elegant solution [21] when a required root sequence of the unknown phylogeny is specified as input, but is NP-complete when the root sequence is not specified [27]. When the missing values can be imputed so that M' has no incompatible site pairs, the sequences in M are consistent with the hypothesis that they originated from a perfect phylogeny; when the values cannot be imputed with zero incompatibilities, the solution to Problem M1 gives a measure of the deviation of M from the Perfect Phylogeny model.

The ILP Formulation for Problem M1. Our ILP formulation for Problem M1 is direct and simple. Its importance is that it generally solves very quickly,

Table 1. Six rows and two columns in the input M to Problem M1

M	p	q
1	0	0
2	?	1
3	1	0
4	?	?
5	?	0
6	0	?

imputing missing values with high accuracy, and that it can be built upon to address more complex problems. More generally, these formulations and computations illustrate that for applied problems whose range of data is known, the fact that a problem is NP-hard does not necessarily imply that exact solutions cannot be efficiently obtained for that data.

The ILP for problem M1 has one binary variable $Y(i,j)$ for each cell (i,j) in M that is missing a value; the value given to $Y(i,j)$ is then the imputed value for $M(i,j)$. The program that creates the ILP for problem M1 identifies all pairs of columns (p,q) of M that are not necessarily incompatible, but can be made incompatible depending on the imputed values are set. We let P be the set of such pairs of columns; for each pair (p,q) in P, the program creates a variable $C(p,q)$ in the formulation, which will be forced to 1 whenever the imputations cause an incompatibility between columns p and q. For each pair in P, the program also determines which of the four binary combinations are not presently found in column pair (p,q); let $d(p,q)$ represent those missing (deficient) binary combinations. The program creates a binary variable $B(p,q,a,b)$ for every ordered binary combination a,b in $d(p,q)$; $B(p,q,a,b)$ will be forced to 1 if the combination a,b has been created (through the setting of the Y variables) in *some* row in columns (p,q). The program next creates inequalities that set a binary variable $C(p,q)$ to 1 if $B(p,q,a,b)$ has been set to 1 for *every* combination a,b in $d(p,q)$. Therefore, $C(p,q)$ is set to 1 if (but not only if) the imputations of the missing values in columns (p,q) cause those sites to be incompatible. To explain the formulation in detail, consider the pair of columns shown in Table 1, where a missing entry is denoted by a "?". Then $d(p,q)$ is $\{(0,1),(1,1)\}$.

For each pair a,b in $d(p,q)$, the program will make one inequality involving $B(p,q,a,b)$ for each row r where the pair a,b can be created in columns p,q. The specific inequality for a pair a,b and a row r depends on the specific pair a,b and whether there is a fixed value in row r in sites p or q. The full details are simple and omitted but explained for variable $B(p,q,1,1)$ using the above example. Those inequalities are:

$$Y(2,p) \le B(p,q,1,1) \tag{1}$$

$$Y(4,p) + Y(4,q) - B(p,q,1,1) \le 1, \tag{2}$$

which force the variable $B(p, q, 1, 1)$ to 1 when the missing value in the second row is set to 1 or the two missing values in the fourth row are set to 1; this is the general pattern for this combination.

In the above example, the inequalities to set variable $B(p, q, 0, 1)$ are:

$$Y(2, p) + B(p, q, 0, 1) \geq 1 \tag{3}$$
$$Y(4, q) - Y(4, p) - B(p, q, 0, 1) \leq 0 \tag{4}$$
$$Y(6, q) - B(p, q, 0, l) \leq 0 \tag{5}$$

The ILP for the example has the following inequality to set the value of variable $C(p, q)$ to one, *if* all the combinations in $d(p, q)$ have been created in the column pair (p, q):

$$C(p, q) \geq B(p, q, 1, 1) + B(p, q, 0, 1) - 1$$

In general, the constant on the right-hand side of the inequality is one less than the number of pairs in $d(p, q)$. These inequalities assure that $C(p, q)$ will be forced to 1 if (but not only if) the missing values in columns (p, q) are imputed (by the setting of the Y variables) in a way that makes site pair (p, q) incompatible.

The overall objective function for the ILP is therefore to Minimize $[\|F\| + \sum_{(p,q) \in P} C(p, q)]$, where F is the set of pairs of incompatibilities forced by the initial 0 and 1 entries in matrix M. We include $|F|$ in the objective for continuity in a later section.

Because the objective function calls for minimizing, we do not need inequalities to assure that $C(p, q)$ will be set to 1 *only if* the missing values in columns (p, q) are imputed in a way that makes site pair (p, q) incompatible. However, such inequalities are possible, and would be added to (or used in place of) the above inequalities if we want to solve the the problem of imputing missing values in order to *maximize* the resulting number of incompatible pairs. Details are omitted for lack of space. The solution to the maximization problem, along with the solution to Problem M1, bracket the number of incompatible pairs in the true data from which M was derived.

If M is an n by m matrix, the above ILP formulation for problem M1 creates at most nm Y variables, $2m^2$ B variables, $\frac{m^2}{2}$ C variables, and $O(nm^2)$ inequalities, although all of these estimates are worst case and the numbers are typically much smaller. For example, if I is the expected percentage of missing entries (which is as low as 3% in many applications), then the expected number of Y variables is nmI. The formulation as described can create redundant inequalities, but we have left them in our description for conceptual clarity, and in practice we have found that the preprocessor in CPLEX removes such redundancies as effectively as any of our more refined programs do. There are additional practical reductions that are possible that we cannot discuss here due to limited space.

Empirical Results for Problem M1. We extensively tested the ILPs for Problem M1 to answer two questions: 1) How quickly are the ILPs for problem M1 solved when problem instances are generated using data from a population genetic process; 2) When parts of the data are removed, and the missing data imputed by the ILP solution, how accurately do those imputations reconstruct the original values? In phylogenetic applications, missing data rates of up to 30% are common, but in population genetic applications, rates from one to five percent are more the norm.

We used the program *ms* created by Richard Hudson [15] to generate the binary sequences. That program is the widely-used standard for generating sequences that reflect the population genetic coalescent model of binary sequence evolution. The program allows one to control the level of recombination (defined in Section 2.3) through a parameter r, and a modified version of the program provided by Yun Song, allows one to control the level of *homoplasy* (recurrent or back mutations violating the infinite sites assumption). After a complete dataset was generated, each value in the data was chosen for removal with probability p, varied in the study; we use $I = p \times 100$ to denote the expected percent of missing values. All computations were done on a 1.5 ghz Intel itanium, and the ILPs were solved using CPLEX 9.1. The time needed to generate the ILPs was minimal and only the time used to solve the ILP is reported. We tested our ILP solution to Problem M1 on all combinations of n (# rows) = {30, 60, 90, 120, 150}; m (# columns) = {30, 60, 90}; r = {0, 4, 16, 30}; and I (expected percent missing values) = {5%, 10%, 15%, 20%, 25%, 30%}[1]. We generated and tested fifty datasets for each parameter combination.

The running times were slightly more influenced by m than by n, and mildly influenced by r, but were mostly a function of $n \times m$ and I (increasing with $n \times m$ and I, and decreasing with r). The largest average execution time (averaged over all 50 datasets) occurred when n, m and I, were at their maximum values and r was zero, but that average time was only 3.98 seconds. Figure 1 (a) shows average running times as a function of I. Hence, despite being NP-hard, problem M1 can be solved very efficiently on a wide range of data whose parameters reflect datasets of current interest in phylogenetics and population genetics.

The imputation accuracy was also excellent, when the product $n \times m$ is large. Error is the percentage of missing values that were incorrectly imputed. The case of 5% missing data illustrates this. When $n = m = 30$, and $r = 4$, the missing values were imputed with an average error of 6.4%, but when $n = 120$, $m = 90$ the average error dropped to 1.8%. In general, error increased with increasing r, and fell with increasing $n \times m$. The likely reason for such good results when $n \times m$ is large is the high level of redundancy in the data, and that these redundancies are exploited when the objective is to minimize incompatibilities. Figure 1 (b) shows error rates as a function of $n \times m$.

2.2 Imputing Values to Minimize Site-Removals

In this section we discuss another natural objective function related to Problem $M1$. We first define the **Site-Removal** problem on *complete* data: Given a binary

---

[1] We had earlier studied the case of $I = 1\%$ using a slower ILP formulation, but almost all of the computations took zero recorded time (to three decimal places) on that data, and so we started with a higher percentage of missing data for this study.

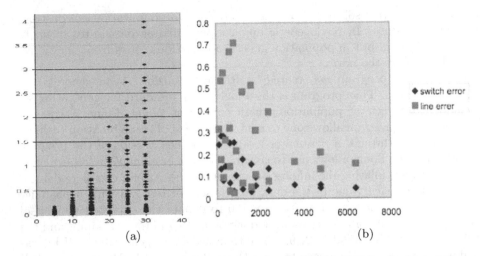

Fig. 1. (a) Average time (in seconds) needed to solve the ILP for Problem M1 as a function of I, the expected percentage of missing data. Each diamond is the average of 50 datasets for a particular combination of the parameters n, m, r and I. (b) Average imputation error rates (percentages) in the solution of the ILP for Problem M1, as a function of $n \times m$. Each diamond is the average of 50 datasets for a particular combination of the parameters n, m, r and I.

matrix M with no missing entries but some pairs of incompatible sites, find the *smallest* set of sites to remove from M so that no remaining pair of sites is incompatible.

Finding a small(est) set of sites to remove so that no remaining pairs are incompatible is often suggested and employed (particularly in phylogenetic studies) as a means to clean up data that does not perfectly conform to the Perfect Phylogeny model. The expectation is that while there may be some sites where homoplasy is observed (due to recurrent or back mutations at that site), a Perfect Phylogeny constructed from the remaining sites will give valid evolutionary information about the taxa. There are similar scenarios in population genetics. Of course, in order to get the most informative phylogeny, we want to remove as few sites as possible, motivating the Site-Removal problem. The Site-Removal Problem is NP-hard and is typically formulated as a Node-Cover problem in a graph where each node represents a site and each edge connects an incompatible pair of sites [25,7].

When M has missing entries, the Site-Removal problem is generalized to:
Problem S1: Over all matrices created by imputing missing values in M, find the matrix M' to minimize the the solution to the Site-Removal problem on M'.

An ILP formulation for Problem S1 is easily obtained from the ILP formulation for Problem M1 as follows: Let $D(i)$ be a binary variable used to indicate whether or not site i will be removed. Then for each pair $(p,q) \in P$, add the inequality $D(p) + D(q) - C(p,q) \geq 0$, which says that if the missing values are imputed so that site pair (p,q) becomes incompatible, then either site p

or site q must be removed. Also, for each pair $(p, q) \in F$, add the inequality $D(p) + D(q) \geq 1$. Finally, change the objective function in the formulation for M1 to Minimize $\sum_{i=1}^{m} D(i)$.

Problem S1 can be solved efficiently on the range of data considered in Section 2.1 (with most computations taking less than one second) but slightly slower than for Problem M1, and with a higher error rate. For example, the average time and error of the 50 datasets with $n = 120, m = 90, r = 16, I = 15$, was 0.53 seconds and 3.1% for Problem M1 and 0.75 seconds and 5.4% for Problem S1.

2.3   Estimating Recombination in Data with Missing Values

Recombination (crossing-over) is a fundamental molecular phenomena, where during meiosis two equal length sequences produce a third sequence of the same length consisting of a prefix of one of the sequences followed by a suffix of the other sequence. A central problem is to determine the *minimum* number of recombinations, denoted $Rmin(M)$, needed to generate a set of binary sequences M from some known or unknown ancestral sequence, when the infinite sites assumption applies [16]. There is a large literature on this problem, but there is no known efficient algorithm to exactly compute $Rmin(M)$. However, there are efficient algorithms that give relatively good *lower bounds* on $Rmin$, and there are biological questions concerning recombination (for example, finding recombination hotspots) that have been successfully addressed using lower bounds on $Rmin$ rather than using $Rmin$ itself [1],[5]. The first published, and most basic lower bound, called the HK bound [16], is obtained as follows: Consider the m sites of M to be integer points $1...m$ on the real line and pick a minimum number of non-integer points R so that for every pair of incompatible sites (p, q) in M, there is at least one point in R (strictly) between p and q. It is easy to show that $|R| \leq Rmin(M)$. When the data in M is complete, the HK bound $|R|$ can be computed in polynomial-time by a greedy-like algorithm. However, under the realistic situation that some entries in M are missing, we have **Problem R1:** Over all matrices created by imputing missing values in M, find the matrix M' to *minimize* the resulting HK lower bound on the $Rmin(M')$.

The reason for minimizing is that the result is then a valid lower bound on the number of recombinations needed to generate the true underlying data, when the infinite sites assumption applies.

Problem R1 is NP-hard [30], but an ILP formulation for it can be easily obtained from the formulation for Problem M1: For each c from 1 to $m - 1$, let $R(c)$ be a binary variable used to indicate whether a point in R should be chosen in the open interval $(c, c + 1)$. Then, for every pair of sites (p, q) in $F \cup P$, add the inequality $\sum_{p \leq c < q} R(c) \geq C(p, q)$ to the ILP for Problem M1, and change the objective function to Minimize $\sum_{c=1}^{m-1} R(c)$.

In our computations (details omitted due to space limitations), we have established that Problem R1 can be solved in practice over the same range of data discussed in Section 2.1; the computation times were longer than for Problem M1, but generally not more than twice as long. Of greater interest is the quality of the imputed values obtained in the solution to Problem R1 on data generated with recombination. Because Problem R1 more explicitly reflects recombination than does Problem M1, we conjectured that the solutions to Problem R1 would impute the original values better than solutions to Problem M1. Surprisingly, the average error for solutions to Problem R1 was somewhat larger than for Problem M1. For example, the average error over all datasets with $n = 150$ was 4% for Problem M1 and 4.75% for problem R1.

3 Haplotyping Problems

One of the key technical problems in the acquisition of variation data in populations is called the "Haplotype Inference (HI) Problem", or the problem of determining the "phase" of unphased genotype data. A very large literature now exists on this problem (see [12] for one survey). Abstractly, input to the HI problem consists of n *genotype* vectors, each of length m, where each value in the vector is either 0,1, or 2. A site with value 2 is called a "heterozygous" site, while the other sites are called "homozygous" sites. In the context of this problem, a vector with only entries of 0 and 1 is called a "haplotype". Given an input set of n genotype vectors, a solution to the HI Problem is a set of n pairs of haplotypes, one pair for each genotype vector. For any genotype vector g, the associated haplotypes v_1, v_2 must both have value 0 (or 1) at any position where g has value 0 (or 1); but for any position where g has value 2, exactly one of v_1, v_2 must have value 0, while the other has value 1. Hence, for an individual with h heterozygous sites there are 2^{h-1} pairs of haplotypes that could appear in a solution to the HI problem. For example, if the observed genotype g is 0212, then the pair of haplotypes 0110, 0011 is one feasible solution out of two feasible solutions. Of course, we want to find the HI solution that is most biologically plausible, and for that we need additional criteria to to guide the algorithm solving the HI problem. The goal is to devise criteria that reflect biological reality and yet allow efficient solution to the HI problem. Criteria have been previously proposed that were encoded as optimization problems with precise objective functions. In this paper, we discuss four additional biologically-motivated optimization problems that have practical ILP solutions.

3.1 Haplotyping Versions of M1, S1, R1

Each of the three problems M1, S1 and R1 has a natural analog as a haplotyping problem, and has biological and historical connections to other haplotyping problems.

Problem HM1: Solve the HI problem so that the number of incompatible pairs of sites in the HI solution is *minimized* over all HI solutions.

Problem HM1 is a natural extension of the following "Perfect Phylogeny Haplotyping (PPH)" problem [10]: Find, if possible, a solution to the HI problem so that the haplotypes in the solution can be derived on a perfect phylogeny; in other words, so that there are *no* incompatible pairs of sites in the HI solution. Such an HI solution is called a "PPH solution", and if there is one, it can be found in linear time [6,22]. See [10] for a discussion of the biological justification of the PPH problem. The PPH model is justified in some applications, but not all, and there are additional applications where the true haplotypes deviate by only a small amount from the PPH model (low recombination "haplotype blocks" are the prime example). In [13], a heuristic approach was developed to handle small deviations from the PPH model. An attempt to more formally model small deviations from the PPH model was explored in [26,23]. Problem HM1 is an alternative way to formalize, and quantify, deviations from the PPH model: a set of genotypes that allow HI solutions with a small number of incompatible pairs deviate less from the PPH model than do genotypes that only allow HI solutions with a large number of incompatible pairs. We were therefore interested in whether the HM1 problem can be solved efficiently in a range of biologically relevant data, and how well the HI solutions obtained this way reconstruct the correct haplotypes.

An ILP formulation for Problem HM1 can be easily obtained by modifying the formulation for Problem M1: First, duplicate each row of M creating matrix $\overline{M}$, and create the ILP for Problem M1 using matrix $\overline{M}$, treating each 2 as a "?". Then for each cell (i, q) where $M(i, q)$ is 2, add the inequality $Y(2i - 1, q) + Y(2i, q) = 1$. This formulation can be further improved, and such improvements have been implemented. For example, if $M(i, p) = 2$ and $M(i, q) = 1$ the binary combinations 0,1 and 1,1 will definitely be generated in columns (p, q) and that information may reduce the elements in $d(p, q)$, and reduce the size of the ILP formulation.

In the same way, we can modify the ILP for Problem S1 to obtain an ILP for **Problem HS1:** Remove the minimum number of columns in the input *genotypes*, so that there is a PPH solution to the HI problem on the remaining data. That is another way to formalize, and quantify, deviation from the PPH model. The same kind of modification also extends the ILP for Problem R1 to an ILP for **Problem HR1:** Solve the HI problem in order to minimize the HK bound on the haplotypes in the solution. That problem has been proposed as a way to search for recombination hotspots in genotypic data rather than haplotypic data [29]. It is interesting to note that Problem HR1 has a polynomial time solution [29], as does the problem of solving the HI problem in order to *maximize* the HK bound [31], even though Problem R1 is NP-hard.

Empirical Results for Problem HM1. We extensively tested instances of Problems HM1, HS1 and HR1 using datasets with $\frac{n}{2}$ genotypes created by pairing n haplotypes output by the program *ms* described in Section 2.1. We tested 50 datasets for each combination of n, m and r that we examined. We observed that we could solve these problems in practical time on a wide range of data, but not as extensive as for Problem M1. Further, the computation times were longer (considerably so for larger instances) for the same parameter combinations of n, m and r. In general, the HI solutions given by Problem HM1 were better than for HS1 and HR1, and so we will only discuss those here, although the running times for HM1 were larger than for HR1. In our experiments, we stopped any computations that exceeded three hours, and considered only combinations of

$n = \{10, 20, 30, 60, 80\}$ haplotypes and $m = \{30, 60, 90\}$ sites. As an example, when $n = 80, m = 60, r = 16$, 94% of the datasets terminated within the three hour limit. However, for most of the parameter choices we examined, all of the datasets terminated within the time limit, and most terminated well below that limit.

In addition to the solution time, we were interested in the quality of the haplotypes produced and how that quality varied depending on whether the deviation from the PPH model was due to recombination or to homoplasy. Datasets with homoplasy were generated with 5, 10 or 20 sites where additional mutations were forced to occur. Over these ranges of of n, m and r, we did not see a significant difference between the quality of the haplotypes produced from datasets generated with recombination and those with homoplasy, and so we will discuss only the recombination case.

To assess the quality of the solutions, we used the standard *switch error* [17,19] and the *line error*, comparing the haplotype pairs obtained from solving Problem HM1 with the original pairs used to generate the genotype data. The switch error is the minimum number of runs (blocks) of contiguous sites that need to be exchanged between the computed haplotype pairs in order to make the resulting haplotype pairs agree with the correct pairs, divided by the number of hetrozygous sites in the data. The line error is simply the number of haplotype pairs in the solution that do not agree completely with the corresponding correct pair, divided by the number of genotypes. The ILP executions that were terminated after three hours all found HI solutions, and so we could test their quality also. Hence, no datasets were excluded from our accuracy analysis. We also compared the accuracy of the HM1-computed haplotypes with the haplotypes found by program FastPhase [24], the successor program of the widely-used program PHASE [28].

We observed switch errors for the HI solutions produced by solving Problem HM1 that were very good in some parameter ranges, often superior (by a small amount) to the switch error of the solutions produced by FastPhase; in other parameter ranges the observed switch errors were somewhat inferior to those from FastPhase, and to accuracies reported for simulations using HapMap data [20] (although in line with some real data [17]). The line errors of solutions from both FastPhase and Problem HM1 were relatively large, but quite similar to each other. Unlike imputation error, switch accuracy is highly influenced by r, the recombination parameter; consistent with imputation error, all accuracies improved with larger data sets, particularly as the number of rows increase. We should note that in our tests we did not require that the minor allele appear above a minimum frequency as is commonly done; it is well known [19] that accuracies are improved by imposing that requirement.

Figure 2 (a) summarizes the switch and line errors of the HI solutions from obtained Problem HM1 compared to solutions given by FastPhase; Figure 2 (b) shows the time needed to solve Problem HM1.

As an illustration of the influence of n, when $n = 20, m = 30, r = 4$, the average switch errors from HM1 and FastPhase were 0.1189 and 0.136 which are

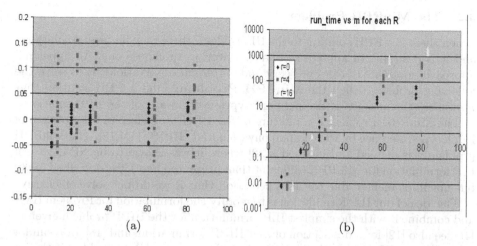

Fig. 2. (a) Comparison of switch and line errors the HI solutions from Problem HM1 and FastPhase, as a function of n. In each simulated dataset the switch and line errors of the HM1 solution were subtracted from switch and line errors of the FastPhase solution. A positive result shows the HM1 solution superior to the FastPhase solution. Each diamond (square) is the average difference of switch (line) errors from the 50 datasets for a particular combination of the parameters n, m and r. (b) Solution times (over the terminating datasets) as a function of m, for three values of parameter r. Each object is the average of 50 datasets with the same values of n.

relatively large, but when n increased to 60, the errors declined to 0.0532 and 0.068, and when n increased to 80, the errors were 0.0452 and 0.062. To see the influence of the recombination parameter r, consider the case of $n = 60, m = 30$, where the switch errors for HM1 were 0.046, 0.0532, 0.0984 with $r = 0, 4$ and 16 respectively, and the errors for FastPhase were 0.063, 0.068, and 0.083.

The comparison of *individual* HI solutions obtained from Problem HM1 and from FastPhase often gave contradictory results, so we averaged the results over all the data examined: the HI solutions from Problem HM1 had an average switch error of 0.13, an average line error of 0.34 and required an average computation time of 186.3 seconds for the terminating computations, while 4% of the computations did not terminate in three hours. The FastPhase solutions had an average switch error of 0.128, an average line error of 0.36 and required an average of 38 seconds to compute. These results suggest that the qualities of the two approaches are very similar. While the time needed to solve Problem HM1 is greater than the time for FastPhase, the main goal in solving HM1 was to see how well this natural extension of perfect phylogeny haplotyping solves the HI problem (although we would have been pleased to report that it solved faster than FastPhase). Having a solvable, simple-to-state objective function allows one to assess the biological fidelity of the model reflected by the objective function, giving much cleaner and clearer semantics compared to more black-box methods whose semantics may be very unclear. We consider the results based on HM1 to be positive and informative.

3.2 The MinPPH Problem

When there is a PPH solution to the HI problem, there may be several solutions, and it is desirable to apply a secondary criterion to choose one. An appealing approach, motivated both by theory and empirical observations, is to solve the following problem called the **MinPPH Problem:** Find a PPH solution that *minimizes* the number of *distinct* haplotypes used in any of the PPH solutions.

The MinPPH Problem is a mixture of the PPH problem and the problem of Haplotype Inference by Pure Parsimony (denoted HIPP) [11,4,18]. The MinPPH problem was defined and justified in [2] where it was shown to be NP-hard. An ILP formulation for MinPPH (different than presented here) was described in [3], but not implemented due to the expectation that it would not solve efficiently.

The idea of our ILP formulation is to modify the formulation for Problem HM1 and combine it with the simplest ILP formulation for the HIPP problem given in [4] (see also [12] for a description of that HIPP formulation, and [18] for a similar formulation). We start with the HIPP formulation from [4], but add to it the inequalities from the HM1 formulation along with the equality $\sum_{(p,q)\in P} C(p,q) = 0$. The end result is an ILP formulation that solves the HI problem using the minimum number of distinct haplotypes possible, subject to the constraint that the HI solution is a PPH solution (assuming a PPH solution exists).

We extensively tested this ILP formulation for solution speed and haplotype accuracy, using genotypic data (created from ms with $r = 0$) where PPH solutions were assured. We obtained two striking empirical results. The first result is that the ILPs solve extremely fast (generally less than one second) over a range of data up to 80 rows and 80 columns. This speed is even more notable considering that the HIPP formulation from [4] requires hours or days to solve the HI problem on the smaller instances, and cannot solve the larger instances

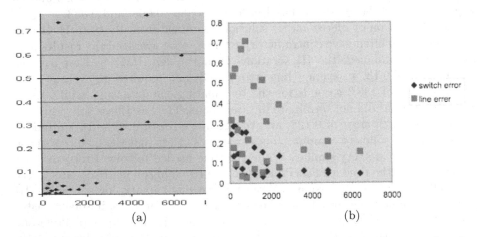

(a)   (b)

Fig. 3. (a) Average time needed to solve the ILP for Problem MinPPH, as a function of $n \times m$. Each diamond is the average of 50 datasets for a particular combination of the parameters n, m. (b) Average switch and line errors as a function of $n \times m$.

in practical time. Hence, it is the PPH constraints added to the HIPP formulation that makes the resulting formulation solve so quickly. We also examined the question of how the running times were influenced by the number of PPH solutions, and we saw no clear pattern.

The second striking empirical result is that the accuracy of the HI solutions given by the MinPPH solution is notably better than solutions obtained by FastPhase (except for instances with a very small number of rows), while running considerably faster. For example, when $n = m = 80$, the average MinPPH solution time was 0.59 seconds with a switch-error of 0.045, while the run time for FastPhase was 163 seconds with a switch-error of 0.074. In general, the MinPPH switch and line errors decrease with increasing problem size, as measured either by n or $m \times n$. Figures 3 (a) and 3 (b) show the MinPPH runtime and switch-error and line-error as a function of $n \times m$.

Acknowledgements

We thank Yun Song for the use of the code to generate SNP sequences with homoplasy events and for helpful conversations and suggestions. We thank Chuck Langley for helpful conversations and suggestions. The research was supported by NSF grants CCF 0515278, IIS 0513910 and REU 0434759.

References

1. Bafna, V., Bansal, V.: Improved recombination lower bounds for haplotype data. In: McLysaght, A., Huson, D.H. (eds.) RECOMB 2005. LNCS (LNBI), vol. 3678, Springer, Heidelberg (2005)
2. Bafna, V., Gusfield, D., Hannenhalli, S., Yooseph, S.: A note on efficient computation of haplotypes via perfect phylogeny. Journal of Computational Biology 11(5), 858–866 (2004)
3. Brown, D., Harrower, I.: A new formulation for haplotype inference by pure parsimony. report cs-2005-03. Technical report, University of Waterloo, School of Computer Science (2005)
4. Brown, D.G., Harrower, I.M.: Integer Programming Approaches to Haplotype Inference by Pure Parsimony. IEEE/ACM Transactions on Computational Biology and Bioinformatics 3(2), 141–154 (2006)
5. International HapMap Consortium. A haplotype map of the human genome. Nature, 437 1299–1320 (2005)
6. Ding, Z., Filkov, V., Gusfield, D.: A linear-time algorithm for the perfect phylogeny haplotyping problem. In: Miyano, S., Mesirov, J., Kasif, S., Istrail, S., Pevzner, P., Waterman, M. (eds.) RECOMB 2005. LNCS (LNBI), vol. 3500, pp. 585–600. Springer, Heidelberg (2005)
7. Felsenstein, J.: Inferring Phylogenies. Sinauer, Sunderland, MA (2004)
8. Gusfield, D.: Efficient algorithms for inferring evolutionary history. Networks 21, 19–28 (1991)
9. Gusfield, D.: Algorithms on Strings, Trees and Sequences: Computer Science and Computational Biology. Cambridge University Press, Cambridge (1997)

10. Gusfield, D.: Haplotyping as Perfect Phylogeny: Conceptual Framework and Efficient Solutions (Extended Abstract). In: Proceedings of RECOMB 2002: The Sixth Annual International Conference on Computational Biology, pp. 166–175 (2002)
11. Gusfield, D.: Haplotype inference by pure parsimony. In: Baeza-Yates, R.A., Chávez, E., Crochemore, M. (eds.) CPM 2003. LNCS, vol. 2676, pp. 144–155. Springer, Heidelberg (2003)
12. Gusfield, D., Orzack, S.: Haplotype inference. In: Aluru, S. (ed.) Handbook of Computational Molecular Biology, vol. 18, pp. 1–25. Chapman and Hall/CRC, Boca Raton (2005)
13. Halperin, E., Eskin, E.: Haplotype reconstruction from genotype data using Imperfect Phylogeny. Bioinformatics 20, 1842–1849 (2004)
14. Hein, J., Schierup, M., Wiuf, C.: Gene Genealogies, Variation and Evolution: A primer in coalescent theory. Oxford University Press, Oxford (2005)
15. Hudson, R.: Generating samples under the Wright-Fisher neutral model of genetic variation. Bioinformatics 18(2), 337–338 (2002)
16. Hudson, R., Kaplan, N.: Statistical properties of the number of recombination events in the history of a sample of DNA sequences. Genetics 111, 147–164 (1985)
17. Kimmel, G., Shamir, R.: GERBIL: Genotype resolution and block identification using likelihood. PNAS 102, 158–162 (2005)
18. Lancia, G., Pinotti, C., Rizzi, R.: Haplotyping populations by pure parsimony: Complexity, exact and approximation algorithms. INFORMS J. on Computing, special issue on Computational Biology 16, 348–359 (2004)
19. Lin, S., Cutler, D., Zwick, M., Chakravarti, A.: Haplotype inference in random population samples. Am. J. of Hum. Genet. 71, 1129–1137 (2002)
20. Marchini, J., Donnelly, P., et al.: A comparison of phasing algorithms for trios and unrelated individuals. Am. J. of Human Genetics 78, 437–450 (2006)
21. Pe'er, I., Pupko, T., Shamir, R., Sharan, R.: Incomplete directed perfect phylogeny. SIAM J. on Computing 33, 590–607 (2004)
22. Satya, R.V., Mukherjee, A.: An optimal algorithm for perfect phylogeny haplotyping. In: Proceedings of 4th CSB Bioinformatics Conference, IEEE Computer Society Press, Los Alamitos (2005)
23. Satya, R.V., Mukherjee, A., Alexe, G., Parida, L., Bhanot, G.: Constructing near-perfect phylogenies with multiple homoplasy events. Bioinformatics 22, 514–522 (2006) Bioinformatics Suppl., Proceedings of ISMB 2006
24. Scheet, P., Stephens, M.: A fast and flexible statistical model for large-scale population genotype data: applications to inferring missing genotypes and haplotypic phase. Am. J. Human Genetics 78, 629–644 (2006)
25. Semple, C., Steel, M.: Phylogenetics. Oxford University Press, Oxford (2003)
26. Song, Y.S., Wu, Y., Gusfield, D.: Haplotyping with one homoplasy or recombination event. In: Casadio, R., Myers, G. (eds.) WABI 2005. LNCS (LNBI), vol. 3692, Springer, Heidelberg (2005)
27. Steel, M.: The complexity of reconstructing trees from qualitative characters and subtrees. J. of Classification 9, 91–116 (1992)
28. Stephens, M., Smith, N., Donnelly, P.: A new statistical method for haplotype reconstruction from population data. Am. J. Human Genetics 68, 978–989 (2001)
29. Wiuf, C.: Inference of recombination and block structure using unphased data. Genetics 166, 537–545 (2004)
30. Wu, Y.: Personal Communication
31. Wu, Y., Gusfield, D.: Efficient computation of minimum recombination over genotypes (not haplotypes). In: Proceedings of Life Science Society Computational Systems Bioinformatics (CSB) 2006, pp. 145–156 (2006)

Improved Exact Algorithms for Counting 3- and 4-Colorings

Fedor V. Fomin[1], Serge Gaspers[1], and Saket Saurabh[1,2]

[1] Department of Informatics, University of Bergen,
N-5020 Bergen, Norway
{fomin,serge,saket}@ii.uib.no
[2] The Institute of Mathematical Sciences,
Chennai 600 113, India
saket@imsc.res.in

Abstract. We introduce a generic algorithmic technique and apply it on decision and counting versions of graph coloring. Our approach is based on the following idea: either a graph has nice (from the algorithmic point of view) properties which allow a simple recursive procedure to find the solution fast, or the pathwidth of the graph is small, which in turn can be used to find the solution by dynamic programming. By making use of this technique we obtain the fastest known exact algorithms
 – running in time $\mathcal{O}(1.7272^n)$ for deciding if a graph is 4-colorable and
 – running in time $\mathcal{O}(1.6262^n)$ and $\mathcal{O}(1.9464^n)$ for counting the number of k-colorings for $k = 3$ and 4 respectively.

1 Introduction

The graph coloring problem is one of the oldest and most intensively studied problems in Combinatorics and Algorithms. The problem is to color the vertices of a graph such that no two adjacent vertices are assigned the same color. The smallest number of colors needed to color a graph G is called the *chromatic number*, $\chi(G)$, of G. The corresponding decision version of the coloring problem is k-COLORING, where for a given graph G and an integer k we are asked if $\chi(G) \leq k$. The k-COLORING problem is one of the classical NP-complete problems [12]. In fact it is known to be NP complete for every $k \geq 3$. A lot of effort was also put in designing efficient approximation algorithms for the optimization version of the problem, namely, given a k-colorable graph to try to color it with as few colors as possible. Unfortunately, it has been shown that if certain reasonable complexity conjectures hold then k-COLORING is hard to approximate within $n^{1-\epsilon}$ for any $\epsilon > 0$ [10,13].

The history of exponential time algorithms for graph coloring is rich. Christofides obtained the first non-trivial algorithm computing the chromatic number of a graph on n vertices running in time $n!n^{O(1)}$ in 1971 [6]. In 1976, Lawler [15] devised an algorithm with running time $\mathcal{O}^*(2.4423^n)$ based on dynamic programming over subsets and enumeration of maximal independent sets. Eppstein [7]

G. Lin (Ed.): COCOON 2007, LNCS 4598, pp. 65–74, 2007.
© Springer-Verlag Berlin Heidelberg 2007

reduced the bound to $\mathcal{O}(2.4151^n)$ and Byskov [5] to $\mathcal{O}(2.4023^n)$. In two break-through papers last year, Björklund & Husfeldt [3] and Koivisto [14] independently devised $2^n n^{\mathcal{O}(1)}$ algorithms based on a combination of inclusion-exclusion and dynamic programming.

Apart from the general chromatic number problem, the problem of k-COLORING for small values of k like 3, 4 has also attracted a lot of attention. The fastest algorithm deciding if a the chromatic number of a graph is at most 3 runs in time $\mathcal{O}(1.3289^n)$ and is due to Beigel & Eppstein [2]. For 4-COLORING Byskov [5] designed the fastest algorithm, running in time $\mathcal{O}(1.7504^n)$.

The counting version of the k-COLORING problem, #k-COLORING, is to count the number of all possible k-colorings of a given graph. #k-COLORING (and its generalization known as *Chromatic Polynomial*) are among the oldest counting problem. Recently Björklund & Husfeldt [3] and Koivisto [14] have shown that the chromatic polynomial of a graph can be computed in time $2^n n^{\mathcal{O}(1)}$.

For small k, #k-COLORING was also studied in the literature. Angelsmark et al. [1] provide an algorithm for #3-COLORING with running time $\mathcal{O}(1.788^n)$. Fürer and Kashiviswanathan [11] show how to solve #3-COLORING with running time $\mathcal{O}(1.770^n)$. No algorithm faster than $2^n n^{\mathcal{O}(1)}$ for #4-COLORING was known in the literature [1,3,11,14].

Our Results. In this paper we introduce a generic technique to obtain exact algorithms for coloring problems and its different variants. This technique can be seen as a generalization of the technique introduced in [8] for a different problem. The technique is based on the following combinatorial property which is proved algorithmically and which is interesting in its own: *Either* a graph G has a nice "algorithmic" property which (very sloppily) means that when we apply branching or a recursive procedure to solve a problem then the branching procedure on subproblems of a smaller size works efficiently, *or* (if branching is not efficient) the pathwidth of the graph is small. This type of technique can be used for a variety of problems (not only coloring and its variants) where sizes of the subproblems on which the algorithm is called recursively decrease significantly by branching on vertices of high degrees.

In this paper we use this technique to obtain exact algorithms for different coloring problems. We show that #3-COLORING and #4-COLORING can be solved in time $\mathcal{O}(1.6262^n)$ and $\mathcal{O}(1.9464^n)$ respectively. We also solve 4-COLORING in time $\mathcal{O}(1.7272^n)$. These improve the best known results for each of the problems.

2   Preliminaries

In this paper we consider simple undirected graphs. Let $G = (V, E)$ be a graph and let n denote the number of vertices and m the number of edges of G. We denote by $\Delta(G)$ the maximum vertex degree in G. For a subset $V' \subseteq V$, $G[V']$ is the graph induced by V', and $G - V' = G[V \setminus V']$. For a vertex $v \in V$ we denote the set of its neighbors by $N(v)$ and its *closed neighborhood* by $N[v] = N(v) \cup \{v\}$.

Similarly, for a subset $D \subseteq V$, we define $N[D] = \cup_{v \in D} N[v]$. An *independent set* in G is a subset of pair-wise non-adjacent vertices. A subset of vertices $S \subseteq V$ is a *vertex cover* in G if for every edge e of G at least one endpoint of e is in S.

Major tools of our paper are tree and path decompositions of graphs. A *tree decomposition* of G is a pair $(\{X_i : i \in I\}, T)$ where each X_i, $i \in I$, is a subset of V, called a *bag* and T is a tree with elements of I as nodes such that we have the following properties :

1. $\cup_{i \in I} X_i = V$;
2. for all $\{u, v\} \in E$, there exists $i \in I$ such that $\{u, v\} \subseteq X_i$;
3. for all $i, j, k \in I$, if j is on the path from i to k in T then $X_i \cap X_k \subseteq X_j$.

The width of a tree decomposition is equal to $\max_{i \in I} |X_i| - 1$. The *treewidth* of a graph G is the minimum width over all its tree decompositions and it is denoted by $\mathbf{tw}(G)$. We speak of a *path decomposition* when the tree T in the definition of a tree decomposition is restricted to be a path. The *pathwidth* of G is defined similarly to its *treewidth* and is denoted by $\mathbf{pw}(G)$.

We need the following bound on the pathwidth of graphs with small vertex degrees.

Proposition 1 ([8]). *For any $\varepsilon > 0$, there exists an integer n_ε such that for every graph G with $n > n_\varepsilon$ vertices,*

$$\mathbf{pw}(G) \leq \frac{1}{6}n_3 + \frac{1}{3}n_4 + \frac{13}{30}n_5 + \frac{23}{45}n_6 + n_{\geq 7} + \varepsilon n,$$

where n_i is the number of vertices of degree i in G for any $i \in \{3, \ldots, 6, \geq 7\}$. Moreover, a path decomposition of the corresponding width can be constructed in polynomial time.

In our algorithms we also use the following results.

Proposition 2 ([7]). *The number of maximal independent sets of size k in a graph on n vertices is at most $3^{4k-n}4^{n-3k}$ and can be enumerated with polynomial time delay.*

Our $\mathcal{O}^*$ notation suppresses polynomial terms. Thus we write $\mathcal{O}^*(T(x))$ for a time complexity of the form $\mathcal{O}(T(x) \cdot |x|^{\mathcal{O}(1)})$ where $T(x)$ grows exponentially with $|x|$, the input size.

3 Framework for Combining Enumeration and Pathwidth Arguments

Let us assume that we have a graph problem for which

(a) we know how to solve it by enumerating independent sets, or maximal independent sets, of the input graph (for an example to check whether a graph G is 3-colorable, one can enumerate all independent sets I of G and for each independent set I can check whether $G - I$ is bipartite), and

(b) we also know how to solve the problem using dynamic programming over the path decomposition of the input graph.

For some instances, the first approach might be faster and for other instances, the path decomposition algorithm might be preferable. One method to get the best of both algorithms would be to compute a path decomposition of the graph using Proposition 1, and choose one of the two algorithms based on the width of this path decomposition. Unfortunately, this method is not very helpful in obtaining better worst case bounds on the running time of the algorithm.

Here in our technique we start by enumerating (maximal) independent sets and based on the knowledge we gain on the graph by this enumeration step, we prove that either the enumeration algorithm is fast, or the pathwidth of the graph is small. This means that either the input graph has a good algorithmic property, or it has a good graph-theoretic property.

To enumerate (maximal) independent sets of the input graph G, we use a very standard approach. Two sets I and C are constructed by a recursive procedure, where I is the set of vertices in the independent set and C the set of vertices not in the independent set. Let v be a vertex of maximum degree in $G - I - C$, the algorithm makes one recursive call where it adds v to I and all its neighbors to C and another recursive call where it adds v to C. This branching into two subproblems decreases the number of vertices in $G - I - C$ according to the following recurrence

$$T(n) \leq T(n - d(v) - 1) + T(n - 1).$$

From this recurrence, we see that the running time of the algorithm depends on how often it branches on a vertex of high degree. This algorithmic property is reflected by the size of C: frequent branchings on vertices of high degree imply that $|C|$ grows fast (in one branch).

On the other hand we can exploit a graph-theoretic property if C is small and there are no vertices of high degree in $G - I - C$. Based on the work of Monien and Preis on the bisection width of 3-regular graphs [16], small upper bounds on the pathwidth of the input graph G depending on their maximum degree have been obtained [8,9] (also see Proposition 1). If a path decomposition of $G - I - C$ of size $\beta_d |V(G) - I - C|$ can be computed, then a path decomposition of G of size $\beta_d |V(G) - I - C| + |C|$ can be computed easily. Here β_d is a constant strictly less than 1 depending on the maximum degree of the graph. If it turns out that a path decomposition of small width can be computed, the algorithm enumerating (maximal) independent sets is completely stopped without any further backtracking and an algorithm based on this path decomposition is executed on the original input graph.

In the rest of this section, we give a general framework combining

- algorithms based on the enumeration of maximal independent sets, and
- algorithms based on path decompositions of small width,

Input: A graph G, an independent set I of G and a set of vertices C such that $N(I) \subseteq C \subseteq V(G) - I$.
Output: An optimal solution which has the problem-dependent properties.
if $(\Delta(G - I - C) \geq$ a$)$ *or*
  $(\Delta(G - I - C) =$ a $- 1$ *and* $|C| > \alpha_{a-1}|V(G)|)$ *or*
  $(\Delta(G - I - C) =$ a $- 2$ *and* $|C| > \alpha_{a-2}|V(G)|)$ *or*
  $\cdots$ *or*
  $(\Delta(G - I - C) = 3$ *and* $|C| > \alpha_3|V(G)|)$
then
  $\quad$ choose a vertex $v \in V(G) - I - C$ of maximum degree in $G - I - C$
  $\quad S_1 \leftarrow$ enumISPw$(G, I \cup \{v\}, C \cup N(v))$ $\qquad$ R1
  $\quad S_2 \leftarrow$ enumISPw$(G, I, C \cup \{v\})$ $\qquad$ R2
  $\quad$ **return** combine(S_1, S_2)
else if $\Delta(G - I - C) = 2$ *and* $|C| > \alpha_2|V(G)|$ **then**
  $\quad$ **return** enumIS(G, I, C)
else
  $\quad$ Stop this algorithm and **run** Pw(G, I, C) instead.

Fig. 1. Algorithm enumISPw(G, I, C)

and discuss the running time of the algorithms based on this framework. This framework is not problem-dependent and it relies on two black boxes that have to be replaced by appropriate procedures to solve a specific problem.

Algorithm enumISPw(G, I, C) in Figure 1 is invoked with the parameters $(G, \emptyset, \emptyset)$, where G is the input graph, and the algorithms enumIS and Pw are problem-dependent subroutines. The function combine takes polynomial time and it is also a problem-dependent subroutine. The values for a, $\alpha_a, \cdots$, and α_2 $(0 = \alpha_a \leq \alpha_{a-1} \leq \cdots \alpha_2 < 1)$ are carefully chosen constants to balance the time complexities of enumeration and path decomposition based algorithms and to optimize the overall running time of the combined algorithm.

Algorithm enumIS(G, I, C) is problem-dependent and returns an optimal solution *respecting* the choice for I and C, where I is an independent set and C is a set of vertices not belonging to the independent set (set of discarded vertices). The sets I and C are supposed to be completed into a (maximal) independent set and a (minimal) vertex cover for G by enumerating (maximal) independent sets of $G - I - C$, before the problem-specific treatment is done.

Algorithm Pw(G, I, C) first computes a path decomposition based on G, I and C and the maximum degree of $G - I - C$. It then calls a problem-dependent algorithm based on this path decomposition of G.

Let n denote the number of vertices of G, $T(n)$ be the running time of Algorithm enumISPw on G, $T_e(n, i, c)$ be the running time of Algorithm enumIS and $T_p(n, i, c)$ be the running time of Algorithm Pw with parameters G, I, C where $i = |I|$ and $c = |C|$. We also assume that for any graph with n vertices and maximum degree d, a path decomposition of width at most $\beta_d n$ can be computed.

The following lemma is used by Algorithm Pw to compute a path decomposition of G of small width.

Lemma 1. *Suppose that given a graph H with $\Delta(H) \le d$, a path decomposition of width at most $\beta_d |H|$ can be computed where $\beta_d < 1$ is a constant depending on d alone. Then a path decomposition of width at most $\beta_d |V(G) - I - C| + |C|$ can be computed for a graph G if I is an independent set in G, $N(I) \subseteq C \subseteq V(G)$ and $\Delta(G - I - C) \le d$.*

Proof. As I is an independent set in G and C separates I from $G - I - C$, every vertex in I has degree 0 in $G - C$. Thus, a path decomposition of $G - C$ of size at most $\beta_d |V(G) - I - C|$ can be computed. Adding C to each bag of this path decomposition gives a path decomposition of width at most $\beta_d |V(G) - I - C| + |C|$ of G. □

Given the conditions under which Pw is executed, the following lemma upper bounds its running time.

Lemma 2. *If the considered problem can be solved on G in time $O^*(t_{pw}^\ell)$, given a path decomposition of width ℓ of G, then*

$$T_p(n, i, c) = O^* \left(\max_{d \in \{2, 3, \cdots, a-1\}} \left(t_{pw}^{(\beta_d + (1 - \beta_d)\alpha_d)n} \right) \right).$$

Proof. The proof follows from Lemma 1 and the conditions on $|C|$ and $\Delta(G - I - C)$ under which Algorithm Pw is executed. □

To estimate the size of the search tree we assume that Algorithm Pw is not executed. We denote $(\alpha_{d-1} - \alpha_d)n$ by $\Delta\alpha_d n$. Let t_n, t_i and t_c be constants such that $T_e(n, i, c) = O^*(t_n^n t_i^i t_c^c)$. The next lemma bounds the size of the search tree when the algorithm based on path decomposition is not used.

Lemma 3. *If Algorithm Pw is not executed, then*

$$T(n) = O^* \left(t_n^n t_c^{\alpha_2 n} \prod_{d=3}^{a} t_d^{\Delta\alpha_d n} \right)$$

where $t_d = (1 + r_d)$ and r_d is the minimum positive root of

$$(1 + r)^{-(d-1)} \cdot r^{-1} \cdot t_i - 1.$$

Proof. We divide $T(n)$ into $T_d(n, i, c)$ for $d \in \{2, 3, \cdots, a\}$ where d corresponds to the maximum degree of $G - I - C$ if $d < a$ and $T_a(n, 0, 0) = T(n)$ if Algorithm Pw is not executed. Clearly, $T_2(n, i, c) = T_e(n, i, c)$. Let us now express $T_d(n, i, c)$ in terms of $T_{d-1}(\cdot, \cdot, \cdot)$ for $d \in \{3, \cdots, a\}$. Consider the part of the search tree with branchings on vertices of degree d (or at least d if $d = a$). Observe that

$|C|$ increases in the worst case by at most $(\alpha_{d-1} - \alpha_d)n = \Delta\alpha_d n$ in this part of the search tree. In each branch of the type **R1**, $|C|$ increases by d and in each branch of the type **R2**, $|C|$ increases by 1. Let $r \in [0, \Delta\alpha_d n/d]$ be the number of times the algorithm branches according to **R1**, then it branches $\Delta\alpha_d n - dr$ times according to **R2**. We get that

$$T_d(n, i, c) = O^* \left(\sum_{r=0}^{\Delta\alpha_d n/d} \binom{\Delta\alpha_d n - (d-1)r}{r} T_{d-1}(n, i+r, c+\Delta\alpha_d n) \right).$$

In general situation the degree d may not change to $d-1$ but rather jump to something smaller. But the worst case bounds on the size of the search tree are achieved when d decreases progressively as considered above.

To prove the lemma, it is sufficient to expand $T_a(n, 0, 0)$ and to prove that

$$\sum_{r=0}^{\Delta\alpha_d n/d} \binom{\Delta\alpha_d n - (d-1)r}{r} t_i^r \leq t_d^{\Delta\alpha_d n}.$$

The sum over binomial coefficients $\sum_{r=0}^{\Delta\alpha_d n/d} \binom{\Delta\alpha_d n - (d-1)r}{r} t_i^r$ is bounded by $(\Delta\alpha_d n/d)B$ where B is the maximum term in this sum. Let us assume that $B = \binom{\Delta\alpha_d n - (d-1)j}{j} t_i^j$ for some $j \in \{0, 1, \ldots, \Delta\alpha_d n/d\}$.

$$B = \binom{\Delta\alpha_d n - (d-1)j}{j} t_i^j \leq \frac{(1+r_i)^{\Delta\alpha_d n - (d-1)j}}{r_i^j} t_i^j.$$

Here we use the well known fact that for any $x > 0$ and $0 \leq k \leq n$,

$$\binom{n}{k} \leq \frac{(1+x)^n}{x^k}.$$

By choosing r_i to be the minimum positive root of $\frac{(1+r)^{-(d-1)}}{r} t_i - 1$, we arrive at $B < (1+r_i)^{\Delta\alpha_d n} = t_d^{\Delta\alpha_d n}$. $\qquad\square$

The following theorem combines Lemmas 2 and 3 to upper bound the overall running time of the algorithms resulting from this framework.

Theorem 1. *The running time of Algorithm* enumISPw *on a graph on n vertices is*

$$T(n) = O^* \left(t_n^n t_c^{\alpha_2 n} \prod_{d=3}^{a} t_d^{\Delta\alpha_d n} + \max_{d \in \{2,3,\cdots,a-1\}} \left(t_{pw}^{(\beta_d + (1-\beta_d)\alpha_d)n} \right) \right)$$

where $t_d = (1 + r_d)$ and r_d is the minimum positive root of

$$(1+r)^{-(d-1)} \cdot r^{-1} \cdot t_i - 1.$$

The current best values for $\beta_d, d = \{2, \cdots, 6\}$ are obtained from Proposition 1.

4   Applications

In this section we use the framework of the previous section to derive improved algorithms for #3-COLORING, #4-COLORING and 4-COLORING.

4.1   Counting 3-Colorings

We first describe the problem-dependent subroutines we need to use in our Algorithm `enumISPw` to solve #3-COLORING in time $\mathcal{O}(1.6262^n)$.

Algorithm `enumISPw` returns here an integer, I corresponds to the color class C_1 and C to the remaining two color classes C_2 and C_3. Algorithm `enumIS` with parameters G, I, C enumerates all independent sets of $G - I - C$ and for each, adds this independent set to I, then checks if $G - I$ is bipartite. If $G - I$ is bipartite, then a counter counting the independent sets is incremented. This takes time $T_e(n, i, c) = 2^{n-i-c}$. Thus, $t_n = 2, t_i = 1/2$ and $t_c = 1/2$.

The function `combine` corresponds in this case to the plus-operation. The running time of Algorithm `Pw` is based on the following lemma.

Lemma 4. *Given a graph $G = (V, E)$ with a path decomposition of G of width ℓ, #k-COLORING can be solved in time $\mathcal{O}(k^\ell n^{O(1)})$.*

Now we use Theorem 1 and Proposition 1 to evaluate the overall complexity of our #3-COLORING algorithm.

Theorem 2. *The #3-COLORING problem can be solved in time $\mathcal{O}(1.6262^n)$ for a graph on n vertices.*

Proof. We use Theorem 1 and Lemma 1 with $a = 5, \alpha_2 = 0.44258, \alpha_3 = 0.33093$ and $\alpha_4 = 0.16387$. The pathwidth part of the algorithm takes time

$$O^*\left(\max\left(3^{\alpha_2 n}, 3^{(1+5\alpha_3)n/6}, 3^{(1+2\alpha_4)n/3}\right)\right)$$

$$= \mathcal{O}(1.62617^n).$$

The branching part of the algorithm takes time

$$O^*\left(2^n \cdot (1/2)^{\alpha_2 n} \cdot 1.29716^{(\alpha_2 - \alpha_3)n} \cdot 1.25373^{(\alpha_3 - \alpha_4)n} \cdot 1.22329^{\alpha_4 n}\right)$$

$$= \mathcal{O}(1.62617^n). \qquad \square$$

4.2   Counting 4-Colorings

To solve #4-COLORING, Algorithm `enumIS` with parameters G, I, C enumerates all independent sets of $G - I - C$ and for each, adds this independent set to I, then counts the number of 3-colorings of $G - I$ using the previous algorithm. This takes time $T_e(n, i, c) = \sum_{\ell=0}^{n-i-c} 1.62617^{\ell+c}$. Thus, $t_n = 2.62617, t_i = 1/2.62617$ and $t_c = 1.62617/2.62617$. We evaluate the running time as previously.

Theorem 3. *The #4-COLORING problem can be solved in time $\mathcal{O}(1.9464^n)$ for a graph on n vertices.*

Proof. We use Theorem 1 and Lemma 1 with $a = 6, \alpha_2 = 0.480402, \alpha_3 = 0.376482, \alpha_4 = 0.220602$ and $\alpha_5 = 0.083061$. The pathwidth part of the algorithm takes time

$$O^* \left(\max \left(4^{\alpha_2 n}, 4^{(1+5\alpha_3)n/6}, 4^{(1+2\alpha_4)n/3} 4^{(13+17\alpha_5)n/30} \right) \right)$$

$$= \mathcal{O}(1.9464^n).$$

The branching part of the algorithm takes time

$$O^* \big(\, 2.62617^n \cdot (1.62617/2.62617)^{\alpha_2 n} \cdot 1.24548^{(\alpha_2 - \alpha_3)n} \cdot 1.21324^{(\alpha_3 - \alpha_4)n} \cdot$$
$$1.18993^{(\alpha_4 - \alpha_5)n} \cdot 1.17212^{\alpha_5 n} \big) = \mathcal{O}(1.9464^n). \qquad \square$$

4.3   4-Coloring

A well known technique [15] to check if a graph is k-colorable is to check for all maximal independent sets I of size at least $\lceil n/k \rceil$ whether $G - I$ is $(k - 1)$-colorable. In the analysis, we use Proposition 2 to bound the number of maximal independent sets of a given size.

We also need the current best algorithm deciding 3-COLORING.

Theorem 4 ([2]). *The 3-COLORING problem can be solved in time $O(1.3289^n)$ for a graph on n vertices.*

In Algorithm `enumISPw`, which returns here a boolean, I corresponds to the color class C_1 and C to the remaining three color classes C_2, C_3 and C_4. Algorithm `enumIS` with parameters G, I, C enumerates all maximal independent sets of $G - I - C$ of size at least $\lceil n/k \rceil - |I|$ and for each, adds this independent set to I, then checks if $G - I$ is 3-colorable using Theorem 4. If yes, then G is 4-colorable. This takes time

$$T_e(n, i, c) = \sum_{\ell = \lceil n/4 \rceil - i}^{n-i-c} 3^{4\ell - n + c + i} 4^{n - c - i - 3\ell} 1.3289^{n - i - \ell}.$$

As $\sum_{\ell=0}^{\lfloor 3n/4 \rfloor - c} 3^{4\ell} 4^{-3\ell} 1.3289^{-\ell}$ is upper bounded by a constant, $t_n = 4^{1/4} 1.3289^{3/4}$, $t_i = 4^2/3^3$ and $t_c = 3/4$.

Theorem 5. *The 4-COLORING problem can be solved in time $\mathcal{O}(1.7272^n)$ for a graph on n vertices.*

Proof. We use Theorem 1 and Lemma 1 with $a = 5, \alpha_2 = 0.39418, \alpha_3 = 0.27302$ and $\alpha_4 = 0.09127$, and the pathwidth algorithm of Lemma 4. $\qquad \square$

References

1. Angelsmark, O., Jonsson, P.: Improved Algorithms for Counting Solutions in Constraint Satisfaction Problems. In: Rossi, F. (ed.) CP 2003. LNCS, vol. 2833, pp. 81–95. Springer, Heidelberg (2003)
2. Beigel, R., Eppstein, D.: 3-coloring in time $O(1.3289^n)$.. Journal of Algorithms 54(2), 168–204 (2005)
3. Björklund, A., Husfeldt, T.: Inclusion–Exclusion Algorithms for Counting Set Partitions. In: The Proceedings of FOCS 2006, pp. 575–582 (2006)
4. Byskov, J.M.: Exact Algorithms for Graph Colouring and Exact Satisfiability. PhD Dissertation (2004)
5. Byskov, J.M.: Enumerating Maximal Independent Sets with Applications to Graph Colouring. Operations Research Letters 32(6), 547–556 (2004)
6. Christofides, N.: An algorithm for the chromatic number of a graph. Computer J. 14, 38–39 (1971)
7. Eppstein, D.: Small Maximal Independent Sets and Faster Exact Graph Coloring. J. Graph Algorithms Appl. 7(2), 131–140 (2003)
8. Fomin, F.V., Gaspers, S., Saurabh, S., Stepanov, A.A.: On Two Techniques of Combining Branching and Treewidth. Report No 337, Department of Informatics, University of Bergen, Norway (December 2006)
9. Fomin, F.V., Høie, K.: Pathwidth of cubic graphs and exact algorithms. Information Processing Letters 97(5), 191–196 (2006)
10. Feige, U., Kilian, J.: Zero Knowledge and the Chromatic Number. Journal of Computer and System Sciences 57(2), 187–199 (1998)
11. Fürer, M., Kasiviswanathan, S.P.: Algorithms for counting 2-SAT solutions and colorings with applications. In: ECCC 33 (2005)
12. Garey, M.R., Johnson, D.S.: Computer and Intractability: A Guide to the Theory of NP-Completeness. W.H. Freeman, San Francisco, CA (1979)
13. Khot, S., Ponnuswami, A.K.: Better Inapproximability Results for MaxClique, Chromatic Number and Min-3Lin-Deletion. In: Bugliesi, M., Preneel, B., Sassone, V., Wegener, I. (eds.) ICALP 2006. LNCS, vol. 4051, pp. 226–237. Springer, Heidelberg (2006)
14. Koivisto, M.: An $O(2^n)$ algorithm for graph coloring and other partitioning problems via inclusion-exclusion. In: Proceedings of FOCS 2006, pp. 583–590 (2006)
15. Lawler, E.L.: A Note on the Complexity of the Chromatic Number. Information Processing Letters 5(3), 66–67 (1976)
16. Monien, B., Preis, R.: Upper bounds on the bisection width of 3- and 4-regular graphs. Discrete Algorithms 4(3), 475–498 (2006)

Connected Coloring Completion for General Graphs: Algorithms and Complexity*

Benny Chor[1], Michael Fellows[2,3], Mark A. Ragan[4], Igor Razgon[5], Frances Rosamond[2], and Sagi Snir[6]

[1] Computer Science Department, Tel Aviv University, Tel Aviv, Israel
benny@cs.tau.ac.il
[2] University of Newcastle, Callaghan NSW 2308, Australia
{michael.fellows,frances.rosamond}@newcastle.edu.au
[3] Durham University, Institute of Advanced Study,
Durham DH1 3RL, United Kingdom
[4] Institute for Molecular Biosciences, University of Queensland, Brisbane, QLD 4072 Australia
m.ragan@imb.uq.edu.au
[5] Computer Science Department, University College Cork, Ireland
i.razgon@cs.ucc.ie
[6] Department of Mathematics, University of California, Berkeley, USA
ssagi@math.berkeley.edu

Abstract. An *r-component connected coloring* of a graph is a coloring of the vertices so that each color class induces a subgraph having at most r connected components. The concept has been well-studied for $r = 1$, in the case of trees, under the rubric of *convex coloring*, used in modeling perfect phylogenies. Several applications in bioinformatics of connected coloring problems on general graphs are discussed, including analysis of protein-protein interaction networks and protein structure graphs, and of phylogenetic relationships modeled by splits trees. We investigate the r-COMPONENT CONNECTED COLORING COMPLETION (r-CCC) problem, that takes as input a partially colored graph, having k uncolored vertices, and asks whether the partial coloring can be completed to an r-component connected coloring. For $r = 1$ this problem is shown to be NP-hard, but fixed-parameter tractable when parameterized by the number of uncolored vertices, solvable in time $O^*(8^k)$. We also show that the 1-CCC problem, parameterized (only) by the treewidth t of the graph, is fixed-parameter tractable; we show this by a method that is of independent interest. The r-CCC problem is shown to be $W[1]$-hard, when parameterized by the treewidth bound t, for any $r \geq 2$. Our proof also shows that the problem is NP-complete for $r = 2$, for general graphs.

Topics: Algorithms and Complexity, Bioinformatics.

1 Introduction

The following two problems concerning colored graphs can be used to model several different issues in bioinformatics.

---

* This research has been supported by the Australian Research Council through the Australian Centre in Bioinformatics. The second and fifth authors also acknowledge the support provided by a William Best Fellowship at Grey College, Durham, while the paper was in preparation.

r-COMPONENT CONNECTED RECOLORING (r-CCR)

Instance: A graph $G = (V, E)$, a set of colors $\mathcal{C}$, a coloring function $\Gamma : V \to$
$\mathcal{C}$, and a positive integer k.

Parameter: k

Question: Is it possible to modify Γ by changing the color of at most k vertices, so that the modified coloring Γ' has the property that each color class induces a subgraph with at most r components?

In the case where G is a tree and $r = 1$, the problem is of interest in the context of maximum parsimony approaches to phylogenetics [17,13]. A connected coloring corresponds to a perfect phylogeny, and the recoloring number can be viewed as a measure of distance from perfection. The problem was introduced by Moran and Snir, who showed that CONVEX RECOLORING FOR TREES (which we term 1-CCR) is NP-hard, even for the restriction to colored paths. They also showed that the problem is fixed-parameter tractable, and described an FPT algorithm that runs in time $O(k(k/\log k)^k n^4)$ for colored trees [17]. Subsequently, Bodlaender *et al.* have improved this to an FPT algorithm that runs in linear time for every fixed k, and have described a polynomial-time kernelization to a colored tree on at most $O(k^2)$ vertices [3].

Here we study a closely related problem.

r-COMPONENT CONNECTED COLORING COMPLETION (r-CCC)

Instance: A graph $G = (V, E)$, a set of colors $\mathcal{C}$, a coloring partial function
$\Gamma : V \to \mathcal{C}$ where there are k uncolored vertices.

Parameter: k

Question: Is it possible to complete Γ to a total coloring function Γ' such that each color class induces a subgraph with at most r components?

The problem is of interest in the following contexts.

(1) Protein-protein interaction networks. In a protein-protein interaction network the vertices represent proteins and edges model interactions between that pair of proteins [22,7,20,21]. Biologists are interested in analyzing such relationship graphs in terms of cellular location or function (either of which may be represented by vertex coloring) [16]. Interaction networks colored by cellular location would be expected to have monochrome subgraphs representing localized functional subnetworks. Conversely, interaction networks colored by function may also be expected to have monochrome connected subgraphs representing cellular localization. The issue of error makes the number of recolorings (corrections) needed to attain color-connectivity of interest [17], and the issue of incomplete information may be modeled by considering uncolored vertices that are colored to attain color-connectivity, the main combinatorial problem we are concerned with here. Protein-protein interaction graphs generally have bounded treewidth.

(2) Phylogenetic networks. Phylogenetic relationships can be represented not only as trees, but also as networks, as in the *splits trees* models of phylogenetic relationships that take into account such issues as evidence of lateral genetic transfer, inconsistencies in the phylogenetic signal, or information relevant to a specific biological hypothesis, e.g., host-parasite relationships [15,14]. Convex colorings of splits trees have essentially the same modeling uses and justifications as in the case of trees [17,4,13]. Splits trees for natural datasets have small treewidth.

Our main results are summarized as follows:

1. 1-CCC is NP-hard for general colored graphs, even if there are only two colors.
2. 1-CCC for general colored graphs, parameterized by the number k of uncolored vertices, is fixed-parameter tractable, and can be solved in time $O^*(8^k)$.
3. 1-CCC is in XP for colored graphs of treewidth at most t, parameterized by t. (That is, it is solvable in polynomial time for any fixed t. Note that under this parameterization the number of uncolored vertices is unbounded.)
4. 1-CCC is fixed-parameter tractable when parameterized by treewidth.
5. For all $r \geq 2$, r-CCC, parameterized by a treewidth bound t, is hard for $W[1]$.

For basic background on parameterized complexity see [10,11,19].

2 Connected Coloring Completion Is NP-Hard

Theorem 1. *The 1-CCC problem is NP-hard, even if there are only two colors.*

Proof. (Sketch.) The reduction from 3SAT can be inferred from Figure 1 (the details are omitted due to space limitations). The two colors are T and F.

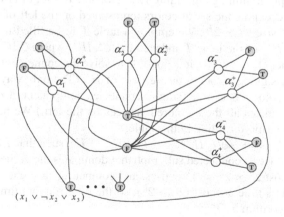

Fig. 1. The reduction from 3SAT

3 1-CCC for k Uncolored Vertices Is Fixed-Parameter Tractable

The input to the problem is a partially colored graph $G = (V, E)$, and the parameter is the number of uncolored vertices.

Soundness for the following reduction rules is easy to verify.

Rule 1. A maximal connected monochromatic subgraph (of colored vertices) can be collapsed to a single vertex. The parameter is unchanged.

Rule 2. If a color occurs on only a single vertex, then that vertex can be deleted. The parameter is unchanged.

Rule 3. An edge between two colored vertices of different color can be deleted. The parameter is unchanged.

Suppose that a partially colored graph G is reduced with respect to the above three reduction rules. The situation can be represented by a bipartite *model graph* that on one side (let us say, the *left side*), has vertices representing the vertices created by Rule 1, but not deleted by Rule 2. On the *right side* are the k uncolored vertices (and their adjacencies), and between the two sides are edges that represent an incidence relationship. Clearly, if in this representation, there are more than k vertices on the left, then the answer is NO. Thus, there are at most k colors represented on the left, and an FPT algorithm that runs in time $O^*(k^k)$ follows by exploring all possibilities of coloring the k uncolored vertices (on the right) with the k colors represented on the left of the model. We can do better than this.

Theorem 2. *1-CCC parameterized by the number of uncolored vertices is fixed-parameter tractable, solvable in time $O^*(8^k)$.*

Proof. We use the *model graph* described above. Instead of the brute-force exploration of all possibilities of coloring the right-side vertices with the left-side colors, we employ a dynamic programming algorithm. Let H denote the set of at most k uncolored vertices, and let $\mathcal{C}$ denote the set of colors represented on the left of the model graph. Create a table of size $2^{\mathcal{C}} \times 2^H$. We employ a table T to be filled in with 0/1 entries. The entry $T(\mathcal{C}', H')$ of the table T indexed by $(\mathcal{C}', H')$, where $\mathcal{C}' \subseteq \mathcal{C}$ and $H' \subseteq H$, represents whether (1) or not (0), it is possible to solve the connectivity problem for the colors in $\mathcal{C}'$ by assigning colors to the vertices of H'. (The only way this can happen, is that, for each color $c \in \mathcal{C}'$, the (single) connected component of c-colored vertices on the right, dominates all the c-colored vertices on the left.) We have the following recurrence relationship for filling in this table:

$$T(\mathcal{C}', H') = 1 \iff \exists(c, H''), \ c \in \mathcal{C}', \ H'' \subset H', \text{ such that } T(\mathcal{C}' - c, H'') = 1$$

and $H' - H''$ induces a connected subgraph that dominates the vertices of color c. The table T has at most $2^k \times 2^k = 4^k$ entries, and computing each entry according to the recurrence requires time at most $O(k \cdot 2^k)$, so the total running time of the dynamic programming algorithm is $O^*(8^k)$.

4 Bounded Treewidth

Most natural datasets for phylogenetics problems have small bounded treewidth. 1-CCR is NP-hard for paths (and therefore, for graphs of treewidth one) [17]. Bodlaender and Weyer have shown that 1-CCR parameterized by (k, t), where k is the number of vertices to be recolored, and t is a treewidth bound, is fixed-parameter tractable, solvable in linear time for all fixed k [6].

4.1 1-CCC Parameterized by Treewidth is Linear-Time FPT

We describe an algorithm for the 1-CCC problem that runs in linear time for any fixed treewidth bound t, and we do this by using the powerful machinery of Monadic Second Order (MSO) logic, due to Courcelle [9] (also [1,5]). At first glance, this seems either surprising or impossible, since MSO does not provide us with any means of describing families of colored graphs, where the number of colors is unbounded. We employ a "trick" that was first described (to our knowledge) in a paper in these proceedings [3]. Further applications of what appears to be more a useful new method, rather than just a trick, are described in [12].

The essence of the trick is to construct an auxiliary graph that consists of the original input, augmented with additional *semantic vertices*, so that the whole ensemble has — or can safely be assumed to have — bounded treewidth, and relative to which the problem of interest *can* be expressed in MSO logic.

Let $G = (V, E)$ be a graph of bounded treewidth, and $\Gamma : V' \to \mathcal{C}$ a vertex-coloring function defined on a subset $V' \subseteq V$. (Assume each color in $\mathcal{C}$ is used at least once.) We construct an auxiliary graph G' from G in the following way: for each color $c \in \mathcal{C}$, create a new *semantic vertex* v_c (these are all of a second *type* of vertex, the vertices of V are of the first type). Connect v_c to every vertex in G colored c by Γ.

Consider a tree decomposition Δ for G, witnessing the fact that it has treewidth at most t. This can be computed in linear time by Bodlaender's algorithm.

Say that a color $c \in \mathcal{C}$ is *relevant* for a bag B of Δ if either of the following holds:
(1) There is a vertex $u \in B$ such that $\Gamma(u) = c$. (When this holds, say that c is *present* in B.)
(2) There are bags B' and B'' of Δ such that B is on the unique path from B' to B'' relative to the tree that indexes Δ, and there are vertices $u' \in B'$ and $u'' \in B''$ such that $\Gamma(u') = \Gamma(u'') = c$, and furthermore, c is not present in B. (When this holds, say that c is *split* by the bag B.)

Lemma 1. *If the colored graph G is a yes-instance for 1-CCC, then for any bag B, there are at most $t + 1$ relevant colors.*

Proof. Suppose that a bag B has more than $t + 1$ relevant colors, and that p of these are present in B. If s denotes the number of colors split by B, then $s > t + 1 - p$. Since B contains at most $t + 1$ vertices, the number of colors split by B exceeds the number of uncolored vertices in B, and because each bag is a cutset of G, it follows that G is a no-instance for 1-CCC.

Lemma 2. *If the colored graph G is a yes-instance for 1-CCC, then the auxiliary graph G' has treewidth at most $2t + 1$.*

Proof. Consider a tree decomposition Δ for G witnessing that the treewidth of G is at most t. By the above lemma, if we add to each bag B of Δ all those vertices v_c for colors c that are relevant to B, then (it is easy to check) we obtain a tree-decomposition Δ' for G' of treewidth at most $2t + 1$.

Theorem 3. *The 1-CCC problem, parameterized by the treewidth bound t, is fixed-parameter tractable, solvable in linear time for every fixed t.*

Proof. The algorithm consists of the following steps.

Step 1. Construct the auxiliary graph G'.

Step 2. Compute in linear time, using Bodlaender's algorithm, a tree-decomposition for G' of width at most $2t + 1$, if one exists. (If not, then correctly output NO.)

Step 3. Otherwise, we can express the problem in MSO logic. That this is so, is not entirely trivial, and is argued as follows (sketch).

The vertices of G' can be considered to be of three types: (i) the original colored vertices of G (that is, the vertices of V'), (ii) the uncolored vertices of G (that is, the vertices of $V - V'$), and (iii) the color-semantic vertices added in the construction of G'. (The extension of MSO Logic to accomodate a fixed number of vertex types is routine.)

If G is a yes-instance for the problem, then this is witnessed by a set of edges F between vertices of G (both colored and uncolored) that provides the connectivity for the color classes. In fact, we can choose such an F so that it can be partitioned into classes F_c, one for each color $c \in C$, such that the classes are disjoint: no vertex $v \in V$ has incident edges $e \in F_c$ and $e' \in F_{c'}$ where $c \neq c'$.

The following are the key points of the argument:

(1) Connectivity of a set of vertices, relative to a set of edges, can be expressed by an MSO formula.

(2) We assert the existence of a set of edges F of $G \subseteq G'$, and of a set of edges F' between uncolored vertices of G and color-semantic vertices of G' such that:

- Each uncolored vertex of G has degree 1 relative to F'. (The edges of F' thus represent a coloring of the uncolored vertices of G.)
- If u and v are colored vertices of G that are connected relative to F, then there is a unique color-semantic vertex v_c such that both u and v are adjacent to v_c.
- If u is a colored vertex of G and v is an uncolored vertex of G that are connected via edges in F, then there is a unique color-semantic vertex v_c such that v is adjacent to v_c by an edge of F', and u is adjacent to v_c by an edge of G'.
- If u is an uncolored vertex of G and v is an uncolored vertex of G that are connected via edges in F, then there is a unique color-semantic vertex v_c such that v is adjacent to v_c by an edge of F', and u is adjacent to v_c by an edge of F'.
- If u and v are colored vertices of G that are both adjacent to some color-semantic vertex v_c, then u and v are connected relative to F.

4.2 r-CCC Parameterized by Treewidth is $W[1]$-Hard for $r \geq 2$

In view of the fact that 1-CCC is fixed-parameter tractable for bounded treewidth, it may be considered surprising that this does not generalize to r-CCC for any $r \geq 2$.

Theorem 4. *The 2-CCC problem, parameterized by the treewidth bound t, is hard for $W[1]$.*

Proof. (Sketch.) The proof is an FPT Turing reduction, based on color-coding [2]. We reduce from the $W[1]$-hard problem of k-CLIQUE. Let $(G = (V, E), k)$ be an instance of the parameterized CLIQUE problem. Let $\mathcal{H}$ be a suitable family of hash functions $h : V \rightarrow \mathcal{A} = \{1, ..., k\}$.

If G is a yes-instance for the k-CLIQUE problem, then for at least one $h \in \mathcal{H}$, the coloring function h is injective on the vertices of a witnessing k-clique in G (that is, each vertex of the k-clique is assigned a different color).

We describe a Turing reduction to instances $G'(h)$ of the 2-CCC problem, one for each $h \in \mathcal{H}$, such that G is a yes-instance for the k-CLIQUE problem if and only at least one $G'(h)$ is a yes-instance for the 2-CCC problem. Each $G'(h)$ has treewidth $t = O(k^2)$.

The construction of $G'(h)$ is based on an *edge-representation of the clique* strategy. We will describe the construction of $G'(h)$ in stages, building up in a modular fashion. A *module* of the construction will be a subgraph that occurs in $G'(h)$ as an induced subgraph, except for a specified set of boundary vertices of the module. These boundary sets will be identified as the various modules are "plugged together" to assemble $G'(h)$. In the figures that illustrate the construction, each kind of module is represented by a symbolic *schematic*, and the modules are built up in a hierarchical fashion. Square vertices represent uncolored vertices.

Figure 2 illustrates the *Choice Module* that is a key part of our construction, and its associated schematic representation. A Choice Module has four "output" boundary vertices, labeled $c_1, ..., c_4$ in the figure. It is easy to see that the module admits a partial solution coloring that "outputs" any one of the (numbered) colors occuring in the module depicted, in the sense that the vertices $c_1, ..., c_4$ are assigned the output color (which is *unsolved*, that is, this color class is not connected in the module), and that the other colors are all solved internally to the module, in the sense that there is (locally) only one connected component of the color class.

A *Co-Incident Edge Set Module* is created from the disjoint union of $k - 1$ Choice Modules, as indicated in in Figure 3. The boundary of the module is the union of the boundaries of the constituent Choice Modules.

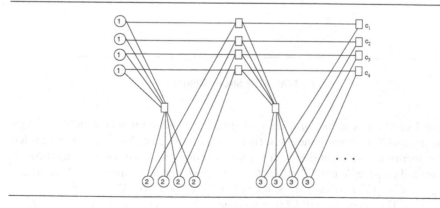

Fig. 2. A *Choice Module* of size 3

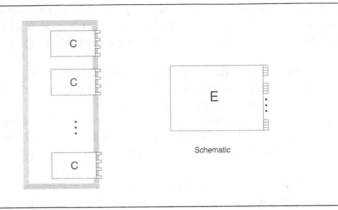

Fig. 3. A Co-Incident Edge Choice Module of size s is the disjoint union of s Choice Modules

An *XOR Stream Module* is shown in Figure 4. This has two "input" boundary sets, each consisting of four uncolored vertices, and one "output" boundary set of four uncolored vertices.

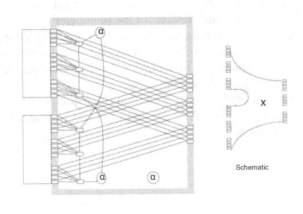

Fig. 4. An XOR Stream Module

Figure 5 shows how a *Tree of Choice Module* is assembled from Co-Incident Edge Set modules and XOR Stream modules. By the *size* of a Tree of Choice module we refer to the number of Co-Incident Edge Set modules occuring as leaves in the construction.

The overall construction of $G'(h)$ for $k = 5$ is illustrated in Figure 6. Suppose the instance graph $G = (V, E)$ that is the source of our reduction from CLIQUE has $|V| = n$ and $|E| = m$. Then each Tree of Choice module $T(h, i)$ has size $n(h, i)$, where $n(h, i)$

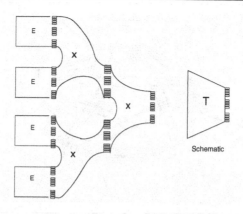

Fig. 5. A Tree of Choice Module of size 4

is the number of vertices colored $i \in \{1, ..., k\}$ by h. Let $V(h, i)$ denote the subset of vertices of V colored i by h.

In the coloring of G by h, if it should happen that a vertex v colored i has no neighbors of color j, $j \neq i$, then v cannot be part of a multicolored k-clique in G, and can be deleted. We consider only colorings of G that are *reduced* in this sense.

The "leaves" of $T(h, i)$ consist of Co-Incident Edge Set modules $E(h, i, u)$, one for each $u \in V(h, i)$. The Co-Incident Edge Set module $E(h, i, u)$ consists of $k - 1$ Choice modules $C(h, i, u, j)$, $j \in \{1, ..., k\}$ and $j \neq i$, and each of these has size equal to the number of edges uv incident to u in G, where v is colored j by h.

The colors used in the construction of $E(h, i, u)$ are in 1:1 correspondence with the edges incident to v in G. Overall, the *colors* of the colored vertices in G' occuring in the Choice modules represent, in this manner, *edges* of G. Each XOR module M has three vertices colored $\alpha = \alpha(M)$ where this color occurs nowhere else in G' (see Figure 4). One of these vertices colored α is an isolated vertex.

Verification that if G has a k-clique, then G' admits a solution to the CCC problem is relatively straightforward. It is important to note that if uv is an edge of G, where $h(u) = i$ and $h(v) = j$, then the color corresponding to uv occurs in exactly two Choice modules: in $C(h, i, u, j)$ and in $C(h, j, v, i)$. If uv is "not selected" (with respect to a coloring completion) then the two local connectivities yield two components of that color.

The argument in the other direction is a little more subtle. First of all, one should verify that the gadgets enforce some restrictions on any solution for G':

(1) Each Choice module necessarily "outputs" one unsolved color (that is, a color not connected into a single component locally), and thus a Co-Incident Edge Set module $E(v)$ "outputs" $k - 1$ colors representing edges incident on v.

(2) Each XOR module forces the "output" stream of unsolved colors to be one, or the other, but not a mixture, of the two input streams of unsolved colors.

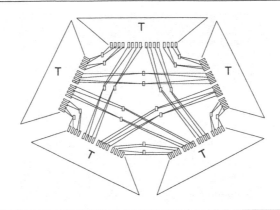

Fig. 6. The modular design of G' for $k = 5$

(3) The unsolved colors that are presented to the central gadget (see Figure 7) can be solved only if these unsolved colors occur in pairs.

The treewidth of G' is easily seen to be $O(k^2)$.

References

1. Arnborg, S., Lagergren, J., Seese, D.: Easy problems for tree-decomposable graphs. J. Algorithms 12, 308–340 (1991)
2. Alon, N., Yuster, R., Zwick, U.: Color-coding. Journal of the ACM 42, 844–856 (1995)
3. Bodlaender, H.L., Fellows, M., Langston, M., Ragan, M.A., Rosamond, F., Weyer, M.: Quadratic kernelization for convex recoloring of trees. Proceedings COCOON 2007, these proceedings (2007)
4. Bar-Yehuda, R., Feldman, I., Rawitz, D.: Improved approximation algorithm for convex recoloring of trees. In: Erlebach, T., Persinao, G. (eds.) WAOA 2005. LNCS, vol. 3879, pp. 55–68. Springer, Heidelberg (2006)
5. Borie, R.B., Parker, R.G., Tovey, C.A.: Automatic generation of linear-time algorithms from predicate calculus descriptions of problems on recursively generated graph families. Algorithmica 7, 555–581 (1992)
6. Bodlaender, H.L., Weyer, M.: Convex anc connected recolourings of trees and graphs. Manuscript (2005)
7. Bu, D., Zhao, Y., Cai, L., Xue, H., Zhu, X., Lu, H., Zhang, J., Sun, S., Ling, L., Zhang, N., Li, G., Chen, R.: Topological structure analysis of the protein-protein interaction network in budding yeast. Nucleic Acids Res. 31(9), 2443–2450 (2003)
8. Chen, J., Chor, B., Fellows, M., Huang, X., Juedes, D., Kanj, I., Xia, G.: Tight lower bounds for certain parameterized NP-hard problems. In: Proceedings of the 19th Annual IEEE Conference on Computational Complexity, pp. 150–160. IEEE Computer Society Press, Los Alamitos (2004)
9. Courcelle, B.: The monadic second-order logic of graphs I: Recognizable sets of finite graphs. Information and Computation 85, 12–75 (1990)
10. Downey, R.G., Fellows, M.R.: Parameterized Complexity. Springer, Heidelberg (1999)

11. Flum, J., Grohe, M.: Parameterized Complexity Theory. Springer, Heidelberg (2006)
12. Fellows, M., Giannopoulos, P., Knauer, C., Paul, C., Rosamond, F., Whitesides, S., Yu, N.: The lawnmower and other problems: applications of MSO logic in geometry, Manuscript (2007)
13. Gramm, J., Nickelsen, A., Tantau, T.: Fixed-parameter algorithms in phylogenetics. Manuscript (2006)
14. Huson, D.H., Bryant, D.: Application of phylogenetic networks in evolutionary studies. Mol. Biol. E 23, 254–267 (2006)
15. Huson, D.H.: SplitsTree: a program for analyzing and visualizing evolutionary data. Bioinfomatics 14, 68–73 (1998)
16. Kelley, B.P., Sharan, R., Karp, R.M., Sittler, T., Root, D.E., Stockwell, B.R., Ideker, T.: Conserved pathways within bacteria and yeast as revealed by global protein network alignment. Proc. Natl. Acad. Sci. USA 100, 11394–11399 (2003)
17. Moran, S., Snir, S.: Convex recolorings of strings and trees: definitions, hardness results and algorithms. To appear in Journal of Computer and System Sciences. In: Dehne, F., López-Ortiz, A., Sack, J.-R. (eds.) WADS 2005. LNCS, vol. 3608, pp. 218–232. Springer, Heidelberg (2005) A preliminary version appeared
18. Moran, S., Snir, S., Sung, W.: Partial convex recolorings of trees and galled networks. Manuscript (2006)
19. Niedermeier, R.: Invitation to Fixed Parameter Algorithms. Oxford University Press, Oxford (2006)
20. Ramadan, E., Tarafdar, A., Pothen, A.: A hypergraph model for the yeast protein complex network. Fourth IEEE International Workshop on High Performance Computational Biology, Santa Fe, NM, April 26, 2004. IEEE Computer Society Press, Los Alamitos (2004)
21. Rual, J.F., Venkatesan, K., Hao, T., Hirozane-Kishikawa, T., Dricot, A., Li, N., Berriz, G.F., Gibbons, F.D., Dreze, M., Ayivi-Guedehoussou, N., Klitgord, N., Simon, C., Boxem, M., Milstein, S., Rosenberg, J., Goldberg, D.S., Zhang, L.V., Wong, S.L., Franklin, G., Li, S., Albala, J.S., Lim, J., Fraughton, C., Llamosas, E., Cevik, S., Bex, C., Lamesch, P., Sikorski, R.S., Vandenhaute, J., Zoghbi, H.Y., Smolyar, A., Bosak, S., Sequerra, R., Doucette-Stamm, L., Cusick, M.E., Hill, D.E., Roth, F.P., Vidal, M.: Nature, 437, 1173–1178 (2005)
22. Schwikowski, B., Uetz, P., Fields, S.: A network of protein-protein interactions in yeast. Nature Biotechnology 18(12), 1257–1261 (2000)
23. Viveshwara, S., Brinda, K.V., Kannan, N.: Protein structure: insights from graph theory. J. Theoretical and Computational Chemistry 1, 187–211 (2002)

Quadratic Kernelization for Convex Recoloring of Trees*

Hans L. Bodlaender[1], Michael R. Fellows[2,3], Michael A. Langston[4],
Mark A. Ragan[3,5], Frances A. Rosamond[2,3], and Mark Weyer[6]

[1] Department of Information and Computing Sciences, Utrecht University,
Utrecht, The Netherlands
hansb@cs.uu.nl
[2] Parameterized Complexity Research Unit, Office of the DVC(Research),
University of Newcastle, Callaghan NSW 2308, Australia
{michael.fellows,frances.rosamond}@newcastle.edu.au
[3] Australian Research Council Centre in Bioinformatics
[4] Department of Computer Science, University of Tennessee,
Knoxville TN 37996-3450 and Computer Science and Mathematics Division,
Oak Ridge National Laboratory, Oak Ridge, TN 37831-6164 USA
langston@cs.utk.edu
[5] Institute for Molecular Bioscience, University of Queensland,
Brisbane, QLD 4072 Australia
m.ragan@imb.uq.edu.au
[6] Institut für Informatik, Humboldt-Universität zu Berlin, Berlin, Germany
mark.weyer@informatik.hu-berlin.de

Abstract. The CONVEX RECOLORING (CR) problem measures how far
a tree of characters differs from exhibiting a so-called "perfect phylogeny". For input consisting of a vertex-colored tree T, the problem
is to determine whether recoloring at most k vertices can achieve a convex coloring, meaning by this a coloring where each color class induces
a connected subtree. The problem was introduced by Moran and Snir,
who showed that CR is NP-hard, and described a search-tree based FPT
algorithm with a running time of $O(k(k/\log k)^k n^4)$. The Moran and Snir
result did not provide any nontrivial kernelization. Subsequently, a kernelization with a large polynomial bound was established. Here we give
the strongest FPT results to date on this problem: (1) We show that in
polynomial time, a problem kernel of size $O(k^2)$ can be obtained, and (2)

* This research has been supported by the Australian Research Council Centre in
Bioinformatics, by the U.S. National Science Foundation under grant CCR–0311500,
by the U.S. National Institutes of Health under grants 1-P01-DA-015027-01, 5-U01-
AA-013512-02 and 1-R01-MH-074460-01, by the U.S. Department of Energy under
the EPSCoR Laboratory Partnership Program and by the European Commission
under the Sixth Framework Programme. The second and fifth authors have been
supported by a Fellowship to the Institute of Advanced Studies at Durham University, and hosted by a William Best Fellowship to Grey College during the preparation
of the paper.

We prove that the problem can be solved in linear time for fixed k. The technique used to establish the second result appears to be of general interest and applicability for bounded treewidth problems.

Topics: Algorithms and Complexity.

1 Introduction

The historically first and most common definition of *fixed-parameter tractability* for parameterized problems is solvability in time $O(f(k)n^c)$, where n is the input size, k is the parameter, and c is a constant independent of n and k. Background and motivation for the general subject of parameterized complexity and algorithmics can be found in the books [9,10,14]. This basic view of the subject is by now well-known.

Less well-known is the point of view that puts an emphasis on FPT kernelization. Kernelization is central to FPT as shown by the lemma:

Lemma 1. *A parameterized problem Π is in FPT if and only if there is a transformation from Π to itself, and a function g, that reduces an instance (x, k) to (x', k') such that:*

1. *the transformation runs in time polynomial in $|(x, k)|$,*
2. *(x, k) is a yes-instance of Π if and only if (x', k') is a yes-instance of Π,*
3. *$k' \leq k$, and*
4. *$|(x', k')| \leq g(k)$.*

The lemma is trivial, but codifies a shift of perspective. To see how this works, consider the Moran and Snir FPT result for CONVEX RECOLORING, with the running time of $O^*((k/\log k)^k)$. When $n > (k/\log k)^k$, then the Moran and Snir algorithm runs in polynomial time, in fact, in time $O(n^5)$. (So we can run the algorithm and determine the answer, and "transform" the input to either a canonical small NO instance or a canonical small YES instance accordingly.) If $n \leq (k/\log k)^k$ then we simply do nothing; we declare that the input is "already" kernelized to the bound $g(k) = (k/\log k)^k$. If a parameterized problem is FPT, that is, solvable in time $f(k)n^c$, then the lemma above gives us the trivial P-time kernelization bound of $g(k) = f(k)$. Normally for FPT problems, $f(k)$ is some exponential function of k, so the subclasses of FPT

$$Lin(k) \subseteq Poly(k) \subseteq FPT$$

of parameterized problems that admit P-time kernelization to kernels of size bounded by *linear* or *polynomial* functions of k would seem to be severe restrictions of FPT.

It is surprising that so many parameterized problems in FPT belong to these much more restrictive subclasses. The connection to heuristics goes like this: *FPT kernelization* is basically a systematic approach to *polynomial-time* preprocessing. Preprocessing plays a powerful role in practical approaches to solving hard problems. For any parameterized problem in FPT there are now essentially two different algorithm design competitions with independent payoffs:

(1) to find an FPT algorithm with the best possible exponential cost $f(k)$, and
(2) to find the best possible polynomial-time kernelization for the problem.

A classic result on FPT kernelization is the VERTEX COVER $2k$ kernelization due to Nemhauser and Trotter (see [14]). The linear kernelization for PLANAR DOMINATING SET is definitely nontrivial [1]. The question of whether the FEED-BACK VERTEX SET (FVS) problem for undirected graphs has a $Poly(k)$ kernelization was a noted open problem in parameterized algorithmics, only recently solved. The first $Poly(k)$ kernelization for FVS gave a bound of $g(k) = O(k^{11})$ [6]. Improved in stages, the best kernelization bound is now $O(k^3)$ [4].

We prove here a polynomial time kernelization for the CONVEX RECOLORING problem, to a reduced instance that has $O(k^2)$ vertices. The basic problem is defined as follows.

CONVEX RECOLORING
Instance: A tree $F = (V, E)$, a set of colors $\mathcal{C}$, a coloring function Γ : $V \to \mathcal{C}$, and a positive integer k.
Parameter: k
Question: Is it possible to modify Γ by changing the color of at most k vertices, so that the modified coloring Γ' is *convex* (meaning that each color class induces a single monochromatic subtree)?

The CONVEX RECOLORING problem is of interest in the context of maximum parsimony approaches to evaluating phylogenetic trees [12,11]. Some further applications in bioinformatics are also described in [12]; see also [13].

2   Preliminaries

Our kernelization algorithm is shown for the following generalized problem.

ANNOTATED CONVEX RECOLORING
Instance: A forest $F = (V, E)$ where the vertex set V is partitioned into two types of vertices, $V = V_0 \cup V_1$, a set of colors $\mathcal{C}$, a coloring function $\Gamma : V \to \mathcal{C}$, and a positive integer k.
Parameter: k
Question: Is it possible to modify Γ by changing the color of at most k vertices in V_1, so that the modified coloring Γ' is *convex*?

A *block* in a colored forest is a maximal set of vertices that induces a monochromatic subtree. For a color $c \in \mathcal{C}$, $\beta(\Gamma, c)$ denotes the number of blocks of color c with respect to the coloring function Γ. A color c is termed a *bad color* if $\beta(\Gamma, c) > 1$, and c is termed a *good color* if $\beta(\Gamma, c) = 1$.

A recoloring Γ' of coloring Γ is *expanding*, if for each block of Γ', there is at least one vertex in the block that keeps its color, i.e., that has $\Gamma'(v) = \Gamma(v)$.

Moran and Snir [12] have shown that there always exists an optimal expanding convex recoloring. It is easy to see that this also holds for the variant where we allow annotations and a forest, as above.

As in [3], we define for each $v \in V$ a set of colors $S(v)$: $c \in S(v)$, if and only if there are two distinct vertices w, x, $w \neq v$, $x \neq v$, such that w and x have color c (in Γ), and v is on the path from w to x.

We introduce another special form of convex recoloring, called *normalized*. For each vertex $v \in V$, we add a new color c_v. We denote $\mathcal{C}' = \mathcal{C} \cup \{c_v \mid v \in V\}$. A recoloring $\Gamma' : V \to \mathcal{C}'$ of coloring $\Gamma : V \to \mathcal{C}$ is *normalized*, if for each $v \in V$: $\Gamma'(v) \in \{\Gamma(v), c_v\} \cup S(v)$.

Lemma 2. *There is an optimal convex recoloring that is normalized.*

Proof. Take an arbitrary optimal convex recoloring Γ'. Define Γ'' as follows. For each $v \in V$, if $\Gamma'(v) \in \{\Gamma(v)\} \cup S(v)$, then set $\Gamma''(v) = \Gamma'(v)$. Otherwise, set $\Gamma''(v) = c_v$. One can show that $\Gamma''(v)$ is convex. Clearly, it is normalized, and recolors at most as many vertices as $\Gamma'(v)$. □

3 Kernelizing to a Linear Number of Vertices Per Color

In this section, we summarize a set of rules from [6] that ensure that for each color, there are at most $O(k)$ vertices with that color. There are two trivial rules: if F is an empty forest and $k \geq 0$, then we return YES; if $k < 0$, then we return NO if and only if the given coloring is not already convex.

Rule 1. *Suppose there are $\alpha \geq 2k + 3$ vertices with color c. Let v be a vertex, such that each connected component of $F - v$ contains at most $\frac{\alpha}{2}$ vertices of color c. Assume that $v \notin V_0$, or $\Gamma(v) \neq c$. Now*

  - *If v has a color, different from c, and $v \in V_0$, then return NO.*
  - *If v has a color, different from c, and $v \notin V_0$, then set $\Gamma(v) = c$ and decrease k by one.*
  - *Fix the color of v: put $v \in V_0$.*

Rule 2. *Suppose $v \in V_0$ has color c. Suppose one of the connected components of F contains at least $k + 1$ vertices with color c, but does not contain v. Then return NO.*

Rule 3. *Let v be a vertex of color c. Suppose $F - v$ has a component with vertex set W that contains at least $k + 1$ vertices with color c, that contains a neighbor w of v. Assume that $w \notin V_0$ or $\Gamma(w) \neq c$. Then*

  - *If w has a color, different from c, and $w \in V_0$, then return NO.*
  - *If w has a color, different from c, and $w \notin V_0$, then give w color c, and decrease k by one.*
  - *Fix the color of w: put $w \in V_0$.*

Rule 4. *If there is a tree T in the colored forest that contains two distinct vertices u and v, both of which have fixed color c, then modify T by contracting the path between u and v (resulting in a single vertex of fixed color c). If r denotes the number of vertices of this path that do not have color c, then $k' = k - r$.*

Rule 5. *If there are two distinct trees in the colored forest that both contain a vertex of fixed color c, then return NO.*

Let $v \in V_0$ have color c. A connected component of $F - v$ with vertex set W is said to be *irrelevant*, if both of the following two properties hold: (1) The vertices with color c in W form a connected set that contains a neighbor of v, and (2) For each color $c' \neq c$, if there are vertices with color c' in W, then c' is not bad. If a component is not irrelevant, it is said to be *relevant*.

Rule 6. *Let $v \in V_0$. Let W be the vertex set of an irrelevant connected component of $F - v$. Then remove W, and all edges incident to vertices in W, from the tree.*

Rule 7. *Suppose $F - v$ contains at least $2k + 1$ components, and each component of $F - v$ is relevant. Then return NO.*

Rule 8. *Let $v \in V_0$ be a vertex with fixed color c. Let W be the vertex set of a relevant component of $F - v$ that contains vertices with color c. Suppose there are vertices with color c, not in $\{v\} \cup W$. Then*

- *Take a new color c', not yet used, and add it to C.*
- *Take a new vertex v', and add it to F.*
- *Set $v' \in V_0$. Color v' with c'.*
- *For all $w \in W$ with color c, give w color c'. (Note that k is not changed.)*
- *For each edge $\{v, w\}$ with $w \in W$: remove this edge from F and add the edge $\{v', w\}$.*

Theorem 1. *An instance that is reduced with respect to the above Rules has at most $2k + 2$ vertices per color.*

4   Kernelizing to a Quadratic Number of Vertices

In this section, we work in a series of steps towards a kernel with $O(k^2)$ vertices. A number of new structural notions are introduced.

4.1   Bad Colors

Rule 9. *If there are more than $2k$ bad colors, then return NO.*

The colors that are not bad are distinguished into three types: gluing, stitching, and irrelevant.

4.2 Gluing Colors

A vertex v is *gluing* for color c, if it is adjacent to two distinct vertices x and y, both of color c, but the color of v is unequal to c. Note that if v is gluing for color c, then v separates (at least) two blocks of color c: x and y belong to different blocks. When we recolor v with color c, then these blocks become one block (are 'glued' together). A vertex v is *gluing*, if v is gluing for some color, and if the color of v is not bad. A color c is *gluing*, if there is a gluing vertex with color c, and c is not bad. (Note that the vertex will be gluing for some other color $c' \neq c$.)

For a color c, let W_c be the set of vertices that have color c or are gluing for color c. A *bunch* of vertices of color c is a maximal connected set of vertices in W_c, i.e., a maximal connected set of vertices that have either color c or are gluing for color c. The following lemma is not difficult, and establishes the soundness of the next Rule.

Lemma 3. *Suppose W is a bunch of color c that contains ℓ vertices that are gluing for color c. Then in any convex recoloring, at least ℓ vertices in W are recolored, and for each, either the old, or the new color is c.*

Rule 10. *If there are more than $2k$ vertices that are gluing, then return NO.*

As a result, there are $O(k)$ colors that are gluing, and thus $O(k^2)$ vertices that have a gluing color. We next bound the number of bunches.

Lemma 4. *Suppose there are ℓ bunches with color c. Then the number of recolored vertices whose old or new color equals c, is at least $\ell - 1$.*

Proof. Omitted due to space limitations.   □

Rule 11. *Suppose that for each bad color c, there are ℓ_c bunches. If the sum over all bad colors c of $\ell_c - 1$ is at least $2k + 1$, then return NO.*

4.3 Stitching Colors

We now define the notion of stitching vertices. To arrive at an $O(k^2)$ kernel, we have to define this notion with some care.

We first define the *cost* of a path between two vertices with the same color that belong to different bunches. Let v and w be two vertices with color c, in the same subtree of F, but in different bunches of color c. The *cost* of the path from v to w is the sum of the number of vertices with color unequal to c on this path, and the sum over all colors $c' \neq c$ such that

- c' is not bad and c' is not gluing, and
- there is at least one vertex with color c' on the path from v to w in T

of the following term: consider the blocks with color c' in the forest, obtained by removing the path from v to w from T. If one of these blocks contains a vertex in V_0 (i.e., it has fixed color c'), then sum the sizes of all other blocks. Otherwise, sum the sizes of the all blocks except one block with the largest number of vertices.

A *stitch* of color c is a path between two vertices v, w, both of color c, such that: (1) v and w are in different bunches, (2) all vertices except v and w have a color, different from c, and do not belong to V_0, and (3) the cost of the path is at most k. A vertex v is *stitching* for color c, if: (1) the color of v is different from c, (2) v is not gluing for color c, and (3) v is on a stitch of color c. A vertex is *stitching*, if it is stitching for at least one color. (Note that this vertex will be stitching for a bad color.) A color c is *stitching* if there is at least one vertex with color c that is stitching, and it is not bad or gluing. A color is *irrelevant*, if it is neither bad, gluing, or stitching. Summarizing, we have the following types of vertices: (a) vertices with a bad color, (b) vertices with a gluing color, (c) vertices with a stitching color that belong to a stitch, (d) vertices with a stitching color, that do not belong to a stitch — these will be partitioned into pieces in Section 4.4, and (e) vertices with an irrelevant color.

Proofs of the next two lemmas can be found in the full paper, and establish the soundness of the next two rules.

Lemma 5. *Suppose there is a convex recoloring with at most k recolored vertices. Then there is an optimal convex recoloring such that for each bad color c, all vertices that receive color c have color c in the original coloring or are gluing for c or are stitching for c.*

Lemma 6. *Suppose there is a convex recoloring with at most k recolored vertices. Then there is an optimal convex recoloring such that no vertex with an irrelevant color is recolored.*

Rule 12. *Let $v \in V_1$ be a vertex with an irrelevant color. Put v into V_0 (i.e., fix the color of v.)*

Rule 13. *Let $v \in V_1$ be a vertex with a stitching color. Suppose v is not vulnerable. Then put v into V_0.*

The following rule, in combination with earlier rules, helps us to get a reduced instance without any vertex with an irrelevant color.

Rule 14. *Let $v \in V_0$ be the only vertex with some color c. Then remove v and its incident edges.*

Soundness is easy to see: the same recolorings in the graph with and without v are convex. The combination of Rules 4, 12, and 14 causes the deletion of all vertices with an irrelevant color: all such vertices first get a fixed color, then all vertices with the same irrelevant color are contracted to one vertex, and then this vertex is deleted. So, we can also use instead the following rule.

Rule 15. *Suppose c is an irrelevant color. Then remove all vertices with color c.*

4.4 Pieces of a Stitching Color

For a kernel of size $O(k^2)$, we still can have too many vertices with a stitching color. In order to reduce the number of such vertices we introduce the concept of a *piece of color c*. Consider a stitching color c, and consider the subforest of the forest, induced by the vertices with color c. If we remove from this subtree all vertices that are on a stitch, then the resulting components are the pieces, i.e., a *piece of color c* is a maximal subtree of vertices with color c that do not belong to a stitch. Assume that we have exhaustively applied all of the rules described so far, and therefore we do not have vertices with an irrelevant color. The next lemma, given here without proof, shows that the next Rule is sound.

Lemma 7. *Suppose W is the vertex set of a piece of stitching color c. Suppose there is a vertex $v \in W \cap V_0$. Then if there is a convex recoloring with at most k recolored vertices, then there is a convex recoloring that does not recolor any vertex in W.*

Rule 16. *Let W be a vertex set of a piece of stitching color c, that contains at least one vertex in V_0. Then put all of the vertices of W into V_0.*

As a result of this rule and Rule 4, a piece containing a vertex with a fixed color will contain only one vertex. We omit the soundness proof for the following rule from this version. If there is a large piece, found by Rule 17, then it will be contracted to a single vertex by Rule 4.

Rule 17. *Suppose c is a stitching color, and there are α vertices with color c. Suppose there is no vertex in V_0 with color c. If W is the vertex set of a piece of color c, and $|W| > \alpha/2$, then put all of the vertices of W into V_0.*

4.5 Kernel Size

We now *tag* some vertices that have a stitching color. A tag is labeled with a bad color. Basically, when we have a stitch for a bad color c, we tag the vertices that count for the cost of the stitch with color c. Tagging is done as follows. For each stitch for bad color c, and for each stitching color c' with a vertex on the stitch with color c', consider the blocks of color c' obtained by removing the vertices on the stitch. If there is no vertex with color c' in V_0, then take a block with vertex set W such that the number of vertices in this piece is at least as large as the number of vertices in any other piece. Tag all vertices in $Q - W$ with color c. If there is a vertex v with color c' in V_0, then tag all vertices with color c', except those that are in the same block as v.

Comparing the tagging procedure with the definition of the cost of a stitch, we directly note that the number of vertices that is tagged equals the cost of the stitch, i.e., for each stitch of bad color c, we tag at most k vertices with c.

We now want to count the number of vertices with a stitching color. To do so, we first count the number of vertices with a stitching color that are tagged. To do this, we consider a bad color c, and count the number of vertices, with a stitching color, that are tagged with c.

The following three lemmas are proved in the full paper.

Lemma 8. *Let c be a bad color with ℓ_c bunches. A reduced instance has at most $2k(\ell_c - 1)$ vertices that are tagged with c.*

Lemma 9. *Let c be a stitching color. There is at most one piece of color c that contains a vertex that is not tagged with a bad color.*

Lemma 10. *In a reduced instance, there are at most $8k^2$ vertices with a stitching color.*

Theorem 2. *In time polynomial in n and k, a kernel can be found with $O(k^2)$ vertices.*

Proof. With standard techniques, one can observe that all rules can be carried out in time polynomial in n and k.

In the reduced instance, there are at most $2k$ bad colors, and at most $2k$ gluing colors. For each of these colors, there are at most $2k + 2$ vertices with that color. So, in total at most $4k^2 + 4k$ vertices have a bad or gluing color. There are at most $8k^2$ vertices with a stitching color, and no vertices with an irrelevant color, so in total we have at most $12k^2 + 4k$ vertices.   □

5   Linear Time FPT with Treewidth and MSOL

We describe an algorithm that solves the CONVEX RECOLORING problem in $O(f(k) \cdot n)$ time. There are four main ingredients of the algorithm: (1) the construction of an auxiliary graph, the *vertex-color* graph, (2) the observation that for yes-instances of the CONVEX RECOLORING problem this graph have bounded treewidth, (3) a formulation of the CONVEX RECOLORING problem for fixed k in Monadic Second Order Logic, and (4) the use of a famous result of Courcelle [8], see also [2,5]. For ease of presentation, we assume that there are no vertices with a fixed color, and that the input is a tree. The "trick" of using such an auxiliary graph seems to be widely applicable. For another example of this new technique, see [7].

Suppose we have a tree $T = (V, E)$, and a coloring $\Gamma : V \to \mathcal{C}$. The *vertex-color graph* has as vertex set $V \cup \mathcal{C}$, i.e., we have a vertex for each vertex in T, and a vertex for each color. The edge set of the vertex-color graph is $E \cup \{\{v, c\} \mid \Gamma(v) = c\}$, i.e., there are two types of edges: the edges of the tree T, and an edge between a vertex of T and the vertex representing its color. Recall that $S(v)$ denotes the set of colors c for which there are vertices w and x, distinct from v, with v on the path from w to x in T, and $\Gamma(w) = \Gamma(x) = c$.

Lemma 11. *Suppose there is a convex recoloring with at most k recolored vertices. Then for each $v \in V$, $|S(v)| \le k + 1$.*

Proof. Suppose Γ' is a convex recoloring with at most k recolored vertices of Γ. Consider a vertex $v \in V$, and a color $c \in S(v)$. Suppose $c \ne \Gamma'(v)$. There are vertices w, x with $\Gamma(w) = \Gamma(x) = c$, and the path from w to x uses v. As v is not recolored to c, either w or x must be recolored. So, there can be at most k colors $\ne \Gamma'(v)$ that belong to $S(v)$.   □

Lemma 12. *Suppose there is a convex recoloring with at most k recolored vertices. Then the treewidth of the vertex-color graph is at most $k + 3$.*

Proof. Without loss of generality, we suppose that Γ is surjective, i.e., for all $c \in \mathcal{C}$, there is a $v \in V$ with $\Gamma(v) = c$. (If Γ is not surjective, then colors c with $\Gamma^{-1}(c) = \emptyset$ are isolated vertices in the vertex-color graph, and removing these does not change the treewidth.)

Take an arbitrary vertex $r \in V$ as root of T. For each $v \in V$, set $X_v = \{v, p(v), \Gamma(v)\} \cup S(v)$, where $p(v)$ is the parent of v. For $v = r$, we take $X_v = \{v, \Gamma(v)\} \cup S(v)$.

It is easy to verify directly that $(\{X_v \mid v \in V\}, T)$ is a tree decomposition of the vertex-color graph of width at most $k + 3$. □

Theorem 3. *For each fixed k, there is a linear time algorithm that, given a vertex-colored tree T, decides if there is a convex recoloring of T with at most k recolored vertices.*

Proof. We show that for a fixed number of recolored vertices k, the CONVEX RECOLORING problem can be formulated as a property of the vertex-color graph that can be stated in Monadic Second Order Logic. The result then directly follows by the result of Courcelle [8], that each such property can be decided in linear time on graphs with bounded treewidth. (See also [2,5]. We use Lemma 12.)

We assume we are given the vertex-color graph $(V \cup \mathcal{C}, E')$, and have sets V and $\mathcal{C}$ distinguished. A recoloring Γ' that differs from Γ at most k vertices can be represented by these vertices $v_1, \ldots, v_k$ and by the corresponding values $c_1 := \Gamma'(v_1), \ldots, c_k := \Gamma'(v_k)$. Then, for a given color c, consider the set V_c defined as follows:

- If $v = v_i$ for some $1 \leq i \leq k$, then $v \in V_c$ if and only if $c_i = c$.
- If $v \notin \{v_1, \ldots, v_k\}$, then $v \in V_c$ if and only if (v, c) is an edge of the vertex-color graph.

Note, that V_c is the color class of c in the coloring Γ'. Hence Γ' is convex if and only if V_c is connected for each $c \in \mathcal{C}$. It is a standard fact, that MSOL can express connectedness, say by a formula $\varphi(X)$, where the set variable X represents V_c. Furthermore, the definition of V_c given above can be expressed even in first-order logic, say by a formula $\psi(x_1, \ldots, x_k, y_1, \ldots, y_k, z, X)$, where additionally $x_1, \ldots, x_k$ represent $v_1, \ldots v_k$, $y_1, \ldots, y_k$ represent $c_1, \ldots c_k$, and z represents c. Then, for this k, the CONVEX RECOLORING problem can be expressed by the formula

$$\exists x_1, \ldots, x_k \exists y_1, \ldots, y_k \left(\bigwedge_{1 \leq i \leq k} V x_i \wedge \bigwedge_{1 \leq i < j \leq k} x_i \neq x_j \wedge \bigwedge_{1 \leq i \leq k} C y_i \wedge \forall z \forall X (\psi \to \varphi) \right).$$

The algorithms works as follows. Given a colored tree (T, Γ), it constructs the vertex-color graph and computes a tree-decomposition of width at most $k + 3$. If this fails, the algorithm returns NO, in accordance with Lemma 12. Otherwise, using the tree-decomposition, it evaluates the above formula on the vertex-color graph. Each step uses at most linear time. □

References

1. Alber, J., Fellows, M.R., Niedermeier, R.: Polynomial-time data reduction for dominating sets. J. ACM 51, 363–384 (2004)
2. Arnborg, S., Lagergren, J., Seese, D.: Easy problems for tree-decomposable graphs. J. Algorithms 12, 308–340 (1991)
3. Bar-Yehuda, R., Feldman, I., Rawitz, D.: Improved approximation algorithm for convex recoloring of trees. In: Erlebach, T., Persinao, G. (eds.) WAOA 2005. LNCS, vol. 3879, pp. 55–68. Springer, Heidelberg (2006)
4. Bodlaender, H.L.: A cubic kernel for feedback vertex set. In: Thomas, W., Weil, P. (eds.) STACS 2007. LNCS, vol. 4393, pp. 320–331. Springer, Heidelberg (2007)
5. Borie, R.B., Parker, R.G., Tovey, C.A.: Automatic generation of linear-time algorithms from predicate calculus descriptions of problems on recursively constructed graph families. Algorithmica 7, 555–581 (1992)
6. Burrage, K., Estivill-Castro, V., Fellows, M.R., Langston, M.A., Mac, S., Rosamond, F.A.: The undirected feedback vertex set problem has a poly(k) kernel. In: Bodlaender, H.L., Langston, M.A. (eds.) IWPEC 2006. LNCS, vol. 4169, pp. 192–202. Springer, Heidelberg (2006)
7. Chor, B., Fellows, M., Ragan, M., Rosamond, F., Razgon, I., Snir, S.: Connected coloring completion for general graphs: algorithms and complexity. In: Proceedings COCOON (2007)
8. Courcelle, B.: The monadic second-order logic of graphs I: Recognizable sets of finite graphs. Information and Computation 85, 12–75 (1990)
9. Downey, R.G., Fellows, M.R.: Parameterized Complexity. Springer, Heidelberg (1999)
10. Flum, J., Grohe, M.: Parameterized Complexity Theory. Springer, Heidelberg (2006)
11. Gramm, J., Nickelsen, A., Tantau, T.: Fixed-parameter algorithms in phylogenetics. To appear in The Computer Journal.
12. Moran, S., Snir, S.: Convex recolorings of strings and trees: Definitions, hardness results, and algorithms. In: Dehne, F., López-Ortiz, A., Sack, J.-R. (eds.) WADS 2005. LNCS, vol. 3608, pp. 218–232. Springer, Heidelberg (2005)
13. Moran, S., Snir, S.: Efficient approximation of convex recolorings. In: Chekuri, C., Jansen, K., Rolim, J.D.P., Trevisan, L. (eds.) APPROX 2005 and RANDOM 2005. LNCS, vol. 3624, pp. 192–208. Springer, Heidelberg (2005)
14. Niedermeier, R.: Invitation to fixed-parameter algorithms. Oxford Lecture Series in Mathematics and Its Applications. Oxford University Press, Oxford (2006)

On the Number of Cycles in Planar Graphs

Kevin Buchin[1], Christian Knauer[1], Klaus Kriegel[1], André Schulz[1], and
Raimund Seidel[2]

[1] Freie Universität Berlin, Institute of Computer Science,
Takustr. 9, 14195 Berlin, Germany
{buchin,knauer,kriegel,schulza}@inf.fu-berlin.de
[2] Universität des Saarlandes, Institute of Computer Science, PO Box 151150,
66041 Saarbrücken, Germany
rseidel@cs.uni-sb.de

Abstract. We investigate the maximum number of simple cycles and
the maximum number of Hamiltonian cycles in a planar graph G with
n vertices. Using the transfer matrix method we construct a family of
graphs which have at least 2.4262^n simple cycles and at least 2.0845^n
Hamilton cycles.

Based on counting arguments for perfect matchings we prove that
2.3404^n is an upper bound for the number of Hamiltonian cycles. Moreover, we obtain upper bounds for the number of simple cycles of a given
length with a face coloring technique. Combining both, we show that
there is no planar graph with more than 2.8927^n simple cycles. This
reduces the previous gap between the upper and lower bound for the
exponential growth from 1.03 to 0.46.

1 Introduction

In this paper we consider the following question:

> *How many simple cycles and how many Hamiltonian cycles can there be
> in a planar graph with n vertices?*

Since the determination of the exact numbers seems to be out of reach, our goal
is to learn more about the asymptotic behavior of these numbers. Denoting by
$C_s(G)$ and $C_h(G)$ the numbers of simple cycles and of Hamiltonian cycles in a
graph G we define

$$C_s(n) = \max\left\{C_s(G) \mid G \text{ is a planar graph on } n \text{ vertices}\right\}, \text{ and}$$
$$C_h(n) = \max\left\{C_h(G) \mid G \text{ is a planar graph on } n \text{ vertices}\right\}.$$

It is easy to observe that both $C_s(n)$ and $C_h(n)$ grow exponentially and thus we
are interested in describing this *exponential growth rate* by constants $c, d \in \mathbb{R}$
such that $c^n \leq C_s(n) \leq d^n$, and analogously for $C_h(n)$.

The lower bound $C_s(n) = \Omega(2.259^n)$ was obtained in [1] and is based on
counting the number of simple paths connecting two adjacent vertices in a special planar graph on 29 vertices and joining $n/28$ copies of it in a cyclic way.

G. Lin (Ed.): COCOON 2007, LNCS 4598, pp. 97–107, 2007.

An $O(3.363^n)$ upper bound was proved in the same paper by a probabilistic argument. Here we extend the original problem setting to Hamiltonian cycles.

The problem has gained new attention by some recent results of Sharir and Welzl [2], [3]. They investigate the numbers of several geometric objects on a point set in the plane, among them triangulations and crossing-free spanning cycles. In particular they note that an upper bound on the number of crossing-free spanning cycles can be obtained by combining an upper bound on the number of triangulations with an upper bound on the number of Hamiltonian cycles in planar graphs. Here, we will present the proof to the $\sqrt{6}^n$ upper bound for the number of Hamiltonian cycles, which is quoted in [2] as a personal communication, along with an improvement to $\sqrt[4]{30^n}$.

In [2] Sharir and Welzl prove a bound of $O(86.81^n)$ for the number of crossing-free spanning cycles on n points with an alternative approach. This bound is better than the combined bound and it remains better even if the improved bound for Hamiltonian cycles presented in our paper is used. In fact, the lower bound on the number of Hamiltonian cycles presented in our paper shows that a better combined bound cannot be obtained without improving the bound on the number of triangulations.

The paper is organized as follows: In Section 2 we present new lower bounds for $C_s(n)$ and for $C_h(n)$. We prove $C_s(n) = \Omega(2.4262^n)$ and $C_h(n) = \Omega(2.0845^n)$. Both bounds are based on the so-called transfer matrix method applied on a twisted cylinder.

In Section 3 we prove first the $O(\sqrt[4]{30^n})$ upper bound on $C_h(n)$. Next we present a new technique for upper bounds on the number of simple cycles with a given length k in planar graphs on n vertices. Combining both we will obtain a new $O(2.8928^n)$ upper bound for $C_s(n)$.

2   Lower Bounds

We will present a new lower bound for $C_s(n)$ by counting cycles on the *twisted cylinder*. We use the technique of the *transfer matrix method* (see [4,5,6]).

The twisted cylinder describes a graph which is parametrized by a *width* w and a *length* l. We will describe the graph by the following construction: Consider an $\lfloor l/w \rfloor \times w$ integer lattice with the upper leftmost point $(0,0)$ and the lower rightmost point (r, w). Furthermore we attach $(l \bmod w)$ squares at the right end, starting from the top. As a next step we triangulate each square of the lattice by adding diagonals $((x,y),(x+1,y+1))$ for all appropriate values x and y. Finally we identify all edges $((x,w),(x+1,w))$ with the edges $((x+1,0),(x+2,0))$ for all x smaller than $\lfloor l/w \rfloor$. Observe that this graph is planar since it can be embedded as the graph of a 3-polytope. Figure 1 shows a twisted cylinder of length 41 and width 5, and Figure 2 shows a planar embedding of a twisted cylinder of length 12 and width 6. To count the cycles, we construct the cylinder by increasing its length consecutively. We name the cylinder of width w after k rounds Z_k^w and call the last inserted $w + 1$ points its *border*. During the construction we have to deal with unfinished cycles. These cycles are represented

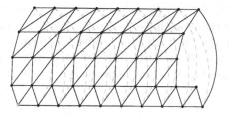

Fig. 1. The graph of a twisted cylinder

as non-intersecting paths which start and end at the border of Z_k^w. To complete
a cycle we need the information which of the points at the border belong to the
same path. We will store this information in a string of length $w + 1$ which we
call the *signature*. The last point introduced corresponds to the first character of
the signature, its predecessor to the second character and so on. Every path has
a start and an end point on the border. The point that was introduced later is
considered as the start point, the other as the end point of a path. We associate a
start point at the border with an A. The position of the end point of a path will be
marked in the signature as B. Interior points of paths at the border are denoted
by X. A point at the border that is not used from any path will be represented
as O in the signature. Thus we get as signature a string from $\{A, B, X, O\}^{w+1}$.
Figure 2 shows an example which has the signature $AXOAOBB$. Notice that

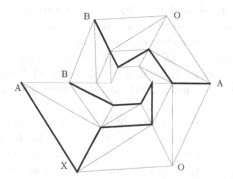

Fig. 2. Partially constructed cycle with signature $AXOAOBB$

the signatures can be represented as 2-Motzkin paths [7, Exercise 6.38.], which
are one of the numerous incarnations of Catalan structures.

We come back to the counting procedure. During the construction we trace
the number of completed and uncompleted cycles. We count the different ways
of generating an uncompleted cycle by a variable indexed by its signature. The
completed cycles are stored in a distinct variable. All variables are stored in a
vector which we call the *state vector* S_k.

Going from cylinder Z_k^w to Z_{k+1}^w will change the state vector. We introduce three new edges and therefore have at most 7 ways to continue uncompleted cycles (choosing all 3 edges will not give a valid successor state). Not all of these choices will produce a valid signature for the successor state. For example the signature $AXOAOBB$ (Figure 2) has four successors, depicted in Figure 3.

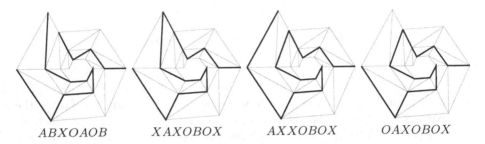

$ABXOAOB$ $XAXOBOX$ $AXXOBOX$ $OAXOBOX$

Fig. 3. The successor states from Figure 2

It is not hard to see that every entry of the "new" state vector S_{k+1} is a non-negative linear combination of the entries of the "old" state vector S_k. Thus $S_{k+1} = T \cdot S_k$, where T is a square matrix with non-negative entries, which is called the *transfer matrix*. By construction this matrix T does not depend on k, and thus for every $k \geq 0$ we have $S_k = T^k \cdot S_0$. The entry in S_k for the completed cycles in Z_k^w is then a lower bound for $C_s(k + w)$.

This entry can be represented as $e^t \cdot S_k$ for the appropriate unit basis vector e. Thus we are interested in the asymptotic behavior of a sequence with elements $p_k = a^t \cdot T^k \cdot b$, where a and b are vectors and T is a square matrix. By considering the Jordan canonical form $T = X^{-1} \cdot J \cdot X$ it is easy to see that $p_k = O(q(k)c^k)$, where c is the largest absolute value of any eigenvalue of T and q is a polynomial of degree smaller than the multiplicity of any eigenvalue of maximal absolute value.

If in our case we remove all unreachable configurations from consideration then the resulting transfer matrix T will be primitive in the sense that for some $\ell > 0$ all entries of T^ℓ are strictly positive. In this case the Perron-Frobenius Theorem [8] guarantees that the eigenvalue of largest absolute value is real and unique. Thus $S_k = O(c^k)$, where c is the largest real eigenvalue of the transfer matrix T.

The generation of T and the computation of its eigenvalues was done with the help of computer programs. We omit the details of the computation of T. The correctness of the calculations was checked by two independent implementations. We observed that larger widths will result in larger growth. The largest width our implementations could handle is 13. The largest absolute eigenvalue for T was computed as 2.4262. The computation was done on a AMD Athlon 64 with 1.8 GHz and 1 GB of RAM. It required 5 days of CPU-time and 550 MB of memory. The implementation can be found under [9]. The results of the computation are listed in Table 1.

Theorem 1. *The maximal number of simple cycles in a planar graph G with n vertices is bounded from below by $\Omega(2.4262^n)$.*

We can also construct a lower bound for Hamiltonian cycles with the method from above. To this end we restrict the state transitions in such a way that if a vertex vanishes from the border, it is guaranteed to be on some path. We forbid all sequences which contain a O as character and calculate the modified transfer matrix.

The largest eigenvalue for the modified transfer matrix is 2.0845. It is obtained for a twisted cylinder of width 13. See Table 1 for the results of the computation.

Table 1. Eigenvalues λ_T of the transfer matrix T, generated for Hamiltonian cycles (H. cyc.) and simple cycles (simple cyc.) depending on the width of the twisted cylinder

w	λ_T H. cyc.	λ_T simple cyc.	w	λ_T H. cyc.	λ_T simple cyc.
2	1.8124	1.9659	8	2.0688	2.4078
3	1.9557	2.2567	9	2.0740	2.4139
4	2.0022	2.3326	10	2.0777	2.4183
5	2.0335	2.3654	11	2.0805	2.4217
6	2.0507	2.3858	12	2.0827	2.4242
7	2.0614	2.3991	13	2.0845	2.4262

Theorem 2. *The maximal number of Hamiltonian cycles in a planar graph G with n vertices is bounded from below by $\Omega(2.0845^n)$.*

3 Upper Bounds

3.1 Hamiltonian Cycles

In this section G denotes a planar graph with n vertices, e edges, and f faces. Since additional edges cannot decrease the number of cycles, we focus on triangulated planar graphs. In this case we have $3f = 2e$, which leads to $e = 3n - 6$ and $f = 2n - 4$.

Let us assume first that n is even and let $M(G)$ denote the number of perfect matchings in G. By a theorem of Kasteleyn, c.f. [10], there is an orientation of the edges of G such that the corresponding skew symmetric adjacency matrix A characterizes $M(G)$ in the following way:

$$(M(G))^2 = |\det(A)|.$$

Note that all but $6n - 12$ entries of A are zero and the nonzero entries are 1 or -1. In this situation we can apply the Hadamard bound for determinants and we obtain $|\det(A)| \leq \sqrt{6}^n$.

In this way we obtain an $\sqrt[4]{6}^n$ upper bound on the number of perfect matchings in G, which improves the $O(\sqrt{3}^n)$ bound from [11]. Moreover, our bound can be improved for graphs with few edges.

Theorem 3. *The number of perfect matchings in a planar graph G with n vertices is bounded from above by $\sqrt[4]{6}^n$.*
The number of perfect matchings in a planar graph G with n vertices and at most kn edges is bounded from above by $\sqrt[4]{2k}^n$.

Our first bound on the number of Hamiltonian cycles follows from Theorem 3 by an easy observation.

Theorem 4. $C_h(n) = O\left(\sqrt[4]{30}^n\right) = O\left(2.3404^n\right).$

Proof. Any Hamiltonian cycle in a graph G with an even number of vertices splits into two perfect matchings, which implies $C_h(G) \leq (M(G))^2 \leq \sqrt{6}^n$. The following modification of the arguments above results in a slight improvement of that bound:

Splitting a Hamiltonian cycle into two perfect matchings, we fix the matching with the lexicographically smallest edge as the first matching m_1 and the other one as the second matching m_2. It follows that if m_1 is fixed, m_2 is a matching in a graph with $2.5\,n - 6$ edges. Repeating the above observations for both matchings, we get

$$M_1(G) \leq \sqrt[4]{6}^n, \qquad M_2(G) \leq \sqrt[4]{5}^n \quad \text{and together} \quad C_h(G) \leq \sqrt[4]{30}^n.$$

Finally we study the case that n is odd. We choose in G a vertex v of degree at most 5, and for each e incident to v we consider the Graph G_e obtained from G by contracting e. Any Hamiltonian cycle in G contains two edges e and e' incident to v and hence induces a Hamiltonian cycle in G_e and $G_{e'}$. On the other hand, any Hamiltonian cycle in some G_e may be extended in up to two ways to a Hamiltonian cycle in G. Thus we obtain an upper bound on $C_h(G)$ by adding the number of Hamiltonian cycles in the at most five planar graphs G_e, leading to a bound of $C_h(G) \leq 5\sqrt[4]{30}^{n-1}$.   □

3.2   Simple Cycles

We start with a new upper bound for the number of cycles in planar graphs and successively improve the bound.

Instead of counting cycles we count paths on G, which can be completed to a simple cycle. We call these paths *cycle-paths*. Their number is an upper bound for the number of cycles. The number of cycle-paths is maximized when G is triangulated. Therefore we assume that G is triangulated.

There exist n paths of length 0. The number of all cycle-paths in G of nonzero length is at most the number of edges e times the maximum number of cycle-paths starting from an arbitrary edge. Thus the exponential growth of the number of cycle-paths is determined by the number of cycle-paths starting from an edge.

Lemma 1. *The maximum number of cycle-paths on G starting from an edge is bounded by $O(n) \cdot 3^n$.*

Proof. We give the starting edge an orientation. We consider only paths in the direction induced by this orientation. The total number of cycle-paths starting from this edge is at most twice the number of cycle-paths with the chosen orientation.

We associate cycle-paths with the nodes of a tree. The root of the tree contains the path of length one corresponding to the starting edge. The children of a tree node contain paths starting with the path stored in the predecessor plus an additional edge. Every cycle-path is only stored in one tree node.

Every cycle-path in G corresponds to a partial red-blue coloring of the faces of G. The coloring is defined as follows: The faces right of the oriented path will be colored blue the faces left of the oriented path red (see Figure 4). We color all faces incident to an inner vertex or the starting edge of the path. The coloring is consistent, because we consider only paths which can be extended to cycles. Therefore the colors correspond to a part of the interior or exterior region induced by the cycle.

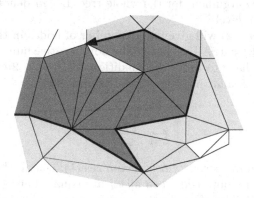

Fig. 4. Example of an induced red-blue coloring by a path (light gray corresponds to red, dark gray to blue)

We construct the tree top down. When we enter a new tree node, the color of at least two faces incident to the last vertex v_i of the path is given. It might be that other faces incident to v_i have been colored before. In that case we color the faces incident to v_i which lie in between two red faces red. The faces which are located in between two blue faces will be colored blue. Observe that at most one non-colored connected region incident to v_i remains. Otherwise it is not possible to extend the path to a cycle and therefore the path stored in this tree node is not a cycle-path. Figure 5 illustrates this procedure. Let k_v be the number of faces of the non-colored region incident to v. We have $k_v + 1$ different ways to continue the path and therefore $k_v + 1$ children of its tree node. No matter which child we choose, we will color all faces incident to v.

It remains to analyze the number of nodes in the tree. A bound on the number of nodes can be expressed by the following recurrence:

$$P(n, f) \leq (k_v + 1)P(n - 1, f - k_v) + 1.$$

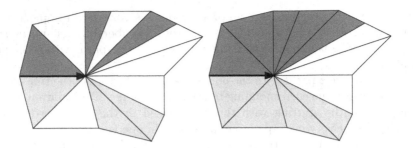

Fig. 5. Completing the red-blue coloring when entering a new vertex (light gray corresponds to red, dark gray to blue)

Because we want to maximize the number of nodes in the tree, we can assume that the k_vs for all v within a level l of the tree are equal. This holds for the root and by an inductive argument for the whole tree. Let κ_l denote the number k_v for the vertices v on level l.

$P := P(n-2, 2n+2)$ will give us the number of nodes in the tree. We know that $P(0, \cdot) = P(\cdot, 0) = 1$. All κ_ls have to be non-negative numbers. The κ_l along a path of length L have to fulfill the condition $\sum_{l \leq L} \kappa_l \leq 2n + 2$. A path is of length at most $n-2$, and therefore we can bound P by

$$1 + \sum_{L=1}^{n-2} \prod_{l \leq L} (\kappa_l + 1). \tag{1}$$

We are interested in a set κ_l which maximizes (1). Due to the convexity of (1) the maximum will be obtained when all κ_l are equal. Thus (1) is bounded by $1 + \sum_{i \leq n} (\frac{2n+2}{n-2} + 1)^i$. Therefore the exponential growth of the maximum number of cycle-paths is $O(n) \cdot 3^n$. □

This already yields an improvement of the best known upper bound for cycles in planar graphs.

Observation 1. *The number of simple cycles on a planar graph with n vertices is bounded from above by $O(n) \cdot 3^n$.*

We improve the obtained upper bound further. For this we go back to the proof of Lemma 1. Instead of considering cycle-paths of length n, we focus on shorter cycle-paths of length αn, where $\alpha \in [0, 1]$.

Lemma 2. *Let $C_s^\alpha(n)$ be the number of simple cycles of length αn in a planar graph with n vertices and f faces. Then we have*

$$C_s^\alpha(n) \leq \left(\frac{f}{\alpha n} + 1 \right)^{\alpha n}. \tag{2}$$

Proof. We reconsider the argumentation which led to Observation 1 and notice that $P(k, f)$ will be maximized by equally distributed values of κ_l. Therefore we set $\kappa_l = f/(\alpha n)$, which proves the Lemma.   □

As final step we combine the result from Lemma 2 with the results of Section 3.1.

Theorem 5. *The number of simple cycles in a planar graph G with n vertices is bounded from above by $O(2.89278^n)$.*

Proof. An upper bound ν^n for the number of Hamiltonian cycles will always imply an upper bound for $C_s(G)$ since every simple cycle is an Hamiltonian cycle on a subgraph of G. This leads to

$$C_s(n) \leq \sum_{t \leq n} \binom{n}{t} \nu^t = (1 + \nu)^n.$$

Plugging in our bound of $\sqrt[4]{30}$ for ν yields $C_s(n) \leq 3.3404^n$, which is larger than 3^n. Responsible for this are cycles with small length. When choosing a small subset of vertices, it is unlikely that they are connected in G. Therefore the Hamiltonian cycles counted for this subset will not correspond to cycles in G. Thus we overestimate the number of small cycles.

We modify the upper bound induced by the Hamiltonian cycles such that they can express $C_s^\alpha(n)$. Every αn-cycle is a Hamiltonian cycle on a subgraph of size αn. Thus

$$C_s^\alpha(n) \leq \binom{n}{\alpha n} \nu^{\alpha n} \leq 5 \binom{n}{\alpha n} (\sqrt[4]{30})^{\alpha n}$$

Since $\sum_i \binom{n}{i} \alpha^i (1 - \alpha)^{n-i} = 1$, for $0 \leq \alpha \leq 1$ every summand of this sum is at most 1. Considering the summand for $i = \alpha n$ yields $\binom{n}{\alpha n} \leq (1/(\alpha^\alpha (1-\alpha)^{1-\alpha}))^n$ and therefore

$$C_s^\alpha(n) = O\left(\left(\frac{\sqrt[4]{30}\alpha}{\alpha^\alpha (1 - \alpha)^{(1-\alpha)}}\right)^n\right). \tag{3}$$

So far we know two upper bounds for $C_s^\alpha(n)$. The two bounds are shown in Figure 6. The graph of (3) is represented as dashed gray curve, whereas the graph of (2) is depicted solid black. Clearly the maximum of the lower envelope of the two functions induces an upper bound for the exponential growth of cycles in G.

One can observe that the two functions intersect in the interval $[0, 1]$ in only one point, which is approximately $\tilde{\alpha} = 0.91925$. The maximal exponential growth is realized for this α. Evaluating $C_s^{\tilde{\alpha}}(n)$ yields a bound of 2.89278^n on the number of cycles.   □

At its core the bound of $\left(\frac{f}{\alpha n} + 1\right)^{\alpha n}$ in Lemma 2 comes from *consuming* f faces in αn steps. A similar bound can be obtained by consuming edges instead. In this case we do not need a coloring scheme. In each step we get as many ways to continue the cycle-path as the number of edges consumed in the step. This yields a bound of $\left(\frac{e}{\alpha n}\right)^{\alpha n}$. For $e = 3n - 6$, $f = 2n - 4$, and $\alpha < 1$ the bound obtained by considering faces is stronger.

For this counting argument the graph does not need to be planar. With $\alpha = 1$ it yields a bound of $\left(\frac{e}{n}\right)^n$ on the number of cycles in the graph. This bound has been independently observed by Sharir and Welzl [2].

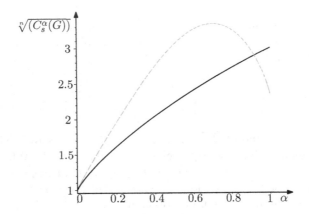

Fig. 6. Plot of the two bounds for $C_s^\alpha(G)$

4   Discussion

We improved the lower and upper bounds for the number of simple cycles in planar graphs. This reduces the gap between the upper and lower bound for the exponential growth from 1.03 to 0.46. We believe that the truth is closer to the lower bound. This is indicated by the technique sketched in the following which might further improve the upper bound.

For Observation 1 the worst case scenario is the situation where there are three possible ways to continue the cycle-path. However it is clear that this situation will not constantly occur during the construction of the cycle-paths. To use this fact we compute recurrences for the number of cycle-paths by simultaneously analyzing two or more consecutive levels of the tree which stores the cycle-paths. A careful analysis reveals other effects in this setting which also reduce the number of cycle-paths. In particular, vertices and faces will be surrounded and absorbed by the colored regions. The main part of the analysis is an intricate case distinction for which we have not checked all cases yet.

Furthermore we used the transfer matrix approach on the twisted cylinder to obtain lower bounds for other structures (for instance perfect matchings) on planar graphs. Moreover we adapted the counting procedure for sub-classes of planar graphs (for instance grid graphs). The results of these computations have not yet been double-checked and we therefore do not include them in this extended abstract.

Acknowledgments

We would like to thank Günter Rote for fruitful discussions on the subject and Andreas Stoffel for providing his implementation [9] of the transfer matrix method.

References

1. Alt, H., Fuchs, U., Kriegel, K.: On the number of simple cycles in planar graphs. Combinatorics, Probability & Computing 8(5), 397–405 (1999)
2. Sharir, M., Welzl, E.: On the number of crossing-free matchings (cycles, and partitions). In: Proc. 17th ACM-SIAM Sympos. Discrete Algorithms, pp. 860–869. ACM Press, New York (2006)
3. Sharir, M., Welzl, E.: Random triangulations of planar point sets. In: 22nd Annu. ACM Sympos. Comput. Geom., pp. 273–281. ACM Press, New York (2006)
4. Barequet, G., Moffie, M.: The complexity of Jensen's algorithm for counting polyominoes. In: Proc. 1st Workshop on Analytic Algorithmics and Combinatorics, pp. 161–169 (2004)
5. Conway, A., Guttmann, A.: On two-dimensional percolation. J. Phys. A: Math. Gen. 28(4), 891–904 (1995)
6. Jensen, I.: Enumerations of lattice animals and trees. J. Stat. Phys. 102(3–4), 865–881 (2001)
7. Stanley, R.P.: Enumerative combinatorics, vol. 2. Cambridge University Press, Cambridge (1999)
8. Graham, A.: Nonnegative Matrices and Applicable Topics in Linear Algebra. John Wiley & Sons, New York (1987)
9. Stoffel, A.: Software for Counting Cycles on the Twisted Cylinder, http://page.mi.fu-berlin.de/~schulza/cyclecount/
10. Lovasz, L., Plummer, M.: Matching theory. Elsevier, Amsterdam (1986)
11. Aichholzer, O., Hackl, T., Vogtenhuber, B., Huemer, C., Hurtado, F., Krasser, H.: On the number of plane graphs. In: Proc. 17th ACM-SIAM Sympos. Discrete Algorithms, pp. 504–513. ACM Press, New York (2006)

An Improved Exact Algorithm for Cubic Graph TSP

Kazuo Iwama* and Takuya Nakashima

School of Informatics, Kyoto University, Kyoto 606-8501, Japan
{iwama,tnakashima}@kuis.kyoto-u.ac.jp

Abstract. It is shown that the traveling salesman problem for graphs of degree at most three with n vertices can be solved in time $O(1.251^n)$, improving the previous bound $O(1.260^n)$ by Eppstein.

1   Introduction

The CNF satisfiability (SAT) and the traveling salesman (TSP) problems are probably the two most fundamental NP-hard problems. However, there are major differences between those two problems, which seems to be especially important when we develop their exact algorithms. For example, SAT is a so-called *subset* problem whose basic search space is 2^n for n variables, while TSP is a *permutation* problem whose search space is $n!$ for n vertices. For SAT, it is enough to find *any* feasible solution (satisfiable assignment), while it is not enough to find a Hamiltonian cycle in the case of TSP; we have to find an *optimal* one. Thus TSP is intuitively much harder than SAT. Not surprisingly, compared to the rich history of exact SAT algorithms (see , e.g. [6]), TSP has a very small amount of literature. For more than four decades, we had nothing other than the simple dynamic programming by Held and Karp [1], which runs in time $O(2^n)$.

In 2003, Eppstein considered this problem for *cubic graphs*, graphs with maximum degree three [8]. His basic idea is to use the fact that if the degree is three then a selection of an edge as a part of a Hamiltonian cycle will implicationally force several edges to be in the cycle or not. He also maintained several nontrivial algorithmic ideas, e.g., the one for how to treat 4-cycles efficiently. As a result, his algorithm runs in $O(2^{n/3}) \approx 1.260^n$ time. He also showed that there is a cubic graph having $O(2^{n/3})$ different Hamiltonian cycles. Hence his algorithm cannot be improved if one tries to enumerate all Hamiltonian cycles.

Our Contribution. In this paper we give an improved time bound, $O(2^{31n/96})$ $\approx 1.251^n$. First of all, the above lower bound for the number of Hamiltonian cycles is not harmful since such a graph includes a lot of 4-cycles for which the special treatment is already considered in [8].

Our new algorithm is similar to Eppstein's and is based on a branching search which tries to enumerate all Hamiltonian cycles (and even more, i.e., some cycle

---

* Supported in part by Scientific Research Grant, Ministry of Japan, 1609211 and 16300003.

Fig. 1. Treatment of 3-cycles

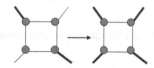

Fig. 2. Treatment of 4-cycles

decompositions). Here is our idea of the improvement: Suppose that at each branch there are two possibilities and the corresponding recurrence relations which look like

$$T(n) \le 2T(n-1) \text{ and } T(n) \le 3T(n-2).$$

The second equation is obviously more desirable for us, but we often have to assume the worst case (i.e., the first equation and $T(n) = 2^n$ as a result) due to the lack of specific information. If we do have useful information, for example, the information that the first case can happen at most $3n/4$ time, then it obviously helps. Note that such information can be incorporated into analysis by introducing new parameters and relations among them. In the above case, we have

$$T(n, m) \le 2T(n-1, m-1),$$
$$T(n, m) \le 3T(n-2, m),$$
$$m \le 3n/4$$

and it is easy to conclude $T(n) \approx 1.931^n$ by setting $T(n, m) = 2^{\alpha n + \beta m}$ and solving equations on α and β. In our analysis, we need to introduce three new parameters other than the main parameter n.

Related Work. If a given graph has a small separator, then as with many other NP-hard problems, TSP has efficient algorithms. For example the planar graphs have a $O(\sqrt{n})$ size separator [3] and the planar graph TSP can be solved in time $O(2^{10.8224\sqrt{n}} n^{3/2} + n^3)$ [9]. Euclidean TSP can also be solved in time $O(n^{\sqrt{n}})$ [5] similarly.

2   Eppstein's Algorithm

For a given cubic graph G of n vertices, we first remove all 3-cycles by merging corresponding three vertices into a single vertex and add the cost of each triangle edge to the opposite non-triangle edge. Since a Hamiltonian cycle must go

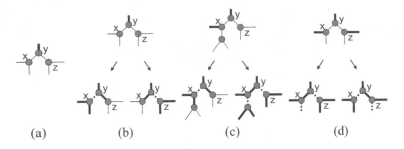

Fig. 3. Single branching of the algorithm

like Fig. 1 (a), (b), or (c) on a 3-cycle, this preprocessing works. As mentioned previously, 4-cycles also receive a special treatment: As shown in Fig. 2, if two diagonal edges attached to the cycle are selected, then the other two are automatically selected and then further decisions (whether two horizontal edges or vertical edges are selected) are postponed until the last step of the algorithm (see [8] for details).

The main part of the algorithm is illustrated in Fig. 3. Suppose that we have a vertex y such that one of three neighboring edges is already selected as a part of the solution (see Fig. 3 (a) where the selected edge is denoted by a thick line) and we try to extend the solution by selecting an edge yx or yz. Then there are three cases (b), (c) and (d) due to the state of the vertices x and z. Suppose that both x and z are *free*, i.e., neither is adjacent to already selected edges as in Fig. 3 (b). Then either selecting yx or yz to extend the solution, we can force at least three more edges to be in the solution. Suppose that one of them, say x, is not free as in (c). Then by a similar extension, we can force at least two and five edges. If neither is free as in (d), we can force at least four and four edges. In this last case, however, we have to be careful since we can only force three edges if this portion is a part of a 6-cycle as shown in Fig. 4.

Thus our recurrence equation can be written as follows using the number n of edges which should be selected from now until the end of the algorithm.

$$T(n) \leq 2T(n-3) \tag{1}$$
$$T(n) \leq T(n-2) + T(n-5) \tag{2}$$
$$T(n) \leq 2T(n-4) \tag{3}$$

It is easy to see that (1) is the worst case and by solving it we have $T(n) = 2^{n/3}$. Because the length of a hamiltonian cycle of a n vertices graph is n, the number of edges which should be selected is also n. Thus, TSP for a cubic graph with n vertices can be sloved in time $O(2^{n/3})$, which is the main result of [8].

3   New Algorithm

Recall that one of the worst cases happens for the case of Fig. 3 (b). Notice, however, that if this case happens once then the number of free vertices decreases

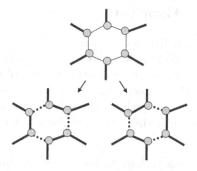

Fig. 4. 6-cycles

by four. Therefore, in each path of the backtrack tree from the root to a leaf, this worst case can happen at most $n/4$ times. Without considering this property, we could only say that the case can happen up to $n/3$ times, since three edges are selected at each step and the backtrack path ends when n edges have been selected. Thus the number of worst-case branches reduces by the new property, which should contribute to a better time complexity.

Unfortunately this idea does not work by itself, since we also have to consider the case of a 6-cycle in Fig. 4: if the six edges attaching (radiating from) the 6-cycle are all selected, then the further extension selects only three edges (every other ones in the six cycle edges) in a single branch. To make things worse, this 6-cycle case and the worst case of Fig. 3 (b) can happen alternatively without any other in between.

Now here is a natural idea. Since a 6-cycle plays a bad role only when all six attaching edges are selected *before* the cycle edges, we should select (some of) the cycle edges before the attaching edges have been all selected. This can be done by placing a priority to edges constituting 6-cycles when we extend the solution.

A 6-cycle is called *live* if none of its six cycle edges are selected. A 6-cycle which is not live is called *dead*. Fig. 9 shows the original algorithm in [8] where F shows the set of edges already selected. Our algorithm differs only in 3 (b), which is replaced by the following (b1) and (b2):

(b1) If there is no such 4-cycle and if $G\backslash F$ contains a live 6-cycle with a vertex y which has a neighboring edge in F (that is not a cycle edge but an attaching one), let z be one of y's neighboring vertices (on the cycle). If two or more such live 6-cycles exist, then select a 6-cycle such that most attaching edges are already selected.

(b2) If there is no such 4-cycle or 6-cycle, but F is nonempty, then let uy be any edge in F such that y is adjacent to at least one vertex which is not free. Let yz be an edge in $G\backslash F$ between y and such a vertex ($= z$). If there is no such uy, then let uy be any edge in F and yz be any adjacent edge in $G\backslash F$.

4   Analysis of the Algorithm

Let $C(i), 0 \leq i \leq 6$, be a live 6-cycle such that i edges of its six attached edges have already been selected. Also, we call the branch shown in Fig. 3 (b) and Fig. 4 *A-branch* and *B-branch*, respectively, and all the other branches *D-branch*. Recall that A- and B-branches are our worst cases. (Note that Fig. 3 (b) may select more than three new edges if there are already selected edges near there. If that is the case, then this branch is not an A-branch any more but a D-branch.) Here is our key lemma:

Lemma 1. *Let P be a single path of the backtrack tree and suppose that we have a $C(6)$, say Q, somewhere on P. Then at least three attached edges of Q have been selected by D-branches.*

Proof. Q is of course $C(0)$ at the beginning of the algorithm, so it becomes $C(6)$ by changing its state like, for instance, $C(0) \rightarrow C(3) \rightarrow C(5) \rightarrow C(6)$. If it once becomes $C(3)$, then the lemma is true since the change from $C(3)$ to $C(6)$ is caused by a branch associated with other 6-cycle(s) which is at least $C(3)$ (see 3 (b1) of our algorithm given in the previous section). One can easily see that only D-branches can be applied to $C(3), C(4)$ and $C(5)$. Furthermore, if Q changes from $C(2)$ to $C(6)$, then it must be done by a D-branch and the lemma is true. Hence we can assume that Q becomes $C(6)$ through $C(4)$ or $C(5)$. In the following we only discuss the case that Q once becomes $C(4)$; the other case is much easier and omitted.

Now Q is $C(4)$. Its previous state is $C(0), C(1)$ or $C(2)$ (not $C(3)$ as mentioned above). If Q changes from $C(0)$ to $C(4)$ directly, then that branch must be a D-branch (A-branch can select only three edges) and we are done. The case that Q changes from $C(1)$ to $C(4)$ directly is also easy and omitted. As a result, we shall prove the following statement: Suppose that Q changes from $C(2)$ to $C(4)$ directly and finally becomes $C(6)$. Then the change from $C(2)$ to $C(4)$ must be by a D-branch and hence the lemma is true.

Suppose for contradiction that Q changes from $C(2)$ to $C(4)$ by other than D-branch. Since B-branch is obviously impossible (it newer increase selected attached edges), the branch must an A-branch. Furthermore, we can assume without loss of generality that when Q was $C(2)$, there were no 6-cycles which were already $C(4)$ or more. (Namely, we are now considering such a moment that a "bad" $C(4)$ has appeared for the first time in the path of the algorithm. $C(4)$'s that previously appeared should have been processed already and should be dead at this moment.) Under such a situation, our first claim is that when Q becomes $C(4)$, there must be at lent one other 6-cycle, Q', which becomes also $C(4)$ at the same time. The reason is easy: If Q would be only one $C(4)$, then it must be processed at the next step and would become dead, meaning it can never be $C(6)$.

Now Suppose that a single A-branch has created two $C(4)$'s Q and Q'. First of all, this is impossible if Q and Q' are *disjoint* or do not share cycle edges for the following reason. If they are disjoint, we need to select four attached edges by a single A-branch. Since A-branch selects only three new edges, this would

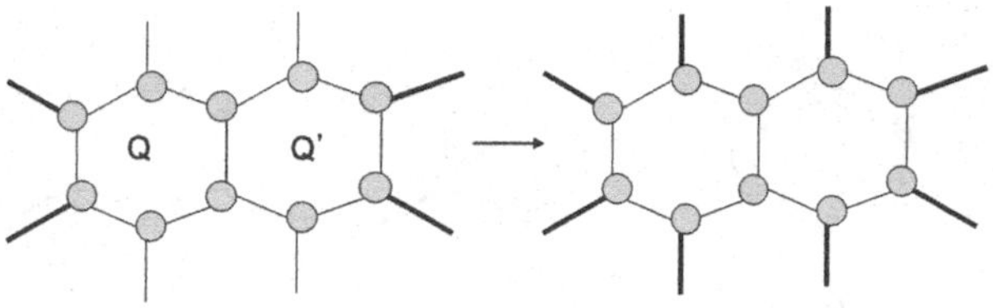

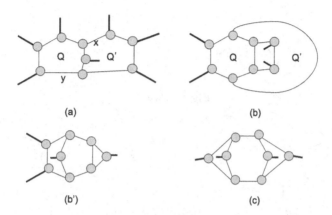

Fig. 5. Case 1

Fig. 6. Case 2

be possible only if some single edge is an attached edge of both Q and Q'. This means that this common edge was selected but no adjacent edges (i.e., cycle edges of Q and Q') are selected, which is obviously impossible. Thus Q and Q' must share some cycle edges.

Case 1. Q and Q' share one cycle edge. See Fig. 5. To change both Q and Q' from $C(2)$ to $C(4)$, we need to select four edges at a single step (recall that we have no common attaching edges), which is impossible by an A-branch.

Case 2. Q and Q' share two cycle edges. See Fig. 6 (a), (b), (b') and (c). In the case of (a), after making two $C(4)'s$, neither can be $C(6)$. (One can see that if we select x then y must be selected also.) In the case (b), we even cannot make both Q and Q' $C(4)$. (b') includes a 3-cycle, which cannot happen in our algorithm. In (c), we have a 4-cycle whose diagonal attaching edges are selected. Then, in the next step, 3 (a) of the algorithm applies and both Q and Q' become dead.

Case 3. Q and Q' share three cycle edges. See Fig. 7 (a), (b), (b') and (c), where shared edges are given by x, Q edges by straight lines and Q' edges by doted lines. The case (a) is obviously impossible since this subgraph is completely disconnected from the rest of the graph. In the case (b), it is impossible to make Q and Q' both $C(4)$. (b') includes a 3-cycle. (c) shows the situation before the branch is made, i.e., Q and Q' are both $C(2)$. Notice that we have the third 6-cycle, say Q'' other than Q and Q'. Now Q and Q' are $C(2)$. But to be so, one can see that at least one of Q, Q' and Q'' has two attached edges such that

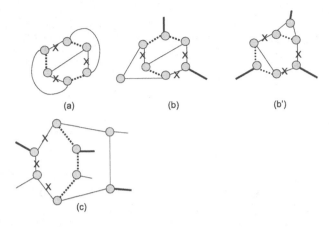

Fig. 7. Case 3

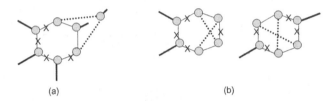

Fig. 8. Case 4

their end points apart only at most distance two on the cycle. Then, by 3 (b2) of our algorithm (see section 3), that cycle must be processed in the next step, which is not an A-branch.

Case 4. Q and Q' share four cycle edges. See Fig. 8 (a) and (b). (a) is only one possibility to make two $C(4)$'s. Then notice that we have a 4-cycle whose two diagonal attached edges are selected, which is the same as (c) of Case 2. In (b), it is obviously impossible to make two $C(4)$'s.

Case 5. Q and Q' share five cycle edges. This is impossible since our graph does not have parallel edges.

Thus we have contradictions in all the cases.   □

Now we are ready to prove our main theorem. As for the correctness of the algorithm, note that our modification from [8] is very small and it is not hard to verify that the modification does not affect its correctness. Therefore we can use the analysis of [8] as it is, and so, we are only interested in its time complexity.

Theorem 1. *Our algorithm runs in time* $O(n^{31n/96})$.

Proof. Since the recurrence relation is most important, we once again summarize several different cases. (All other parts of the algorithm including the treatment of 4-cycles, runs in a polynomial time, see [8] for details.).

1. Repeat the following steps until one of the steps returns or none of them applies:
   (a) If G contains a vertex with degree zero or one, return None.
   (b) If G contains a vertex with degree two, add its incident edges to F.
   (c) If F consists of a Hamiltonian cycle, return the cost of this cycle.
   (d) If F contains a non-Hamiltonian cycle, return None.
   (e) If F contains three edges meeting at a vertex, return None.
   (f) If F contains exactly two edges meeting at some vertex, remove from G that vertex and any other edge incident to it; replace the two edges by a single forced edge connecting their other two endpoints, having as its cost the sum of the costs of the two replaced edges costs.
   (g) If G contains two parallel edges, at least one of which is not in F, and G has more than two vertices, then remove from G whichever of the two edges is unforced and has larger cost.
   (h) If G contains a self-loop which is not in F, and G has more than one vertex, remove the self-loop from G.
   (i) If G contains a triangle xyz, then for each non-triangle edge e incident to a triangle vertex, increase the cost of e by the cost of the opposite triangle edge. Also, if the triangle edge opposite e belongs to F, add e to F. Remove from G the three triangle edges, and contract the three triangle vertices into a single supervertex.
   (j) If G contains a cycle of four unforced edges, two opposite vertices of which are each incident to a forced edge outside the cycle, then add to F all non-cycle edges that are incident to a vertex of the cycle.
2. If $G \backslash F$ forms a collection of disjoint 4-cycles, perform the following steps.
   (a) For each 4-cycle C_i in $G \backslash F$, let H_i consist of two opposite edges of C_i, chosen so that the cost of H_i is less than or equal to the cost of $C_i \backslash H_i$.
   (b) Let $H = \cup_i H_i$. Then $F \cup H$ is a degree-two spanning subgraph of G, but may not be connected.
   (c) Form a graph $G' = (V', E')$, where the vertices of V' consist of the connected components of $F \cup H$. For each set H_i that contains edges from two different components K_j and K_k, draw an edge in E' between the corresponding two vertices, with cost equal to the difference between the costs of C_i and of H_i.
   (d) Compute the minimum spanning tree of (G', E').
   (e) Return the sum of the costs of $F \cup H$ and of the minimum spanning tree.
3. Choose an edge yz according to the following cases:
   (a) If $G \backslash F$ contains a 4-cycle, two vertices of which are adjacent to edges in F, let y be one of the other two vertices of the cycle and let yz be an edge of $G \backslash F$ that does not belong to the cycle.
   (b) If there is no such 4-cycle, but F is nonempty, let xy be any edge in F and yz be an adjacent edge in G F.
   (c) If F is empty, let yz be any edge in G.
4. Call the algorithm recursively on $G, F \cup \{yz\}$.
5. Call the algorithm recursively on $G \backslash \{yz\}, F$.
6. Return the minimum of the set of at most two numbers returned by the two recursive calls.

Fig. 9. Eppstein's algorithm

(1) Fig. 3 (b) happens. In either branch, three new edges are selected and four free vertices disapper. (As mentioned before, it may happen that more than three edges are selected due to the existence of nearby selected edges. But this is obviously more desirable for us and can be omitted.)

(2) Fig. 4 happens. In either branch, three new edges are selected and the number of free vertices does not change at all.

(3) Fig. 3 (c) happens. In one branch, two new edges are selected and in the other branch five. At least two free vertices disappear in either branch.

(4) Fig. 3 (d) happens. In either branch, four new edges are selected and at least two free vertices disappear. However, we have to be careful here again. Now we have to consider an 8-cycle such that its attaching edges have been all selected and none of its cycle edges have been selected. In this case, by selecting one edge in the cycle, four edges are forced to be selected in either branch. However, the number of free vertices does not change and so this case is worse than Fig. 3 (d).

Now let $T(n, a, b, f)$ be the number of nodes that appear in the backtrack tree for a graph G such that in each path starting from this node to a leaf, (i) we further need to select n edges, (ii) the extension of Fig. 3 (b) type happens a times from now on, (iii) the extension of Fig. 4 happens b times from now on, and (iv) G has f free vertices. Then by the case (1) to case (4) above, we have the following recurrence relation:

$$T(n, a, b, f) \leq \max \begin{cases} 2T(n-3, a-1, b, f-4) \\ 2T(n-3, a, b-1, f) \\ T(n-5, a, b, f-2) + T(n-2, a, b, f-2) \\ 2T(n-4, a, b, f) \end{cases}$$

Note that in the case of (1), we have now used this type of extension once, we can decrease the value of a by one, and similarly for (2). Also, at each leaf, we have T(0,0,0,0) = 1.

Now let $T(n, a, b, f) = 2^{\frac{n+\frac{1}{2}(a+2b)+f/8}{4}}$. Then as shown below, this function satisfies all the recurrence formulas above:

$$2T(n-3, a-1, b, f-4) = 2^{\frac{(n-3)+\frac{1}{2}((a-1)+2b)+(f-4)/8}{4}+1} = 2^{\frac{n+\frac{1}{2}(a+2b)+f/8}{4}}$$

$$2T(n-3, a, b-1, f) = 2^{\frac{(n-3)+\frac{1}{2}(a+2(b-1))+f/8}{4}+1} = 2^{\frac{n+\frac{1}{2}(a+2b)+f/8}{4}}$$

$$2T(n-4, a, b, f) = 2^{\frac{(n-4)+\frac{1}{2}(a+2b)+f/8}{4}+1} = 2^{\frac{n+\frac{1}{2}(a+2b)+f/8}{4}}$$

$$T(n-5, a, b, f-2) + T(n-2, a, b, f-2) = 2^{\frac{(n-5)+\frac{1}{2}(a+2b)+(f-2)/8}{4}} + 2^{\frac{(n-2)+\frac{1}{2}(a+2b)+(f-2)/8}{4}}$$

$$= (2^{\frac{-5-2/8}{4}} + 2^{\frac{-2-2/8}{4}})2^{\frac{n+\frac{1}{2}(a+2b)+f/8}{4}}$$

$$> (1.07)2^{\frac{n+\frac{1}{2}(a+2b)+f/8}{4}} > 2^{\frac{n+\frac{1}{2}(a+2b)+f/8}{4}}$$

$$T(0,0,0,0) = 2^0 = 1$$

Here we use Lemma 1, which says that if type (2) happens b times in the path, then at least $3b$ edges are selected by neither type (1) or type (2). Since we select $3a + 3b$ edges by types (1) and (2) and the total number of selected edges is n, we have

$$3a + 3b + 3b \leq n,$$

which means $a + 2b \leq n/3$. Also $f \leq n$ obviously. Using these two inequalities, we have

$$2^{\frac{n+\frac{1}{2}(a+2b)+n/8}{4}} \leq 2^{\frac{n+\frac{1}{2}n/3+n/8}{4}} = 2^{31n/96}. \qquad \square$$

References

1. Held, M., Karp, R.M.: A dynamic programming approach to sequencing problems. SIAM Journal on Applied Mathematics 10, 196–210 (1962)
2. Garey, M.R., Johnson, D.S.: Computers and Intractability: A Guide to the Theory of NP-Completeness, W.H. Freeman (1979)
3. Lipton, R.J., Tarjan, R.E.: A separator theorem for planar graphs. SIAM Journal on Applied Mathematics 36, 177–189 (1979)
4. Lipton, R.J., Tarjan, R.E.: Applications of a planar separator theorem. SIAM Journal on Computing 9, 615–627 (1980)
5. Hwang, R.Z., Chang, R.C., Lee, R.C.T.: The Searching over Separators Strategy To Solve Some NP-Hard Problems in Subexponential Time. Algorithmica 9, 398–423 (1993)
6. Iwama, K., Tamaki, S.: Improved Upper Bounds for 3-SAT, 15th annual ACM-SIAM Symposium on Discrete Algorithms. In: Proc. SODA, January 2004, pp. 328–329 (2004)
7. Woeginger, G.J.: Exact Algorithms for NP-Hard Problems: A Survey. In: Jünger, M., Reinelt, G., Rinaldi, G. (eds.) Combinatorial Optimization - Eureka, You Shrink! LNCS, vol. 2570, pp. 185–207. Springer, Heidelberg (2003)
8. Eppstein, D.: The Traveling Salesman Problem for Cubic Graphs. In: Dehne, F., Sack, J.-R., Smid, M. (eds.) WADS 2003. LNCS, vol. 2748, pp. 307–318. Springer, Heidelberg (2003)
9. Dorn, F., Penninkx, E., Bodlaender, H.L., Fomin, F.V.: Efficient Exact Algorithms on Planar Graphs:Exploiting Sphere Cut Branch Decompositions. In: Brodal, G.S., Leonardi, S. (eds.) ESA 2005. LNCS, vol. 3669, pp. 95–106. Springer, Heidelberg (2005)

Geometric Intersection Graphs:
Do Short Cycles Help?
(Extended Abstract)

Jan Kratochvíl* and Martin Pergel**

Department of Applied Mathematics, Charles University, Malostranské nám. 25,
118 00 Praha 1, Czech Republic
{honza,perm}@kam.mff.cuni.cz

Abstract. Geometric intersection graphs are intensively studied both for their practical motivation and interesting theoretical properties. Many such classes are hard to recognize. We ask the question if imposing restrictions on the girth (the length of a shortest cycle) of the input graphs may help in finding polynomial time recognition algorithms. We give examples in both directions. First we present a polynomial time recognition algorithm for intersection graphs of polygons inscribed in a circle for inputs of girth greater than four (the general recognition problem is NP-complete). On the other hand, we prove that recognition of intersection graphs of segments in the plane remains NP-hard for graphs with arbitrarily large girth.

1 Introduction

Intersection graphs are defined as graphs with representations by set-systems of certain types. Each set corresponds to a vertex and two vertices are adjacent iff the corresponding sets have nonempty intersection. Any graph can be represented by some set-system, but interesting and nontrivial graph classes are obtained when further restrictions are imposed on the sets representing the vertices. Especially geometrical representations are popular and widely studied. They stem from practical applications, have many interesting structural and algorithmic properties, and often serve as a method of graph visualization. Numerous examples include interval graphs, circle graphs, circular arc graphs, boxicity two graphs, and many others, cf. e.g., [16,22]. (In all examples that are mentioned in this paper, representations are assumed in the Euclidean plane.)

Many of these classes are NP-hard to recognize (e.g., boxicity two graphs [12], intersection graphs of segments, convex sets or curves in the plane [11], intersection graphs of unit and general disks in the plane [1,13], contact graphs of curves and unit disks [8,1], and others, see [2]). It has been observed in [13] that triangle-free disk graphs are planar and hence recognizable in polynomial

---

* Support of Institute for Theoretical Computer Science (ITI), a project of Czech Ministry of Education 1M0021620808 is gladly acknowledged.
** Supported by Project 201/05/H014 of the Czech Science Foundation (GAČR).

G. Lin (Ed.): COCOON 2007, LNCS 4598, pp. 118–128, 2007.
© Springer-Verlag Berlin Heidelberg 2007

time (in fact, the observation was done for a larger class of pseudodisk graphs), but that forbidding triangles does not help recognizing intersection graphs of curves (so called string graphs). In the present paper we propose a more thorough study of this phenomenon and we prove two results showing that forbidding short cycles does help sometimes, but not always. We first introduce the relevant graph classes in more detail.

Polygon-circle graphs (shortly *PC-graphs*) are intersection graphs of convex polygons inscribed into a circle. They extend e.g., interval graphs, circular-arc graphs, circle graphs, and chordal graphs. M. Fellows observed that this class is closed under taking induced minors [personal communication, 1988]. These graphs are also interesting because the Clique and Independent Set problems can be solved in polynomial time [6]. On the other hand, determining their chromatic number is NP-hard, since PC-graphs contain circle graphs [7], but the graphs are near-perfect in the sense that their chromatic number is bounded by a function of their clique number [10]. The recognition problem for PC-graphs is in the class NP, since every PC-graph with n vertices has a representation whose all polygons have at most n corners each, and an asymptotically tight bound $n - \log n + o(\log n)$ on the maximum number of corners needed in a representation is presented in [15]. A polynomial time recognition algorithm was announced in [9], but a full paper containing the algorithm was never published. On the contrary, the recognition has recently been shown NP-complete [18]. We show that restricting the girth (the length of a shortest cycle) of the input graphs does make the recognition easier.

Theorem 1. *PC-graphs of girth greater than four can be recognized in polynomial time.*

Segment (or SEG-) graphs are intersection graphs of segments in the plane. These graphs were first considered in [5] where it is shown that determining the chromatic number of these graphs is NP-hard. The near-perfectness of these graphs is a well known open problem going back to Erdős. The recognition of SEG-graphs is shown NP-hard in [11] in a uniform reduction which also shows NP-hardness of recognition of *string graphs* (intersection graphs of curves), *1-string graphs* (intersection graphs of nontangent curves such that each two curves intersect in at most one point) and *CONV-graphs* (intersection graphs of convex sets). SEG- and CONV-graphs have been further studied in [14] where their recognition is shown to be in the class PSPACE. Membership in the class NP is still open. On the contrary, a surprising development occurred for the most general class of string graphs. This class, also mentioned in [5], was introduced already in 1966 in [21]. For decades no recursive algorithm for its recognition was known, until 2001 when two bounds on the number of crossing points needed in their representations were proved independently in [17,19] (yielding recursive recognition algorithms), followed by the proof of NP-membership in 2002 in [20]. The complexity of recognition is connected to the question of maximum size of a representation. The construction of [14] showing that there are SEG-graphs requiring representations of double exponential size (the size of a representation by segments with integral coordinates of endpoints is the maximum absolute value of the endpoint coordinates) shows that the would-be-straightforward 'guess-and-verify' algorithm is not in NP. Another long-standing open problem is whether every planar graph is a SEG-graph (it is easy to show that every planar graph is a string graph, and this was recently improved in [3] showing that every planar graph is a 1-string graph). Regarding the question of recognizing graphs of large girth, it was noted in [13] that triangle-free string graphs are NP-hard to recognize, but all known reductions hinge on the presence of cycles of length 4. We show that for SEG-graphs, this is not the case, and in fact no girth restriction would help.

Theorem 2. *For every* k, *recognition of SEG-graphs of girth greater than* k *is NP-hard.*

2   Polygon-Circle Graphs

2.1   Preliminaries

We first explain the terminology. All graphs considered are finite and undirected. For a graph G and two disjoint subsets U and H of its vertex set, we denote by $G[U]$ the subgraph induced by U, and we denote by $E_G(U, H)$ the edges of G between U and H, i. e., $E_G(U, H) = E(G[U \cup H]) \setminus (E(G[U]) \cup E(G[H]))$.

Given a PC-graph G and a PC-representation, we refer to *vertices* of the graph and to *corners* of the polygons (these are their vertices lying on the geometrical bounding circle). For a vertex v of the graph, we use P_v to denote the corresponding polygon representing v (but sometimes and namely in figures, we avoid multiple subscripts by denoting also the polygon simply by v). Given a representation R of G with $G \subseteq H$, we say that R *is extended into* a representation S of H if S is obtained from R by adding new corners to existing polygons and by adding new polygons representing the vertices of $V(H) \setminus V(G)$.

In a PC representation, we describe the relative position of a set of polygons in terms of "visibility" from one polygon to another. The corners of any polygon R divide the bounding circle into circular arcs referred to as *R-segments*. If P, Q, R are disjoint polygons and Q lies in a different P-segment than R, we say that Q is *blocked* from R by P. A set A of polygons is said to *lie around the circle* if for no triple of disjoint polygons $P, Q, R \in A$, the polygon Q is blocked by R from P.

It is easy to check that we may assume that the corners of all polygons are distinct. By cutting the bounding circle in an arbitrary point and listing the names of the vertices in the order as the corners of their polygons appear along the circle we obtain the so called *alternating sequence* of the representation. To be more precise, we say that two symbols a and b *alternate* in a sequence S, if S contains a subsequence of the form $...a...b...a...b...$ or $...b...a...b...a....$ It is a well-known fact that $G = (\{v_1, \ldots, v_n\}, E)$ is a PC-graph if and only if there exists an alternating sequence S over the alphabet $v_1, \ldots, v_n$ such that $v_i v_j \in E(G)$ if and only if v_i and v_j alternate in S.

Though polygon-circle representations lie behind the original geometric definition of PC-graphs and are visually more accessible, precise proofs of several observations and auxiliary technical lemmas are easier to formulate in the language of alternating sequences. Thus in the paper we often switch between these two equivalent descriptions of PC-representations.

2.2 PC-Graphs of Low Connectivity and Their Decompositions

We first observe that we may restrict our attention to bi-connected (vertex-2-connected) graphs.

Lemma 1. *A graph G is a PC-graph if and only if all its connected components are PC-graphs.*

Proof. We just take alternating representations of individual components and place one after another.

Lemma 2. *A graph G is a PC-graph if and only if all its biconnected components are PC-graphs.*

Proof. Similar to the proof of Lemma 1.

From now on we assume that our input graphs are bi-connected. It turns out that vertex cuts containing two nonadjacent vertices do not create any problems, but those containing two adjacent vertices do. We formalize this in two definitions of decompositions in the next subsection.

Definition 1. *Let $G = (V, E)$ be a graph. An edge ab whose endpoints form a cut in G is called a* cutting edge. *Let $C_1, C_2, ..., C_k$ be the connected components of $G[V \setminus \{a, b\}]$ for a cutting edge ab. The* pair-cutting decomposition *of G based on ab is the collection of graphs $G_i^{ab} = (C_i \cup \{a, b, c\}, E(G[C_i \cup \{a, b\}]) \cup \{ac, bc\})$, where c is a new extra vertex, $i = 1, 2, \ldots, k$.*

A graph is called pc-prime *(pair-cutting-prime) if it has no nontrivial pair-cutting decomposition (i.e., if every cutting edge divides the graph into one connected component and a single vertex).*

Proposition 1. *A biconnected graph G with a cutting edge ab is a PC-graph if and only if all the graphs in the pair-cutting decomposition based on ab are PC-graphs.*

Proof. We start with alternating representations $S_1, ..., S_k$ of the graphs $G_1^{ab}, ..., G_k^{ab}$ in the pair-cutting decomposition. Without loss of generality we may assume that each of them begins with the newly added vertex c. Let $\tilde{S}_i$ be obtained from S_i by removing all occurences of c and adding an occurence of a to the beginning and an occurence of b to the end of the sequence. Then the concatenation $\tilde{S}_1 \ldots \tilde{S}_k$ is an alternating representation of G.

Obviously a and b alternate. By our transformation we could not remove any alternation among vertices of G. Thus it suffices to check that we did not add new ones. For $i \neq j$, any symbol from $\tilde{S}_i$ distinct from a and b cannot alternate with any symbol from $\tilde{S}_j$ (distinct from a and b). Suppose that by replacing the first occurence of c by a we have created a new alternation of a and $x \neq b$ in $\tilde{S}_i$. Thus in S_i all occurences of a lie between two occurences of x. But a and c alternated in S_i,

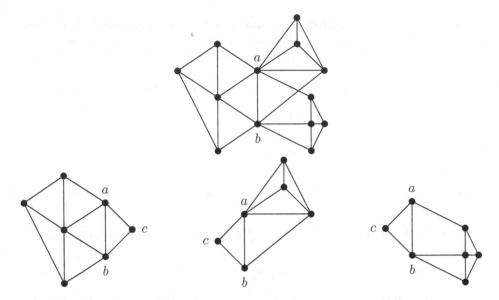

Fig. 1. An example of the pair-cutting decomposition. At the top of the figure is the original graph with vertices a and b forming a cutting edge. Below are depicted the three elements of its pair-cutting decomposition based on ab. Note that the left one and the right one are pc-primes, but the middle one is not.

thus between these two occurences of x there is an occurence of c and thus x and c alternated. Hence $x = b$ (c alternated only with a and b). A contradiction.

The converse is clear from the fact that the class of PC-graphs is closed under taking induced minors and the fact that each G_i^{ab} is an induced minor of G. Indeed, for some $j \neq i$, contract all vertices of C_j into a single vertex c and contract all vertices of all remaining C_l's for $l \neq i, j$ into the vertex a. Since G was biconnected and vertices a and b formed a cut, c is adjacent exactly to a and b.

The previous proposition is general in the sense that it does not require large girth of G. However, as we aim at graphs of large girth, the reduction is inconvenient because it creates triangles in the blocks of the decomposition. For this sake we introduce the following adjustment. In each element of the pair-cutting decomposition $G[V(C_i) \cup \{a, b\}]$ we mark the edge ab "red" instead of adding the triangle abc. We will talk about the *red-edged decomposition* and each element of this decomposition will be called a *red-edged graph*. It is not true anymore that a biconnected graph is in PC if and only if each red-edged component of the decomposition is a PC-graph. However, we can still control how to glue representations together if G has no short cycles.

In the sense of Definition 1 we denote by G^e the graph obtained from G by adding a new vertex adjacent to the endvertices of e. Similarly, $G^{e_1, \dots, e_k}$ is the graph obtained by adding k new vertices, each connected to the endpoints of one of the edges e_i.

Lemma 3. *Let R be an alternating representation of a graph G. If this representation can be extended to a representation of G^{ab}, it can be done so by adding the subsequence cabc between some two consecutive symbols in R.*

Proof. Will be presented in the journal version.

Proposition 2. *Let G be a connected red-edged graph of girth at least 4 with $Q = \{e_1, ...e_k\}$ being the set of its red edges. Let R be a PC-representation of G. If for any $e \in Q$, the representation R can be extended into a representation of G^e, then R can be extended into a representation of $G^{e_1, ... e_k}$.*

Proof. Will be presented in the journal version.

2.3   Pseudoears

A well-known characterization of biconnected graphs is given by the *ear-decomposition lemma*:

Lemma 4. *A biconnected graph can be constructed from any of its cycles by consecutive addition of paths (with endpoints in the already constructed subgraph, so called* ears) *and/or single edges.*

However, we need a version where the constructed graph is an induced subgraph, and therefore we need to avoid adding edges. Thus we define the technical notion of a pseudoear and introduce a special version of this lemma for K_3-free graphs.

Definition 2. *Let H be an induced subgraph of a graph G, and let $U = a_1 \ldots a_3$ be an induced path in H of length at least 2. A pseudoear attached along U is an induced path $P = p_1...p_k$ in $G \setminus H$ of length at least 1 such that either*

 I. H has length at least 3 and $E(P, H) = \{p_1 a_1, p_k a_3\}$, or
 II. $U = a_1 a_2 a_3$ has length 2 and $E(p_1, H) = \{p_1 a_1\}$, $E(p_k, H) = \{p_k a_3\}$, and $E(\{p_2, \ldots, p_{k-1}\}, H) \subseteq \{p_i a_2 | i = 2, \ldots, k-1\}$.

Lemma 5. *Any biconnected K_3-free graph without cutting edges can be constructed from any of its induced cycles by consecutive addition of pseudoears and/or single vertices adjacent to at least two vertices of the so far constructed*

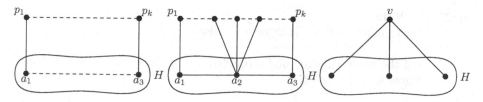

Fig. 2. The first two pictures show pseudoears of type I and II, the third one is an example of a vertex we operate with in our version of ear-decomposition lemma. Solid lines denote edges, dashed lines represent paths.

subgraph. On the other hand, any K_3-free graph constructed in this way is bi-connected and has no cutting edges.

Proof. Will be presented in the journal version. We only note here that the proof is constructive and yields a polynomial time algorithm.

2.4   Minimal Representations of Biconnected Pc-Prime Graphs

We call an alternating representation of a PC-graph G *minimal* if deleting any occurrence of any symbol from the sequence results in a sequence that does not represent G anymore. For instance, in every minimal alternating representation of a cycle of length n, every symbol occurs exactly twice and the representation is unique (up to rotation and symmetrical flip). The core theorem our algorithm is based on says that this is also true for bi-connected pc-prime graphs of large girth.

Theorem 3. *Every bi-connected pc-prime graph with girth at least 5 has at most 1 minimal PC-representation. This representation (if it exists) can be found in polynomial time.*

Proof. Will be presented in journal version. It uses the Pseudoear lemma and by induction on the number of pseudoears shows how to find this unique representation (when it exists).

2.5   Algorithms

In this subsection we formalize our algorithm and estimate its running time.

Algorithm 1:
Input: A biconnected graph G
Output: Decision whether G has a PC-representation.
Auxiliary variables: $\mathcal{H}$ – set of graphs, $\mathcal{S}$ – set of representations, initially empty

```
Add G to set H.
while exists F ∈ H with a cutting-edge ab do
        remove F from H and replace it by the elements of the red-edged
        decomposition of F with respect to ab.
done
forall F ∈ H do
     if Algorithm2(F)='false' then return 'false'
     else
          add the representation of F provided by
          Algorithm2(F) into S
done
forall R ∈ S do
     for e ∈ red_edges_of(graph_of(R)) do
     if Algorithm3(R,e)='false' then return 'false';
return 'true';
```

The correctness of Algorithm 1 follows from Lemmas 1, 2 and Propositions 1, 2.

Next we present the algorithm finding a representation of a pc-prime graph as the rest is either brute-force or obvious (and was described above). In Algorithm 2 we call an arc of a bounding circle between two corners of a polygon a *sector*.

Algorithm 2:
Input: pc-prime graph G.
Output: PC-representation or `false`.
Auxiliary variables: Graph H, initially empty.

```
while (|V(H)| < |V(G)|) do
    find a pseudoear or a single vertex P attachable to H;
    by brute force represent it;
    if the representation is impossible then return false;
    add P to H;
done
```

When looking for pseudoears (or single vertices), we implement the (constructive) proof of Lemma 5, while finding a suitable representation follows the proof of Proposition 3 (none of these proofs is included in this extended abstract). Note that looking for a pseudoear or a single vertex is (naively) possible in $O(n^2)$, representing a pseudoear is possible in $O(n^4)$ (looking for feasible sector, trying different placements of p_1 and p_k), representing of single vertex is possible in $O(n^5)$. Pseudoears (or single vertices) are added only linearly many times. Thus the complexity of Algorithm 2 is $O(n^6)$.

Algorithm 3:
Input: Alternating representation $\mathcal{R}$ of a graph G and a red edge $ab = e$ in G (c is not a vertex of G)
Output: Decision whether G^e has PC-representation.

By brute force try to add $cabc$ between all consecutive pairs of symbols into the alternating representation and check whether a correct representation is obtained.
If we at least once succeed, `return representation; else return` 'false';

The algorithm is just brute force, but polynomial ($O(n^4)$ as the length of representation is at most n^2). Its correctness is obvious from Observation 1. The total complexity of all algorithms together is therefore $O(n^7)$.

3   Segment Intersection Graphs

To prove the desired NP-hardness result, we show a polynomial reduction of the problem P3CON3SAT(4) to the recognition problem. P3CON3SAT(4) is the SATISFIABILITY problem restricted to formulas Φ with variables $v_1, \ldots, v_n$ and clauses $C_1, \ldots, C_m$ in CNF such that each clause contains exactly 3 literals, each variable occurs in Φ at most 4 times (positively or in negation) and the bipartite graph

$$G(\Phi) = (\bigcup_{i=1}^{n}\{v_i\} \cup \bigcup_{j=1}^{m}\{C_j\}, \{\{v_i, C_j\}|v_i \in C_j\})$$

is planar and (vertex) 3-connected. This variant of SATISFIABILITY is shown NP-complete in [12] and we will use it in the proof of Theorem 2.

We use a construction similar to [11]. Given a formula Φ, we use $G(\Phi)$ to construct a graph H with girth at least k such that H is a SEG-graph if and only if Φ is satisfiable. We modify the graph $G(\Phi)$ in the following way. Each vertex corresponding to a variable is replaced by a circle of length $\max\{16, k\}$. We replace each edge of $G(\Phi)$ by a pair of long-enough paths, and pairs of paths leaving the same variable gadget are linked by so called cross-over gadgets. Each vertex corresponding to a clause is replaced by a clause gadget. The assignment to variables is obtained from the orientation of the circle representing the particular variable. The "orientation" of pair of paths from the vertex gadget to a clause gadget describes whether the respective occurrence in the clause is positive or negative. By "orientation" of a pair of paths we mean that one path is "to the left" to the other. This has now become a standard trick used in NP-hardness reductions of geometric flavor.

The variable gadgets do not differ much from those used in [11], only the bounding cycle of the variable gadget may need to be artificially extended to make it meet the girth requirement. The clause gadget is slightly modified and it is depicted in Fig. 4. The main novelty of the present proof is the cross-over gadget depicted in Fig. 3.

The arguments are based on the fact that two intersecting segments share only one common point. The details of the proofs will be presented in the journal version.

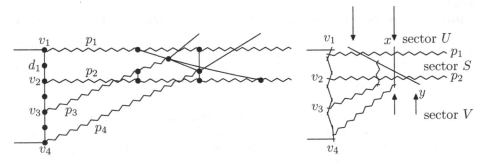

Fig. 3. The cross-over gadget and a representation starting from the variable gadget. Broken lines denote sequences of arbitrarily many segments (depending on the required girth). Arrows in the representation propose two possibilities for the respective paths – either to cross or to touch only. Note that the clause gadget is surrounded by a cycle, therefore a *sector S* is well-defined as a place surrounded by p_1, p_2, v_1, d_1, v_2 and the corresponding boundary-part of the clause gadget. As $G(\Phi)$ is 3-connected, even the sectors U and V are well-defined as the place above p_1 (distinct from S) and below p_2, respectively. Moreover S, U and V describe disjoint regions.

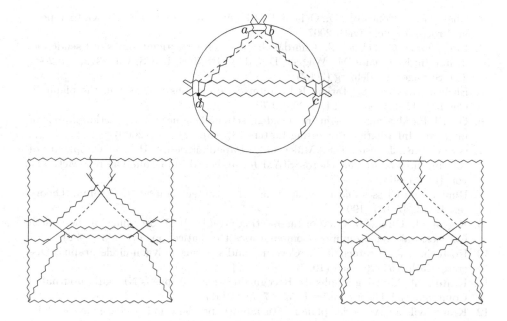

Fig. 4. The clause gadget (at the top, depicting the "all-true" position of the literals involved in it). Dashed and broken lines denote paths of lengths depending on k. The bottom two figures are examples of representations, the left one corresponds to the "false-true-false" valuation of the literals, the right one corresponds to "true-false-true" (the order of literals being left-top-right). Other cases can be represented similarly. The "all-false" position has no representation, even by curves with 1 crossing per pair.

4   Conclusion

We have shown that restricting the girth of the input graph helps with recognizing polygon-circle graphs, but that recognition of segment intersection graphs remains NP-hard. We believe that this phenomenon is worth of studying for other graph classes and that interesting results may be obtained. Also our results leave some room for strengthening. Several particular presently open cases are

- String graphs of girth at least k (known NP-hard only for $k = 4$).
- Polygon-circle graphs of girth 4.
- Intersection graphs of convex sets of girth at least k.

References

1. Breu, H., Kirkpatrick, D.G.: Unit Disk Graph Recognition is NP-Hard. Comput. Geom. Theory and Applications 9(1–2), 3–24 (1998)
2. Brandstaedt, A., Le, V.B., Spinrad, J.: Graph Classes: A Survey. SIAM (1999)

3. Chalopin, J., Goncalves, D., Ochem, P.: Planar graphs are in 1-STRING, to appear in: Proceedings of SODA 2007
4. Dangelmayr, C., Felsner, S.: Chordal Graphs as Intersection Graphs of Pseudosegments. In: Kaufmann, M., Wagner, D. (eds.) GD 2006. LNCS, vol. 4372, pp. 208–219. Springer, Heidelberg (2007)
5. Ehrlich, G., Even, S., Tarjan, R.E.: Intersection graphs of curves in the plane. J. Combin. Theory Ser. B. 21, 8–20 (1976)
6. Gavril, F.: Maximum weight independent sets and cliques in intersection graphs of filaments. Information Processing Letters 73(5–6), 181–188 (2000)
7. Garey, M.R., Johnson, D.S., Miller, G.L., Papadimitriou, C.H.: The complexity of coloring circular arcs and chords. SIAM Journal of Algebraic and Discrete Methods, vol. 1(2) (1980)
8. Hliněný, P.: Classes and recognition of curve contact graphs. J. Combin. Theory ser. B. 74, 87–103 (1998)
9. Koebe, M.: On a New Class of Intersection Graphs. In: Proceedings of the Fourth Czechoslovak Symposium on Combinatorics Prachatice, pp. 141–143 (1990)
10. Kostochka, A., Kratochvíl, J.: Covering and coloring polygon-circle graphs. Discrete Math. 163, 299–305 (1997)
11. Kratochvíl, J.: String graphs. II. Recognizing string graphs is NP-hard. Journal of Combinatorial Theory, Series B 52, 67–78 (1991)
12. Kratochvíl, J.: A special planar satisfiability problem and a consequence of its NP-completeness. Discr. Appl. Math. 52, 233–252 (1994)
13. Kratochvíl, J.: Intersection Graphs of Noncrossing Arc-Connected Sets in the Plane. In: North, S.C. (ed.) GD 1996. LNCS, vol. 1190, pp. 257–270. Springer, Heidelberg (1997)
14. Kratochvíl, J., Matoušek, J.: Intersection Graphs of Segments. Journal of Combinatorial Theory, Series B 62, 289–315 (1994)
15. Kratochvíl, J., Pergel, M.: Two Results on Intersection Graphs of Polygons. In: Liotta, G. (ed.) GD 2003. LNCS, vol. 2912, pp. 59–70. Springer, Heidelberg (2004)
16. McKee, T.A., McMorris, F.R.: Topics on Intersection Graphs, SIAM (1999)
17. Pach, J., Tóth, G.: Recognizing string graphs is decidable. In: Mutzel, P., Jünger, M., Leipert, S. (eds.) GD 2001. LNCS, vol. 2265, pp. 247–260. Springer, Heidelberg (2002)
18. Pergel, M.: Recognition of Polygon-circle Graphs and Graphs of Interval Filaments is NP-complete, in preparation
19. Schaefer, M., Štefankovič, D.: Decidability of string graphs, In: STOC, pp. 241–246 (2001)
20. Schaefer, M., Sedgwick, E., Štefankovič, D.: Recognizing String Graphs in NP. In: STOC 2002, pp. 1–6 (2002)
21. Sinden, F.W.: Topology of thin film RC-circuits. Bell System Technological Journal, pp. 1639–1662 (1966)
22. Spinrad, J.: Efficient Graph Representations, Fields Institute Monographs 19, American Mathematical Society (2003)

Dimension, Halfspaces, and the Density of Hard Sets[*]

Ryan C. Harkins and John M. Hitchcock

Department of Computer Science, University of Wyoming

Abstract. We use the connection between resource-bounded dimension and the online mistake-bound model of learning to show that the following classes have polynomial-time dimension zero.

1. The class of problems which reduce to nondense sets via a majority reduction.
2. The class of problems which reduce to nondense sets via an iterated reduction that composes a bounded-query truth-table reduction with a conjunctive reduction.

As corollary, all sets which are hard for exponential time under these reductions are exponentially dense. The first item subsumes two previous results and the second item answers a question of Lutz and Mayordomo. Our proofs use Littlestone's Winnow2 algorithm for learning r-of-k threshold functions and Maass and Turán's algorithm for learning halfspaces.

1 Introduction

Recent work has found applications of computational learning theory to the resource-bounded measure [10] and dimension [12] of complexity classes. Lindner, Schuler, and Watanabe [8] studied connections between computational learning theory and resource-bounded measure [10], primarily focusing on the PAC (probably approximately correct) model. They also observed that any admissible subclass of P/poly that is learnable in Littlestone's online mistake-bound model [9] has p-measure 0. This observation was later developed into a general tool for resource-bounded dimension [6]. To show that a class has p-dimension 0, it suffices to show that it is reducible to a learnable concept class family. This idea was used to show that the following classes have p-dimension 0.

(1) $P_{ctt}(DENSE^c)$.
(2) $P_{dtt}(DENSE^c)$.
(3) $P_{n^\alpha-T}(DENSE^c)$, for all $\alpha < 1$.

Here $P_r(DENSE^c)$ is the class of all problems which reduce to nondense sets under $\leq_r^p$ reductions, where a problem is nondense if its census function is subexponential. The result for (3) improved previous work [4,13,15] and solved one of Lutz and Mayordomo's twelve problems in resource-bounded measure [14].

---

[*] This research was supported in part by NSF grant 0515313.

The classes in (2) and (3) were reduced to disjunctions, which can be learned by Littlestone's Winnow algorithm [9]. We obtain further results in this direction using more sophisticated learning algorithms and concept classes that generalize disjunctions.

In our first result we show that the class

(4) $P_{maj}(DENSE^c)$

of problems which reduce to nondense sets via majority reductions has p-dimension 0. Our proof gives a reduction to r-of-k threshold functions and applies Littlestone's Winnow2 algorithm. This subsumes the results about (1) and (2) above and answers a question of Fu [5].

Our second result concerns iterated reductions and answers the following question of Lutz and Mayordomo [13]:

(Q) Does the class $P_{btt}(P_{ctt}(DENSE^c))$ have measure 0 in E?

Agrawal and Arvind [1] showed that $P_{btt}(P_{ctt}(SPARSE)) \subseteq P_m(LT_1)$, where LT_1 is the class of problems that have a nonuniform family of depth-1 weighted linear threshold circuits. Equivalently, LT_1 is the class of problems where each input length is a halfspace. We use their technique to reduce

(5) $P_{\alpha \log n-tt}(P_{ctt}(DENSE^c))$, for all $\alpha < 1$

to a subexponential-size family of halfspaces. We then apply the online learning algorithm of Maass and Turán [16] to learn these halfspaces and conclude that the classes in (5) have p-dimension 0. This strongly answers (Q) in the affirmative.

This paper is organized as follows. Section 2 contains preliminaries about halfspaces, learning, and dimension. The majority reductions result is in section 3 and the iterated reductions result is in section 4. Section 5 concludes with some observations for NP and directions for further work.

2   Preliminaries

A language L is a subset of $\{0, 1\}^*$. For the length of a string x, we write $|x|$. By $L_{=n}$ we denote the set all strings in L of length n, and by $L_{\leq n}$ we denote the set of all strings in L with length at most n. Let L be a language.

- L is *sparse* if for all $n \in \mathbb{N}$, $|L_{\leq n}| \leq p(n)$, where $p(n)$ is a polynomial.
- L is *dense* if for some $\epsilon > 0$, for all but finitely many n, $|L_{\leq n}| > 2^{n^{\epsilon}}$.
- L is *io-dense* if for some $\epsilon > 0$, for infinitely many n, $|L_{\leq n}| > 2^{n^{\epsilon}}$.

We write SPARSE, DENSE, and DENSE$_{i.o.}$ for the classes of sparse, dense, and io-dense languages, respectively. Note that $L \in DENSE^c$ if for all $\epsilon > 0$, for infinitely many n, $|L_{\leq n}| < 2^{n^{\epsilon}}$, and $L \in DENSE^c_{i.o.}$ if for all $\epsilon > 0$, for all sufficiently large n, $|L_{\leq n}| < 2^{n^{\epsilon}}$.

We assume the reader is familiar with the various notions of polynomial-time reductions. If a reduction $g(x)$ produces a single query, then $|g(x)|$ refers to the size of that query. If it produces multiple queries, then $|g(x)|$ is the number of queries produced.

2.1   Threshold Circuits

A *weighted linear threshold gate* with n inputs is determined by a weight vector $\hat{w} \in \mathbb{Q}^n$ and a threshold $T \in \mathbb{Q}$ such that on inputs $x \in \{0, 1\}^n$, where x is considered an n-valued vector $(x_1, x_2, \ldots, x_n)$, the gate will output 1 if and only if $\sum_{1 \leq i \leq n} w_i x_i > T$. An *exact weighted linear threshold gate* is defined similarly, except that the gate will output 1 if and only if $\sum_{1 \leq i \leq n} w_i x_i = 0$. As this is an inner product on vectors, we use the notation $\hat{w} \cdot \hat{x}$ for $\sum_{1 \leq i \leq n} w_i x_i$.

A linear threshold circuit has a linear threshold gate at its root. A language L is in the class LT_1 if there exists a family of nonuniform, depth-1 weighted linear threshold circuits defined by a family of weight vectors $\{\hat{w}_n\}_{n \geq 0}$ such that for all $x \in \{0, 1\}^*$, $x \in L$ if and only if $\hat{w}_{|x|} \cdot \hat{x} > 0$. Similarly, L is in the class ELT_1 if there exists a family of nonuniform, depth-1 exact weighted linear threshold circuits defined by a family of weight vectors $\{\hat{w}_n\}_{n \geq 0}$ such that for all $x \in \{0, 1\}^*$, $x \in L$ if and only if $\hat{w}_{|x|} \cdot \hat{x} = 0$.

Topologically, a linear threshold gate on n inputs describes a halfspace S in $\{0, 1\}^n$, and an exact linear threshold gate describes a hyperplane H in $\{0, 1\}^n$, where strings in $\{0, 1\}^n$ are viewed as binary vectors.

For more information on LT_1 and ELT_1, we refer the reader to Agrawal and Arvind [1], from which we will make several useful extensions in Section 4.

2.2   Dimension and Learning

Resource-bounded dimension was introduced by Lutz [12] as a refinement of resource-bounded measure [10]. Each class X of languages has a p-dimension $\dim_\mathrm{p}(X) \in [0, 1]$, and if $\dim_\mathrm{p}(X) < 1$, then X has p-measure 0. In this paper we do not use original definition of p-dimension but instead the result that if X reduces to a learnable concept class family, then $\dim_\mathrm{p}(X) = 0$ [6]. For more information on measure and dimension we refer to [3,7,11,14].

A concept is a set $C \subseteq U$ for some universe U. A concept class is a set $\mathcal{C}$ of concepts. An online learner, given a concept class $\mathcal{C}$ and a universe U, attempts to learn a target concept $C \in \mathcal{C}$. Given a sequence of examples $x_1, x_2, \ldots$ in U, the learner must predict whether $x_i \in C$. The answer is then revealed, the learner adjusts its strategy, and the next concept is presented for classification. The learner makes a mistake if it incorrectly classifies an example. The *mistake bound* of a learning algorithm for a concept class $\mathcal{C}$ is the maximum over all $C \in \mathcal{C}$ of the number of mistakes made when learning C, over all possible sequences of examples. The running time of the learner is the time required to predict the classification of an example.

Let $L \subseteq \{0, 1\}^*$ and let $\mathcal{C} = (\mathcal{C}_n | n \in \mathbb{N})$ be a sequence of concept classes. For a time bound $r(n)$, we say L reduces to $\mathcal{C}$ in $r(n)$ time if there is a reduction f computable in $O(r(n))$ time such that for infinitely many n, there is a concept $C_n \in \mathcal{C}_n$ such that for all $x \in \{0, 1\}^{\leq n}$, $x \in L$ if and only if $f(0^n, x) \in C_n$. Note that the reduction is not required to hold for all n, but only infinitely many n.

Let $\mathcal{L}(t, m)$ be the set of all sequences of concept classes $\mathcal{C}$ such that for each $\mathcal{C}_n \in \mathcal{C}$, there is an algorithm that learns $\mathcal{C}_n$ in $O(t(n))$ time with mistake bound

$m(n)$. Then the class $\mathcal{RL}(r, t, m)$ is the class of languages that reduce to some sequence of concept classes in $\mathcal{L}(t, m)$ in $r(n)$ time.

Theorem 2.1 (Hitchcock [6]). *For every $c \in \mathbb{N}$, the class $\mathcal{RL}(2^{cn}, 2^{cn}, o(2^n))$ has p-dimension 0.*

Because $X \subseteq Y$ implies $\dim_p(X) \leq \dim_p(Y)$, the task of proving that a class has p-dimesion 0 can be reduced to showing the class is a subset of $\mathcal{RL}(2^{cn}, 2^{cn}, o(2^n))$ for some constant c.

2.3 Learning Algorithms

We make use of two learning algorithms. The first is the second of Littlestone's Winnow algorithms [9], which can be used to learn Boolean r-of-k functions on n variables. In an r-of-k function there is a subset V of the n variables with $|V| = k$ such that the function evaluates to 1 if at least r of the variables in V are set to 1. Winnow2 has two parameters α, a weight update multiplier, and θ, a threshold value. Initially, each of the variables x_i has a weight $w_i = 1$. Winnow2 operates by predicting that an example x is in the concept if and only if $\sum_i w_i x_i > \theta$. The weights are updated following each mistake by the following rubric:

- If Winnow2 incorrectly predicts that x is in the target concept, then for each x_i such that $x_i = 1$, set $w_i = w_i/\alpha$.
- If Winnow2 incorrectly predicts that x is not in the target concept, then for each x_i such that $x_i = 1$, set $w_i = \alpha \cdot w_i$.

Littlestone showed that for $\alpha = \frac{1}{2r}$ and $\theta = n$, Winnow2 has a mistake bound on learning r-of-k functions of $8r^2 + 5k + 14kr \ln n$. Winnow2 also classifies examples in polynomial time.

The second learning algorithm we use is Maass and Turán's [16] algorithm for learning halfspaces. They first describe the Convex Feasability Problem: given a separation oracle and a guarantee r for an unknown convex body P, find a point in P. By a guarantee, they mean a number such that the volume of the convex body P (in d dimensions) within the ball of radius r around $\hat{0}$ is at least r^{-d}.

Theorem 2.2 (Maass and Turán [16]). *Assume that there is an algorithm A^* solving the Convex Feasability Problem with query complexity $q(d, \log r)$ (where q is a function of both the dimension d and the guarantee r) and time complexity $t(d, \log r)$. Then there is a learning algorithm A for learning a halfspace in d dimensions and n values such that the mistake bound of A is $q(d, 4d(\log d + \log n + 3)) + 1$ and the running time is at most $t(d, 4d(\log d + \log n + 3)) + q(d, 4d(\log d + \log n + 3)) \cdot p(d, \log n)$ for some polynomial p.*

Using Vaidya's algorithm for learning convex bodies [17], which is an algorithm for the Convex Feasability Problem, they show that learning a halfspace on d dimensions and n values (in our case, with the binary alphabet, $n = 2$) has a mistake bound of $O(d^2(\log d + \log n))$ and a polynomial running time.

3   Majority Reductions

We say that $A \leq_{\mathrm{maj}}^{\mathrm{P}} B$ if there is a polynomial-time computable function $f : \{0,1\}^* \to \mathcal{P}(\{0,1\}^*)$ such that for all $x \in \{0,1\}^*$, $x \in A$ if and only if

$$|f(x) \cap B| \geq \frac{|f(x)|}{2}.$$

The following lemma says that if B is nondense, then we can assume that the majority reduction makes the same number of queries for all inputs of each length.

Lemma 3.1. *Let* $A \in \mathrm{P}_{\mathrm{maj}}(\mathrm{DENSE}^c)$. *Then there exists a* $B \in \mathrm{DENSE}^c$, *a majority reduction* f *computable in polynomial time, and a polynomial* q *such that for all* $x \in \{0,1\}^*$, $|f(x)| = q(|x|)$.

We now prove our first main result.

Theorem 3.2. $\mathrm{P}_{\mathrm{maj}}(\mathrm{DENSE}^c)$ *has* p-*dimension 0*.

Proof. It suffices to show that there is a concept class family $\mathcal{CF} \in \mathcal{L}(2^{cn}, o(2^n))$ and a reduction g computable in 2^{cn} time such that for all $A \in \mathrm{P}_{\mathrm{maj}}(\mathrm{DENSE}^c)$, A reduces to $\mathcal{CF}$ by g.

Let $A \in \mathrm{P}_{\mathrm{maj}}(\mathrm{DENSE}^c)$. Then there is a $p(n)$-time-bounded majority reduction f that makes exactly $r(n)$ queries for each n, and a set $B \in \mathrm{DENSE}^c$ such that for all $x \in \{0,1\}^*$, $x \in A$ if and only if $|f(x) \cap B| \geq \frac{|f(x)|}{2}$.

Let $Q_n = \bigcup_{|x| \leq n} f(x)$ be the set of all queries made by f up through length n. Then $|Q_n| \leq 2^{n+1} p(n)$. Enumerate Q_n as $q_1, \ldots, q_N$. Then each subset $R \subseteq Q_n$ can be identified with its characteristic string $\chi_R \in \{0,1\}^N$ according to this enumeration.

Let $\delta \in (0,1)$. Then $|B_{\leq p(n)}| < 2^{n^\delta}$ for infinitely many n because B is nondense. Thus $M = |Q_n \cap B| \leq 2^{n^\delta}$. Since for all $x \in \{0,1\}^*$, $f(x)$ makes exactly $r(|x|)$ number of queries and $x \in A$ if and only if $|f(x) \cap B| \geq \frac{r(|x|)}{2}$, our target concept is a $\frac{r(n)}{2}$-of-M threshhold function h, such that $x \in A$ if and only if $h(\chi_{f(x)}) = 1$, which can be learned by Winnow2.

Given x, $\chi_{f(x)}$ can be computed in $O(2^{2n})$ time, and thus Winnow2 can classify examples in $O(2^{2n})$ time, making only $2r^2(n) + 5 \cdot 2^{n^\delta} + 7 \cdot 2^{n^\delta} r(n) \ln 2^{n+1} p(n)$ mistakes, which is $o(2^n)$. Thus $\mathrm{P}_{\mathrm{maj}}(\mathrm{DENSE}^c) \subseteq \mathcal{RL}(2^{2n}, 2^{2n}, o(2^n))$ and the theorem follows by Theorem 2.1. □

We remark that as r-of-k threshold functions are a special case of halfspaces, we could also use the halfspace learning algorithm instead of Winnow2 to prove Theorem 3.2. As $\mathrm{P}_{\mathrm{dtt}}(\mathrm{DENSE}^c) \subseteq \mathrm{P}_{\mathrm{maj}}(\mathrm{DENSE}^c)$ and $\mathrm{P}_{\mathrm{ctt}}(\mathrm{DENSE}^c) \subseteq \mathrm{P}_{\mathrm{maj}}(\mathrm{DENSE}^c)$, Theorem 3.2 subsumes two results from [6]. We also have the following corollary about hard sets for exponential time.

Corollary 3.3. $\mathrm{E} \not\subseteq \mathrm{P}_{\mathrm{maj}}(\mathrm{DENSE}^c)$. *That is, every* $\leq_{\mathrm{maj}}^{\mathrm{P}}$-*hard set for* E *is dense.*

4   Iterated Reductions

Our proof that $P_{\alpha \log n-tt}(P_{ctt}(\text{DENSE}^c))$ has p-dimension zero follows the proof technique of Agrawal and Arvind [1] that

$$P_{btt}(P_{ctt}(\text{SPARSE})) \subseteq P_m(\text{LT}_1)$$

to reduce the class to a family of halfspaces. We then use Maass and Turán's [16] learning algorithm to learn these halfspaces. As long as the reduction runs in 2^{n^α} time for some $\alpha < 1$, the halfspaces have subexponential size and the mistake bound is $2^{o(n)}$. Therefore by Theorem 2.1,

$$\dim_p(R_m^{2^{n^\alpha}}(\text{LT}_1)) = 0, \tag{1}$$

where $R_m^{t(n)}$ denotes many-one reductions that run in time $t(n)$.

Instead of $P_{\alpha \log n-tt}(P_{ctt}(\text{DENSE}^c))$ we will focus on the smaller class

$$P_{\alpha \log n-tt}(P_{ctt}(\text{DENSE}_{i.o.}^c)).$$

The benefit is that the nondense set will be small almost everywhere rather than infinitely often, which simplifies the arguments. Our proofs can be adapted to the infinitely-often case. First we need to extend some of Agrawal and Arvind's results. They proved the following technical lemma.

Lemma 4.1. *Let $A \in P_m(\text{ELT}_1)$ (resp. $A \in P_m(\text{LT}_1)$). Then there exist $L \in \text{ELT}_1$ ($L \in \text{LT}_1$), an FP function f, and a polynomial r, such that for every x, for every $n \geq r(|x|)$, $x \in A$ iff $f(x, 1^n) \in H(\hat{w}_n)$ ($f(x, 1^n) \in S_+(\hat{w}_n)$), where $\hat{w}_n$ are the weight vectors associated with L.*

We use the following extension.

Lemma 4.2. *Let Δ be a family of computable functions that is closed under multiplication and composition with polynomials, and let $F\Delta$ be the functional class with bounds in Δ. Then Lemma 4.1 holds with $f \in F\Delta$ and $r \in \Delta$.*

We say that a time bound $t(n)$ is *subexponential* if for all $\epsilon > 0$, $t(n) < 2^{n^\epsilon}$ for all sufficiently large n. We write se for the class of all subexponential time bounds.

Corollary 4.3. *Let $A \in R_m^{se}(\text{ELT}_1)$. Then there is a subexponential-time function f and a language $B \in \text{ELT}_1$ such that for every $x \in \{0,1\}^*$, for all $q_i, q_j \in f(x)$, $|q_i| = |q_j|$.*

Agrawal and Arvind showed that SPARSE $\subseteq P_m(\text{ELT}_1)$. We extend this in the following lemma.

Lemma 4.4. $\text{DENSE}_{i.o.}^c \subseteq R_m^{se}(\text{ELT}_1)$.

Proof. Let $S \in \mathrm{DENSE}^c_{\mathrm{i.o.}}$. Then for all but finitely many n, $|S_{=n}| \leq 2^{n^\epsilon}$ for all $\epsilon > 0$. Let $S_{=n} = \{s_{n,1}, s_{n,2}, \ldots, s_{n,m(n)}\}$, where $m(n) \leq f(n)$ and f is a subexponential function. Let the string $s_{n,i}$ also stand for the natural number representing the lexicographic rank of the string $s_{n,i}$ in $\{0,1\}^n$ for every n and $1 \leq i \leq m(n)$. Define $T_n(z)$ to be the polynomial $\prod_{i=1}^{m(n)} (z - s_{n,i})$. Clearly, $T_n(z)$ is a polynomial in z of degree bounded by $f(n)$. Letting z represent both a string in $\{0,1\}^n$ as well as the lexicographic rank of z in $\{0,1\}^n$, then $z \in S$ iff $T_n(z) = 0$.

Rewriting $T_n(z)$, we have $T_n(z) = \sum_{1 \leq j \leq f(n)} a_j z^j$. For $1 \leq j \leq f(n)$, we can write z^j as $\sum_{1 \leq r \leq n \cdot f(n)} 2^r y_{j,r}$, where the $y_{j,r}$ essentially denotes the bits in the binary representation of z^j. Thus it follows that $T_n(z)$ can be rewritten as a linear combination $\sum_{1 \leq j \leq f(n)} \sum_{1 \leq r \leq n \cdot f(n)} w_{j,r} y_{j,r}$ of the bits $y_{j,r}$ defined above.

Now we can define a language $L \in \mathrm{ELT}_1$ using these linear functions to define the corresponding weighted exact threshold gates in the circuit family accepting L. As there will be $n \cdot f(n)^2$ variables, which is subexponential, we have that $S \leq^{se}_m L$. $\qquad\square$

Agrawal and Arvind showed that $\mathrm{P_{ctt}}(\mathrm{ELT}_1) = \mathrm{P_m}(\mathrm{ELT}_1)$ and $\mathrm{P_{btt}}(\mathrm{ELT}_1) \subseteq \mathrm{P_m}(\mathrm{LT}_1)$. This allows them to show $\mathrm{P_{btt}}(\mathrm{P_{ctt}}(\mathrm{SPARSE})) \subseteq \mathrm{P_m}(\mathrm{LT}_1)$ through the following steps:

$$
\begin{aligned}
\mathrm{P_{btt}}(\mathrm{P_{ctt}}(\mathrm{SPARSE})) &\subseteq \mathrm{P_{btt}}(\mathrm{P_{ctt}}(\mathrm{P_m}(\mathrm{ELT}_1))) \\
&\subseteq \mathrm{P_{btt}}(\mathrm{P_{ctt}}(\mathrm{ELT}_1)) \\
&\subseteq \mathrm{P_{btt}}(\mathrm{P_m}(\mathrm{ELT}_1)) \\
&\subseteq \mathrm{P_{btt}}(\mathrm{ELT}_1) \\
&\subseteq \mathrm{P_m}(\mathrm{LT}_1).
\end{aligned}
$$

We will adapt this proof to our setting. Agrawal and Arvind made use of the following technical lemma.

Lemma 4.5. *Let* $\{F_n(\hat{x})\}_{n \geq 1}$, $F_n(\hat{x})$ *defined over* $\mathbb{Q}^n$, *be a family of degree k multinomials (for a constant $k > 0$). Let the family of weight vectors* $\{\hat{c}_n\}_{n > 0}$ *and the FP function f be such that for every $\hat{x} \in \mathbb{Q}^n$, $F_n(\hat{x}) = \hat{c}_m \cdot f(\hat{x})$ where $f(\hat{x}) \in \mathbb{Q}^m$. Then the function f reduces the set*

$$
A = \bigcup_{n \geq 1} \{x \in \{0,1\}^* \mid F_n(x) = 0\}
$$

to the set in ELT_1 *defined by weight vectors* $\{\hat{c}_n\}_{n > 0}$. *Also, f reduces the set*

$$
B = \bigcup_{n \geq 1} \{x \in \{0,1\}^* \mid F_n(x) > 0\}
$$

to the set in LT_1 *defined by weight vectors* $\{\hat{c}_n\}_{n > 0}$ *(where a string x of length n is interpreted as an n-dimensional 0-1 vector when it is an argument to F_n).*

Lemma 4.6. $R_{ctt}^{se}(ELT_1) = R_m^{se}(ELT_1)$.

Proof. Let A be a set that is conjuctively reducible to some set $B \in ELT_1$. Then there is an se-computable function f such that for every $x \in \{0,1\}^*$, $f(x)$ is a list of queries such that $x \in A$ iff for every q in $f(x)$, $q \in B$. Using Corollary 4.3, there exist $B' \in ELT_1$ defined by a family of weight vectors $\{\hat{c}_n\}_{n \geq 1}$, an se-computable function g, and a subexponential function r such that for every x, for every $j \geq r(|x|)$, $x \in A$ iff $g(x, 1^j) \subseteq H(\hat{c}_j)$.

Since f is a conjunctive reduction, there is a subexponential p such that for every x, $g(x, 1^{r(|x|)})$ has exactly $p(|x|)$ queries (this can be achieved simply by repeating the last query a suitable number of times). Define

$$F_{p(n)r(n)}(\hat{q}_1, \hat{q}_2, \ldots, \hat{q}_{p(n)}) = \sum_{i=1}^{p(n)} (\hat{c}_{r(n)} \cdot \hat{q}_i)^2$$

where $\hat{q}_i \in \{0,1\}^{r(n)}$ for $1 \leq i \leq p(n)$. The set L is defined as

$$L = \bigcup_{n \geq 1} \left\{ x \in \{0,1\}^* p(n)r(n) \mid F_{p(n)r(n)}(x) = 0 \right\}.$$

Note that as an argument to $F_{p(n)r(n)}$, x is interpreted as a 0-1 vector.

It is easy to see that $x \in A$ iff $(\hat{q}_1, \hat{q}_2, \ldots, \hat{q}_{p(|x|)}) \in L$ where $g(x, 1^{r(|x|)}) = \{\hat{q}_1, \ldots, \hat{q}_{p(|x|)}\}$. Lemma 4.5 implies that L is in $R_m^{se}(ELT_1)$. $\square$

To show $P_{btt}(ELT_1) \subseteq P_m(LT_1)$, Agrawal and Arvind divide a k-tt reduction into each separate condition τ, and then note that

$$P_\tau(A) \subseteq P_{b\oplus}(P_{bc}(P_{1-tt}(A))),$$

where $P_{b\oplus}$ is the closure under the bounded parity reduction and P_{bc} is the closure under the bounded conjunctive reduction. They then show that

$$P_{b\oplus}(P_{bc}(P_{1-tt}(ELT_1))) \subseteq P_m(LT_1),$$

and finish their proof by showing $P_m(LT_1)$ is closed under the join operation, i.e. it is possible to create a linear threshold circuit from all 2^{2^k} k-tt conditions in polynomial time. We proceed in the same fashion.

Lemma 4.7. $R_{b\oplus}^{se}(LT_1) = R_m^{se}(LT_1)$.

Lemma 4.8. $R_{bd}^{se}(ELT_1) = R_m^{se}(ELT_1)$.

Lemma 4.9. $R_\tau^{se}(LT_1) \subseteq R_m^{se}(LT_1)$.

Lemma 4.10. $P_{bc}(R_{1-tt}^{se}(ELT_1)) \subseteq R_\tau^{se}(ELT_1)$.

Proof. For $A \in P_{bc}(R_{1-tt}^{se}(ELT_1))$, there is a reduction f to $B \in ELT_1$ with k queries such that $x \in A$ iff m of the k queries are in B and $k - m$ are not in B. We label these queries $\hat{q}_1, \hat{q}_2, \ldots, \hat{q}_m$ and $\hat{r}_1, \hat{r}_2, \ldots, \hat{r}_{k-m}$. Thus $x \in A$ iff $\hat{q}_i \in B$ for all $1 \leq i \leq m$ and no $\hat{r}_j \in B$ for all $1 \leq j \leq k - m$. We can look at this as the combination of a bounded conjunctive reduction and a bounded disjunctive reduction, both requiring subexponential time. By Lemma 4.6 and Lemma 4.8, we can alter these reductions to a single query each to a language $B' \in \text{ELT}_1$. Call these single queries $\hat{q}$ and $\hat{r}$. Then $x \in A$ iff $\hat{q} \in B'$ and $\hat{r} \notin B'$. This transformation can be carried out in subexponential time, so the lemma follows. $\qquad\square$

We are ready to prove the simplified version of our main result.

Theorem 4.11. $P_{\alpha \log n - \text{tt}}(P_{\text{ctt}}(\text{DENSE}_{\text{i.o.}}^c))$ *has* p-*dimension 0.*

Proof. Through Lemma 4.4, Lemma 4.6, Lemma 4.10, and Lemma 4.7 respectively, the following holds for each truth-table condition τ:

$$
\begin{aligned}
P_\tau(P_{\text{ctt}}(\text{DENSE}_{\text{i.o.}}^c)) &\subseteq P_{\text{b}\oplus}(P_{\text{bc}}(P_{1-\text{tt}}(P_{\text{ctt}}(\text{DENSE}_{\text{i.o.}}^c)))) \\
&\subseteq P_{\text{b}\oplus}(P_{\text{bc}}(P_{1-\text{tt}}(R_{\text{m}}^{\text{se}}(\text{ELT}_1)))) \\
&\subseteq P_{\text{b}\oplus}(P_{\text{bc}}(P_{1-\text{tt}}(R_{\text{ctt}}^{\text{se}}(\text{ELT}_1)))) \\
&\subseteq P_{\text{b}\oplus}(P_{\text{bc}}(P_{1-\text{tt}}(R_{\text{m}}^{\text{se}}(\text{ELT}_1)))) \\
&\subseteq P_{\text{b}\oplus}(P_{\text{bc}}(R_{1-\text{tt}}^{\text{se}}(\text{ELT}_1))) \\
&\subseteq P_{\text{b}\oplus}(R_\tau^{\text{se}}(\text{ELT}_1)) \\
&\subseteq R_{\text{m}}^{\text{se}}(\text{LT}_1).
\end{aligned}
$$

With $\alpha \log n$ queries for $|x| = n$, there are 2^{n^α} truth-table conditions. Through the reduction above, each condition corresponds to a different set of weights $\hat{c}_{n,j}$, $1 \leq j \leq 2^{n^\alpha}$, defining a threshold circuit that is subexponential in size. Let us say this size is $s(n)$. Thus we can create a single linear threshhold circuit with weights $\hat{d}_n = (\hat{c}_{n,1}, \hat{c}_{n,2}, \ldots, \hat{c}_{n,2^{n^\alpha}})$ as the join of all these individual circuits. The size of this new circuit is $2^{n^\alpha} \cdot s(n) \leq 2^{n^\delta}$ for some δ such that $0 < \delta < 1$. Let $L \in \text{LT}_1$ be the set defined by the weight vectors $\{\hat{d}_n\}$.

Let $A \in P_{\alpha \log n - \text{tt}}(P_{\text{ctt}}(\text{DENSE}_{\text{i.o.}}^c))$. Let g be the $\alpha \log n$-tt reduction, and suppose that $g(x)$ uses the condition which corresponds to $\hat{c}_{|x|,j}$. Let $\hat{q}_x$ be the many-one query corresponding to that condition produced by the reduction above. Then the reduction f mapping A to L is defined by

$$
f(x) = (\hat{0}_{(j-1)s(|x|)}, \hat{q}_x, \hat{0}_{(2^{|x|^\alpha} - (j+1))s(|x|)}).
$$

Then $x \in A$ iff $f(x) \in L$. It follows that

$$
P_{\alpha \log n - \text{tt}}(P_{\text{ctt}}(\text{DENSE}_{\text{i.o.}}^c)) \subseteq R_{\text{m}}^{2^{n^\alpha}}(\text{LT}_1),
$$

which yields the theorem by (1). $\qquad\square$

A similar argument yields our main result.

Theorem 4.12. $P_{\alpha \log n - \text{tt}}(P_{\text{ctt}}(\text{DENSE}^c))$ *has* p-*dimension 0.*

Corollary 4.13. $E \not\subseteq P_{\alpha \log n\text{-tt}}(P_{ctt}(\text{DENSE}^c))$.

Theorem 4.12 gives the answer to Lutz and Mayordomo's question [13].

Corollary 4.14. $P_{btt}(P_{ctt}(\text{DENSE}^c))$ *has* p-*measure 0.*

5 Conclusion

We conclude with a brief remark about the density of hard sets for NP. If NP has positive p-dimension, then it follows from our results that

$$NP \not\subseteq P_{maj}(\text{DENSE}^c)$$

and

$$NP \not\subseteq P_{\alpha \log n\text{-tt}}(P_{ctt}(\text{DENSE}^c))$$

for all $\alpha < 1$. These conclusions are stronger than what is known from the hypothesis $P \neq NP$. If $P \neq NP$, then $NP \not\subseteq P_{btt}(P_{ctt}(\text{SPARSE}))$ [2], but nothing is known about majority reductions.

One direction for further research is to improve the $\alpha \log n$-tt bound in Theorem 4.12. Can the bound be improved to n^α-tt? Or ideally, to subsume the main result in [6], can it be improved to n^α-T? A more basic direction is to find more applications of learning algorithms in resource-bounded measure and dimension.

References

1. Agrawal, M., Arvind, V.: Geometric sets of low information content. Theoretical Computer Science 158(1–2), 193–219 (1996)
2. Arvind, V., Han, Y., Hemachandra, L., Köbler, J., Lozano, A., Mundhenk, M., Ogiwara, A., Schöning, U., Silvestri, R., Thierauf, T.: Reductions to sets of low information content. In: Ambos-Spies, K., Homer, S., Schöning, U. (eds.) Complexity Theory: Current Research, pp. 1–45. Cambridge University Press, Cambridge (1993)
3. Athreya, K.B., Hitchcock, J.M., Lutz, J.H., Mayordomo, E.: Effective strong dimension in algorithmic information and computational complexity. SIAM Journal on Computing (to appear)
4. Fu, B.: With quasilinear queries EXP is not polynomial time Turing reducible to sparse sets. SIAM Journal on Computing 24(5), 1082–1090 (1995)
5. Fu, B.: Personal communication (2006)
6. Hitchcock, J.M.: Online learning and resource-bounded dimension: Winnow yields new lower bounds for hard sets. SIAM Journal on Computing 36(6), 1696–1708 (2007)
7. Hitchcock, J.M., Lutz, J.H., Mayordomo, E.: The fractal geometry of complexity classes. SIGACT News 36(3), 24–38 (2005)
8. Lindner, W., Schuler, R., Watanabe, O.: Resource-bounded measure and learnability. Theory of Computing Systems 33(2), 151–170 (2000)
9. Littlestone, N.: Learning quickly when irrelevant attributes abound: A new linear-threshold algorithm. Machine Learning 2(4), 285–318 (1987)

10. Lutz, J.H.: Almost everywhere high nonuniform complexity. Journal of Computer and System Sciences 44(2), 220–258 (1992)
11. Lutz, J.H.: The quantitative structure of exponential time. In: Hemaspaandra, L.A., Selman, A.L. (eds.) Complexity Theory Retrospective II, pp. 225–254. Springer, Heidelberg (1997)
12. Lutz, J.H.: Dimension in complexity classes. SIAM Journal on Computing 32(5), 1236–1259 (2003)
13. Lutz, J.H., Mayordomo, E.: Measure, stochasticity, and the density of hard languages. SIAM Journal on Computing 23(4), 762–779 (1994)
14. Lutz, J.H., Mayordomo, E.: Twelve problems in resource-bounded measure. Bulletin of the European Association for Theoretical Computer Science 68, 64–80 (1999). Current Trends in Theoretical Computer Science: Entering the 21st Century, pp. 83–101. World Scientific Publishing, Singapore (2001)
15. Lutz, J.H., Zhao, Y.: The density of weakly complete problems under adaptive reductions. SIAM Journal on Computing 30(4), 1197–1210 (2000)
16. Maass, W., Turán, G.: How fast can a threshold gate learn? In: Hanson, S.J., Drastal, G.A., Rivest, R.L. (eds.) Computational Learning Theory and Natural Learning Systems. Constraints and Prospects, vol. I, pp. 381–414. MIT Press, Cambridge (1994)
17. Vaidya, P.M.: A new algorithm for minimizing convex functions over convex sets. In: Proceedings of the 30th IEEE Symposium on Foundations of Computer Science, pp. 338–349. IEEE Computer Society Press, Los Alamitos (1989)

Isolation Concepts
for Enumerating Dense Subgraphs

Christian Komusiewicz*, Falk Hüffner**, Hannes Moser***,
and Rolf Niedermeier

Institut für Informatik, Friedrich-Schiller-Universität Jena,
Ernst-Abbe-Platz 2, D-07743 Jena, Germany
{ckomus,hueffner,moser,niedermr}@minet.uni-jena.de

Abstract. In a graph $G = (V, E)$, a vertex subset $S \subseteq V$ of size k is called c-isolated if it has less than $c \cdot k$ outgoing edges. We repair a nontrivially flawed algorithm for enumerating all c-isolated cliques due to Ito et al. [European Symposium on Algorithms 2005] and obtain an algorithm running in $O(4^c \cdot c^4 \cdot |E|)$ time. We describe a speedup trick that also helps parallelizing the enumeration. Moreover, we introduce a more restricted and a more general isolation concept and show that both lead to faster enumeration algorithms. Finally, we extend our considerations to s-plexes (a relaxation of the clique notion), pointing out a W[1]-hardness result and providing a fixed-parameter algorithm for enumerating isolated s-plexes.

1 Introduction

Finding and enumerating cliques and clique-like structures in graphs has many applications ranging from technical networks [9] to social and biological networks [1–3]. Unfortunately, clique-related problems are known to be notoriously hard for exact algorithms, approximation algorithms, and fixed-parameter algorithms [7,5]. Ito et al. [9] introduced an interesting way out of this quandary by restricting the search to *isolated* cliques. Herein, given a graph $G = (V, E)$, a vertex subset $S \subseteq V$ of size k is called *c-isolated* if it has less than $c \cdot k$ outgoing edges. As their main result, Ito et al. [9] claimed an $O(4^c \cdot c^5 \cdot |E|)$ time algorithm for enumerating all c-isolated cliques in a graph. In particular, this means linear time for constant c and fixed-parameter tractability with respect to the parameter c. Unfortunately, the algorithm proposed by Ito et al. [9] suffers from serious deficiencies[1].

---

* Partially supported by the Deutsche Forschungsgemeinschaft, project OPAL (optimal solutions for hard problems in computational biology), NI 369/2.

** Supported by the Deutsche Forschungsgemeinschaft, Emmy Noether research group PIAF (fixed-parameter algorithms), NI 369/4.

*** Supported by the Deutsche Forschungsgemeinschaft, project ITKO (iterative compression for solving hard network problems), NI 369/5.

[1] A later manuscript [8] does not fundamentally resolve the problem.

G. Lin (Ed.): COCOON 2007, LNCS 4598, pp. 140–150, 2007.

We start with describing the algorithm of Ito et al. [9] and show that it does not fulfill the claimed running time bound. Then, we present some new results that eventually help us to repair Ito et al.'s approach, ending up with an algorithm that enumerates all c-isolated cliques in $O(4^c \cdot c^4 \cdot |E|)$ time. We also observe a speedup trick which seems to have high practical potential and which allows to parallelize the so far purely sequential enumeration algorithm.

Next, inspired by Ito et al.'s isolation concept, we propose two further isolation definitions, a weaker (less demanding) and a stronger concept, both practically motivated. Somewhat surprisingly, we can show that *both* concepts lead to faster enumeration algorithms for isolated cliques, improving the exponential factor from 4^c to 2^c and 2.44^c, respectively.

Finally, we show how to adapt the isolation scenario to the concept of s-plexes, a relaxation of cliques occurring in social networks analysis [15,1]. In a graph $G = (V, E)$, a vertex subset $S \subseteq V$ of size k is called an *s-plex* if the minimum degree in $G[S]$ is at least $k - s$. First, strengthening an NP-hardness result of Balasundaram et al. [1], we point out that the problem of finding s-plexes is W[1]-hard with respect to the parameter k; that is, the problem seems as (parameterized) intractable as CLIQUE is. This motivates our final result, a fixed-parameter algorithm (the parameter is the isolation factor) for constant s that enumerates all of one type of maximal isolated s-plexes. As a side result, here we improve a time bound for a generalized vertex cover problem first studied by Nishimura et al. [12].

Preliminaries. We consider only undirected graphs $G = (V, E)$ with $n := |V|$, $m := |E|$, $V(G) := V$, and $E(G) := E$. Let $N(v) := \{u \in V \mid \{u, v\} \in E\}$ and $N[v] := N(v) \cup \{v\}$. For $v \in V$, let $\deg_G(v) := |N(v)|$. For $A, B \subseteq V, A \cap B = \emptyset$, let $E(A, B) := \{\{u, v\} \mid u \in A, v \in B\}$. For $V' \subseteq V$, let $G[V']$ be the subgraph of G induced by V' and $G \setminus V' := G[V \setminus V']$. For $v \in V$, let $G - v := G[V \setminus \{v\}]$. A set S with property P is called *maximal* if no proper superset of S has property P, and *maximum* if no other set with property P has higher cardinality.

Parameterized complexity [5,11] is an approach to finding optimal solutions for NP-hard problems. The idea is to accept the seemingly inevitable combinatorial explosion, but to confine it to one aspect of the problem, the *parameter*. If for relevant inputs this parameter remains small, then even large problems can be solved efficiently. More precisely, a problem of size n is *fixed-parameter tractable* (FPT) with respect to a parameter k if there is an algorithm solving it in $f(k) \cdot n^{O(1)}$ time.

Due to the lack of space, most proofs will appear in the full version of this paper. Some material also appears in [10].

2   Enumerating Isolated Cliques

We begin with describing Ito et al.'s algorithm for enumerating maximal c-isolated cliques [9]. Given a graph $G = (V, E)$ and an isolation factor c, first the vertices are sorted by their degree such that $u < v \Rightarrow \deg(u) \leq \deg(v)$. The *index* of a vertex is its position in this sorted order. For a vertex set $C \subseteq V$, an *outgoing edge* is an edge $\{u, v\}$ with $u \in C$ and $v \notin C$, and for a vertex $v \in C$, its outgoing edges are the outgoing edges of C that are incident on v. Let $N_+[v] := \{u \in N[v] \mid u \geq v\}$ and $N_-(v) := \{u \in N(v) \mid u < v\}$.

In a c-isolated clique, the vertex with the lowest index is called the *pivot* of the clique. Clearly, a pivot has less than c outgoing edges. Since every c-isolated clique has a pivot, we can enumerate all maximal c-isolated cliques of a graph by enumerating all maximal c-isolated cliques with pivot v for each $v \in V$ and then removing those c-isolated cliques with pivot v that are a subset of a c-isolated clique with another pivot.

The enumeration of maximal c-isolated cliques with pivot v for each $v \in V$ is the central part of the algorithm. We call this the *pivot procedure*. It comprises three successive stages.

Trimming stage. In this stage, we build a candidate set C that is a superset of all c-isolated cliques with pivot v. The candidate set C is initialized with $N_+[v]$, and then vertices that obviously cannot be part of a c-isolated clique with pivot v are removed from C. We refer to Ito et al. [9] for details.

Enumeration stage. In this stage, all maximal c-isolated cliques with pivot v are enumerated. Let C be the candidate set after the trimming stage, which deleted d vertices from $N_+[v]$. In total, we can delete only less than c vertices from $N_+[v]$, since otherwise v obtains too many outgoing edges. Therefore, $\tilde{c} := c - d - 1$ is the number of vertices that we may still remove from C. We can enumerate cliques $C' \subseteq C$ of size *at least* $|C| - \tilde{c}$ by enumerating vertex covers of size *at most* $\tilde{c}$ in the complement graph $\overline{G[C]}$. Ito et al. propose to enumerate *all* vertex covers of size at most $\tilde{c}$ [9]. We point to problems with this approach in Sect. 2.1.

Screening stage. In the screening stage, all cliques that are either not c-isolated or that are c-isolated but not maximal are removed. First the c-isolation is checked. Then those cliques that pass the test for isolation are compared pairwise, and we only keep maximal cliques. Finally, we check each clique that is left for pivot v against each clique obtained during calls to pivot(u) with $u \in N_-(v)$, since these are the only cliques that can be superset of a clique obtained for pivot v. The claimed overall running time, in the exponential part dominated by this last step, is then $O(4^c \cdot c^5 \cdot |E|)$ [9].

2.1   Problems with the Algorithm

The crucial part of the algorithm is the enumeration stage, in which the algorithm enumerates *all* vertex covers of size less than $\tilde{c}$. The authors argued that for a graph of size $|C|$, this can be done in $O(1.39^{\tilde{c}} \cdot \tilde{c}^2 + |C|^2)$ time. In contrast, Fernau [6] showed that the vertex covers of size *exactly* k in a graph can be enumerated in time $O(2^k k^2 + kn)$ if and only if k is the size of a minimum vertex cover of the graph and that otherwise no algorithm of running time $f(k) \cdot n^{O(1)}$ that enumerates all of these vertex covers exists, simply because there are too many. But in the course of the pivot procedure it may happen that we have to do

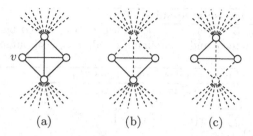

(a)   (b)   (c)

Fig. 1. Example for the enumeration stage with pivot v. *Solid lines* are edges between members of the clique; *dashed lines* are outgoing edges.

just that: enumerate all vertex covers of size $\tilde{c}$ or less, where $\tilde{c}$ is not the size of a minimum vertex cover of $\overline{G[C]}$. Since this cannot be done in time $f(c) \cdot n^{O(1)}$, the algorithm does not yield fixed-parameter tractability with respect to the parameter c.

Figure 1 (a) illustrates such a situation. Consider the case $c = 4$ with v as pivot. No trimming takes place. This means that at the beginning of the enumeration stage, we may still remove up to $\tilde{c} = c - 1 = 3$ vertices from C to obtain a c-isolated clique. Since $C = N_+[v]$ forms a clique, the graph $\overline{G[C]}$ has only one minimum vertex cover, namely the empty set. This means that all subsets of $C \setminus \{v\}$ (v as pivot must not be eliminated from C) of size 3 or less are vertex covers that we would have to enumerate. Clearly, the number of such vertex covers is not only dependent on the size of the covers, but also on the size of $N_+[v]$. In our example, there are 8 such covers, and it is easy to see that we can increase the number of vertex covers simply by increasing the size of $N_+[v]$.

In contrast, enumeration of minimal vertex covers was shown to be *inclusion-minimally fixed parameter enumerable* [4]; in particular, all minimal solutions of size at most c can be enumerated in $O(2^c c^2 + m)$ time. So running time is not a problem here; however, we miss some c-isolated cliques when only considering minimal vertex covers. This is because we cannot simply discard a maximal clique that violates the isolation condition; it might have some subsets that *are* c-isolated. As an example, in Fig. 1 (a), the clique has 4 vertices and 16 outgoing edges and is thus not 4-isolated. However, two subsets ((b) and (c)) are cliques with 3 vertices and 11 outgoing edges, and thus are 4-isolated.

2.2   Repairing the Enumeration Stage

To cope with the problems described in Sect. 2.1, we propose a two-step approach for enumerating all maximal c-isolated cliques. First, we enumerate all *minimal* vertex covers and thus obtain maximal cliques in the candidate set C. Then, to also capture c-isolated cliques that are subsets of non-c-isolated cliques enumerated this way, for each of these cliques, we enumerate all maximal subsets that fulfill the isolation condition. The problem ISOLATED CLIQUE SUBSET of finding these c-isolated subsets is then: given a graph $G = (V, E)$ and a clique $C \subseteq V$, find a set $C' \subseteq C$ that forms a c-isolated vertex set, that is, a set

	procedure isolated-subset$(C, c, x_{\min})$
Input:	A clique $C = \{v_1, v_2, \ldots, v_k\}$ with vertices sorted by degree, an isolation factor c and a minimum number $x_{\min}$ of outgoing edges from each vertex.
Output:	The set of maximal c-isolated cliques $\mathcal{C}$ in C.

```
1:      foreach v ∈ C: x(v) := deg(v) − |C| − 1 − x_min
2:      ĉ := c − x_min
3:      e(C) := (∑_{v∈C} x(v)) − ĉ · |C| + 1
4:      D := {∅}, C := ∅
5:      repeat ĉ times
6:          foreach D ∈ D
7:              if C \ D is a c-isolated clique then C := C ∪ {C \ D}
8:              else
9:                  if D = ∅ then i := k + 1 else i := min_{v_l ∈ D}{l}
10:                     D := D ∪ {D ∪ {v_j} | k − ĉ < j < i}
11:                 D := D \ {D}
12:     return C
```

Fig. 2. Algorithm for enumerating maximal c-isolated subsets of a clique C

with less than $c \cdot |C'|$ outgoing edges. The difficulty is in doing this fast enough, in particular with the running time depending only polynomially on $|C|$. For this (Theorem 1), the key is the following lemma, which reduces the choices of which vertices to omit from C.

Lemma 1. *Given a clique C with $|C| = k$, every maximal c-isolated subset of C is a superset of C^{k-c+1}, where C^{k-c+1} is the set of the $k - c + 1$ vertices with lowest index in C.*

According to Lemma 1, we may only remove vertices from the $c - 1$ vertices in $C \setminus C^{k-c+1}$ to obtain maximal c-isolated subsets of C. Hence, there are 2^{c-1} subsets of $C \setminus C^{k-c+1}$, and we enumerate maximal c-isolated subsets of C by generating the subsets of $C \setminus C^{k-c+1}$ in order of increasing cardinality and testing for each generated set whether its removal from C yields a maximal c-isolated subset. In this way, we can avoid examining supersets of removal sets for which a c-isolated clique was already output, since they would yield non-maximal cliques. The algorithm is shown in Fig. 2. Note that in lines 1–2 we compute an equivalent instance of ISOLATED CLIQUE SUBSET with isolation factor $\hat{c}$ by decreasing the number of outgoing edges of each vertex by $x_{\min}$, where $x_{\min}$ is the minimum number of outgoing edges from each vertex in C. The computation of line 3 derives the number of outgoing edges above the threshold allowed by the isolation condition. With this, the condition in line 7 can be tested in constant time. This is needed to obtain the running time as claimed by the following theorem.

Theorem 1. *Given an instance of ISOLATED CLIQUE SUBSET with at least $x_{\min}$ outgoing edges from each vertex, all of the at most $O(2^{c-x_{\min}})$ maximal solutions can be enumerated in $O(2^{c-x_{\min}} + |C|)$ time.*

We now describe how to use Theorem 1 to obtain a correct pivot procedure. Our modified pivot procedure differs from the original procedure only in the enumeration stage and in the screening stage. The enumeration stage is divided into two steps: the enumeration of maximal cliques and the enumeration of maximal subsets that fulfill the isolation condition for each of those cliques. Using Theorem 1, we can upper-bound the running time of the enumeration stage.

Lemma 2. *Given a graph $G = (V, E)$, a vertex $v \in V$, a set $C \subseteq N_+[v]$, and an isolation factor c, there are at most $2^{c-1} \cdot c$ maximal c-isolated cliques with pivot v, and they can be enumerated in $O(2^c \cdot c^2 \cdot m(C))$ time, where $m(C)$ is the number of edges in $G[C]$.*

In the screening stage, we filter non-maximal cliques by $O(4^c \cdot c^3)$ pairwise comparisons. Since the cliques obtained in the enumeration stage have size at most $\deg_G(v)$, these comparisons can be performed in $O(4^c \cdot c^3 \cdot \deg_G(v))$ time. With the running times of the stages of the pivot procedure for pivot v we can upper-bound the running time of the whole algorithm:

Theorem 2. *All maximal c-isolated cliques of a graph can be enumerated in $O(4^c \cdot c^3 \cdot m)$ time.*

2.3 Improved Screening of Cliques

In addition to fixing Ito et al.'s algorithm [9], we present an improved screening stage. While we can improve the asymptotic running time derived in Theorem 2 only slightly, the improvement facilitates parallelization of the enumeration algorithm and allows an exponential speedup for a variant of isolated clique enumeration to be presented in Sect. 3.2. More precisely, instead of a brute-force all-pairwise comparison, we achieve a simple and efficient test for checking whether an enumerated clique is subset of a clique with a different pivot.

Lemma 3. *A c-isolated clique C with pivot v is subset of a c-isolated clique C' with pivot $u \neq v$ iff $u \in N_-(v)$ and $N(u) \supseteq C$.*

Proof. We prove both directions separately. If $C' \supsetneq C$ is a clique with pivot u, then u must be adjacent to all vertices in C, in particular $u \in N_+[v]$. Since u is the pivot of C', it has lower index than v and thus $u \in N_-(v)$.

If there is a vertex $u \in N_-(v)$ that is adjacent to all vertices in C, then $C \cup \{u\}$ is a clique and a superset of C. It is furthermore c-isolated, since with u we have added a vertex with less than c outgoing edges (because $u < v$). Also, u is its pivot, again because $u < v$. $\qquad\square$

According to Lemma 3, we can replace the pairwise comparisons between cliques enumerated in previous calls of the pivot procedure and those of the current call for pivot v with a simple test that looks for vertices in $N_-(v)$ that are adjacent to all vertices of an enumerated clique. This test takes $O(c \cdot |C|)$ time. Since the enumerations of cliques for different pivots now run completely independent from each other, we can parallelize our algorithm by executing the pivot procedures for different pivot vertices on up to n different processors. Unfortunately, the asymptotic running time derived in Theorem 2 remains largely unchanged, since there are still $O(4^c c^2)$ pairwise comparisons between cliques for a single pivot; however, we save a factor of c and there is also a conceivable speedup in practice since we significantly reduce the number of brute-force set comparisons.

3   Alternative Isolation Concepts

Since isolation is not merely a means of developing efficient algorithms for the enumeration of cliques but also a trait in its own right, it makes sense to consider varying degrees of isolation. For instance, this is useful for the enumeration of isolated dense subgraphs for the identification of communities, which play a strong role in the analysis of biological and social networks [13].

In this context, the definition of c-isolation is not particularly tailored to these applications and we propose two alternative isolation concepts. One of them, min-c-isolation, is a weaker notion than c-isolation and the other, max-c-isolation, is a stronger notion than c-isolation. For both isolation concepts, we achieve a considerable speedup in the exponential part of the running time.

3.1   Minimum Isolation

Min-c-isolation is a weaker concept of isolation than the previously defined c-isolation, since we only demand that a set contains at least one vertex with less than c outgoing edges.

Definition 1. *Given a graph $G = (V, E)$ and a vertex set $S \subseteq V$ of size k, S is* min-c-isolated *when there is at least one vertex in S with less than c neighbors in $V \setminus S$.*

Obviously, every c-isolated set is also min-c-isolated. The enumeration of maximal min-c-isolated cliques consequently yields sets that are at least as large and often larger than c-isolated cliques.

The algorithm for the enumeration of maximal min-c-isolated cliques is mainly a simplification of the algorithm from Sect. 2. However, we lose linear-time solvability in the case of constant isolation factors c—the running time then becomes $O(n \cdot m)$. We use the same pivot definition and enumerate cliques for each possible pivot; from our definition of min-c-isolation it follows directly that the pivot of a min-c-isolated clique must have less than c neighbors outside of the clique. Subsequently, we point out the differences in the three main stages of the pivot procedure.

In the trimming stage, we start with $C := N[v]$ as candidate set. After trimming, we can assume that every vertex u that was not removed has at least $|C| - c$ neighbors in C. In the enumeration stage, we simply enumerate minimal vertex covers in $\overline{G[C]}$ of size at most $\tilde{c}$, where $\tilde{c}$ is the number of vertices that can still be removed from the candidate set C. For each enumerated minimal vertex cover D, the set $C \setminus D$ is a maximal min-c-isolated clique. Hence, we need not test for maximality, but the enumerated cliques might contain a vertex with lower index than v, since we have not necessarily removed all vertices from $N_-(v)$. If a clique C' features a vertex with lower index than v, then C' is removed from the output. Compared to Theorem 2, the fact that we do not perform any maximality test results in an improved exponential part of the running time.

Theorem 3. *All maximal min-c-isolated cliques of a graph can be enumerated in $O(2^c \cdot c \cdot m + n \cdot m)$ time.*

3.2   Maximum Isolation

Compared to c-isolation, max-c-isolation is a stronger notion. This results in most cases in the enumeration of smaller cliques for equal values of c.

Definition 2. *Given a graph $G = (V, E)$ and a vertex set $S \subseteq V$ of size k, S is max-c-isolated if every vertex $v \in S$ has less than c neighbors in $V \setminus S$.*

This isolation concept is especially useful for graphs where the vertices have similar degrees. Consider for example a graph in which all vertices have the same degree. Here, the notions c-isolation and max-c-isolation become equivalent for cliques, but max-c-isolation allows a better worst-case running time.

We apply the algorithm scheme presented in Sect. 2, that is, for every vertex $v \in V$ we enumerate all maximal max-c-isolated cliques with pivot v.

Trimming Stage. We compute a candidate set $C \subseteq N_+[v]$ by removing every vertex from $N_+[v]$ that cannot be in a maximum max-c-isolated clique with pivot v.

Enumeration Stage. In this stage, we enumerate max-c-isolated cliques $C' \subseteq C$ with pivot v. As in Sect. 2, we first enumerate maximal cliques in C via enumeration of minimal vertex covers of size at most $\tilde{c}$ in $\overline{G[C]}$, where $\tilde{c}$ is the number of vertices that can still be removed from the candidate set C. The cliques thus obtained may violate the isolation condition, since they may contain vertices with too many outgoing edges. We can restore the isolation condition for each enumerated clique by simply removing these vertices. This is done until the resulting clique is either max-c-isolated or we have removed more than $\tilde{c}$ vertices. In the latter case we discard the clique. The remaining enumerated cliques are not necessarily maximal, and therefore non-maximal cliques must be removed from the output in the screening stage.

Screening Stage. There are two possibilities for an enumerated clique C to be non-maximal. First, it can be proper subset of another max-c-isolated clique with pivot v. Second, it can be proper subset of a max-c-isolated clique with pivot $u < v$. For the first possibility, we test whether there is a set of vertices $D \subseteq N_+[v] \setminus C$ such that $C \cup D$ is a max-c-isolated clique. Clearly, D has to form a clique and all its vertices have to be adjacent to all vertices in C. Furthermore, whenever D contains a vertex u with degree $|C| + c + x$, then $|D|$ must have size at least $x + 1$. Otherwise, $C \cup D$ is not max-c-isolated, because u has at least c outgoing edges from $C \cup D$. Hence, we test for all $0 \leq x < c - 1$ whether the set

$$D^x := \{w \in N_+[v] \setminus C \mid C \subseteq N(w) \wedge \deg(w) \leq |C| + c + x\}$$

contains a clique of size at least $x + 1$. If this is not the case for any x, then C is a maximal max-c-isolated clique for pivot v. Otherwise, C is removed from the output.

It remains to check whether C is a proper subset of a clique with another pivot $u < v$. This can be tested in the manner described in Sect. 2.3. The running time of the pivot procedure is dominated by the first maximality test of the screening stage. For each of the $O(2^c)$ enumerated cliques, we have to solve MAXIMUM CLIQUE up to c times. Since $|D^x| < c$ for all $0 \leq x < c - 1$, this can be done in $O(1.22^c)$ time [14]. The overall running time of this test is then

$$O(2^c \cdot 1.22^c \cdot c) = O(2.44^c \cdot c).$$

The running time of the whole enumeration can be bounded in a similar way as in Sect. 2.2.

Theorem 4. *All maximal max-c-isolated cliques of a graph can be enumerated in $O(2.44^c \cdot c \cdot m)$ time.*

4 Enumerating Isolated s-Plexes

In many applications such as social network analysis, cliques have been criticized for their overly restrictive nature or modelling disadvantages. Hence, more relaxed concepts of dense subgraphs such as s-plexes [15, 1] are of interest.

An s-plex is a degree-based relaxation of the clique concept. In a graph $G = (V, E)$, a subset of vertices $S \subseteq V$ of size k is called an s-plex if the minimum degree in $G[S]$ is at least $k - s$. It has been shown that the problem of deciding whether a graph has an s-plex of size k is NP-complete [1]. We strengthen this by the corresponding parameterized hardness result. The parameter-preserving reduction from CLIQUE is given in the full version of this paper. It shows that MAXIMUM s-PLEX is W[1]-hard with respect to the combined parameter (s, k) and thus also if parameterized only by either one of s and k. Therefore, as for CLIQUE, we rather consider isolation as parameter in terms of studying fixed-parameter tractability.

We present an algorithm for the enumeration of maximal min-c-isolated s-plexes that runs in FPT time with respect to parameter c for any constant s. In this paper, we have chosen to consider only min-c-isolation, since the enumeration algorithm is easier to describe with this isolation concept. A min-c-isolated s-plex S contains at least one vertex that has less than c neighbors in $V \setminus S$. Compared to the enumeration of maximal min-c-isolated cliques, we face two obstacles when enumerating maximal min-c-isolated s-plexes. First, we cannot use the algorithm for the enumeration of minimal vertex covers, since an s-plex does not necessarily induce an independent set in the complement graph. Instead, since in an s-plex S of size k every vertex $v \in S$ is adjacent to at least $k - s$ vertices, the subgraph induced by S in the complement graph $\overline{G}[S]$ is a graph with maximum degree at most $s - 1$. Consider therefore the following generalization of a vertex cover:

Definition 3. *Given a graph G and a nonnegative integer d, we call a subset of vertices $S \subseteq V$ a max-deg-d deletion set if $G[V \setminus S]$ has maximum degree at most d.*

The idea is to enumerate maximal s-plexes in G by enumerating minimal max-deg-d deletion sets in $\overline{G}$. We present a fixed-parameter algorithm for the enumeration of minimal max-deg-d deletion sets that uses the size of the solution sets as parameter.

Finding a minimum max-deg-d deletion set was also considered by Nishimura et al. [12], who presented an $O((d + k)^{k+3} \cdot k + n(d + k))$ time algorithm for the decision version. We improve the exponential part of this running time while also covering the enumeration version. The idea is to pick a vertex v with more than d neighbors and then branch into $d + 2$ cases corresponding to the deletion of v or the deletion of one of the $d + 1$ first neighbors of v:

Lemma 4. *Given a graph G and an integer k, all minimal max-deg-d deletion sets of size at most k can be enumerated in $O((d + 2)^k \cdot (k + d)^2 + m)$ time.*

The second obstacle lies in the fact that given a pivot vertex v, maximal min-c-isolated s-plexes with pivot v are not necessarily a subset of $N_+[v]$, since they can contain up to $s - 1$ vertices that are not adjacent to v. We deal with this by enumerating all maximal min-c-isolated s-plexes for a given pivot *set* instead of a single pivot. The *pivot set* of a min-c-isolated s-plex is defined as the set that contains the pivot vertex v of the s-plex and those vertices that belong to the s-plex but are not adjacent to v. The *pivot vertex* is defined as the vertex with lowest index among the vertices with less than c neighbors outside of the s-plex. There has to be at least one such vertex, since otherwise the condition of min-c-isolation would be violated, but it does not necessarily have to be the vertex with the lowest index of all vertices in the s-plex.

The enumeration algorithm also consists of the three stages. In the trimming stage, we build a candidate set C by removing vertices from $N(v)$ that obviously cannot belong to a min-c-isolated s-plex with pivot v. For each possible pivot set P with pivot v, we independently enumerate the maximal min-c-isolated s-plexes. This is done in the enumeration stage by first building the complement graph $\overline{G}[C \cup P]$ and then enumerating minimal max-deg-$(s - 1)$ deletion sets of size at most $c - 1$ in $\overline{G}[C \cup P]$. In the screening stage, we first test whether any of the enumerated min-c-isolated s-plexes contains a vertex $u < v$, where u has less than c neighbors outside of the s-plex. Then this s-plex has pivot u and not v and is therefore removed from the output. Finally we perform a maximality test and remove non-maximal min-c-isolated s-plexes from the output.

Theorem 5. *All maximal min-c-isolated s-plexes of a graph can be enumerated in $O((s + 1)^c \cdot (s + c) \cdot n^{s+1} + n \cdot m)$ time.*

Thus, for every fixed s, we obtain a fixed-parameter algorithm for enumerating all maximal min-c-isolated s-plexes with respect to the parameter c.

Acknowledgement. We thank H. Ito and K. Iwama (Kyoto) for making their manuscript [8] available to us and J. Guo (Jena) for the idea for Lemma 4.

References

1. Balasundaram, B., Butenko, S., Hicks, I.V., Sachdeva, S.: Clique relaxations in social network analysis: The maximum k-plex problem. Manuscript (2006)
2. Balasundaram, B., Butenko, S., Trukhanovzu, S.: Novel approaches for analyzing biological networks. Journal of Combinatorial Optimization 10(1), 23–39 (2005)
3. Butenko, S., Wilhelm, W.E.: Clique-detection models in computational biochemistry and genomics. European Journal of Operational Research 173(1), 1–17 (2006)
4. Damaschke, P.: Parameterized enumeration, transversals, and imperfect phylogeny reconstruction. Theoretical Computer Science 351(3), 337–350 (2006)
5. Downey, R.G., Fellows, M.R.: Parameterized Complexity. Springer, Heidelberg (1999)
6. Fernau, H.: On parameterized enumeration. In: Ibarra, O.H., Zhang, L. (eds.) CO-COON 2002. LNCS, vol. 2387, pp. 564–573. Springer, Heidelberg (2002)
7. Håstad, J.: Clique is hard to approximate within $n^{1-\epsilon}$. Acta Mathematica 182(1), 105–142 (1999)
8. Ito, H., Iwama, K.: Enumeration of isolated cliques and pseudo-cliques. Manuscript (August 2006)
9. Ito, H., Iwama, K., Osumi, T.: Linear-time enumeration of isolated cliques. In: Brodal, G.S., Leonardi, S. (eds.) ESA 2005. LNCS, vol. 3669, pp. 119–130. Springer, Heidelberg (2005)
10. Komusiewicz, C.: Various isolation concepts for the enumeration of dense subgraphs. Diplomarbeit, Institut für Informatik, Friedrich-Schiller Universität Jena (2007)
11. Niedermeier, R.: Invitation to Fixed-Parameter Algorithms. Oxford University Press, Oxford (2006)
12. Nishimura, N., Ragde, P., Thilikos, D.M.: Fast fixed-parameter tractable algorithms for nontrivial generalizations of vertex cover. Discrete Applied Mathematics 152(1–3), 229–245 (2005)
13. Palla, G., Derényi, I., Farkas, I., Vicsek, T.: Uncovering the overlapping community structure of complex networks in nature and society. Nature 435(7043), 814–818 (2005)
14. Robson, J.M.: Algorithms for maximum independent sets. Journal of Algorithms 7(3), 425–440 (1986)
15. Seidman, S.B., Foster, B.L.: A graph-theoretic generalization of the clique concept. Journal of Mathematical Sociology 6(1), 139–154 (1978)

Alignments with Non-overlapping Moves, Inversions and Tandem Duplications in $O(n^4)$ Time

Christian Ledergerber and Christophe Dessimoz

ETH Zurich, Institute of Computational Science, Switzerland
ledergec@student.ethz.ch, cdessimoz@inf.ethz.ch

Abstract. Sequence alignment is a central problem in bioinformatics. The classical dynamic programming algorithm aligns two sequences by optimizing over possible insertions, deletions and substitution. However, other evolutionary events can be observed, such as inversions, tandem duplications or moves (transpositions). It has been established that the extension of the problem to move operations is NP-complete. Previous work has shown that an extension restricted to non-overlapping inversions can be solved in $O(n^3)$ with a restricted scoring scheme. In this paper, we show that the alignment problem extended to non-overlapping moves can be solved in $O(n^5)$ for general scoring schemes, $O(n^4 \log n)$ for concave scoring schemes and $O(n^4)$ for restricted scoring schemes. Furthermore, we show that the alignment problem extended to non-overlapping moves, inversions and tandem duplications can be solved with the same time complexities. Finally, an example of an alignment with non-overlapping moves is provided.

1 Introduction

In computational biology, alignments are usually performed to identify the characters that have common ancestry. More abstractly, alignments can also be represented as edit sequences that transform one sequence into the other under operations that model the evolutionary process. Hence, the problem of aligning two sequences is to find the most likely edit sequence, or equivalently, under an appropriate scoring scheme, the highest scoring edit sequence.

Historically, the only edit operations allowed were insertions, deletions and substitutions of characters, which we refer to as standard edit operations. The computation of the optimal alignment with respect to standard edit operations is well understood [1], and commonly used. But in some cases, standard edit operations are not sufficient to accurately model gene evolution. To take into account observed phenomena such as inversions, duplications or moves (intragenic transpositions) of blocks of sequences [2], the set of edit operations must be extended correspondingly. Such extensions have been studied in the past and a number of them turned out to be hard [3]. In particular an extension to general move operations was shown to be NP-complete [4]. For the simple greedy

G. Lin (Ed.): COCOON 2007, LNCS 4598, pp. 151–164, 2007.
© Springer-Verlag Berlin Heidelberg 2007

algorithm presented in [4] it was shown by [5] that $O(n^{0.69})$ is an upper bound for the approximation factor. An efficient $O(\log^* n \log n)$ factor approximation algorithm for the general problem is presented in [6].

A number of results for the extension of the standard alignment including block inversions have been achieved. It is shown in [7] that sorting by reversals is NP-hard, but the complexity of alignment with inversions is still unknown. A restricted problem of non-overlapping inversions was proposed by [8] who found an $O(n^6)$ algorithm. This result was then further improved by [9,10,11,12] where [12] obtained an $O(n^3)$ algorithm for a restricted scoring scheme.

In this paper, we show that the alignment problem extended with non-overlapping moves and non-overlapping tandem duplications can be solved exactly in polynomial time, and provide algorithms with time complexity of $O(n^5)$ for general scoring schemes, $O(n^4 \log n)$ for concave scoring schemes and $O(n^4)$ for restricted scoring schemes.

Since the probability that k independent and uniformly distributed moves be non-overlapping decreases very rapidly[1], this restriction is only of practical interest for small k, that is, if such events are very rare. Convincing evidence that this is indeed the case can be found in [13]. They show that protein domain order is highly conserved during evolution. It is established in [13] that most domains cooccur with only zero, one or two domain families. Since a move operation of the more elaborate type such as $ABCD \rightarrow ACBD$ immediately implies that B cooccurs with three other domains, we conclude that move operations have to be rare. Furthermore, exon shuffling is highly correlated to domain shuffling [14,15,16] and hence cannot lead to a large amount of move operations. Finally, a number of move operations can be found in the literature [17,18]. As for tandem duplication events, articles on domain shuffling reveal that the most abundant block edit operations are tandem duplications where the duplicate stays next to the original [19,20].

In the next section, we present a rigorous definition of the two alignment problems solved here: an extension to non-overlapping moves and an extension to non-overlapping moves, inversions and tandem duplications. Then, we provide solutions to both problems. The last section presents the experimental results using the extension to non-overlapping moves.

2   Definition of the Problems and Preliminaries

2.1   Notation and Definitions

In the following, we will denote the two strings to be aligned with $S = s_1 \ldots s_n$ and $T = t_1 \ldots t_m$ where $|S| = n$ and $|T| = m$. The i-th character of S is $S[i]$ and $S[i..j] = s_{i+1} \ldots s_j$ (note the indices). Thus, if $j \leq i$, $S[i..j] = \lambda$. By this definition, $S[i..j]$ and $S[j..k]$ are disjoint. $\overline{S} = s_n \ldots s_1$ denotes the reverse of S and $\overline{S[i..j]} = \overline{S}[n - j..n - i]$ is the reverse of a substring, the substring of the reverse respectively (note the extension of the bar). Let us denote the score of

---

[1] For long sequences, this probability converges to $\frac{1}{(2k-1)!!} = \frac{1}{1\cdot3\cdot5\cdots2k-1}$.

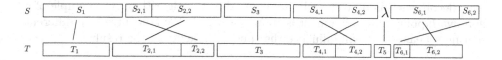

Fig. 1. Example of a non-overlapping move alignment of S with T

the standard alignment of S with T with $\delta(S,T)$. The score for substituting a character a with character b is denoted by an entry in the scoring matrix $\sigma(a,b)$. To simplify the definition of the alignment problems we introduce the concept of d-decompositions:

Definition 1. *Let a d-decomposition of a string S be a decomposition of S in d substrings such that the concatenation of all the substrings is equal to S. E.g. $S = S_1 \cdots S_d$. Let $\mathcal{M}_d(S)$ be the set of all d-decompositions of S.*

Note that S_i denotes a substring of a d-decomposition while s_i denotes a character. Let us further define the cyclic string to string comparison problem as introduced by [21]:

Definition 2. *The cyclic string comparison problem is to find the 2-decomposition $S_1 S_2 \in \mathcal{M}_2(S)$ and $T_1 T_2 \in \mathcal{M}_2(T)$ such that the score $\delta(S_1, T_2) + \delta(S_2, T_1)$ is maximal. The optimal score is denoted by $\delta_c(S, T)$.*

Due to the symmetry of the problem there exists always a two decomposition of $S = S_1 S_2$ such that $\delta_c(S, T) = \delta(S_2 S_1, T)$ as proven by [21].

Finally, we assume that the reader is familiar with the concept of edit graphs as defined for instance in [22] or [23].

2.2   Definition of Alignment with Non-overlapping Moves

Using d-decompositions and the cyclic string to string comparison problem we can now define the alignment with non-overlapping moves as follows.

Definition 3. *The problem of aligning S and T with non-overlapping moves is to find $d \in \mathbb{N}$ and d-decompositions of S and T such that the score $\sum_{i=1}^{d} \max\{\delta(S_i, T_i), \delta_c(S_i, T_i) + \sigma_c(l_{S_{i1}}) + \sigma_c(l_{S_{i2}}) + \sigma_c(l_{T_{i1}}) + \sigma_c(l_{T_{i2}})\}$ is maximal for all $d \in \mathbb{N}, S_1 \ldots S_d \in \mathcal{M}_d(S)$ and $T_1 \ldots T_d \in \mathcal{M}_d(T)$. Where $l_{S_{i1}}, l_{S_{i2}}, l_{T_{i1}}$ and $l_{T_{i2}}$ are the lengths of the blocks involved in the move operation and $\sigma_c(l)$ is a penalty function for move operations. The optimal score is denoted by $\delta_m(S, T)$.*

Note that substrings S_i, T_i may be empty. However, a substring needs to have a length of at least 2 to contain a move. In other words, we align d pairs of substrings of S and T and allow for each aligned pair of substrings at most one swap of a prefix with a suffix as defined by the cyclic string comparison problem. $\sigma_c(l_{S_1}) + \sigma_c(l_{S_2}) + \sigma_c(l_{T_1}) + \sigma_c(l_{T_2})$ is a penalty function for such a move operation and depends on the lengths of the four substrings involved in the move operation.

This decomposition in a sum will be required in the algorithm. An example of a non-overlapping move alignment is shown in Fig. 1. We now introduce different scoring schemes that will influence the time complexity of the results.

Definition 4. *General scoring scheme: the standard alignment of substrings is done with affine gap penalties, $\sigma_c(l)$ is an arbitrary function and the scoring matrix $\sigma(a, b)$ is arbitrary. Concave scoring scheme: the standard alignment of substrings is done with constant indel penalties, $\sigma_c(l)$ is a concave function and the scoring matrix $\sigma(a, b)$ is arbitrary. Restricted scoring scheme: the standard alignment of substrings is done with constant indel penalties and $\sigma_c(l)$ is a constant. The scoring matrix $\sigma(a, b)$ is selected such that the number of distinct values of $DIST[i, j] - DIST[i, j - 1]$ is bounded by a constant ψ. For more details on the restricted scoring scheme, we refer to [24].*

2.3 Definition of Alignment with Non-overlapping Moves, Inversions and Tandem-Duplications

In favor of simplicity, we assume constant indel penalties and constant penalties for block operations in the treatment of this problem. However, the scoring schemes of section 2.2 could be used here as well.

Definition 5. *The problem of aligning S and T with non-overlapping moves, reversals and tandem duplications is to find $d \in \mathbb{N}$ and d-decompositions of S and T such that the score $\sum_{i=1}^{d} \max\{\delta(S_i, T_i), \delta_c(S_i, T_i) + \sigma_c, \delta_d(S_i, T_i) + \sigma_d, \delta_r(S_i, T_i) + \sigma_r\}$ is maximal for all $d \in \mathbb{N}, S_1 \ldots S_d \in \mathcal{M}_d(S)$ and $T_1 \ldots T_d \in \mathcal{M}_d(T)$, where $\delta_d(A, B) = \max\{\delta(AA, B), \delta(A, BB)\}$ and $\delta_r(A, B) = \delta(A, \overline{B})$. Where $\sigma_c, \sigma_d, \sigma_r$ are penalties for move operations, duplications or reversals respectively. The optimal score is denoted by $\delta_{drm}(S, T)$.*

2.4 Other Preliminaries

The notion of $DIST[i, j]$ arrays as used in [24,25] can be defined as follows.

Definition 6. *Let $DIST_{S,T}[i, j], 0 \le i \le j \le m$ denote the score of the optimal alignment of $T[i..j]$ with S.*

Let us further introduce input vectors I, output vectors O and a matrix OUT.

Definition 7. *Let $OUT_{S,T}[i, j] = I[i] + DIST_{S,T}[i, j] + \sigma_c(j - i), 0 \le i \le j \le m$. Then I is an arbitrary vector called input vector and $O[j] = \max_i OUT[i, j]$ is called output vector containing all the column maxima of OUT.*

Lemma 1. *$DIST_{S,T}[i, j]$ arrays are inverse monge arrays.*

The following lemma will become useful in the selection of the parameters.

Lemma 2. *If $f(l)$ is concave then $f_l(j', j) := f(j - j'), 0 \le j' \le m, 0 \le j \le m$ is inverse Monge.*

A proof of these lemmas can be found with the definition of inverse Monge in the *Appendix*.

Corollary 1. $OUT_{S,T}[i,j] = DIST_{S,T}[i,j] + f(j-i) + I[i]$ *is inverse Monge for f concave and constant indel penalties.*

Proof. Due to lemma 1 $DIST_{S,T}$ arrays with constant indel penalties are inverse Monge. The rest follows from definition 8 and lemma 2 (in *Appendix*).

Using our observations and the results from [24,25], we can conclude with the following results: (i) For arbitrary penalty functions σ_c and affine gap penalties as in the general scoring scheme, we can compute $DIST_{S[0..l],T}$ from $DIST_{S[0..l-1],T}$ in $O(m^2)$ as indicated in the *Appendix*. Then we can trivially compute the output vector O as in definition 7 in $O(m^2)$ time by inspecting all entries. (ii) For concave functions σ_c and constant indel penalties as in the concave scoring scheme, we can compute a representation of $DIST_{S[0..l],T}$ from $DIST_{S[0..l-1],T}$ in $O(m \log m)$ time using the data structure of [25]. Then since OUT is inverse Monge, we can compute the output vector O by applying the algorithm of [26] for searching all column maxima in a Monge array to OUT. This algorithm will access $O(m)$ entries of the array and hence the computation of O will take $O(m \log m)$ time since we can access an entry of $DIST$ in the data structure of [25] in $O(\log m)$ time. (iii) For constant functions σ_c, constant indel penalties and a restricted scoring matrix as in the restricted scoring scheme, we can compute a representation of $DIST_{S[0..l],T}$ from $DIST_{S[0..l-1],T}$ in $O(m)$ time due to section 6 of [25] and then compute the output vector O using the algorithm of [24] in $O(m)$ time.

Note that the $O(m \log m)$ and $O(m)$ results rely heavily on the fact that $DIST$ arrays are Monge. Since this is not true for affine gap penalties these results cannot be easily extended to affine gap penalties.

3 Algorithms

3.1 Alignment with Non-overlapping Moves

Let $SCO_{S,T}[i,j]$ be the score of the optimal alignment of $S[0..i]$ and $T[0..j]$ with non-overlapping moves. Then the following recurrence relation and initialization of the table will lead to a dynamic programming solution for the problem.

Base Case: $SCO_{S,T}[i,0] = i \cdot \sigma_I$ and $SCO_{S,T}[0,j] = j \cdot \sigma_I$

$$\text{Recurrence: } SCO_{S,T}[i,j] = \max \begin{cases} SCO_{S,T}[i,j-1] + \sigma_I \\ SCO_{S,T}[i-1,j-1] + \sigma(S[i],T[j]) \\ SCO_{S,T}[i-1,j] + \sigma_I \\ MOVE \end{cases}$$

where

$$MOVE = \max_{0\le i'<i,0\le j'<j}\{\ SCO_{S,T}[i',j'] + \delta_c(S[i'..i],T[j'..j]) +$$

$$\sigma_c(l_{S_{d1}}) + \sigma_c(l_{S_{d2}}) + \sigma_c(l_{T_{d1}}) + \sigma_c(l_{T_{d2}})\}$$

Proof. Let us consider an optimal non-overlapping move alignment of $S[0..i]$ with $T[0..j]$. Let S_d and T_d be the last substrings of the optimal d-composition of $S[0..i]$ and $T[0..j]$. Then there are two cases: (1) S_d and T_d are aligned using the cyclic string comparison or (2) S_d and T_d are aligned by the standard alignment. In case (1), we know that $SCO_{S,T}[i,j] = SCO_{S,T}[i',j'] + \delta_c(S_d,T_d)$ which is considered in $MOVE$. In case (2), we are in the usual standard alignment cases. Hence, we consider all the cases and therefore find the optimal solution.

With the goal of economizing the computation of the table let us rewrite $MOVE$ as

$$\max_{\substack{0\ \le\ i'\ <\ i''\ <\ i \\ 0\ \le\ j'\ <\ j''\ <\ j}}\{SCO_{S,T}[i',j'] + DIST_{S[i''..i],T}[j',j''] + \sigma_c(j''-j') + \sigma_c(i-i'')$$

$$+\ DIST_{S[i'..i''],T}[j'',j] + \sigma_c(j-j'') + \sigma_c(i''-i')\}.$$

To compute $MOVE$ for a given i' and i'' we can first maximize over j' and then over j''. That is, we can first compute the output row of the first $DIST_{S[i''..i],T}$ array and then, given that output, compute the output of the second $DIST_{S[i'..i''],T}$ array. This leads to the following definitions (illustrated in Fig. 2).

$$O_1[j''] = \max_{0\le j'<j''} SCO_{S,T}[i',j']+DIST_{S[i''..i],T}[j',j'']+\sigma_c(j''-j')+\sigma_c(i-i'') \quad (1)$$

$$O_2[j] = \max_{0\le j''<j} O_1[j''] + DIST_{S[i'..i''],T}[j'',j] + \sigma_c(j-j'') + \sigma_c(i''-i') \quad (2)$$

Given $DIST_{S[i''..i],T}[j',j'']$ and $DIST_{S[i'..i''],T}[j'',j]$, $O_1[j'']$ and $O_2[j]$ can be computed efficiently using the results from section 2.4 since both of them are output vectors as in definition 7.

DP_MOVE

1: **for all** i,j such that $0\le i\le n, 0\le j\le m$ **do**
2:     {base case}
3:     $SCO[i,0] := i\cdot\sigma_I$
4:     $SCO[0,j] := j\cdot\sigma_I$
5:     $SCO[i,j] := -\infty$ if $i\ne 0, j\ne 0$
6: **end for**
7: **for** i from 0 to n **do**
8:     **if** $i\ge 1$ **then**
9:         **for** j from 1 to m **do**
10:            {standard alignment recurrence}
11:            $SCO[i,j] := \max\{SCO[i,j], SCO[i-1,j] + \sigma_I, SCO[i,j-1] + \sigma_I,$
               $SCO[i-1,j-1] + \sigma(S[i],T[j])\}$

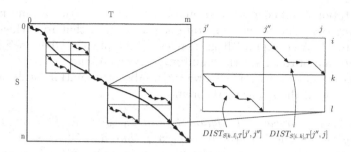

Fig. 2. An illustration of the computation of a move operation in DP_MOVE. Since the scores are additive: $SCO[l, j] = SCO[i, j'] + DIST_{S[k..l],T}[j', j''] + DIST_{S[i..k],T}[j'', j]$. In DP_MOVE this is maximized for all $i < k < l, j' < j'' < j$.

```
12:      end for
13:    end if
14:    for k from i to n do
15:       {move operations}
16:       DIST_{S[i..k],T} := calcDist(DIST_{S[i..k-1],T})
17:       for l from k to n do
18:          DIST_{S[k..l],T} := calcDist(DIST_{S[k..l-1],T})
19:          O_1 := calcOutput(OUT[j', j''] = SCO[i, j'] + DIST_{S[k..l],T}[j', j''] +
                 σ_c(j'' - j') + σ_c(l - k))
20:          O_2 := calcOutput(OUT[j'', j] = O_1[j''] + DIST_{S[i..k]}[j'', j] + σ_c(j -
                 j'') + σ_c(k - i))
21:          for j from 0 to m do
22:             SCO[l, j] := max{SCO[l, j], O_2[j]}
23:          end for
24:       end for
25:    end for
26: end for
```

Where $calcDist(DIST_{S[0..l-1],T})$ computes $DIST_{S[0..l],T}$ from $DIST_{S[0..l-1],T}$ and $calcOutput(OUT[i, j])$ computes O as in definition 7.

Correctness. To show the correctness of the algorithm it suffices to show that we process all edges in the edit graph and whenever we process an edge $(u, v) \in E$ we have completed the computation of the score of u and any of its predecessors in topological order [22]. The computation of the score of a node u is completed iff all the incoming edges of u have been processed. This can be proven by induction. In our edit graph, the only edges are either due to the standard alignment, or due to move operations, as can be seen in the recurrence. Assuming that when computing the i-th row of SCO, all edges due to move operations starting in a row $i' < i$ have already been processed and the computation of any node (i', j) with $i' < i$ has been completed, we can see that the processing of the edges due to the standard alignment recurrence ending in the i-th row as done on line 9 to 12 is legitimate. After having processed those edges, we have completed the computation of all edges ending on any node in the i-th row and hence can compute any edge due to move operations starting on that row which is done on line 14 to 24. We compute all such edges. Consequently, when we advance to the computation of row $i + 1$ the assumption is again true.

Using the results from section 2.4 we can analyze the runtime of the algorithm and conclude with the following theorem.

Theorem 1. *The problem of aligning S and T, $|S| = n, |T| = m$, with non-overlapping moves can be solved in $O(n^3 m^2)$ time and $O(nm + m^2)$ space for general scoring schemes, in $O(n^3 m \log m)$ time and $O(nm + m^2)$ space for concave scoring schemes and in $O(n^3 m)$ time and $O(nm)$ space for restricted scoring schemes.*

3.2   Alignment with Non-overlapping Moves, Inversions, and Tandem Duplications

The dynamic programming recurrence of non-overlapping move operations extends nicely to this problem.

Base Case: $SCO_{S,T}[i, 0] = i \cdot \sigma_I$ and $SCO_{S,T}[0, j] = j \cdot \sigma_I$

$$\text{Recurrence: } SCO_{S,T}[i,j] = \max \begin{cases} SCO_{S,T}[i, j-1] + \sigma_I \\ SCO_{S,T}[i-1, j-1] + \sigma(S[i], T[j]) \\ SCO_{S,T}[i-1, j] + \sigma_I \\ MOVE + \sigma_c \\ S\_DUPLICATE + \sigma_d \\ T\_DUPLICATE + \sigma_d \\ REVERSE + \sigma_r \end{cases}$$

where

$$MOVE = \max_{0 \le i' < i, 0 \le j' < j} \{ SCO_{S,T}[i', j'] + \delta_c(S[i'..i], T[j'..j]) \}$$

$$S\_DUPLICATE = \max_{0 \le i' < i, 0 \le j' < j'' < j} \{ SCO_{S,T}[i', j'] + \delta(S[i'..i], T[j'..j'']) $$
$$+ \delta(S[i'..i], T[j''..j]) \}$$

$$T\_DUPLICATE = \max_{0 \le i' < i'' < i, 0 \le j' < j} \{ SCO_{S,T}[i', j'] + \delta(S[i'..i''], T[j'..j]) $$
$$+ \delta(S[i''..i], T[j'..j]) \}$$

$$REVERSE = \max_{0 \le i' < i, 0 \le j' < j} \{ SCO_{S,T}[i', j'] + \delta(\overline{S[i'..i]}, T[j'..j]) \}$$

A proof of this recurrence is analogous to the proof for non-overlapping moves and is omitted.

We have split tandem duplication into tandem duplication of a substring of S and tandem duplication of a substring of T. We have already shown how $MOVE$ can be treated and in [11] it is shown how to handle $REVERSE$. $S\_DUPLICATE$ can be done as follows. We calculate $DIST_{S[i..k]}$. Then, we

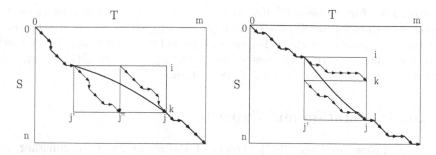

Fig. 3. An illustration on how duplications are treated

first use $SCO_{S,T}[i,j']$ as input vector for $DIST_{S[i..k]}$ to get $O_1[j']$ and then use $O_1[j']$ as input for $DIST_{S[i..k]}$ to get $O_2[j]$. $T\_DUPLICATE$ can be computed by computing the output vector of $DIST_{S[i..k],T}[j',j] + DIST_{S[k..l],T}[j',j]$ for the input vector $SCO_{S,T}[i,j']$. Note, that this array is well defined and is again inverse Monge because it is a sum of two inverse Monge arrays. Using these observations which are illustrated in Fig. 3 we can now present our algorithm for this problem.

DP_MOVE_INV_DUPL

1: {initialize the table as in DP_MOVE}
2: **for** i from 0 to n **do**
3:    {compute REVERSE as done in [12]}
4:    {compute the standard alignment recurrence as in DP_MOVE}
5:    {treat MOVE as done in DP_MOVE}
6:    **for** k from i to n **do**
7:       {duplication}
8:       $DIST_{S[i..k],T} := calcDist(DIST_{S[i..k-1],T})$
9:       $O_1 := calcOutput(OUT[j',j''] = SCO[i,j'] + DIST_{S[i..k],T}[j',j''])$
10:       $O_2 := calcOutput(OUT[j'',j] = O_1[j''] + DIST_{S[i..k],T}[j'',j])$
11:       **for** l from k to n **do**
12:          $DIST_{S[k..l],T} := calcDist(DIST_{S[k..l-1],T})$
13:          $O := calcOutput(OUT[j',j] = SCO[i,j'] + DIST_{S[i..k],T}[j',j] + DIST_{S[k..l],T}[j',j])$
14:          **for** j from 0 to m **do**
15:             $SCO[l,j] := \max\{SCO[l,j], O[j] + \sigma_d, O_2[j] + \sigma_d\}$
16:          **end for**
17:       **end for**
18:    **end for**
19: **end for**

A proof of the correctness of the algorithm is analogous to the proof for DP_MOVE. This proof however reveals that it is important to process edges due to REVERSE *before* the standard alignment recurrence.

Theorem 2. *The problem of aligning S and T with non-overlapping moves, reversals and tandem duplications can be solved in $O(n^3m^2)$ time and $O(nm + m^2)$ space for general scoring schemes, in $O(n^3m\log m)$ time and $O(mn + m^2)$ space for concave scoring schemes and in $O(n^3m)$ time and $O(nm)$ space for restricted scoring schemes, where $n = |S|$ and $m = |T|$.*

4   Implementation and Experiments

We have implemented an $O(n^5)$ version of the algorithm with constant gap penalties in C^2. This implementation has proven useful for aligning sequences of up to about 400 AA, taking a few hours to compute the alignment. We have tested the algorithm on real data and were able to confirm a number of examples found in [2]. In addition we run the algorithm on an example found in [18]. This alignment is shown in figure 4 and is compared with a standard alignment obtained from Darwin [27].

```
DP_MOVE
Seq1: RPSTVPLP_NTQ__A_LAMA_[GTAYKGYVKVP_KPTGVK_KGWQRAYAVVCDCKLFLYDLPEGK_STQPGVIASQVLDLRDDEFAVSSVLA
Seq2: LSSADNDPEDSQHSSLLSLTQ[DSVFEGWLSVPNKQNRRRGHGWKRQYVIVSSRKIIFYNSDIDKHNTTDAVL___ILDL_SKVYHVRSVTQ

Seq1: SDVIHATRRDIPCIFRVT_ASLLG_S__PSKTSSL_L_ILTENENEKRK|GP_KPKA
Seq2: GDVIRADAKEIPRIFQLLYAGE_GASHRPDEQSQLDVSVLHGNCNEERP|GTIVHKG

Seq1:                          PIPPE_QSKRP___LGVDVQ_RGI]WVGILEGLQAILHKNRLRSQVV_HVAQEAYD_S_SLPLI
Seq2:                          KLNHDPRSARDMLLLAATPEDQSL]WVARL__LKRI_QKSGYKAASYNNNSTDGSKISPSQSTR

DARWIN
Seq1: RPSTVPLPNTQALAMAGPKPKA                                                     PIPPEQSKRPLGVDVQRGIG
Seq2:           LSSADNDPEDSQHS__SLLSLTQ                                                             D

Seq1: TAYKGYVKVPKPTGVKK__GWQRAYAVVCDCKLFLY__DLPEGKSTQPGVIASQVLDLRDDEFAVSSVLASDVIHATRRDIPCIFRV_____
Seq2: SVFEGWLSVPNKQNRRRGHGWKRQYVIVSSRKIIFYNSDIDKHNTTD_____AVLILDL_SKVYHVRSVTQGDVIRADAKEIPRIFQLLYAGE

Seq1: _____TASLLGS
Seq2: GASHRPDEQSQLDVSVLHGNCNEERPGTIVHKG_____KLNHD

Seq1: PSKTSSLLILTENENEKRKWVGIL_____EGLQAILHKNRLRSQVVHVAQEAYDSSLPLI
Seq2: PRSARDMLLLAATPEDQSLWVARLLKRIQKSGYKAASYNNN_____STDGSKISPSQSTR
```

Fig. 4. An example found in [18]. In this figure we compare an alignment computed with our new algorithm and an alignment done with Darwin [27]. The brackets '[' and ']' indicate the boundary of the substrings containing a move operation and '|' marks the position of the split. Seq1: Q7TT49 AA 1005-1241; Seq2: Q9VXE3 AA 1112-1367. Using the annotation of SMART we have marked the domains involved in the move operation. Pleckstrin homology phospholipid binding domain is shown in blue, protein kinase C-type diacylglycerol binding domain is shown in yellow.

5   Conclusions

In this paper, we have presented a number of new alignment problems extending the notion of non-overlapping inversions to non-overlapping moves and tandem duplications. For all of them we found algorithms that solve the problems exactly and can be implemented to run in $O(n^2)$ space and $O(n^5)$, $O(n^4 \log n)$ or $O(n^4)$

---

^{2} Available from the authors upon request.

time depending on the scoring scheme used. We believe that this approach may yield new insights by finding the best alignment of two sequences, and think that it is justifiable due to the rarity of such events in nature. Using the implementation of the $O(n^5)$ variant of the algorithm, we were able to align previously identified cases of pairs of sequences with move operations. Furthermore, these experiments also showed the necessity of an $O(n^4 \log n)$ implementation to be applicable to large sequences, which are more likely to contain a move.

Acknowledgments

We thank Manuel Gil, Gaston H. Gonnet, Gina M. Cannarozzi and three anonymous referees for helpful critiques and discussions.

References

1. Needleman, S.B., Wunsch, C.D.: A general method applicable to the search for similarities in the amino acid sequence of two proteins. J. Mol. Biol. 48(3), 443–453 (1970)
2. Fliess, A., Motro, B., Unger, R.: Swaps in protein sequences. Proteins. 48(2), 377–387 (2002)
3. Lopresti, D., Tomkins, A.: Block edit models for approximate string matching. Theor. Comput. Sci. 181(1), 159–179 (1997)
4. Shapira, D., Storer, J.A.: Edit distance with move operations. In: Apostolico, A., Takeda, M. (eds.) CPM 2002. LNCS, vol. 2373, pp. 85–98. Springer, Heidelberg (2002)
5. Chrobak, M., Kolman, P., Sgall, J.: The greedy algorithm for the minimum common string partition problem. ACM Trans. Algorithms 1(2), 350–366 (2005)
6. Cormode, G., Muthukrishnan, S.: The string edit distance matching problem with moves. In: SODA '02. Proceedings of the thirteenth annual ACM-SIAM symposium on Discrete algorithms, Philadelphia, PA. Society for Industrial and Applied Mathematics, pp. 667–676. ACM Press, New York (2002)
7. Caprara, A.: Sorting by reversals is difficult. In: RECOMB '97. Proceedings of the first annual international conference on Computational molecular biology, pp. 75–83. ACM Press, New York (1997)
8. Schoeninger, M., Waterman, M.S.: A local algorithm for dna sequence alignment with inversions. Bull. Math. Biol. 54(4), 521–536 (1992)
9. Chen, Z.Z., Gao, Y., Lin, G., Niewiadomski, R., Wang, Y., Wu, J.: A space-efficient algorithm for sequence alignment with inversions and reversals. Theor. Comput. Sci. 325(3), 361–372 (2004)
10. do Lago, A.P., Muchnik, I.: A sparse dynamic programming algorithm for alignment with non-overlapping inversions. Theoret. Informatics Appl. 39(1), 175–189 (2005)
11. Alves, C.E.R., do Lago, A.P., Vellozo, A.F.: Alignment with non-overlapping inversions in $o(n^3 \log n)$ time. In: Proceedings of GRACO 2005. Electronic Notes in Discrete Mathematics, vol. 19, pp. 365–371. Elsevier, Amsterdam (2005)
12. Vellozo, A.F., Alves, C.E.R., do Lago, A.P.: Alignment with non-overlapping inversions in $o(n^3)$-time. In: Bücher, P., Moret, B.M.E. (eds.) WABI 2006. LNCS (LNBI), vol. 4175, pp. 186–196. Springer, Heidelberg (2006)

13. Apic, G., Gough, J., Teichmann, S.A.: Domain combinations in archaeal, eubacterial and eukaryotic proteomes. J. Mol. Biol. 310(2), 311–325 (2001)
14. Kaessmann, H., Zöllner, S., Nekrutenko, A., Li, W.H.: Signatures of domain shuffling in the human genome. Genome Res. 12(11), 1642–1650 (2002)
15. Liu, M., Walch, H., Wu, S., Grigoriev, A.: Significant expansion of exon-bordering protein domains during animal proteome evolution. Nucleic Acids Res. 33(1), 95–105 (2005)
16. Vibranovski, M.D., Sakabe, N.J., de Oliveira, R.S., de Souza, S.J.: Signs of ancient and modern exon-shuffling are correlated to the distribution of ancient and modern domains along proteins. J. Mol. Evol. 61(3), 341–350 (2005)
17. Bashton, M., Chothia, C.: The geometry of domain combination in proteins. J. Mol. Biol. 315(4), 927–939 (2002)
18. Shandala, T., Gregory, S.L., Dalton, H.E., Smallhorn, M., Saint, R.: Citron kinase is an essential effector of the pbl-activated rho signalling pathway in drosophila melanogaster. Development. 131(20), 5053–5063 (2004)
19. Andrade, M.A., Perez-Iratxeta, C., Ponting, C.P.: Protein repeats: structures, functions, and evolution. J. Struct. Biol. 134(2-3), 117–131 (2001)
20. Marcotte, E.M., Pellegrini, M., Yeates, T.O., Eisenberg, D.: A census of protein repeats. J. Mol. Biol. 293(1), 151–160 (1999)
21. Maes, M.: On a cyclic string-to-string correction problem. Inf. Process. Lett. 35(2), 73–78 (1990)
22. Myers, E.W.: An overview of sequence comparison algorithms in molecular biology. Technical Report 91-29, Univ. of Arizona, Dept. of Computer Science (1991)
23. Gusfield, D.: Algorithms on Strings, Trees, and Sequences: computer science and computational biology. Press Syndicate of the University of Cambridge, Cambridge (1997/1999)
24. Landau, G.M., Ziv-Ukelson, M.: On the common substring alignment problem. J. Algorithms 41(2), 338–354 (2001)
25. Schmidt, J.P.: All highest scoring paths in weighted grid graphs and their application to finding all approximate repeats in strings. SIAM J. Comput. 27(4), 972–992 (1998)
26. Aggarwal, A., Klawe, M.M., Moran, S., Shor, P., Wilber, R.: Geometric applications of a matrix-searching algorithm. Algorithmica 2(1), 195–208 (1987)
27. Gonnet, G.H., Hallett, M.T., Korostensky, C., Bernardin, L.: Darwin v. 2.0: An interpreted computer language for the biosciences. Bioinformatics 16(2), 101–103 (2000)
28. Monge, G.: Déblai et remblai. Mémoires de l'Académie Royale des Sciences (1781)

Appendix

Monge Properties

For a proof of lemma 1 and lemma 2 we have to define the notion of Monge arrays [28] first:

Definition 8. *A matrix $M[0\ldots n; 0\ldots m]$ is Monge if for all $i = 1\ldots n, j = 1\ldots m$*

$$M[i,j] + M[i-1,j-1] \leq M[i-1,j] + M[i,j-1]$$

and it is called inverse Monge if for all $i = 1\ldots n, j = 1\ldots m$

$$M[i,j] + M[i-1,j-1] \geq M[i-1,j] + M[i,j-1]$$

Then the proof for lemma 2 goes as follows.

Proof

$$f_l(j' - 1, j - 1) + f_l(j', j) = 2f(j - j')$$
$$\geq 2\frac{f(j - j' - 1) + f(j - j' + 1)}{2}$$
$$= f_l(j' - 1, j) + f_l(j', j - 1)$$

Where the inequality follows from the definition of concave. A function f is concave iff $f(tx + (1-t)y) \geq tf(x) + (1-t)f(y)$ holds for all $x, y \in \mathbb{R}, t \in [0, 1]$. In other words every point on a secant is below the function. In the proof we used $t = 1/2$ as shown in Fig. 5.

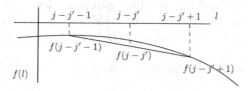

Fig. 5. For concave functions any point on a secant is below the function

Note that if any three points $f(j - 1), f(j), f(j + 1), 0 < j < m$ are not in concave position the resulting array will not be inverse Monge. That is, lemma 2 holds with equivalence if we restrict the definition of concave to values in $\{0 \ldots m + 1\} \subseteq \mathbb{N}$.

The proof for Lemma 1 is analogous to the proof in [25].

Proof The paths represented by $DIST_{S,T}[i - 1, j]$ and $DIST_{S,T}[i, j - 1]$ have to cross properly in a vertex v as shown in figure 6. Therefore, we have

$$DIST_{S,T}[i - 1, j] + DIST_{S,T}[i, j - 1] = (a + b) + (c + d)$$
$$= (a + d) + (b + c)$$
$$\leq g + f$$
$$= DIST_{S,T}[i, j] + DIST_{S,T}[i - 1, j - 1]$$

where the inequality follows from $a + d \leq f$ and $b + c \leq g$. Where $a + d \leq f$ holds since $a + d$ is the length of a path from $(0, i - 1)$ to $(n, j - 1)$ and the optimal path of length f can only be longer and analogously $b + c \leq g$.

In figure 6 2) a counter example for affine gap penalties is given. This is a counter example because the total number of gaps on the paths from $(0, i)$ to $(n, j - 1)$ and $(0, i - 1)$ to (n, j) is smaller than the total number of gaps on the paths from $(0, i)$ to (n, j) and $(0, i - 1)$ to $(n, j - 1)$. Hence $DIST_{S,T}$ arrays cannot be monge for large initial penalties.

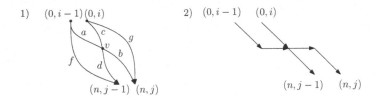

Fig. 6. 1) illustrates that the path from vertex $(0, i)$ to $(n, j - 1)$ and the path from vertex $(0, i - 1)$ to (n, j) in the grid graph have to cross in a common vertex v. 2) gives a counter example for affine gap penalties.

Extension of *DIST* Arrays

This simple algorithm is inspired by [25]. It is repeated here to provide an idea on how to extend our algorithms to affine gap penalties.

For the base cases we observe that $DIST_{S,T}[i, j] = B_{S,T[i..j]}[n, j - i]$ in particular for constant indel penalties $DIST_{S[0..0],T}[i, j] = (j - i) \cdot \sigma_I$ and $DIST_{S[0..l],T}[i, i] = l \cdot \sigma_I$.

By mapping the standard alignment recurrence to *DIST* arrays we obtain:

$$DIST_{S[0..l],T}[i, j] = \max \begin{cases} DIST_{S[0..l-1],T}[i, j] + \sigma_I \\ DIST_{S[0..l-1],T}[i, j - 1] + \sigma(S[l], T[j]) \\ DIST_{S[0..l],T}[i, j - 1] + \sigma_I \end{cases}$$

Therefore, we can compute $DIST_{S[0..l],T}$ given $DIST_{S[0..l-1],T}$ in $O(m^2)$ time. This recurrence can be extended to include affine gap penalties by mapping the more complicated recurrence for the standard alignment with affine gap penalties to *DIST* arrays.

Counting Minimum Weighted Dominating Sets

Fedor V. Fomin and Alexey A. Stepanov

Department of Informatics, University of Bergen, PO Box 7803, 5020 Bergen, Norway
fomin@ii.uib.no, ljosha@ljosha.org

Abstract. We show how to count all minimum weighted dominating sets of a graph on n vertices in time $\mathcal{O}(1.5535^n)$. Our algorithm is a combination of branch and bound approach along with dynamic programming on graphs with bounded treewidth. To achieve $\mathcal{O}(1.5535^n)$ bound we introduce a technique of measuring running time of our algorithm by combining *measure and conquer* approach with linear programming.

1   Introduction

The story of exact (exponential-time) algorithms for hard problems dates back to the sixties and seventies but especially the last decade has seen a growing interest in new techniques for such type of algorithms. We refer to the following recent surveys [8,21,23] for an overview of the field.

Counting problem is a natural extension of a decision problem, where instead of deciding on the existence of a solution, the task is to find the number of solutions. Valiant [22] defined the class #P and showed that computing the permanent is #P-complete. Many NP-complete as well as some problems in P can have their counting versions to be #P-hard. In particular, counting minimum dominating set is #P-hard [14] even when restricted to planar instances. There is a lot of research going on counting problems and the complexity class #P. (See the book by Jerrum [15] for an introduction.)

While exact algorithms for many NP-complete decision problems were studied quite intensively, there are not so many results on exact algorithms for #P-complete problems in the literature. For a long time the only known exact algorithm for an #P-complete problem was time $2^n \cdot n^{\mathcal{O}(1)}$ counting perfect matching in bipartite graph algorithm due to Ryser [20]. There are known exact algorithms for different counting versions of satisfiability problem like counting maximum weighted models for 2SAT and 3SAT [5,11,24], CSP [1], and counting maximum weighted independent sets [4,6].

The principle of inclusion and exclusion was used by Karp [16] to solve many counting problems such as Hamiltonian path, sequencing and bin packing. Lately, Bax and Franklin [2] generalized this technique to the finite-difference approach and obtained algorithms for counting paths and cycles of a given length in a directed graph. Very recently time $2^n \cdot n^{\mathcal{O}(1)}$ algorithm computing chromatic polynomial (counting all k-colorings for each k) was obtained in [3,17].

G. Lin (Ed.): COCOON 2007, LNCS 4598, pp. 165–175, 2007.

Previous results. The dominating set problem is one of the classical NP-complete graph optimization problem which fits into the broader class of domination and covering problems. Hundreds of papers have been written on them (see e.g. the survey [13] by Haynes et al.). Recently, several groups of authors independently obtained exact algorithms that solve minimum dominating set problem in a graph on n vertices faster than the trivial $2^n \cdot n^{O(1)}$-time brute force algorithm [10,12,19]. The fastest known algorithm computes a minimum dominating set of a graph in time $\mathcal{O}(1.5137^n)$ [7]. The algorithm from [7] cannot be used to compute a dominating set of minimum weight (in a weighted graph), however, as it was observed in [9], the problem can be solved in time $\mathcal{O}(1.5780^n)$ by similar techniques. In the same paper it was shown that all minimal dominating sets can be listed in time $\mathcal{O}(1.7697^n)$ (later improved to $\mathcal{O}(1.7170^n)$), which implies that minimum dominating sets can be counted in this time.

Our results. In this paper we give an algorithm that counts minimum weight dominating sets in a weighted graph on n vertices in time $\mathcal{O}(1.5535^n)$. The basic idea is as follows: First we turn the instance of the dominating set problem to the instance of a set cover problem (this is the idea used by Grandoni [12] for decision problem) and perform branching on large sets and sets of size three containing elements of high degree. When branching is complete, we turn the instance of set cover into an instance of red/blue domination on bipartite graphs and use dynamic programming to count all solutions.

The novel and the most difficult part of the paper is the analysis of the algorithm. To analyze the running time we need to investigate the behavior of the pathwidth of a graph as a function of the measure of the corresponding set cover instance. The difficulty here is to find the measure of the problem that "balances" branching and dynamic programming parts of the algorithm. To choose the right measure we express the bounds on pathwidth as a linear program.

Combining branching with dynamic programming is a general approach that works for many decision and counting problems [6,18] and our technique can be used to improve the analysis of many algorithms of this type.

Finally, let us remark that the running time $\mathcal{O}(1.5535^n)$ of our algorithm counting weight dominating sets is even (slightly) better than the running time of the minimum weighted dominating set algorithm from [9].

2   Preliminaries

Let $G = (V, E)$ be an n-vertex undirected, simple graph without loops. We denote by $\Delta(G)$ the maximum vertex degree in G. For a vertex $v \in V$ we denote the set of its neighbors by $N(v)$ and its *closed neighborhood* by $N[v] = N(v) \cup \{v\}$.

A set $D \subseteq V$ is called a *dominating set* for G if every vertex from V is either in D, or adjacent to some vertex in D. Given a weight function $w : V \longrightarrow \mathbb{R}^+$ the *Minimum Weighted Dominating Set* problem (MWDS) asks to find a dominating set $D \subseteq V$ of minimum weight $w(D) = \sum_{v \in D} w(v)$. We denote by #MWDS the counting version of MWDS where the objective is to count all dominating sets of minimum weight.

Let $\mathcal{U}$ be a set of elements and $\mathcal{S}$ be a collection of non-empty subsets of $\mathcal{U}$. Given a weight function $w : \mathcal{S} \longrightarrow \mathbb{R}^+$ the *Minimum Weighted Set Cover* problem (MWSC) asks to find a subset $\mathcal{S}^* \subseteq \mathcal{S}$ of minimum weight $w(\mathcal{S}^*) = \sum_{S \in \mathcal{S}^*} w(S)$ which *covers* $\mathcal{U}$; that is,

$$\bigcup_{S \in \mathcal{S}^*} S = \mathcal{U}.$$

We denote by #MWSC the problem of counting set covers of minimum weight.

The *frequency* of an element $u \in \mathcal{U}$ is the number of sets $S \in \mathcal{S}$ in which u is contained. We denote it by $\mathrm{freq}(u)$.

#MWDS can be reduced to #MWSC by imposing $\mathcal{U} = V$ and $\mathcal{S} = \{N[v] \mid v \in V\}$. Given a weight function $w(v)$ for MWDS we define weight for $S \in \mathcal{S}$ as follows.

$$w(S) = w(S = \{N[v] \mid v \in V\}) = w(v).$$

Thus D is a dominating set of G if and only if $\{N[v] \mid v \in D\}$ is a set cover of $\{N[v] \mid v \in V\}$.

We also need a reduction from MWSC to a version of weighted dominating set problem called *minimum red/blue weighted dominating set* (RBWDS) problem. For a bipartite graph $G = (V, E)$ with a bipartition $V = V_{Red} \cup V_{Blue}$ and a weight function $w : V \longrightarrow \mathbb{R}^+$, a subset $D \subseteq V_{Red}$ is *red-blue dominating set* if every vertex in V_{Blue} is adjacent to a vertex of D. RBWDS problem is to determine the minimum weight of a red/blue dominating set in G.

With an instance $(\mathcal{U}, \mathcal{S}, w)$ of MWSC one can associate an *incidence graph* $G_{\mathcal{S}}$, which is a bipartite graph on a vertex set $\mathcal{S} \cup \mathcal{U}$ with a bipartition $(\mathcal{S}, \mathcal{U})$, and vertices $S \in \mathcal{S}$ and $u \in \mathcal{U}$ are adjacent if and only if u is an element of S. Let us observe, that $\mathcal{S}$ has a cover of weight k if and only if its incidence graph has a red-blue (with $V_{Red} = \mathcal{S}$ and $V_{Blue} = \mathcal{U}$) dominating set of weight k.

Let $G = (V, E)$ be a graph. A *tree decomposition* of G is a pair $\langle \{X_i \mid i \in I\}, T \rangle$, where each X_i is a subset of V, called a *bag*, and T is a tree with the elements of I as vertices. The following three properties must hold:

1. $\bigcup_{i \in I} X_i = V$.
2. For every edge $(u, v) \in E$, there is an $i \in I$ such that $\{u, v\} \subseteq X_i$.
3. For all $i, j, k \in I$, if j lies on the path between i and k in T, then $X_i \cap X_k \subseteq X_j$.

The *width* of $\langle \{X_i \mid i \in I\}, T \rangle$ equals $\max\{|X_i| \mid i \in I\} - 1$. The *treewidth* of G is the minimum k such that G has a tree decomposition of width k. A tree decomposition is called a *path decomposition* if T is a path. Accordingly, the *pathwidth* of G is the minimum k such that G has a path decomposition of width k. We denote by $\mathbf{pw}(G)$ the pathwidth of G.

We need the following result which can be obtained by a standard dynamic programming techniques.

Lemma 1. *All minimum red/blue weighted dominating sets of a bipartite graph G on n vertices with bipartition (V_{Red}, V_{Blue}) given together with its path decomposition of width at most p can be counted in time $\mathcal{O}(2^p n)$.*

3 Algorithm for Counting Minimum Weighted Set Covers

We consider a recursive algorithm countMWSC for solving #MWSC. The algorithm depends on the following observation.

Lemma 2. *For a given instance of $(\mathcal{S}, w)$, if there is an element $u \in \mathcal{U}(\mathcal{S})$ that belongs to a unique $S \in \mathcal{S}$, then S belongs to every set cover.*

Input: A collection on sets $\mathcal{S}$ and a weight function $w : \mathcal{S} \to \mathbb{R}^+$.
Output: A couple (*weight,num*) where *weight* is the minimum weight of a set cover of $\mathcal{S}$ and *num* is the number of different set covers of this weight.

1 **if** $|\mathcal{S}| = 0$ **then**
2   **return** (0,1);

3 **if** $\exists u \in \mathcal{U}(\mathcal{S}) : \mathrm{freq}(u) = 1$, Let $u \in S'$ **then**
4   **return** countMWSC(remove($S', \mathcal{S}$), w);

5 Pick $S \in \mathcal{S}$ of maximum cardinality;

6 **if** $|S| \le 3$ *and for every $S \in \mathcal{S}$ the degree of all its elements is at most 6* **then**
7   **return** countPW($\mathcal{S}, w$);
8 **else**
9      $(w_{in}, n_{in}) = $ countMWSC(remove($S, \mathcal{S}$), w);
10      $(w_{out}, n_{out}) = $ countMWSC($\mathcal{S} \setminus \{S\}, w$);
11      $w_{in} = w_{in} + w(S)$;
12   **if** $w_{in} < w_{out}$ **then**
13   **return** (w_{in}, n_{in});
14   **else if** $w_{in} = w_{out}$ **then**
15   **return** $(w_{in}, n_{in} + n_{out})$;
16   **else**
17   **return** (w_{out}, n_{out});

Fig. 1. countMWSC($\mathcal{S}, w$)

Let us go through the algorithm countMWSC; see Figure 1. First, if $|\mathcal{S}| = 0$ then the size of MWSC is 0 and the number of such set covers is 1. Otherwise (lines 3–4), the algorithm tries to reduce the size of the problem by checking whether condition of Lemma 2 is applicable. Specifically, if there is an element $u \in S'$ of frequency one, then we should put S' into minimum set cover. Thus we remove S' and all its elements from the other sets $S \in \mathcal{S}$. Namely remove($S', \mathcal{S}$) = $\{Z \mid Z = S \setminus S', S \in \mathcal{S}\}$.

If the condition of Lemma 2 does not apply, then we take a set $S \in \mathcal{S}$ of maximum cardinality. If $|S| \le 3$ and for every $S \in \mathcal{S}$ the degree of all its elements is at most 6 then we solve the problem with the algorithm countPW. We discuss this algorithm and its complexity later.

Otherwise we branch on the following two subproblems. First subproblem (remove($S, \mathcal{S}$),w) corresponds to the case where S belongs to the minimum set cover. Whereas in $(\mathcal{S} \setminus \{S\}, w)$ subproblem S does not belong to the minimum set cover.

And finally we compare the weights of two subproblems and return total weight and number (lines 12–17).

The function `countPW` computes a minimum set cover for a specific instance $(\mathcal{S}, w)$. Namely, $\mathcal{S}$ consists of sets with cardinalities at most 3 and for every $S \in \mathcal{S}$ the degree of all its elements is at most 6. For such a set $\mathcal{S}$, the function `countPW` does the following.

- Constructs the incidence graph $G_\mathcal{S}$ of $\mathcal{S}$;
- Counts the number of minimum red/blue dominating sets in $G_\mathcal{S}$ (to perform this step, we construct a path decomposition of $G_\mathcal{S}$ and perform dynamic programming algorithm described in Lemma 1);
- Counts the number of minimum set covers of $\mathcal{S}$ from the number of red/blue dominating sets of $G_\mathcal{S}$.

4 Analysis of `countMWSC` Algorithm

In this section we show that the running time of the algorithm `countMWSC` is $\mathcal{O}^*(1.2464^{|\mathcal{S}|+|\mathcal{U}|})$. The analysis is based on *Measure & Conquer* technique [8,7]. The analysis of the branching part of the algorithm is quite similar to analysis from [7].

Let n_i be the number of subsets $S \in \mathcal{S}$ of cardinality i and let m_j be the number of elements $u \in \mathcal{U}$ of frequency j. We use the following measure $k = k(\mathcal{S})$ of the size of $\mathcal{S}$:

$$k(\mathcal{S}) = \sum_{i \geq 1} w_i n_i + \sum_{j \geq 1} v_j m_j,$$

where the weights $w_i, v_j \in (0, 1]$ will be fixed later. Note that $k \leq |\mathcal{S}| + |\mathcal{U}|$. Let

$$\Delta w_i = w_i - w_{i-1}, \text{if } i \geq 2 \quad \text{and} \quad \Delta v_i = \begin{cases} v_i - v_{i-1}, & \text{if } i \geq 3, \\ v_2, & \text{if } i = 2. \end{cases}$$

Intuitively, $\Delta w_i (\Delta v_i)$ is a reduction of the size of the problem corresponding to the reduction of the cardinality of a set (the frequency of an element) from i to $i - 1$. Note that this also holds for Δv_2, because the new element of frequency one introduced is removed before next branching. And thus we get the total reduction $1 - (1 - v_2) = v_2$.

Theorem 1. *Algorithm* `countMWSC` *solves* #MWSC *in time* $\mathcal{O}^*(1.2464^{|\mathcal{S}|+|\mathcal{U}|})$.

Proof. In order to simplify the running time analysis, we make the following assumptions:

- $v_1 = 1$;
- $w_i = 1$ for $i \geq 6$ and $v_i = 1$ for $i \geq 6$;
- $0 \leq \Delta w_i \leq \Delta w_{i-1}$ for $i \geq 2$.

Note that this implies $w_i \geq w_{i-1}$ for every $i \geq 2$.

Let $P_h(k)$ be the number of subproblems of size $h \in \{0, \ldots, k\}$, solved by countMWSC to solve a problem of size k. Clearly, $P_k(k) = 1$. Consider the case $h < k$ (which implies $|\mathcal{S}| \neq 0$). If one of the condition in line 3 of the algorithm holds, we remove one set from $\mathcal{S}$. Thus the reduction of size of the problem is at least w_1 (worst case, $|S| = 1$) and $P_h(k) \leq P_h(k - w_1)$. Otherwise, let S be the subset selected in line 5. If $|S| \leq 3$ and for every $S \in \mathcal{S}$ the degree of all its elements is at most 6 (line 6), no subproblem is generated. Otherwise, we branch on two subproblems $S_{out} = (w_{out}, n_{out})$ and $S_{in} = (w_{in}, n_{in})$.

Consider subproblem S_{out}. It corresponds to the case where S does not belong to the set cover. The size of S_{out} decreases by $w_{|S|}$ because of the removal of S. Let m_i be the number of elements of S with frequency i. Note that there cannot be elements of frequency 1. Consider an element $u \in S$ with frequency $i \geq 2$. When we remove S, the frequency of u decreases by one. Thus, the size of the subproblem decreases by Δv_i. The overall reduction due to the reduction of the frequencies is at least

$$\sum_{i \geq 2} m_i \Delta v_i = \sum_{i=2}^{6} m_i \Delta v_i.$$

Finally, the total reduction of the size of S_{out} is

$$w_{|S|} + \sum_{i=2}^{6} m_i \Delta v_i.$$

Now consider the subproblem S_{in}. The size of S_{in} decreases by $w_{|S|}$ because of the removal of S. Since we also remove all elements from S, we also get the reduction of size

$$\sum_{i \geq 2} m_i v_i = \sum_{i=2}^{6} m_i v_i + m_{\geq 7}.$$

Here $m_{\geq 7}$ is the number of elements with frequency at least 7. Let S' be the set sharing element u with S ($S' \cap S \neq \emptyset$). Note that $|S'| \leq |S|$. When we remove u, the cardinality of S' is reduced by one. This implies the reduction of size S_{in} by $\Delta w_{|S'|} \geq \Delta w_{|S|}$. Thus the overall reduction of the size of S_{in} due to the reduction of the cardinalities of the sets S' is at least

$$\Delta w_{|S|} \sum_{i \geq 2} (i-1) m_i \geq \Delta w_{|S|} \left(\sum_{i=2}^{6} (i-1) m_i + 6 \cdot m_{\geq 7} \right).$$

Finally, the total reduction of the size of S_{in} is

$$w_{|S|} + \sum_{i=2}^{6} m_i v_i + m_{\geq 7} + \Delta w_{|S|} \left(\sum_{i=2}^{6} (i-1) m_i + 6 \cdot m_{\geq 7} \right).$$

Putting all together, for all possible values of $|S| \geq 3$ and for all values m_i such that

$$\sum_{i=2}^{6} m_i + m_{\geq 7} = |S|,$$

(except if $|S| = 3$, then we choose only those sets of m_i where $m_{\geq 7} \neq 0$) we have the following set of recursions

$$P_h(k) \leq P_h(k - \Delta k_{out}) + P_h(k - \Delta k_{in}),$$

where

$$\Delta k_{out} = w_{|S|} + \sum_{i=2}^{6} m_i \Delta v_i,$$

$$\Delta k_{in} = w_{|S|} + \sum_{i=2}^{6} m_i v_i + m_{\geq 7} + \Delta w_{|S|} (\sum_{i=2}^{6} (i-1)m_i + 6 \cdot m_{\geq 7}).$$

Since $\Delta w_{|S|} = 0$ for $|S| \geq 7$, we have that each recursion with $|S| \geq 8$ is "dominated" by some recurrence with $|S| = 7$. Thus we restrict our attention only to the cases $3 \leq |S| \leq 7$. We need to consider a large number of recursions (1653). For every fixed 9-tuple $(w_1, w_2, w_3, w_4, w_5, v_2, v_3, v_4, v_5)$ the number $P_h(k)$ is upper bounded by α^{k-h}, where α is the largest number from the set of real roots of the set of equations

$$\alpha^k = \alpha^{k - \Delta k_{out}} + \alpha^{k - \Delta k_{in}}$$

corresponding to the different combinations of values $|S|$ and m_i. Thus to estimate $P_h(k)$ we need to choose the weights w_i and v_j minimizing α.

Let K denote the set of the possible sizes of the subproblems solved. Note that $|K|$ is polynomially bounded. Thus the total number $P(k)$ of subproblems is

$$P(k) \leq \sum_{h \in K} P_h(k) \leq \sum_{h \in K} \alpha^{k-h}.$$

After performing branching the algorithm calls countPW algorithm. Thus the total running time of the algorithm on an instance of measure k is

$$\mathcal{O}(\sum_{h \in K} \alpha^{k-h} \cdot \beta^h) = \mathcal{O}(\max\{\alpha, \beta\}^k).$$

Here $\mathcal{O}(\beta^h)$ is the running time of countPW algorithm on a problem of size h. So we need to choose the weights w_i and v_j minimizing both α and β. To estimate the running time of countPW algorithm we use the idea of measure and conquer applied to linear programming.

Let us remind that countPW is called on an instance of MWSC problem with all sets of size at most 3 and elements of frequency at most 6. There are no elements of frequency 1. Let S be an instance of set cover of measure k and let G_S be its incidence graph. Then G_S is a bipartite graph with the bipartition

$(\mathcal{X} = \mathcal{S}, \mathcal{Y} = \mathcal{U}(\mathcal{S}))$. By Lemma 1, the running time of dynamic programming algorithm on $G_{\mathcal{S}}$ is $\mathcal{O}(\beta^k) = \mathcal{O}(2^{\mathbf{pw}(G_{\mathcal{S}})})$. Let us remind that both constants α and β depend on the choice of the weights in the measure function. In the remaining part of the proof we show how to choose the weights that balance branching and dynamic programming parts of the algorithm.

We denote by $\mathcal{X}_i$ all vertices from $\mathcal{X}$ of degree i. Let $x_i = |\mathcal{X}_i|$. We define $\mathcal{Y}_j \subseteq \mathcal{Y}$ and y_j in the same way for every $j \in \{2, \ldots, 6\}$. We need the following lemma. The proof can be obtained by technique used in [6].

Lemma 3. *For any $\varepsilon > 0$, there exists an integer n_ε such that for every graph G with $n > n_\varepsilon$ vertices and maximum degree at most 6,*

$$\mathbf{pw}(G) \leq \frac{1}{6}n_3 + \frac{1}{3}n_4 + \frac{13}{30}n_5 + \frac{23}{45}n_6 + \varepsilon n,$$

where n_i is the number of vertices of degree i in G for any $i \in \{3, \ldots, 6\}$. Moreover, a path decomposition of such width can be constructed in polynomial time.

We need to evaluate the running time of `countPW` on an instance of $\mathcal{S}$ of measure k. This gives us the following:

$$k = k(\mathcal{S}) = \sum_{i=1}^{3} w_i x_i + \sum_{j=2}^{5} v_j y_j + y_6. \tag{1}$$

Here values w_i and v_j are taken from analysis of `countMWSC` algorithm. By counting edges of $G_{\mathcal{S}}$, we arrive at the second condition

$$x_1 + 2x_2 + 3x_3 = \sum_{j=2}^{6} j \cdot y_j. \tag{2}$$

By Lemma 3,

$$\mathbf{pw}(G_{\mathcal{S}}) \leq \frac{1}{6}(x_3 + y_3) + \frac{1}{3}y_4 + \frac{13}{30}y_5 + \frac{23}{45}y_6. \tag{3}$$

Combining (1),(2), and (3) we conclude that the pathwidth of $G_{\mathcal{S}}$ is at most the maximum of the following linear function

$$\frac{1}{6}(x_3 + y_3) + \frac{1}{3}y_4 + \frac{13}{30}y_5 + \frac{23}{45}y_6 \rightarrow \max$$

subject to the following constraints:

$$\text{measure:} \quad \sum_{i=1}^{3} w_i x_i + \sum_{j=2}^{5} v_j y_j + y_6 = k$$

$$\text{edges:} \quad x_1 + 2x_2 + 3x_3 = \sum_{j=2}^{6} j \cdot y_j$$

$$\text{variables:} \quad x_i \geq 0, i \in \{1, 2, 3\}$$
$$y_j \geq 0, j \in \{2, \ldots, 6\}$$

The running time of countPW is $\mathcal{O}^*(2^{\mathrm{pw}(G)})$. Thus everything boils up to finding the measure that minimizes the maximum of α and maximum of LP obtained from pathwidth bounds. Finding of such weights is an interesting (and nontrivial) computational problem on its own. To find the weights we use a modification of random search with plugged LP solver.

We numerically obtained the following values of the weights.

$$w_i = \begin{cases} 0.1039797, & \text{if } i = 1, \\ 0.4664084, & \text{if } i = 2, \\ 0.8288271, & \text{if } i = 3, \\ 0.9429144, & \text{if } i = 4, \\ 0.9899772, & \text{if } i = 5, \end{cases} \quad \text{and} \quad v_i = \begin{cases} 0.5750176, & \text{if } i = 2, \\ 0.7951411, & \text{if } i = 3, \\ 0.9165365, & \text{if } i = 4, \\ 0.9771879, & \text{if } i = 5. \end{cases}$$

With such weights the optimum of LP is obtained on $x_3 = 0.7525...$ and $y_6 = 0.3762...$ and all other variables equal to 0.

This gives us the total running time $\mathcal{O}(1.2464^{k(\mathcal{S})}) = \mathcal{O}(1.2464^{|\mathcal{U}|+|\mathcal{S}|})$. The exponential space is used by the algorithm during the dynamic programming part and thus is bounded by $\mathcal{O}(1.2464^{|\mathcal{U}|+|\mathcal{S}|})$. This finalizes the proof. $\square$

As we mentioned already, #MWDS can be reduced to #MWSC by imposing $\mathcal{U} = V$ and $\mathcal{S} = \{N[v] \mid v \in V\}$. The size of the #MWSC instance obtained is at most $2n$, where n is the number of vertices in G. Thus, we have

Corollary 1. *The #MWDS problem can be solved in* $\mathcal{O}(1.2464^{2n}) = \mathcal{O}(1.5535^n)$ *time.*

5 Conclusions and Open Problems

In this paper we have used dynamic programming on bounded treewidth techniques to speed up branching algorithm counting weighted dominating sets. The running time of our algorithm can be slightly improved by considering more detailed bounds on pathwidth of bipartite graphs with different coefficients for vertices adjacent to degree two and three vertices. However in this case the arguments become much more technical and we do not include them in this extended abstract.

Another general technique to speed up branching algorithms (by using exponential space) is *memorization*, see [8]. Both techniques have many similarities and it would be interesting to understand in which cases each of the techniques is more efficient.

The best known algorithm for the decision version of minimum dominating set is not significantly faster than our counting algorithms. Similar strange effect was observed by Magnus Wahlström (private communication) for different SAT problems. Thus despite the fact that #P complete problems seems to be much more difficult than NP complete problems, the running time of best known algorithms for many decision and counting problems do not differ too much. Is there any reasonable explanation to this phenomena? Is this because faster exponential algorithms for decision problems are still to be found or this is because in the world of exponential time algorithms the gaps between different complexity classes are small?

Acknowledgement. We thank Serge Gaspers for helpful comments.

References

1. Angelsmark, O., Jonsson, P.: Improved algorithms for counting solutions in constraint satisfaction problems. In: Rossi, F. (ed.) CP 2003. LNCS, vol. 2833, pp. 81–95. Springer, Heidelberg (2003)
2. Bax, E.T., Franklin, J.A: finite-difference sieve to count paths and cycles by length. Inf. Process. Lett. 60(4), 171–176 (1996)
3. Björklund, A., Husfeldt, T.: Inclusion-exclusion algorithms for counting set partitions. In: Proceedings of the 47th Annual IEEE Symposium on Foundations of Computer Science (FOCS 2006), pp. 575–582. IEEE Computer Society Press, Los Alamitos (2006)
4. Dahllöf, V., Jonsson, P.: An algorithm for counting maximum weighted independent sets and its applications. In: Proceedings of the 13th Annual ACM-SIAM Symposium on Discrete Algorithms (SODA 2002). Society for Industrial and Applied Mathematics, pp. 292–298. ACM Press, New York (2002)
5. Dahllöf, V., Jonsson, P., Wahlström, M.: Counting models for 2 SAT and 3 SAT formulae. Theoretical Computer Science 332 332(1-3), 265–291 (2005)
6. Fomin, F.V., Gaspers, S., Saurabh, S.: Branching and treewidth based exact algorithms. In: Deng, X., Du, D.-Z. (eds.) ISAAC 2005. LNCS, vol. 3827, pp. 16–25. Springer, Heidelberg (2005)
7. Fomin, F.V., Grandoni, F., Kratsch, D.: Measure and conquer: Domination – a case study. In: Caires, L., Italiano, G.F., Monteiro, L., Palamidessi, C., Yung, M. (eds.) ICALP 2005. LNCS, vol. 3580, pp. 191–203. Springer, Heidelberg (2005)
8. Fomin, F.V., Grandoni, F., Kratsch, D.: Some new techniques in design and analysis of exact (exponential) algorithms. Bulletin of the EATCS 87, 47–77 (2005)
9. Fomin, F.V., Grandoni, F., Pyatkin, A.V., Stepanov, A.A.: Bounding the number of minimal dominating sets: a measure and conquer approach. In: Deng, X., Du, D.-Z. (eds.) ISAAC 2005. LNCS, vol. 3827, pp. 573–582. Springer, Heidelberg (2005)
10. Fomin, F.V., Kratsch, D., Woeginger, G.J.: Exact (exponential) algorithms for the dominating set problem. In: Hromkovič, J., Nagl, M., Westfechtel, B. (eds.) WG 2004. LNCS, vol. 3353, pp. 245–256. Springer, Heidelberg (2004)
11. Fürer, M., Kasiviswanathan, S.P.: Algorithms for counting 2-SAT solutions and colorings with applications. Electronic Colloquium on Computational Complexity (ECCC) 33 (2005)
12. Grandoni, F.: A note on the complexity of minimum dominating set. Journal of Discrete Algorithms 4(2), 209–214 (2006)
13. Haynes, T.W., Hedetniemi, S.T.: Domination in graphs (Advanced topics). Monographs and Textbooks in Pure and Applied Mathematics, vol. 209. Marcel Dekker Inc., New York (1998)
14. Hunt III, H.B., Marathe, M.V., Radhakrishnan, V., Stearns, R.E.: The complexity of planar counting problems. SIAM Journal on Computing 27(4), 1142–1167 (1998)
15. Jerrum, M.: Counting, sampling and integrating: algorithms and complexity. Lectures in Mathematics ETH Zürich. Birkhäuser Verlag, Basel (2003)

16. Karp, R.M.: Dynamic programming meets the principle of inclusion and exclusion. Operations Research Letters 1 2(1981/82), 49–51 (1981)
17. Koivisto, M.: An $O(2^n)$ algorithm for graph coloring and other partitioning problems via inclusion-exclusion. In: Proceedings of the 47th Annual IEEE Symposium on Foundations of Computer Science (FOCS 2006), pp. 583–590. IEEE Computer Society Press, Los Alamitos (2006)
18. Mölle, D., Richter, S., Rossmanith, P.: Enumerate and expand: Improved algorithms for connected vertex cover and tree cover. In: Grigoriev, D., Harrison, J., Hirsch, E.A. (eds.) CSR 2006. LNCS, vol. 3967, pp. 270–280. Springer, Heidelberg (2006)
19. Randerath, B., Schiermeyer, I.: Exact algorithms for MINIMUM DOMINATING SET. Technical Report zaik-469, Zentrum für Angewandte Informatik Köln, Germany (2004)
20. Ryser, H.J.: Combinatorial mathematics. The Carus Mathematical Monographs, No. 14. Published by The Mathematical Association of America (1963)
21. Schöning, U.: Algorithmics in exponential time. In: Diekert, V., Durand, B. (eds.) STACS 2005. LNCS, vol. 3404, pp. 36–43. Springer, Heidelberg (2005)
22. Valiant, L.G.: The complexity of computing the permanent. Theoretical Computer Science 8(2), 189–201 (1979)
23. Woeginger, G.J.: Exact algorithms for NP-hard problems: A survey. In: Jünger, M., Reinelt, G., Rinaldi, G. (eds.) Combinatorial Optimization - Eureka, You Shrink! LNCS, vol. 2570, pp. 185–207. Springer, Heidelberg (2003)
24. Zhang, W.: Number of models and satisfiability of sets of clauses. Theoretical Computer Science 155(1), 277–288 (1996)

Online Interval Scheduling: Randomized and Multiprocessor Cases[*]

Stanley P.Y. Fung[1], Chung Keung Poon[2], and Feifeng Zheng[3]

[1] Department of Computer Science, University of Leicester, United Kingdom
pyfung@mcs.le.ac.uk
[2] Department of Computer Science, City University of Hong Kong, China
ckpoon@cs.cityu.edu.hk
[3] School of Management, Xi'an JiaoTong University, China
zhengff@mail.xjtu.edu.cn

Abstract. We consider the problem of scheduling a set of equal-length intervals arriving online, where each interval is associated with a weight and the objective is to maximize the total weight of completed intervals. An optimal 4-competitive algorithm has long been known in the deterministic case, but the randomized case remains open. We give the first randomized algorithm for this problem, achieving a competitive ratio of 3.618. We also prove a randomized lower bound of 4/3, which is an improvement over the previous 5/4 result, and a lower bound of 2 for a class of barely random algorithms which include our new algorithm. We also show that the techniques can be carried to the deterministic multiprocessor case, giving a 3.618-competitive 2-processor algorithm, a 5/4 lower bound for any number of processors, and a 2 lower bound for 2 processors.

1 Introduction

We study the problem of scheduling a set of intervals which arrive online. Each interval has a weight and all intervals are of the same length. The objective is to schedule a set of non-overlapping intervals such that the total weight of all these intervals is maximized. Intervals being processed can be interrupted, but the value will be lost. This can also be viewed as a job scheduling problem where each job must be served immediately or else it is lost. This is a fundamental problem in scheduling and has been widely studied, and is also related to a number of online problems such as call control and bandwidth allocation (see e.g. [15,4,1]).

Related Work. For the basic problem where intervals are of the same length and with arbitrary weights, Woeginger [15] gave an optimal deterministic 4-competitive algorithm and a matching lower bound. In the paper, the open

---

[*] The work described in this paper was fully supported by a grant from City University of Hong Kong (SRG 7001969), and NSFC Grant No. 70471035.

G. Lin (Ed.): COCOON 2007, LNCS 4598, pp. 176–186, 2007.

question of whether randomization can help give better algorithms was raised. Miyazawa and Erlebach [12] gave a 3-competitive randomized algorithm for the special case where the weights of the intervals form a non-decreasing sequence. They also gave the first randomized lower bound of 1.25.

Many other variations exist for the problem. One is to allow variable interval lengths: Canetti and Irani [4] gave a randomized lower bound of $\Omega(\sqrt{\log \Delta / \log\log \Delta})$ and a randomized upper bound of $O(\log \Delta)$ where Δ is the ratio of the longest to shortest interval length. The deterministic case (with variable lengths) seems not studied before on its own. However, more general models have been studied. The problem of *online scheduling of broadcasts with restarts* is considered in [8,5,7,16,14]. In this broadcast scheduling problem, a set of pages is stored in a server, and requests for pages arrive online. Each request has a deadline by which the request has to be completed or else no profit is obtained. Different requests for the same page can be served together in one single broadcast. Requests being served can be interrupted but will need to restart from the beginning if it is to be broadcasted again. It can be seen that the interval scheduling problem is a special case of the broadcast scheduling problem (where all requests have tight deadlines and each request asks for a different page). In particular, there are matching upper and lower bounds $\Theta(\Delta / \log \Delta)$ [14,16] for the deterministic case. These bounds also apply to the interval scheduling problem.

Another related problem, the *real-time job scheduling problem*, gives a bound on the interval scheduling problem w.r.t. a different parameter. In this problem, jobs have release times, deadlines and execution times. Preemption is allowed and preempted jobs can be resumed at the point where they were preempted. When all jobs have no laxity (i.e., execution time equals the difference between deadline and release time), this problem reduces to the interval scheduling problem. Matching upper and lower bounds of $(1 + \sqrt{k})^2$ are established in [2,9], where k is the *importance ratio*, defined as the ratio of maximum to minimum weight of the jobs. (Here jobs have different lengths and the weight of a job is the value per unit length of the job.) Both bounds apply to the interval scheduling case.

Yet another case is where intervals can have different lengths and their weights are some function of their lengths. Seiden [13] gave a randomized 3.732-competitive algorithm for the case of *benevolent instances*, where (roughly speaking) the weight of an interval is either a convex or a monotonic decreasing function of its length. If the weight of an interval is equal to its length, the non-preemptive case was considered in [11]. They gave a randomized $O((\log \Delta)^{1+\epsilon})$-competitive algorithm and a $\Omega(\log \Delta)$ lower bound.

Multiprocessor scheduling of intervals were studied in [6], giving an optimal (1-competitive) algorithm when all intervals have unit weight (and not necessarily equal length). Multiprocessor scheduling of jobs with different lengths and weights were studied in [10], with competitive ratio roughly $\Theta(\log k)$ where k is the importance ratio. If the profit is equal to the length, a tight bound of 2 is known for 2 processors [2].

Our Results. In this paper we consider the case where all intervals are of the same length. In fact our algorithm applies to the more general case where intervals may not have equal lengths but have *agreeable deadlines*, i.e. no interval being strictly contained in another interval. This is because of the result that the class of interval graphs for agreeable-deadlines intervals is equal to the class of interval graphs for equal-length intervals (see e.g. [3]). In Section 3 we give a randomized online algorithm which is 3.618-competitive, which is lower than the optimal deterministic bound. The algorithm only uses a constant number of random bits as it only makes a single random choice when it starts. In Section 4 we consider lower bounds, giving an improved randomized lower bound of 4/3 on the competitive ratio. For the class of barely random algorithms that choose between two deterministic algorithms with equal probabilities (which includes our proposed algorithm), we show a lower bound of 2.

There is a close relation between the randomized single-processor case and the deterministic multiprocessor case. We show in Section 5 that (with some modifications) our 3.618-competitive algorithm can be applied in the 2-processor case. The lower bounds also apply: we give a 4/3 lower bound for the competitive ratio of any deterministic or randomized algorithms for any number of processors, and a deterministic 2 lower bound for 2 processors.

Due to space limitations some proofs are omitted and can be found in the full version of the paper.

2   Preliminaries

An interval I is specified by $s(I)$, its arrival time; $\ell(I)$, its length; and $|I|$, its weight. Since we only consider the case where all intervals are of the same length, we can, without loss of generality, assume $\ell(I) = 1$ for all I.

Intervals arrive online and the scheduling algorithm has to make decisions without knowledge of future intervals. When a scheduling algorithm completes an interval, it receives a *profit* equal to the weight of the interval. An interval being scheduled can be aborted (e.g. when an interval of larger weight arrives) but the value of the aborted interval will be lost. The objective of the algorithm is to maximize the total profit obtained by completing the intervals.

Let $A(S)$ denote the profit obtained by algorithm A on input instance S. An online algorithm A is *c-competitive* if for all input instances S, $OPT(S)/A(S) \leq c$ where OPT is the offline optimal algorithm which has knowledge of all future intervals and hence can schedule optimally. When A is a randomized algorithm, the definition of competitive ratio becomes $OPT(S)/E[A(S)] \leq c$ where the expectation is taken over the random choices of A. A randomized algorithm that only uses a constant number of random bits is called *barely random*.

3   A Randomized Algorithm

The Algorithm. Consider the simple deterministic algorithm Greedy$_r$: when an interval I arrives and the algorithm is currently executing another interval I', it aborts I' and starts I if $|I| \geq r|I'|$. If the machine is idle at that time then $|I'| = 0$, meaning that I will always get started. We call r the *abortion ratio*. This algorithm is 4-competitive when $r = 2$ and is the best possible for deterministic algorithms [15].

Fix constants α, β, p where $1 < \alpha < \beta$ and $0 < p < 1$. The randomized algorithm **RGreedy**$_{\alpha,\beta,p}$ chooses to run one of the two deterministic algorithms Greedy$_\alpha$ and Greedy$_\beta$ with probability p and $1 - p$ respectively. It is barely random since the choice is only made in the beginning. Below we will analyze the competitive ratio of this algorithm.

Basic subschedules. Let A and B denote the schedules produced by Greedy$_\alpha$ and Greedy$_\beta$ respectively on a particular input instance. We define a *basic subschedule* to be a sequence of execution of intervals $(I_1, \ldots, I_k)$, for $k \geq 1$, where I_i is aborted by I_{i+1} for $1 \leq i \leq k - 1$ and I_k is completed. That is, each basic subschedule consists of zero or more aborted intervals followed by a completed interval. Each of the two online schedules A and B can then be partitioned into a sequence of basic subschedules. In a basic subschedule $(I_1, \ldots, I_k)$ with abortion ratio r, we have $|I_i| \leq |I_{i+1}|/r$ because this is the condition for I_{i+1} to abort I_i.

Profit amortization. Consider a basic subschedule $(I_1, \ldots, I_k)$ with abortion ratio r. Only I_k is completed. Therefore the profit received for the whole basic subschedule is $|I_k|$. For the purpose of analysis, we will 'amortize' the profit of a basic subschedule to the individual intervals (not just the completed one) as follows. I_k transfers a profit of $|I_{k-1}|$ to I_{k-1} and keep the rest of $|I_k| - |I_{k-1}|$ profit to itself. Inductively, for $i = k - 1, \ldots, 2$, I_i receives an amortized profit of $|I_i|$ from I_{i+1}; it then transfers a profit of $|I_{i-1}|$ to I_{i-1} and keep the rest of $|I_i| - |I_{i-1}|$ to itself. For I_1 it keeps all its received profit $|I_1|$. Obviously, the total profit remains the same. From now on, unless explicitly stated otherwise, we will refer to the amortized profits.

Schedule segments. Consider a basic subschedule, $(X_1, \ldots, X_k)$, of A. Let $x_1, x_2, \ldots, x_k$ be the weights of these intervals. Let t_i be the time when X_i is started, and define $t_{k+1} = t_k + 1$. During $[t_i, t_{i+1})$, OPT can start at most one interval, Z_i, with weight z_i. If there is no interval started by OPT during that time period, we simply skip this interval X_i from our consideration; thus X_i and t_i only refer to those intervals in A which have corresponding Z_i's. This will only underestimate the profit of the online algorithm. By the property of basic subschedules we have $x_i \geq \alpha x_{i-1}$ for all $1 < i \leq k$. (Note that X_i and X_{i-1} may not be consecutive intervals in the basic subschedule because of the skipping just mentioned. Nevertheless the inequality remains true.)

Let u_i be the time when Z_i starts. At time u_i, Greedy$_\alpha$ must be serving some intervals with weight larger than z_i/α or else Greedy$_\alpha$ would abort what it is serving and start Z_i instead (which also has weight $> z_i/\alpha$). Thus $x_i > z_i/\alpha$ for all i. Similarly, at time u_i, the other algorithm Greedy$_\beta$ (if it is chosen instead) must be serving some interval, Y_i, of weight larger than z_i/β, or else it will abort what it is serving and start Z_i instead. Denote by y_i the weight of this interval that Greedy$_\beta$ is serving at time u_i. We have $y_i > z_i/\beta$.

We make two observations about these y_i's. First, any two Z_i's must correspond to different Y_i's. This is because each interval in OPT is completed and thus takes 1 unit of time, so $u_{i+1} \geq u_i + 1$ and hence Y_i and Y_{i+1} cannot be the same interval. Second, two consecutive Y_{i-1} and Y_i may or may not belong to the same basic subschedule in B. If they do, then we have $y_i \geq \beta y_{i-1}$. Note that even if they are in the same basic subschedule, they may not be consecutive intervals (there may be a number of abortions in-between), but even so the previous inequality remains true. If they are not in the same basic subschedule, we cannot say anything about y_i and y_{i+1}.

Therefore, we further split the basic subschedules in A and B into *segments* as follows. Let X_1 and Y_1 be the first interval in A and B respectively after the previous segment (initially they are simply the first intervals). A segment is then the set of intervals $(X_1, \ldots, X_n)$ from A and $(Y_1, \ldots, Y_n)$ from B such that at least one of X_n or Y_n is completed, and all other X_i and Y_i is aborted (directly or indirectly) by X_{i+1} and Y_{i+1} respectively. For a pair of corresponding segments, at least one of the two ending intervals is completed. In effect, intervals in a segment satisfy $x_i \leq x_{i+1}/\alpha$ (for those in A) and $y_i \leq y_{i+1}/\beta$ (for those in B).

Note that X_i's and Y_i's may not be consecutive intervals in a basic subschedule, as explained before. In the analysis we will ignore all those skipped abortions, e.g. in the profit amortization we treat Y_i and Y_{i+1} as if they are consecutive without giving any profit to any aborted intervals in-between, if any.

Bounding the expected profit. We now consider each such segment, where $(X_1, \ldots, X_n)$ is a segment from A, $(Y_1, \ldots, Y_n)$ is a segment from B, and $(Z_1, \ldots, Z_n)$ is the corresponding sequence of intervals in OPT.

The total profit of OPT in this segment is $\sum_{i=1}^{n} z_i$. As for the online algorithm, A has an amortized profit of at least $x_n - x_1/\alpha$ for this segment: the last interval has profit x_n and subsequently transferred to other intervals in this segment, except x_1 may transfer profit to a previously aborted interval, which has weight at most x_1/α. Similarly B receives an amortized profit of $y_n - y_1/\beta$ for this segment. The expected profit is thus at least $p(x_n - x_1/\alpha) + (1-p)(y_n - y_1/\beta)$. Note that the terms x_1/α and y_1/β would not be there if they are the first interval in a basic subschedule. But at least one of x_1 and y_1 must be such a first interval, since otherwise the segment would be extended to the front. Therefore, we can remove the smaller of these two terms from the expression. Thus the expected profit of the online algorithm is at least $px_n + (1-p)y_n - \max(px_1/\alpha, (1-p)y_1/\beta)$. We call the $\max(px_1/\alpha, (1-p)y_1/\beta)$ term the *amortized term*. The ratio R of optimal profit to the expected online profit in this segment is at most

$$\frac{\sum_{i=1}^{n} z_i}{px_n + (1-p)y_n - \max(px_1/\alpha, (1-p)y_1/\beta)} \quad (1)$$

We want to upper bound the above ratio, under the following constraints:

$$z_i < \min(\alpha x_i, \beta y_i), \quad x_i \leq x_{i+1}/\alpha, \quad y_i \leq y_{i+1}/\beta$$

Each interval served by OPT must belong to exactly one segment. Therefore, if we can upper bound the ratio of the total OPT profit to the expected online profit, for all such segments, this gives a bound on the competitive ratio of the randomized algorithm.

For the rest of the paper, we fix $\alpha = \phi \approx 1.618$, $\beta = \phi^2 \approx 2.618$ and $p = 1/2$, where $\phi = (1+\sqrt{5})/2$ is the golden ratio. We first state a technical lemma which will be required later.

Lemma 1. *Suppose* $x_i = 1/\alpha^{n-i}$ *and* $y_i = y/\beta^{n-i}$ *for all* i. *Then the function*

$$F(y) = \frac{\sum_{i=1}^n \min(\alpha x_i, \beta y_i)}{1 + y - 1/\alpha^n}$$

is increasing in y *for* $0 \le y < \alpha/\beta$, *and decreasing in* y *for* $y > \alpha/\beta$.

Theorem 1. *The competitive ratio of* $RGreedy_{\alpha,\beta,p}$ *is* $\phi + 2 \approx 3.618$ *when* $\alpha = \phi$, $\beta = \phi^2$ *and* $p = 1/2$.

Proof. Consider a segment with $(X_1, \ldots, X_n)$, $(Y_1, \ldots, Y_n)$, and $(Z_1, \ldots, Z_n)$. Without loss of generality, assume $x_n = 1$ and denote y_n simply as y. To maximize (1), observe that (for a fixed y) we should make x_i and y_i as large as possible, so that z_i's are large and also x_1 and y_1 are large. This means $x_i = 1/\alpha^{n-i}$ and $y_i = y/\beta^{n-i}$. Together with $p = 1/2$, (1) becomes at most

$$\frac{2\sum_{i=1}^n \min(\alpha x_i, \beta y_i)}{1 + y - \max(1/\alpha^n, y/\beta^n)} \tag{2}$$

We consider these cases:

Case 1: $y \le (\beta/\alpha)^n$. In this case the amortized term is $1/\alpha^n$. The ratio (2) is equal to $2F(y)$ in Lemma 1, which we know is maximum when $y = \alpha/\beta$. At this value of y, all min terms in the numerator are βy_i terms and the ratio has maximum value

$$\frac{2\beta y(1 + 1/\beta + \cdots + 1/\beta^{n-1})}{1 + y - 1/\alpha^n} = \frac{2\beta y(1 - 1/\beta^n)/(1 - 1/\beta)}{1 + y - 1/\alpha^n}$$

$$= \frac{\frac{2\alpha\beta}{\beta-1}(1 - 1/\beta^n)}{1 - 1/\alpha^n + \alpha/\beta} = \frac{2\phi^2(1 - 1/\phi^{2n})}{1 + 1/\phi - 1/\phi^n}.$$

This is at most $\phi + 2 \approx 3.618$ for any value of n (maximum occurs when $n = 2$).

Case 2: $y > (\beta/\alpha)^n$. In this case all min terms are the αx_i terms and the amortizing term is y/β^n. So the ratio is

$$\frac{2(\alpha + 1 + \cdots + 1/\alpha^{n-2})}{1 + y - y/\beta^n} = \frac{2\alpha(1 - 1/\alpha^n)/(1 - 1/\alpha)}{1 + y(1 - 1/\beta^n)}$$

$$\le \frac{2\alpha(1 - 1/\alpha^n)/(1 - 1/\alpha)}{1 + (\beta/\alpha)^n - 1/\alpha^n} = \frac{2\phi^3(1 - 1/\phi^n)}{1 + \phi^n - 1/\phi^n}.$$

This is at most ϕ^2 for any value of n.

Therefore in either case the ratio is at most $\phi + 2$.   □

We can show that there is an instance which actually attains the competitive ratio of 3.618 using our algorithm (with these chosen parameters), so that the analysis is tight.

4 Lower Bounds

4.1 Randomized Algorithms

Theorem 2. *No randomized algorithm for interval scheduling has competitive ratio better than* $4/3$.

Proof. We will use Yao's principle which states that the randomized lower bound can be obtained by bounding $E[OPT]/E[A]$ for any deterministic algorithm A over a probability distribution of input instances. (See for example, [12].) Thus we define an input distribution as follows. Let (r, w) denote the release time and weight respectively of an interval. Fix a large even integer n. Define $n + 1$ intervals, $I_0, I_1, \ldots, I_n$, such that for $0 \leq i \leq n - 1$, $I_i = (i/2, v_i)$ where $v_i = 2^i$, and $I_n = (n/2, v_n)$ where $v_n = 2^{n-1}$. Define n sets of intervals, $S_1, S_2, \ldots, S_n$, such that $S_i = \{I_0, I_1, \ldots, I_i\}$ for $1 \leq i \leq n$. Finally, we define our distribution of inputs to be one such that S_i occurs with probability $p_i = 1/2^i$ for $1 \leq i \leq n-1$ and S_n occurs with probability $p_n = 1/2^{n-1}$.

Since any I_i does not overlap with I_{i+2}, we have $OPT(S_i) = 1 + 4 + \cdots + 2^i$ $= \frac{4^{i/2+1}-1}{3}$ if i is even, and $OPT(S_i) = 2 + 8 + \cdots + 2^i = \frac{2(4^{(i+1)/2}-1)}{3}$ if i is odd, and $OPT(S_n) = 1 + 4 + \cdots + 2^{n-2} + 2^{n-1} = \frac{4^{n/2}-1}{3} + 2^{n-1}$. Hence

$$E[OPT] = \sum_{\substack{i=2,i \text{ even}}}^{n-2} \frac{4^{i/2+1}-1}{3 \cdot 2^i} + \sum_{\substack{i=1,i \text{ odd}}}^{n-1} \frac{2(4^{(i+1)/2}-1)}{3 \cdot 2^i} + \frac{4^{n/2}-1}{3 \cdot 2^{n-1}} + 1$$

$$= 4n/3 - o(n).$$

We now derive an upper bound on the expected profit of an arbitrary deterministic algorithm A on our input distribution. More specifically, for $i = 1, \ldots, n$, we let Q_i be the contribution to the expected profit of A on $I_{i-1}, \ldots, I_n$ when the input is one of $S_i, \ldots, S_n$.

Consider the case when the input is S_n. This happens with probability p_n. When I_n arrives at time $n/2$, A may or may not be serving another interval. If it does, it must be serving I_{n-1}. Since we choose $v_{n-1} = v_n$, A will obtain at most a profit of v_n whether it aborts I_{n-1} or not. Thus, $Q_n \leq p_n v_n$.

Now, suppose the input is either S_{n-1} or S_n. When I_{n-1} arrives at time $(n - 1)/2$, A may or may not be serving I_{n-2}. There are two cases.

Case 1: A is serving I_{n-2} and it continues until its completion. Then A gains a profit of v_{n-2} on I_{n-2} whether the input is S_{n-1} or S_n. Further, it can gain an expected profit of at most Q_n on I_{n-1} and I_n when the input is S_n. Hence, $Q_{n-1} \leq (p_{n-1} + p_n)v_{n-2} + Q_n$.

Case 2: A is not serving I_{n-2} or if it aborts I_{n-2}. Then A may have an expected profit of $p_{n-1}v_{n-1}$ on I_{n-1} when the input is S_{n-1} and an expected profit of Q_n on I_{n-1} and I_n when the input is S_n. Note that the input being S_{n-1} and being S_n are two disjoint events. Thus, $Q_{n-1} \leq p_{n-1}v_{n-1} + Q_n$.

Setting $(p_{n-1}+p_n)v_{n-2}+Q_n = p_{n-1}v_{n-1}+Q_n$ (which is satisfied by requiring $v_{n-2} = \frac{p_{n-1}}{p_{n-1}+p_n}v_{n-1}$), we have $Q_{n-1} \leq p_{n-1}v_{n-1} + Q_n$ no matter what A does.

In general, consider the case when the input is one of $S_i, \ldots, S_n$. When I_i arrives at time $i/2$, A may or may not be serving I_{i-1} and we consider the following cases.

Case 1: A is serving I_{i-1} and it continues with it until completion. Then A gains an expected profit of $(p_i + \cdots + p_n)v_{i-1}$ on I_{i-1} (no matter what the true input is) and an expected profit of Q_{i+1} on $I_i, \ldots, I_n$ when the input is $S_{i+1}, \ldots, S_n$. Thus, $Q_i \le (p_i + \cdots + p_n)v_{i-1} + Q_{i+1}$.

Case 2: A is not serving I_{i-1} or if it aborts I_{i-1}. Then A gains an expected profit of $p_i v_i$ on I_i when the input is S_i, and an expected profit of Q_{i+1} on $I_i, \ldots, I_n$ when the input is one of $S_{i+1}, \ldots, S_n$. Hence $Q_i \le p_i v_i + Q_{i+1}$.

Setting $v_{i-1} = \frac{p_i}{p_i + \cdots + p_n} v_i$, we have $(p_i + \cdots + p_n)v_{i-1} + Q_{i+1} = p_i v_i + Q_{i+1}$. So $Q_i \le p_i v_i + Q_{i+1}$.

One can easily check that setting p_i and v_i as mentioned earlier, the conditions $v_{i-1} = \frac{p_i}{p_i + \cdots + p_n} v_i$ for $1 \le i \le n$, are satisfied and the total expected profit of A is $Q_1 \le p_1 v_1 + \cdots + p_n v_n = n$.

Therefore, $E[OPT]/E[A] \to 4/3$ for $n \to \infty$.   □

Remark on benevolent instances. The lower bound construction does not rely on the exact lengths of the intervals. The only requirement on the lengths is that I_i and I_{i+1} intersect while I_i and I_{i+2} do not. Therefore, the lower bound also holds for benevolent instances; we just create the instances with the specified weights and adjust the lengths accordingly.

4.2   Barely Random Algorithms

Our randomized algorithm in Section 3 chooses between two deterministic algorithms with equal probabilities. We next show a lower bound on such algorithms.

Theorem 3. *No barely random algorithms that choose between two deterministic algorithms with equal probabilities can be better than 2-competitive.*

Proof. Suppose on the contrary there exists such a randomized algorithm which is $(2 - \epsilon)$-competitive for some constant $\epsilon > 0$. Let D1 and D2 be the two deterministic algorithms. We construct an adversarial request sequence to show that this results in a contradiction.

Consider a set of a large number of intervals where each interval differs from the previous one by arriving slightly later and having a slightly larger weight (difference in weight being δ). The minimum weight of intervals in this set is 1 and the maximum weight is α. Here δ is a sufficiently small and α a sufficiently large constant to be chosen later. The last interval arrives before the deadline of the first interval, and hence any algorithm can serve at most one interval in this set. (This is the set of intervals used in [15].) Given this set of intervals, let x and y be the weights of intervals chosen by D1 and D2, where without loss of generality, assume $x \le y$. We emphasize that the adversary knows the values of x and y. We consider the following cases.

Case 1: $x = y = 1$. Both D1 and D2 obtains a profit of 1 while OPT schedules the heaviest interval giving a profit of α. So the competitive ratio is α.

Case 2: $x = y \neq 1$. One more interval of weight y is released just before the deadline of the y in the set. Both D1 and D2 either continue with the x or y, or abort and switch to the new y. In either case their profit is at most y. The adversary schedules the interval in the set just before y, together with the new y, giving a profit of $(y - \delta) + y$. Hence the competitive ratio is $2 - \delta/y > 2 - \delta$.

Case 3: $1 = x < y$. D1 and D2 gets a profit of 1 and y respectively while OPT gets α. Thus the competitive ratio is $\alpha/((1+y)/2) \geq 2\alpha/(1+\alpha) = 2 - 2/(1+\alpha)$.

Case 4: $1 < x < y$. The adversary releases another interval with weight y just before the deadline of x in the set. We distinguish two subcases.

If D1 does not abort x in favour of the new y, no more intervals are released. (We remark that the adversary knows the response of D1 and can make requests accordingly.) In this case D1 and D2 get a profit of x and y respectively, while OPT gets $x - \delta + y$. Then the competitive ratio $= (x + y - \delta)/((x + y)/2) = 2 - 2\delta/(x + y) > 2 - 2\delta/(1 + 1) = 2 - \delta$.

If D1 aborts x and serves y, then one more interval of weight y arrives just before the deadline of y in the original set. Then both D1 and D2 gets a profit of y no matter what they do, and OPT gets a profit of $y - \delta + y$. The competitive ratio is $(2y - \delta)/y = 2 - \delta/y > 2 - \delta$.

Considering all cases, the competitive ratio is at least $\min\{\alpha, 2 - \delta, 2 - 2/(1 + \alpha)\}$. By choosing $\delta < \epsilon$ and $\alpha > \max(2 - \epsilon, 2/\epsilon - 1) = 2/\epsilon - 1$, we have the competitive ratio being at least $2 - \epsilon$.   □

5   The Multiprocessor Case

In this section we consider the case of using more than one processor to schedule the intervals. We will see that the cases of randomization and multiple processors are closely related. We first show that the idea of the barely random algorithm in Section 3 can be used to give a deterministic 2-processor algorithm with the same competitive ratio. Then we show that the lower bounds in Section 4 can also be carried to the multiprocessor case; namely, that no deterministic or randomized algorithm can be better than 4/3-competitive for any number of processors m, and no 2-processor deterministic algorithm can be better than 2-competitive.

5.1   A 2-Processor Algorithm

We consider the following deterministic 2-processor algorithm. Call the two processors P1 and P2. In simple terms, P1 runs Greedy$_\alpha$ whereas P2 runs Greedy$_\beta$. Specifically, suppose the two processors are running intervals I_1 and I_2 respectively. When a new interval I arrives, if $|I| < \alpha|I_1|$ and $|I| < \beta|I_2|$ then I is rejected. If one of $|I| \geq \alpha|I_1|$ and $|I| \geq \beta|I_2|$ is true, the corresponding I_1 or I_2 is aborted and I is started on that processor. If I is at least as large as both $\alpha|I_1|$ and $\beta|I_2|$, it aborts I_2 and start I on $P2$. (This is the only difference from the randomization case: since the two processors cannot be doing the same interval, we need some way of tie-breaking.) A processor which has completed its interval will become idle. Note that an idle processor is regarded as executing a weight-0 interval. Therefore if P1 is idle and $|I| \geq \beta |I_2|$, it will still abort I_2 (and P1 remains idle). Again we set $\alpha = \phi$ and $\beta = \phi^2$.

We will separately bound the value of the two optimal offline schedules produced by the two processors, OPT1 and OPT2. As before, we divide the schedule into basic subschedules and segments. With the same notation as in Section 3, consider a segment where OPT1 schedules $(Z_1, \ldots, Z_n)$, P1 schedules $(X_1, \ldots, X_n)$, and P2 schedules $(Y_1, \ldots, Y_n)$. We will show the same bound on the competitive ratio, i.e. $2 \sum z_i / (x_n + y_n - \max(x_1/\alpha, y_1/\beta)) \leq \phi + 2$. Therefore over the whole OPT1, $OPT1/((P1 + P2)/2) \leq \phi + 2$. Here $OPT1$, $P1$ and $P2$ represent both the schedules and their profits. Since OPT2 can be analyzed similarly, we have $OPT2/((P1 + P2)/2) \leq \phi + 2$. Adding these two together, $OPT1 + OPT2 \leq (\phi + 2)(P1 + P2)$ and therefore the algorithm is $(\phi + 2)$-competitive. In the analysis below we only consider OPT1.

We call (z_k, x_k, y_k) in a segment a *triplet*. We first make an observation:

Lemma 2. *For any triplet (z_k, x_k, y_k), one of the following two cases holds: (i) $x_i > z_i/\alpha$ and $y_i > z_i/\beta$, (ii) $z_i = y_i$ and $x_i \leq z_i/\alpha$.*

We call triplets of case (i) *normal triplets* and those of case (ii) *violating triplets*. The main idea of the competitiveness proof is as follows: if there are no violating triplets in a segment, then we are done by the same proof as in the randomized algorithm. If there are violating triplets, we further divide the segment into *subsegments* so that each subsegment has at most one violating triplet at the beginning of the subsegment. We then perform a similar analysis to the randomized algorithm on each subsegment. There is a small difference in the amortized terms: both the x_1/α and y_1/β terms may be subtracted, since the last pair of intervals in the previous subsegment may not be completed, and hence do not have any real profit. So the amortized term may sometimes become $x_1/\alpha + y_1/\beta$ instead of $\max(x_1/\alpha, y_1/\beta)$. We omit the proof to this theorem:

Theorem 4. *The algorithm is $\phi + 2 \approx 3.618$-competitive for 2 processors.*

5.2   Lower Bounds

The proofs of the following theorems use almost identical constructions to that in Theorems 2 and 3, so we omit the proofs.

Theorem 5. *No deterministic or randomized algorithm for online interval scheduling on m processors is better than $4/3$-competitive, for any m.*

Theorem 6. *No deterministic algorithm for online interval scheduling on 2 processors is better than 2-competitive.*

6   Conclusion

In this paper we give the first randomized algorithm and improved lower bounds for the online interval scheduling problem. The gap between the upper and lower bounds remains wide, however. It may be possible to generalize the barely random algorithm to use 3 or more deterministic algorithms but we encounter some technical difficulties in extending the technique here. Algorithms for three or more processors will also yield randomized algorithms for the one-processor case.

References

1. Awerbuch, B., Bartal, Y., Fiat, A., Rosen, A.: Competitive non-preemptive call control. In: Proc. 5th SODA, pp. 312–320 (1994)
2. Baruah, S., Koren, G., Mao, D., Mishra, B., Raghunathan, A., Rosier, L., Shasha, D., Wang, F.: On the competitiveness of on-line real-time task scheduling. Real-Time Systems 4, 125–144 (1992)
3. Bogart, K.P., West, D.B.: A short proof that proper = unit. Discrete Mathematics 201, 21–23 (1999)
4. Canetti, R., Irani, S.: Bounding the power of preemption in randomized scheduling. SIAM Journal on Computing 27(4), 993–1015 (1998)
5. Chan, W.-T., Lam, T.-W., Ting, H.-F., Wong, P.W.H.: New results on on-demand broadcasting with deadline via job scheduling with cancellation. In: Chwa, K.-Y., Munro, J.I.J. (eds.) COCOON 2004. LNCS, vol. 3106, pp. 210–218. Springer, Heidelberg (2004)
6. Faigle, U., Nawijn, W.M.: Greedy k-coverings of interval orders. Technical Report 979, University of Twente (1991)
7. Fung, S.P.Y., Chin, F.Y.L., Poon, C.K.: Laxity helps in broadcast scheduling. In: Coppo, M., Lodi, E., Pinna, G.M. (eds.) ICTCS 2005. LNCS, vol. 3701, pp. 251–264. Springer, Heidelberg (2005)
8. Kim, J.-H., Chwa, K.-Y.: Scheduling broadcasts with deadlines. Theoretical Computer Science 325(3), 479–488 (2004)
9. Koren, G., Shasha, D.: D^{over}: An optimal on-line scheduling algorithm for overloaded uniprocessor real-time systems. SIAM Journal on Computing 24, 318–339 (1995)
10. Koren, G., Shasha, D.: MOCA: A multiprocessor on-line competitive algorithm for real-time system scheduling. Theoretical Computer Science 128(1-2), 75–97 (1994)
11. Lipton, R.J., Tomkins, A.: Online interval scheduling. In: Proc. 5th SODA, pp. 302–311 (1994)
12. Miyazawa, H., Erlebach, T.: An improved randomized on-line algorithm for a weighted interval selection problem. Journal of Scheduling 7(4), 293–311 (2004)
13. Seiden, S.S.: Randomized online interval scheduling. Operations Research Letters 22(4-5), 171–177 (1998)
14. Ting, H.-F.: A near optimal scheduler for on-demand data broadcasts. In: Calamoneri, T., Finocchi, I., Italiano, G.F. (eds.) CIAC 2006. LNCS, vol. 3998, pp. 163–174. Springer, Heidelberg (2006)
15. Woeginger, G.J.: On-line scheduling of jobs with fixed start and end times. Theoretical Computer Science 130(1), 5–16 (1994)
16. Zheng, F., Fung, S.P.Y., Chan, W.-T., Chin, F.Y.L., Poon, C.K., Wong, P.W.H.: Improved on-line broadcast scheduling with deadlines. In: Chen, D.Z., Lee, D.T. (eds.) COCOON 2006. LNCS, vol. 4112, pp. 320–329. Springer, Heidelberg (2006)

Scheduling Selfish Tasks: About the Performance of Truthful Algorithms

George Christodoulou[1], Laurent Gourvès[2], and Fanny Pascual[3]

[1] Max-Planck-Institut für Informatik, Saarbrücken, Germany
gchristo@mpi-inf.mpg.de
[2] LAMSADE, CNRS UMR 7024, Université de Paris-Dauphine, Paris, France
laurent.gourves@lamsade.dauphine.fr
[3] Equipe MOAIS (CNRS-INRIA-INPG-UJF), Grenoble, France
fanny.pascual@imag.fr

Abstract. This paper deals with problems which fall into the domain of selfish scheduling: a protocol is in charge of building a schedule for a set of tasks without directly knowing their length. The protocol gets these informations from agents who control the tasks. The aim of each agent is to minimize the completion time of her task while the protocol tries to minimize the maximal completion time. When an agent reports the length of her task, she is aware of what the others bid and also of the protocol's algorithm. Then, an agent can bid a false value in order to optimize her individual objective function. With erroneous information, even the most efficient algorithm may produce unreasonable solutions. An algorithm is truthful if it prevents the selfish agents from lying about the length of their task. The central question in this paper is: *"How efficient a truthful algorithm can be?"* We study the problem of scheduling selfish tasks on parallel identical machines. This question has been raised by Christodoulou et al [8] in a distributed system, but it is also relevant in centrally controlled systems. Without considering side payments, our goal is to give a picture of the performance under the condition of truthfulness.

Keywords: scheduling, algorithmic game theory, truthful algorithms.

1 Introduction

The Internet is a complex distributed system involving many autonomous entities (*agents*). Protocols organize this network, using the data held by these agents and trying to maximize the social welfare. Agents are often supposed to be trustworthy but this assumption is unrealistic in some settings as they might try to manipulate the protocol by reporting false information in order to maximize their own profit. With false information, even the most efficient protocol may lead to unreasonable solutions if it is not designed to cope with the selfish behavior of the single entities. Then, it is natural to ask the following question: *How efficient a protocol can be if it guarantees that no agent has incentive to lie?*

G. Lin (Ed.): COCOON 2007, LNCS 4598, pp. 187–197, 2007.
© Springer-Verlag Berlin Heidelberg 2007

In this paper, we deal with the problem of scheduling n selfish tasks on m identical parallel machines. We consider two distinct settings in which the aim is to minimize the *makespan*, i.e. the maximum completion time. The first setting is centralized, while the second one is distributed. Both problems share the following characteristics. Each task is owned by an agent[1]. The length l_i of a task i is known to its owner only. The agents, considered as players of a non-cooperative game, want to minimize the completion time of their tasks. The protocol builds the schedule with rules known to all players and fixed in advance. In particular, mixing the execution of two jobs (like round-robin) is not allowed. Before the execution begins, the agents report a value representing the length of their tasks. We assume that every agent behave rationally and selfishly. Each one is aware of the situation the others face and tries to optimize her own objective function. Thus an agent can report a value which is not equal to her real length. Practically, an agent can add "fake" data to artificially increase the length of her task if it decreases her completion time. This selfish behavior can prevent the protocol to produce a reasonable (i.e. close to the social welfare) schedule. Without considering side payments, which are often used with the aim of inciting the agents to report their real value, some algorithmic tools can simultaneously offer a guarantee on the quality of the schedule (its makespan is not arbitrarily far from the optimum) and guarantee that the solution is *truthful* (no agent can lie and improve her own completion time). For both centralized and distributed settings, our goal is to give lower and upper bounds on the performance under the condition of truthfulness. It is important to mention that we do not strictly restrict the study to polynomial time algorithms.

Since the length of a task is private, each agent bids a value which represents the length of her task. We assume that an agent cannot shrink the length of her task (otherwise she will not get her result), but if she can decrease her completion time by bidding a value larger than the real one, then she will do so. We also assume that an agent does not report a distribution on different lengths. A player may play according to a distribution, but she just announces the outcome, so the protocol does not know if she lies.

In the *centralized setting*, the strategy of agent i is a value b_i representing the length of her task. The protocol, called an algorithm, is in charge of indicating when and on which machine a task will be scheduled. An algorithm is *truthful* when no agent has incentive to report a false value. We focus on the performance of truthful algorithms with respect to the makespan of the schedule. In particular, we are interested in giving lower and upper bounds on the *approximation ratio* that a (deterministic or randomized) truthful algorithm can achieve. For example, a truthful algorithm can be obtained by greedily scheduling the tasks following the increasing order of their lengths. This algorithm, known as SPT, produces a $(2 - 1/m)$-approximate schedule [11]. *Are there truthful algorithms with better approximation guarantee for the considered scheduling problem?*

---

[1] We equally refer to a task and its owner since we assume that two tasks cannot be held by the same agent.

In the *distributed setting*, the strategy of agent i is a couple (M_i, b_i), where M_i is the machine which will execute the task and b_i is the length bidden. As opposed to the centralized setting, the agents choose their machine and M_i can be a probability distribution on different machines. The protocol, called a *coordination mechanism* in this context [8], consists in selecting a *scheduling policy* for each machine (e.g. scheduling the tasks in order of decreasing lengths). An important and natural condition is due to the decentralized nature of the problem: the scheduling on a machine should depend only on the tasks assigned to it, and should be independent of the tasks assigned to the other machines. A coordination mechanism is *truthful* when no agent has incentive to lie on the length of her task. Using the *price of anarchy* [14], we study the performance of truthful coordination mechanisms with respect to the makespan. The price of anarchy of a coordination mechanism is, in the context, equal to the largest ratio between the makespan of a schedule where agent's strategies form a *Nash equilibrium*[2] and the optimal makespan.

Interestingly, it is possible to slightly transform the SPT algorithm in a truthful coordination mechanism, as suggested in [8]: each machine P_j schedules its tasks in order of increasing lengths, and adds at the very beginning of the schedule a small delay equal to $(j-1)\varepsilon$ times the length of the first task. By this way, and if ε is small enough, the schedule obtained in a Nash equilibrium is similar to the one returned by the SPT algorithm (excepted the small delays at the beginning of the schedule). When ε is negligible, the price of anarchy of this coordination mechanism is $2 - 1/m$. *Are there truthful coordination mechanisms with better price of anarchy for the considered scheduling problem?*

For both centralized algorithms and coordination mechanisms, we consider the two following execution models:

- **Strong model of execution:** If the owner of task i bids $b_i \geq l_i$, then the execution time will still be l_i (i.e. the task will be completed l_i time units after its start).
- **Weak model of execution:** If the owner of task i bids $b_i \geq l_i$, then the execution time will be b_i (i.e. the task will be completed b_i time units after its start).

The strong execution model corresponds to the case where tasks have to be linearly executed – from their beginning to their end–, whereas the weak execution model corresponds to the case where a task can be executed in any order[3] (and the "fake" part of the task is not anymore necessarily executed at the end), or when the machine returns the result of the task only at the end of its execution. Depending on the applications of the scheduling problem, either the strong or the weak model of execution will be used.

---

[2] Situation in which no agent can unilaterally change her strategy and improve her own completion time. A Nash equilibrium is *pure* if each agent has a pure strategy : each agent chooses only one machine. A Nash equilibrium is *mixed* if the agents give a probability distribution on the machines on which they will go.

[3] Nevertheless, the execution of two jobs is never interlaced.

Related Work

The field of *Mechanism Design* can be useful to deal with the selfishness of the agents. Its main idea is to pay the agents to convince them to perform strategies that help the system to optimize a global objective function. The most famous technique for designing truthful mechanisms is perhaps the Vickrey-Clarke-Groves (VCG) mechanism [20,7,12]. However, when applied to combinatorial optimization problems, this mechanism guarantees the truthfulness under the hypothesis that the objective function is *utilitarian* (i.e. the value of the objective function is equal to the sum of the agents individual objective functions) and that the mechanism is able to compute the optimum. Archer and Tardos introduce in [4] a method which allows to design truthful mechanisms for several combinatorial optimization problems to which the VCG mechanism does not apply. However, both approaches cannot be applied to our problem.

Scheduling selfish agents has been intensively studied these last years, started with the seminal work of Nisan and Ronen [17], and followed by a series of papers [1,2,4,6,9,15,16]. However, all these works differ from ours since in their case, the selfish agents are the machines while here we consider that the agents are the tasks. Furthermore, they use side payments whereas we focus on truthful algorithms without side payments.

A more closely related work is the one of Christodoulou et al [8] who considered the same model but only in the distributed context of coordination mechanisms. They proposed different coordination mechanisms with a price of anarchy better than the one of the SPT coordination mechanism. Nevertheless, these mechanisms are not truthful. In [13], the authors gave coordination mechanisms for the same model for related machines (i.e. machines can have different speeds), but their mechanisms are also not truthful.

In [3], the authors gave a truthful randomized algorithm for the strong model of execution defined before, and they gave, for the weak model of execution, a coordination mechanism which is truthful if there are two machines and if the lengths of the tasks are powers of a certain constant. An optimal (but exponential time) truthful randomized algorithm and a truthful randomized PTAS for the weak model of execution appear in [18,19]. The technique consists in computing an optimal (resp. a $(1 + \varepsilon)$-approximate) schedule and each machine executes its tasks in a random order (the truthfulness is due to the introduction of fictitious tasks which guarantee that all the machines have the same load).

Another related work is the one of Auletta et al. who considered in [5] the problem of scheduling selfish tasks in a centralized case. Their work differs from ours since they considered that each machine uses a round and robin policy and thus that the completion of each task is the completion time of the machine on which the task is (this model is known as the KP model). They considered that the tasks can lie in both directions, and that there are some payments.

Contribution and Organization of the Article

Sections 3 and 4 are devoted to the centralized setting. In particular, we study the strong (resp. weak) model of execution in Section 3 (resp. Section 4). Results on the distributed setting are presented in Section 5 for both execution models.

Table 1 gives a summary of the bounds that we are aware of (those with a † are presented in this article). LB stands for "Lower bound", UB for "Upper bound" and NE for "Nash equilibria". Due to space constraints, some proofs are omitted.

Table 1. Bounds for m identical machines

Strong model of execution:

	Deterministic		Randomized	
	LB	UB	LB	UB
centralized setting	$2 - \frac{1}{m}$ †	$2 - \frac{1}{m}$ [8]	$\frac{3}{2} - \frac{1}{2m}$ †	$2 - \frac{1}{m+1}\left(\frac{5}{3} + \frac{1}{3m}\right)$ [3]
distributed setting	$2 - \frac{1}{m}$ (pure NE) † $\frac{3}{2} - \frac{1}{2m}$ (mixed NE) †	$2 - \frac{1}{m}$ [8]	$\frac{3}{2} - \frac{1}{2m}$ †	$2 - \frac{1}{m}$

Weak model of execution:

	Deterministic		Randomized	
	LB	UB	LB	UB
centralized setting	$m = 2 : 1 + \frac{\sqrt{105}-9}{12} > 1.1$ † $m \geq 3 : \frac{7}{6} > 1.16$ †	$\frac{4}{3} - \frac{1}{3m}$ †	1 [18,19]	1 [18,19]
distributed setting	$\frac{1+\sqrt{17}}{4} > 1.28$ (pure NE) †	$2 - \frac{1}{m}$	$1 + \frac{\sqrt{13}-3}{4} > 1.15$ † (pure NE)	$2 - \frac{1}{m}$

2 Notations

We are given m machines (or processors) $\{P_1, \ldots, P_m\}$, and n tasks $\{1, \ldots, n\}$. Let l_i denote the real execution time (or length) of task i. We use the identification numbers to compare tasks of the same (bidden) lengths: we will say that task i, which bids b_i, is larger than task j, which bids b_j, if and only if $b_i > b_j$ or ($b_i = b_j$ and $i > j$). It is important to mention that an agent cannot lie on her (unique) identification number.

A randomized algorithm can be seen as a probability distribution over deterministic algorithms. We say that a (randomized) algorithm is truthful if for every task the expected completion time when she declares her true length is smaller than or equal to her expected completion time in the case where she declares a larger value. More formally, we say that an algorithm is *truthful* if $E_i[l_i] \leq E_i[b_i]$, for every i and $b_i \geq l_i$, where $E_i[b_i]$ is the expected completion time of task T_i if she declares b_i. In order to evaluate the quality of a randomized algorithm, we use the notion of expected approximation ratio.

We will refer in the sequel to the list scheduling algorithms LPT and SPT, where LPT (resp. SPT) [11] is the algorithm which greedily schedules the tasks, sorted in order of decreasing (resp. increasing) lengths: this algorithm schedules, as soon as a machine is available, the largest (resp. smallest) task which has not yet been scheduled. An LPT (resp. SPT) schedule is a schedule returned by the LPT (resp. SPT) algorithm.

3   About Truthful Algorithms for the Strong Model of Execution

3.1   Deterministic Algorithms

We saw that the deterministic algorithm SPT, which is $(2 - \frac{1}{m})$-approximate, is truthful. Let us now show that there is no truthful deterministic algorithm with a better approximation ratio.

Theorem 1. *Let us consider that we have m identical machines. There is no truthful deterministic algorithm with an approximation ratio smaller than $2 - \frac{1}{m}$.*

Proof. Let us suppose that we have $n = m(m-1) + 1$ tasks of length 1. Let us suppose that we have a truthful deterministic algorithm $\mathcal{A}$ which has an approximation ratio smaller than $(2 - 1/m)$ Let t be the task which has the maximum completion time, C_t, in the schedule returned by $\mathcal{A}$. We know that $C_t \geq m$.

Let us now suppose that task t bids m instead of 1. We will show that the completion time of t is then smaller than m. Let OPT be the makespan of an optimal solution where there are $n - 1 = m(m-1)$ tasks of length 1 and a task of length m. We have: $OPT = m$. Since te approximation ratio of algorithm $\mathcal{A}$ is smaller than $(2 - 1/m)$, the makespan of the schedule it builds with this instance is smaller than $(2 - 1/m)m = 2m - 1$. Thus, the task of length m starts before time $(m-1)$. Thus, if task t bids m instead of 1, it will start before time $m - 1$ and be completed one time unit after, that is before time m. Thus task t will decrease its completion time by bidding m instead of 1, and algorithm $\mathcal{A}$ is not truthful.

Note that we can generalize this result to the case of related machines: we have m machines $P_1, \ldots, P_m$, such that machine P_i has a speed v_i, $v_1 = 1$, and $v_1 \leq \ldots \leq v_m$. By this way, the bound becomes $2 - \frac{v_m}{\sum_{i=1}^{m} v_i}$.

Concerning the strong model of execution, no deterministic algorithm can outperform SPT in the centralized setting. Then, it is interesting to consider randomized algorithms to achieve a better approximation ratio.

3.2   Randomized Algorithms

In [3], the authors present a randomized algorithm which consists in returning a LPT schedule with a probability $1/(m+1)$ and a slightly modified SPT schedule with a probability $m/(m+1)$. They obtain a truthful algorithm whose expected approximation ratio improves $2 - \frac{1}{m}$ but no instance showing the tightness of their analysis is provided. A good candidate should be simultaneously a tight example for both LPT and SPT schedules. We are not aware of the existence of such an instance and we believe in a future improvement of this upper bound. The following Theorem provides a lower bound.

Theorem 2. *Let us consider that we have m identical machines. There is no truthful randomized algorithm with an approximation ratio smaller than $\frac{3}{2} - \frac{1}{2m}$.*

Generalizing this result to the case of related machines, the bound becomes $\frac{3}{2} - \frac{v_m}{2\sum_{i=1}^m v_i}$.

4 About Truthful Algorithms for the Weak Model of Execution

4.1 A Truthful Deterministic Algorithm

We saw in the Section 3 that SPT is a truthful and $(2 - 1/m)$-approximate algorithm for the strong model of execution, and that no truthful deterministic algorithm can have a better approximation ratio. If we consider the weak model of execution, we can design a truthful deterministic algorithm, called LPT_{mirror}, with a better performance guarantee. We are given n tasks $\{1, \ldots, n\}$ which bid lengths $b_1, \ldots, b_n$. Make a schedule σ_{LPT} with the LPT list algorithm. Let C_{max}^{OPT} be the optimal makespan. Let $p(i)$ be the machine on which the task i is executed in σ_{LPT}. Let C_i be date at which the task i ends in σ_{LPT}. LPT_{mirror} returns the schedule in which task i is executed on machine $p(i)$ and starts at time $(4/3 - 1/(3m))C_{max}^{OPT} - C_i$.

Theorem 3. LPT_{mirror} *is a deterministic, truthful and* $(\frac{4}{3} - \frac{1}{3m})$-*approximate algorithm.*

Proof. We are given n tasks with true lengths $l_1, \ldots, l_n$. Let us suppose than each task has bidden a value, and that task i bids $b_i > l_i$. This can make the task i start earlier in σ_{LPT} but never later. In addition, the optimal makespan when i bids $b_i > l_i$ is necessarily larger than or or equal to the optimal makespan when task i reports its true length.

Let S_i be the date at which task i starts to be executed in σ_{LPT}. The completion time of task i in LPT_{mirror} is $(4/3 - 1/(3m))OPT - C_i + b_i = (4/3 - 1/(3m))OPT - S_i$ because $S_i = C_i - b_i$. By bidding $b_i > l_i$, task i can only increase its completion time in the schedule returned by LPT_{mirror} because OPT does not decrease and S_i does not increase. Thus task i does not have incentive to lie.

Since the approximation ratio of the schedule obtained with the LPT list algorithm is at most $(4/3 - 1/(3m))$ [11], the schedule returned by LPT_{mirror} is clearly feasible and its makespan is, by construction, $(4/3 - 1/(3m))$-approximate. Thus LPT_{mirror} is a truthful and $(\frac{4}{3} - \frac{1}{3m})$-approximate algorithm.

Note that LPT_{mirror} is not a polynomial time algorithm, since we need to know the value of the makespan in an optimal solution, which is an NP-hard problem [10]. However, it is possible to have a polynomial time algorithm which is $(4/3 - 1/(3m))$-approximate, even if some tasks do not bid their true values. Consider the following simple algorithm: we first compute a schedule σ_{LPT} with the LPT algorithm. Let $p(i)$ be the machine on which the task i is executed in σ_{LPT}, let C_i be the completion time of task i in σ_{LPT}, and let C_{max} be the makespan of

σ_{LPT}. We then compute the final schedule σ' in which task i is scheduled on $p(i)$ and starts at time $C_{max} - C_i$.

We can show that this algorithm is $(4/3 - 1/(3m))$-approximate (i.e. the schedule returned by this algorithm is at most $(4/3 - 1/(3m))$ times larger than the optimal schedule in which all the tasks bid their true values). We can show this by the following way. We suppose that all the tasks except i have bidden some values. Let $\sigma_{LPT}(b_i)$ be the schedule σ_{LPT} obtained when i bids b_i, let $S_i(b_i)$ be the date at which task i starts to be executed in $\sigma_{LPT}(b_i)$, and let $C_{max}(\sigma_{LPT}(b_i))$ be the makespan of $\sigma_{LPT}(b_i)$. The completion time of task i (which bids b_i) in σ' is equal to $C_{max}(\sigma_{LPT}(b_i)) - S_i(b_i)$. Since with the LPT algorithm, tasks are scheduled in decreasing order of lengths, if $b_i > l_i$ then $S_i(b_i) \leq S_i(l_i)$. Thus, whatever the values bidden by the other tasks are, i has incentive to lie and bid $b_i > l_i$ only if $C_{max}(\sigma_{LPT}(b_i)) < C_{max}(\sigma_{LPT}(l_i))$. Since this is true for each task, no task will unilaterally lie unless this decreases the makespan of the schedule. The makespan of the schedule σ' in which all the tasks bid their true values is $(4/3 - 1/(3m))$-approximate, and then the solution returned by this algorithm will also be $(4/3 - 1/(3m))$-approximate.

4.2 Deterministic Algorithms: Lower Bounds

We suppose that the solution returned by an algorithm depends on the length and the identification number of each task, even those which can be identified with their unique length.

Theorem 4. *Let us consider that we have two identical machines. There is no truthful deterministic algorithm with an approximation ratio smaller than* $1 + (\sqrt{105} - 9)/12 \approx 1.1039$.

Theorem 5. *Let us consider that we have $m \geq 3$ identical machines. There is no truthful deterministic algorithm with an approximation ratio smaller than $7/6$.*

The assumption made to derive Theorems 4 and 5 is, in a sense, stronger than the usual one since we suppose that the solution returned by an algorithm for two similar instances (same number of tasks, same lengths but different identification numbers) can be completely different. If we relax this assumption, i.e. if identification numbers are only required for the tasks which have the same length, the bound presented in Theorem 4 can be improved to $7/6$.

Theorem 6. *Let us consider that we have two identical machines. No truthful deterministic algorithm can be better than $7/6$-approximate if it does not take into account the identification number of tasks whose length is unique.*

5 About Truthful Coordination Mechanisms

Let $\rho \geq 1$. If there is no truthful deterministic algorithm which has an approximation ratio of ρ, then there is no truthful deterministic coordination mechanism which always induce pure Nash equilibria and which has a price of anarchy smaller than or equal to ρ. Indeed, if this was not the case, then the deterministic algorithm which consists in building the schedule obtained in a pure Nash equilibrium with this ρ-approximate coordination mechanism would be a ρ-approximate truthful deterministic algorithm.

Likewise, if there is no truthful (randomized) algorithm which has an approximation ratio of ρ, then there is no truthful coordination mechanism which has a price of anarchy smaller than or equal to ρ. Indeed, if this was not the case, the algorithm which consists in building the schedule obtained in a Nash equilibrium with this ρ-approximate coordination mechanism would be a ρ-approximate truthful algorithm.

This observation leads us to the following results for the strong model of execution. We deduce from Theorem 1 that there is no truthful deterministic coordination mechanism which always induce pure Nash equilibria and which has a price of anarchy smaller than $2 - 1/m$. Thus there is no truthful coordination mechanism which performs better than the truthful SPT coordination mechanism, whose price of anarchy tends towards $2 - 1/m$. We deduce from Theorem 2 that there is no truthful coordination mechanism which has a price of anarchy smaller than $\frac{3}{2} - \frac{1}{2m}$. We now consider the weak model of execution.

Theorem 7. *If we consider the weak model of execution, there is no truthful deterministic coordination mechanism which induces pure Nash equilibria, and which has a price of anarchy smaller than $\frac{1+\sqrt{17}}{4} \approx 1.28$.*

Proof. Let us first prove this result in the case where there are two machines, P_1, and P_2. Let $\varepsilon > 0$. Let us suppose that there exists a truthful coordination mechanism $\mathcal{M}$ with a price of anarchy of $\frac{1+\sqrt{17}}{4} - \varepsilon$. Let us consider the following instance I_1: three tasks of length 1. Since $\mathcal{M}$ is a deterministic coordination mechanism which induces pure Nash equilibria, there is at least a task in I_1 which has a completion time larger than or equal to 2. Let t be such a task.

Let us first consider this instance I_2: we have two tasks of length $\frac{-1+\sqrt{17}}{2} \approx$ 1.56. Since $\mathcal{M}$ is $(\frac{1+\sqrt{17}}{4} - \varepsilon)$-approximate, there is one task on each machine, and each task is completed before time $\frac{-1+\sqrt{17}}{2} \times \frac{1+\sqrt{17}}{4} = 2$. Thus, when it has a task of length $\frac{-1+\sqrt{17}}{2}$, each machine must end it before time 2.

Let us now consider the following instance I_3: two tasks of length 1, and a task of length $\frac{-1+\sqrt{17}}{2}$. Since $\mathcal{M}$ is $(\frac{1+\sqrt{17}}{4} - \varepsilon)$-approximate, the task of length $\frac{-1+\sqrt{17}}{2}$ is necessarily alone on its machine (without loss of generality, on P_2). As we have seen it, P_2 must schedule this task before time 2. Thus, task t of instance I_1, has incentive to bid $\frac{-1+\sqrt{17}}{2}$ instead of 1: by this way it will end before time 2, instead of a time larger than or equal to 2.

We can easily extend this proof in the case where there are more than 2 machines, by having $m + 1$ tasks of length 1 in I_1; m tasks of length $\frac{-1+\sqrt{17}}{2}$ in I_2; and m tasks of length 1 and a task of length $\frac{-1+\sqrt{17}}{2}$ in I_3.

Theorem 8. *If we consider the weak model of execution, there is no truthful coordination mechanism which induces pure Nash equilibria, and which has a price of anarchy smaller than* $1 + \frac{\sqrt{13}-3}{4} \approx 1.15$.

6 Conclusion

We showed that, in the strong model of execution, the list algorithm SPT, which has an approximation ratio of $2 - 1/m$ is the best truthful deterministic algorithm, and that there is no truthful randomized algorithm which has an approximation ratio smaller than $3/2 - 1/(2\,m)$. On the contrary, if we relax the constraints on the execution model, i.e. if the result of a task which bid b is given to this task only b time units after its start, then we can obtain better results. In this model of execution, there is a truthful $4/3 - 1/(3\,m)$-approximate deterministic algorithm and a truthful optimal randomized algorithm. For both execution models, we also gave lower bounds on the approximation ratios that a truthful coordination mechanism can have.

As a future work, it would be interesting to improve the results for which a gap between the lower and the upper bound exists. For example, we believe that the lower bound $\frac{1+\sqrt{17}}{4}$ (lower bound on the performance of a truthful deterministic coordination mechanism for the weak model of execution) can be improved to $3/2$ for two machines.

Another direction would be to restrict the study to truthful algorithms (or coordination mechanisms) which run in polynomial time. Giving improved lower bounds which rely on a computational complexity argument would be very interesting.

Acknowledgments. We thank Elias Koutsoupias for helpful suggestions and discussions on the problem.

References

1. Ambrosio, P., Auletta, V.: Deterministic Monotone Algorithms for Scheduling on related Machines. In: Persiano, G., Solis-Oba, R. (eds.) WAOA 2004. LNCS, vol. 3351, pp. 267–280. Springer, Heidelberg (2005)
2. Andelman, N., Azar, Y., Sorani, M.: Truthful Approximation Mechanisms for Scheduling Selfish Related Machines. In: Diekert, V., Durand, B. (eds.) STACS 2005. LNCS, vol. 3404, pp. 69–82. Springer, Heidelberg (2005)
3. Angel, E., Bampis, E., Pascual, F.: Truthful Algorithms for Scheduling Selfish Tasks on Parallel Machines. In: Deng, X., Ye, Y. (eds.) WINE 2005. LNCS, vol. 3828, pp. 698–707. Springer, Heidelberg (2005)
4. Archer, A., Tardos, E.: Truthful Mechanisms for One-Parameter Agents. In: Proc. of FOCS 2001, pp. 482-491 (2001)
5. Auletta, V., Penna, P., De Prisco, R., Persiano, P.: How to Route and Tax Selfish Unsplittable Traffic. In: Proc. of SPAA 2004, pp. 196-204 (2004)

6. Auletta, V., De Prisco, R., Penna, P., Persiano, P.: Deterministic Truthful Approximation Mechanisms for Scheduling Related Machines. In: Diekert, V., Habib, M. (eds.) STACS 2004. LNCS, vol. 2996, pp. 608–619. Springer, Heidelberg (2004)
7. Clarke, E.: Multipart pricing of public goods. Public Choices, pp. 17-33 (1971)
8. Christodoulou, G., Koutsoupias, E., Nanavati, A.: Coordination Mechanisms. In: Díaz, J., Karhumäki, J., Lepistö, A., Sannella, D. (eds.) ICALP 2004. LNCS, vol. 3142, pp. 345–357. Springer, Heidelberg (2004)
9. Christodoulou, G., Koutsoupias, E., Vidali, A.: A lower bound for scheduling mechanisms. In: Proc. of SODA 2007 (2007)
10. Garey, M., Johnson, D.: Computers and Intractability: A Guide to the Theory of NP-Completeness. W. H. Freeman & Co (1979)
11. Graham, R.: Bounds on multiprocessor timing anomalies. In: SIAM Jr. on Appl. Math. vol. 17(2), pp. 416-429 (1969)
12. Groves, T.: Incentive in teams. Econometrica 41(4), 617–631 (1973)
13. Immorlica, N., Li, L., Mirrokni, V.S., Schulz, A.: Coordination Mechanisms for Selfish Scheduling. In: Deng, X., Ye, Y. (eds.) WINE 2005. LNCS, vol. 3828, pp. 55–69. Springer, Heidelberg (2005)
14. Koutsoupias, E., Papadimitriou, C.: Worst Case Equilibria. In: Meinel, C., Tison, S. (eds.) STACS 99. LNCS, vol. 1563, pp. 404–413. Springer, Heidelberg (1999)
15. Kovács, A.: Fast monotone 3-approximation algorithm for scheduling related machines. In: Brodal, G.S., Leonardi, S. (eds.) ESA 2005. LNCS, vol. 3669, pp. 616–627. Springer, Heidelberg (2005)
16. Mu'alem, A., Schapira, M.: Setting lower bounds on truhfulness. In: Proc. of SODA 2007 (2007)
17. Nisan, N., Ronen, A.: Algorithmic mechanism design. In: Proc. STOC 1999, pp. 129-140 (1999)
18. Pascual, F.: Optimisation dans les réseaux : de l'approximation polynomiale à la théorie des jeux. Ph.D Thesis, University of Evry, France, 2006 (in french).
19. Tchetgnia, A-A.: Truthful algorithms for some scheduling problems. Master Thesis MPRI, École Polytechnique, France (2006)
20. Vickrey, W.: Counterspeculation, auctions and competitive sealed tenders. J. Finance 16, 8–37 (1961)

Volume Computation Using a Direct Monte Carlo Method*

Sheng Liu[1,2], Jian Zhang[1], and Binhai Zhu[3]

[1] State Key Laboratory of Computer Science, Institute of Software,
Chinese Academy of Sciences, Beijing 100080, China
{lius,zj}@ios.ac.cn
[2] Graduate School, Chinese Academy of Sciences, Beijing 100049, China
[3] Department of Computer Science, Montana State University, Bozeman, MT
59717-3880, USA
bhz@cs.montana.edu

Abstract. Volume computation is a traditional, extremely hard but highly demanding task. It has been widely studied and many interesting theoretical results are obtained in recent years. But very little attention is paid to put theory into use in practice. On the other hand, applications emerging in computer science and other fields require practically effective methods to compute/estimate volume. This paper presents a practical Monte Carlo sampling algorithm on volume computation/estimation and a corresponding prototype tool is implemented. Preliminary experimental results on lower dimensional instances show a good approximation of volume computation for both convex and non-convex cases. While there is no theoretical performance guarantee, the method itself even works for the case when there is only a membership oracle, which tells whether a point is inside the geometric body or not, and no description of the actual geometric body is given.

1 Introduction

Volume computation is a highly demanding task in software engineering, computer graphics, economics, computational complexity analysis, linear systems modeling, VLSI design, statistics, etc. It has been studied intensively and lots of progress has been made especially in recent decades [7,2,14,13]. However, so far most of the research work is only concerned with the computational complexity aspect. For instance, some researchers tried to obtain a theoretical lower bound at all costs and neglected the practical feasibility of their algorithms. Therefore, although there are some strong theoretical results on this problem, little progress has been made in putting them into practical use. For this reason, this paper focuses on practically usable tools on computing/estimating the volume of a body.

---

* This work is partially supported by the National Natural Science Foundation (NSFC) under grant number 60673044 and 60633010, and by Montana EPSCOR Visiting Scholar's Program.

Generally speaking, we want to compute an arbitrary body's volume, i.e., in general a body does not even have to be a polyhedron. But for convenience, we will discuss convex polyhedron in most of the paper and in the end we will show some empirical results on some non-convex bodies' volume computation. Specially, we use the *halfspace representation* of a polyhedron. That is to say, a polyhedron is specified by $P = \{X | AX \leq B\}$ for some $m \times n$ matrix A and m-vector B (the pair (A, B) is called a *halfspace representation* of P).

As for two and three dimensional polyhedra, the volume of some basic shape (rectangle for instance) can be computed using some known mathematical formulas. But it is not straightforward to efficiently handle more general and complicated instances using exact analytic method. What is more, as the dimension n increases, the computational effort required arises drastically. Dyer and Frieze show that if P is a polyhedron then it is $\#P$-hard to evaluate the volume of P [4]. But later on, by introducing randomness into volume computation, Dyer, Frieze and Kannan gave a polynomial time randomized algorithm to estimate the volume to arbitrary accuracy in their pathbreaking paper [6]. That paper triggers the following works in this field, which reduce the complexity of the algorithm from n^{23} down to n^4 [10,9,1,5,11,8,12,13]. The method used is to reduce volume computation to sampling from a convex body, using the Multi-phase Monte Carlo method. It first constructs a sequence of incremental convex body $K_0 \subset K_1 \subset \cdots \subset K_m = V$ (the volume of K_0 is easy to compute). Then the volume of K_1 can be estimated by generating sufficiently many independent uniformly distributed random points in K_1 and counting how many of them fall in K_0, using an equation like equation (1). The other K_i's can be computed similarly. At last $V(K_m)$ is obtained. When generating random points, the Markov Chain method is adopted.

However, in contrast with the brisk development on the complexity aspect of the randomized algorithms, little attention is paid to bring the theoretical results into practical use. As mentioned in [3], although these randomized algorithms are very interesting, there does not seem to be any implementation of them. That is why the authors of [3] ignored the randomized algorithm in their practical study on volume computation. In this paper, in contrast, we mainly focus on such practical randomized approximate algorithms. Our algorithm is also based on a Monte Carlo sampling method. But, compared with those Markov chain Monte-Carlo-based algorithms, our method is much simpler and much easier to implement. What is more, preliminary experimental results are very promising.

2   The Framework of the Sampling Algorithm

Assume that the volume of the convex polyhedron to be computed is V, our algorithm tries to estimate the value of V by the random sampling method.

We first generate N uniformly distributed random points in the convex polyhedron P, then at step i we build a probing sphere S with radius r_i in the polyhedron[1]. After that we count the number of points that fall into S, denoted by N_p. Since the volume of the sphere (denoted by W) can be obtained immediately[2], we can easily obtain the volume of the polyhedron V_i from the following formula:

$$\frac{W}{V_i} = \frac{N_p}{N} \tag{1}$$

In particular, for the sake of higher accuracy, we carry out the probing procedure multiple times to obtain the average value (in the algorithm we use an adjustable parameter **Num_Of_Iteration** to denote this number).

Formally, the algorithm can be described as follows:

Algorithm 1. VOL(N)

1: generate N points randomly in P;
2: $sum = 0$;
3: **for** $i = 1$ to **Num_Of_Iteration do**
4:     build a probing sphere S in P with radius r_i;
5:     count the number of points in S;
6:     compute the volume V_i of P using formula (1);
7:     $sum = sum + V_i$;
8: **end for**
9: $V = sum/$**Num_Of_Iteration**;
10: **return** V;

---

3   Implementation

In general, the algorithm framework in section 2 is very simple and is easy to understand. But when putting it into practice, we must handle some difficult technical points and some of them need theoretical analysis.

3.1   Generating Random Points

As shown in Algorithm 1, first of all, we have to generate a lot of uniformly distributed points in the convex polyhedron. Generating points in a given (fat) convex body is easy. But it is hard to generate points that are distributed uniformly. Most previous work adopts a Markov Chain method to obtain theoretically uniformly distributed points. The idea is to divide the space into n-dimensional cubes and perform a random walk on all the cubes that lie within the given convex polyhedron. For each walk, one of the cubes that are orthogonally adjacent to the current cube in the convex body is randomly selected. Thus the random walk is ergodic and the stationary distribution is uniform on cubes in the

---

[1] Formally, an n-dimensional probing sphere S centering at $(o_1, o_2, \cdots, o_n)$ is defined as $S = \{(x_1, x_2, \cdots, x_n) | \sum_{j=1}^{n} (x_j - o_j)^2 \leq r_i^2\}$.

[2] We know that $W = \pi^{n/2} r_i^n / \Gamma(1 + n/2)$, where Γ denotes the gamma function [15].

convex polyhedron. However, this method is too complicated to use in practice. Instead, we use the pseudo-random number generator to generate many points in the polyhedron and assume them to be uniformly distributed. But there are still some uncertainties in our method. For instance, how many random points are needed. Apparently, for the sake of accuracy, the more the better. But on the other hand, more points mean more time and lower speed. So we must take both into consideration and find a proper compromise. In our preliminary implementation, the number of points is defined as an adjustable parameter so that it can be tuned according to different cases.

3.2   On Selecting the Center of the Sphere

Once the sampling problem has been solved, we begin to probe in the polyhedron with a probing sphere. But before probing, an implied prerequisite condition must be fulfilled. That is to say, the probing sphere must perfectly lie in the convex polyhedron. Otherwise, it is easy to see that the result obtained from equation (1) will not be accurate. But how to make sure that the whole probing sphere stays within the convex polyhedron? Strictly speaking, the distance from the center of the sphere to each of the facets of the convex polyhedron should be at least as large as the radius of the sphere. To achieve this goal, we define a new polyhedron contained in the original polyhedron named the *shrunk* polyhedron. The *shrunk* polyhedron has the same number of facets and vertices as the original polyhedron. In fact they should have the same shape except that it is a smaller version of the original polyhedron. Each facet of the *shrunk* convex polyhedron is parallel to its counterpart in the original convex polyhedron and the distance between them should be at least r_i. If we have such a *shrunk* polyhedron, then the problem can be solved easily by restricting the center of the sphere to lie within the *shrunk* polyhedron. If we use the *halfspace representation* of P, the *shrunk* polyhedron can be obtained easily and accurately by simply replacing each linear constraint $\sum_{k=1}^{n} a_{jk}x_k \leq b_j$ with $\sum_{k=1}^{n} a_{jk}x_k \leq b_j - r_i * \sqrt{\sum_{k=1}^{n} a_{jk}^2}$.

However, Algorithm 1 does not state that the body must be represented by linear constraints. In fact, the body can be presented to the algorithm using a very general mechanism called a *membership oracle*, which only tells whether a point is inside the body[3]. That is to say, the algorithm is also applicable to nonlinear constraints and other complicated constraints. But for these representations, it is hard for us to obtain the *shrunk* body accurately. So we have to use some approximate methods or heuristics. For example, we may adopt a random select-and-test heuristic. First, we randomly select a point in the original body as the center of the probing sphere. Then, given a radius r_i, we randomly choose some points on the surface of the probing sphere and test whether all of these points are also contained in the original body. If some point fails the test, we will try another point as the center of the probing sphere and perform this select-and-test procedure again. Formally, it can be formulated in Algorithm 2.

---

[3] Remember that in general a body may not be a polyhedron.

Algorithm 2. ForCenter(r_i)

1: FOUND=0;
2: **for** $j = 1$ to **Num_Try_Center do**
3:     Selecting a point in the body randomly as the center of the sphere;
4:     **for** $k = 1$ to **Num_Try_Surface do**
5:         Generating a point x on the surface of the sphere with radius r_i;
6:         **if** X is not in the body **then**
7:             break;
8:         **end if**
9:     **end for**
10:     **if** $k >$ **Num_Try_Surface then**
11:         FOUND=1;
12:         break;
13:     **end if**
14: **end for**
15: **return** FOUND;

The select-and-test heuristic is easy to carry out but there are still some details that we need to clarify. For example, how many points on the surface of the sphere should be tested in the testing process. In general, the more the better. But again in practice more points mean more resources and more running time. The number of points should also vary with the dimension of the probing sphere. We again use an adjustable parameter **Num_Try_Surface** to represent the number in our experiments and it turns out that the heuristic works very well.

3.3   On Radius Selection

As described above, when building the probing sphere, we should determine the radius of the sphere beforehand. It is easy to understand that the radius should not be too small so that there is no point falling into the probing sphere at all. That is to say, there should be at least some point in the sphere. Otherwise, that probing sphere is useless. Based on the random sampling method, we have the following probabilistic analysis.

Theorem 1. *Given an n-dimensional polyhedron with volume V and the total number of sampling points N, if the radius r of the probing sphere satisfies $r = \Theta(\sqrt[n]{\frac{V * \ln N}{N}})$ then the probability that there is at least one point in the sphere converges to 1 as N grows to infinity.*

Proof. Let W be the volume of the probing sphere. Let C_1 denote $\pi^{n/2}/\Gamma(1 + n/2)$. Then $W = C_1 * r^n$. Assume that $r = C_2 * \sqrt[n]{\frac{V * \ln N}{N}}$. Let E represent the event that the sphere is empty, then $\bar{E}$ represents the event that there is at least one point in the sphere. Then we have:

$$P[\bar{E}] = 1 - P[E]$$

$$\geq 1 - (\frac{V - W}{V})^N$$

$$= 1 - (1 - \frac{C_1 * r^n}{V})^N$$

$$= 1 - (1 - \frac{C_1 * C_2{}^n * \ln N}{N})^N \tag{2}$$

$$= 1 - e^{-C_1 * C_2{}^n * \ln N}$$

$$= 1 - \frac{1}{N^{C_1 * C_2{}^n}}$$

Although C_1 varies with dimension n [15], it is clear that given an n-dimensional instance, when N is big enough, $P[\bar{E}]$ will converge to 1. Thus we complete the proof.   □

On the other hand, in theory, the bigger r is, the better the approximation is.

Claim. Convergence is better when r is bigger.

Proof. Let $p = W/V$ denote the probability that a random point from the polyhedron also falls into the probing sphere. Then the distribution of the random variable N_p will conform to a binomial distribution $\mathbf{B}(N, p)$. Thus we have

$$\mathbf{Var}(N_p) = N * p * (1 - p) \qquad \mathbf{Mean}(N_p) = N * p$$

It follows that

$$\mathbf{Var}(N_p/N) = p * (1 - p)/N \qquad \mathbf{Var}(N_p)/\mathbf{Mean}(N_p) = 1 - p$$

The formula on $\mathbf{Var}(N_p/N)$ reveals that the convergence is better when p is near one[4] than near $1/2$. If $\mathbf{Var}/\mathbf{Mean}$ is used as a measure of convergence, we will also find that the smaller $1 - p$ is, the better the convergence is. While small $1 - p$ means big W, so it is easy to see that the claim holds.   □

The above analyses suggest that we had better find a radius that is as big as possible. Theoretically it is indeed the case but in practice a probing sphere with the largest possible radius may have its weakness in the uniformity of probing. Take a triangle for example, the largest probing sphere inside it may be the inscribed circle of it. If we use the largest probing sphere, we will restrict our probing to the sampling points in the inscribed circle only. However, because all the sampling points are simulated by pseudo-random numbers, we cannot guarantee whether they are absolutely uniformly distributed in any part of the polyhedron. Therefore, we make sure that the probing sphere can visit as many parts of the sampling points as possible, so as to make the probing more general. For this reason, a moderately large but not an extremely large radius may be more suitable.

---

[4] The case of p near 0 is trivial, so we do not consider it.

Theorem 1 also reveals that r depends on V, which is exactly something we need to compute. In theory we can estimate an upper bound of V by building another convex polyhedron, which contains the original convex polyhedron and has a volume that is easier to compute. For example, the smallest axis-parallel bounding box may be enough for that purpose. This method heavily depends on the ratio between the volume of the polyhedron and the volume of the bounding box. But the ratio can be very small. What is more, for instances with nonlinear constraints, it is not always convenient to obtain the smallest bounding box.

To handle this problem, we adopt a self-adaptive heuristic in our implementation. First, we randomly choose two of the sampling points in the polyhedron and let r be the distance between them. With the current radius r, if we fail to find a proper center of the sphere after **Num_Try_Center** tries using the heuristic method given in the previous subsection, we assign $r \leftarrow r/2$. Once we find a proper sphere center, we stop so as to make r as big as possible. Experiments show that this self-adaptive method not only works on polyhedra with normal shapes, but is also competent for polyhedra of long skinny shapes.

4   Experiments and Analysis

Based on the above observations, a prototype tool is implemented. We experiment on many simple instances to examine its performance. For the sake of comparison, we also test it on instances with known volume (named REAL volume) and examine the ratios of the results computed by our program to the REAL volume. Due to the space limit, we only introduce some simple ones.

4.1   Simple Examples

$Example\ 1$ $\begin{cases} -x + 2y \leq 200 \\ -y \leq 0 \\ x - y \leq 0 \\ -x - y \leq 50 \\ x + y \leq 200 \end{cases}$. It is in fact a pentagon with vertices $(0,0)$, $(-50, 0)$, $(-100, 50)$, $(200/3, 400/3)$, and $(100, 100)$ and its REAL volume (the area of the pentagon) is $38750/3$.

$Example\ 2$ $\begin{cases} x + y + z \leq 255 \\ -x \leq 0 \\ -y \leq 0 \\ -z \leq 0 \end{cases}$. It is in fact a tetrahedron defined by the four vertices $(0,0,0)$, $(255, 0, 0)$, $(0, 255, 0)$ and $(0, 0, 255)$. So we can easily obtain the REAL volume $(255 * 255 * 255)/6$ by hand.

We test our tool on these instances and check the ratio of the program results to the REAL volumes. The experimental results are given in Fig. 1 and 2 respectively in detail.

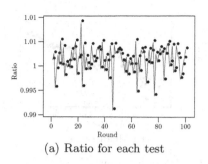

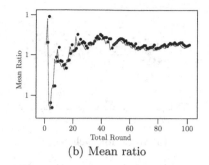

(a) Ratio for each test   (b) Mean ratio

Fig. 1. Experimental results of *Example 1*

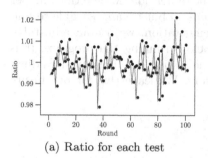

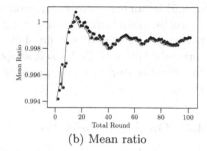

(a) Ratio for each test   (b) Mean ratio

Fig. 2. Experimental results of *Example 2*

4.2 Variance Analysis

Experimental results from both Fig. 1 and Fig. 2 show that our method indeed has a good approximation. The mean values are very close to 1 and they converge well. However, there are still some small differences between Fig. 1(a) and Fig. 2(a). For example, all the ratio values in Fig. 1(a) fall within a small interval while those in Fig. 2(a) fall within a relatively large interval. As far as the variance is concerned, why does the data in Fig. 2(a) have a larger variance compared with those in Fig. 1(a)? To find out the possible reason, we experiment on *Example 2* again with fewer sampling points. Results are presented in Fig. 3. Comparing Fig. 3(a) with Fig. 2(a), we find that given a fixed volume, fewer sampling points result in relatively larger variance.

Theoretically speaking, our sampling algorithm needs sufficiently many independent uniformly distributed points, but in practice we can only generate a finite number of points. How large should this number be? It is a problem to be settled. If we use a fixed number for each dimension, then instances with smaller volume will have larger density compared with those with bigger volume. Given two polyhedra in the same dimension, the volume ratio, however, can be arbitrary large, which may result in sharp difference on density. On the other hand,

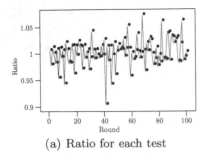

(a) Ratio for each test

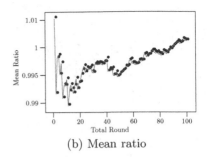

(b) Mean ratio

Fig. 3. Experimental results of *Example 2* (using fewer sampling points)

it is clear that this number should vary with the dimension. Maybe this number should be an exponential function on the dimension but we are not able to find a precise definition yet. In our current implementation, we use an adjustable parameter to denote the number of points before running the program. It is certainly better to find a self-adaptive heuristic, which can dynamically adjust the number during the running of the program. We leave this for our future work.

4.3   On the Mean Ratio

Although the variance varies with different sampling densities, all the mean ratios in Fig. 1(b), Fig. 2(b) and Fig. 3(b) show a very good approximation to 1. The curves reveal a nice trend of convergence to 1 as the total Round number grows from 1 to 100. This is easy to understand. Even if the ratio deviation of one particular case is 100%, it will only contribute 1% up to the total deviation of the mean value because it is divided equally among all the 100 sampling Rounds. So in general it makes sense to run more Rounds.

At last, we want to emphasize that the sampling algorithm is a general method. Although there are some negative results of it when the dimension

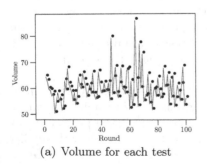

(a) Volume for each test

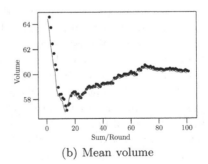

(b) Mean volume

Fig. 4. Experimental results

n grows to infinity [14], our experiments show that the method is still feasible for real-life instances in high dimension. However, as for high dimensional and general non-convex instances, although the method can compute/approximate volume, we cannot always carry out comparisons as we do above, because we have no other means to obtain the REAL volume for general instances. Despite that, for some two-dimensional non-convex cases (whose REAL volume can be computed analytically) our method can obtain a mean ratio which always converges to one. For more complex non-convex cases, we believe that our algorithm also has good approximations in practice. For instance, given an instance $\begin{cases} y \geq x^2 - 2 \\ y \leq \frac{1}{2}x^2 \\ -10 \leq z \leq 5 \\ xyz \leq 1 \end{cases}$, we do not know the REAL volume, but our tool can tell us that it is about 60, as depicted in Fig. 4.[5]

5   Concluding Remarks

Volume computation is a very hard but widely studied problem in mathematics and computer science. Generally speaking, there are two different methods on this problem: the exact method and the randomized approximation method. The exact method itself can be classified into two classes: triangulation methods and signed decomposition methods. Benno Büeler, Andreas Enge and Komei Fukuda present some practical study on these methods [3]. But these methods are only applicable to convex bodies with linear constraints. The randomized approximation method is a more general one. It can cope with almost any constraints presented to it. Although the randomized approximation method is relatively new, a lot of progress has been made since its birth. Our work is also based on the randomized method.

However, most of these efforts on randomized algorithms are on the complexity aspect and little practical studies are given. In this paper some implementation issues of volume computation are studied. Some techniques in randomized volume computation algorithms are quantitatively evaluated. For example, our algorithm is also based on Monte Carlo sampling, but we do not use the Multiphase method as described above. Instead, we use only one phase Monte Carlo method but we run the probing process many times to obtain a more accurate average result. On generating uniformly distributed points, we do not use the Markov Chain method although it does well in simulating the uniformly distribution in theory. Instead, we generate random points directly within the polyhedron and view them as uniformly distributed ones in our algorithm. Techniques and problems on efficient and effective implementation to achieve good performance are also discussed. Preliminary empirical results show that the tool developed by utilizing these results works very well. Of course, there are still some unsolved problems related to some of the manually adjustable parameters. We leave them for future research.

---

[5] See Appendix A for more examples.

References

1. David Applegate and Ravi Kannan. Sampling and integration of near log-concave functions. In: Proc. 23rd annual ACM symp. on Theory of Computing (STOC), pp. 156–163 (1991)
2. Bollobás, B.: Volume estimates and rapid mixing. Flavors of geometry. Math. Sci. Res. Inst. Publ. 31, 151–182 (1997)
3. Büeler, B., Enge, A., Fukuda, K.: Exact volume computation for polytopes: a practical study. Polytopes–combinatorics and computation (1998)
4. Dyer, M., Frieze, A.: On the complexity of computing the volume of a polyhedron. SIAM J. Comput. 17(5), 967–974 (1988)
5. Martin Dyer and Alan Frieze. Computing the volume of convex bodies: A case where randomness provably helps. In: Proc. 44th Symp. in Applied Mathematics (PSAM) (1991)
6. Dyer, M., Frieze, A., Kannan, R.: A random polynomial-time algorithm for approximating the volume of convex bodies. J. ACM 38(1), 1–17 (1991)
7. Gritzmann, P., Klee, V.: On the complexity of some basic problems in computational convexity: II. volume and mixed volumes. Polytopes: abstract, convex and computational (Scarborough, ON, 1993), NATO Adv. Sci. Inst. Ser. C Math. Phys. Sci., pp. 373–466 (1994)
8. Kannan, R., Lovász, L., Simonovits, M.: Random walks and an $O^*(n^5)$ volume algorithm for convex bodies. Random Struct. Algorithms 11(1), 1–50 (1997)
9. ó Lovász, L.: How to compute the volume? Jber. d. Dt. Math.-Verein, Jubiläumstagung, B. G. Teubner, Stuttgart, pp. 138–151 (1990)
10. Lovász, L., Simonovits, M.: The mixing rate of markov chains, an isoperimetric inequality, and computing the volume. In: Proc. 31th IEEE Annual Symp. on Found. of Comp. Sci (FOCS), pp. 482–491 (1990)
11. Lovász, L., Simonovits, M.: Random walks in a convex body and an improved volume algorithm. Random Struct. Algorithms 4(4), 359–412 (1993)
12. Lovász, L., Vempala, S.: Simulated annealing in convex bodies and an $O^*(n^4)$ volume algorithm. In: ó Lovász, L. (ed.) Proc. 44th IEEE Annual Symp. on Found. of Comp. Sci (FOCS), pp. 650–659 (2003)
13. Rademacher, L., Vempala, S.: Dispersion of mass and the complexity of randomized geometric algorithms. In: Proc. 47th IEEE Annual Symp. on Found. of Comp. Sci (FOCS), pp. 729–738 (2006)
14. Simonovits, M.: How to compute the volume in high dimension? Mathematical Programming 97, 337–374 (2003)
15. Weisstein, E.: Ball. From MathWorld – A Wolfram Web Resource (2003), available at http://mathworld.wolfram.com/Ball.html

A   More Examples

Example (a)   Given a 4-dimensional polyhedron below, its REAL volume is about 2/3. Our tool can also give the results depicted on the right of the below instance.

$$
\begin{cases}
-x + y + z - u \leq 1 \\
-x + y + z + u \leq 1 \\
-x + y - z - u \leq 1 \\
-x + y - z + u \leq 1 \\
-x - y - z - u \leq 1 \\
-x - y - z + u \leq 1 \\
-x - y + z + u \leq 1 \\
-x - y + z - u \leq 1 \\
x - y - z + u \leq 1 \\
x - y - z - u \leq 1 \\
x - y + z + u \leq 1 \\
x - y + z - u \leq 1 \\
x + y - z + u \leq 1 \\
x + y - z - u \leq 1 \\
x + y + z + u \leq 1 \\
x + y + z - u \leq 1
\end{cases}
$$

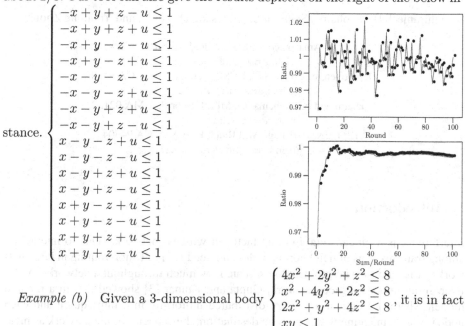

Example (b)   Given a 3-dimensional body $\begin{cases} 4x^2 + 2y^2 + z^2 \leq 8 \\ x^2 + 4y^2 + 2z^2 \leq 8 \\ 2x^2 + y^2 + 4z^2 \leq 8 \\ xy \leq 1 \end{cases}$, it is in fact the intersection of three ellipsoids and another instance denoted by $\{xy \leq 1\}$. We do not know its REAL volume, but our tool can tell us the approximate volume (about 16.2) as depicted below.

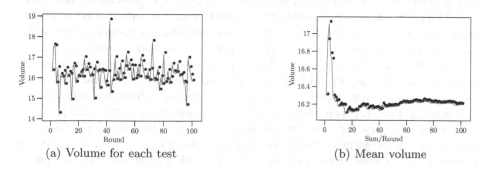

(a) Volume for each test   (b) Mean volume

Fig. 5. Experimental results

Improved Throughput Bounds for Interference-Aware Routing in Wireless Networks*

Chiranjeeb Buragohain[1], Subhash Suri[2], Csaba D. Tóth[3], and Yunhong Zhou[4]

[1] Amazon.com Inc., Seattle, WA 98104
chiran@amazon.com
[2] Department of Comp. Sci., UCSB, Santa Barbara, CA 93106
suri@cs.ucsb.edu
[3] Department of Mathematics, MIT, Cambridge, MA 02139
toth@math.mit.edu
[4] HP Labs, 1501 Page Mill Road, Palo Alto, CA 94304
yunhong.zhou@hp.com

1 Introduction

Interference is a fundamental limiting factor in wireless networks. Due to interaction among transmissions of neighboring nodes and need for multi-hop routing in large networks, it is a non-trivial problem to estimate how much throughput a network can deliver. In an important piece of work, Gupta and Kumar [3] showed that in a random model, where n identical nodes are distributed uniformly in a unit square and each node is communicating with a random destination, the capacity of the network as measured in bit-meters/sec is $O(\sqrt{n})$. This result articulates the packing constraint of the n paths: on average each path is $\Theta(\sqrt{n})$ hops long, and thus in the space of size $O(n)$, only $O(\sqrt{n})$ paths can be accommodated.

The Gupta-Kumar result is quite elegant, but its relevance to practical networks can be questioned because of the *random* source-destination (s–t) pairs assumption. As Li et al. [10] point out, such an assumption may hold in small networks, but as the network scales, it is unlikely that communications patterns will exhibit uniform randomness. Instead, the more relevant question is: given a particular network instance and a set of s–t pairs, what is the maximum throughput that this network can deliver? Motivated by this question, Jain et al. [4], Alicherry et al. [1], and Kumar et al. [8] have investigated the capacity of wireless networks for arbitrary source-destination pairs, and arbitrary networks. All these papers model the problem as a linear program (LP), and provide a computational scheme for estimating the throughput. This is indeed an important direction and, as one of our main results, we show that a novel *node-based* LP formulation combined with a *node ordering* technique yields a 1/3 approximation of the optimal throughput, which improves the previous best lower bound of 1/8. But we first begin with a natural fundamental question.

Is there a generalization of the Gupta-Kumar result for arbitrary networks and arbitrary sets of s–t pairs? In other words, can one estimate the network capacity in broad

* The research of Chiranjeeb Buragohain and Subhash Suri was supported in part by the National Science Foundation under grants CNS-0626954 and CCF-0514738.

G. Lin (Ed.): COCOON 2007, LNCS 4598, pp. 210–221, 2007.

terms, without resorting to computational techniques? And how widely does the capacity vary for different choices of s–t pairs in the network? Recall that in the random model of Gupta and Kumar, the gap between the best-case and worst-case bounds is only a constant factor.

Of course, it is easy to observe that without some additional parameters this question is not particularly meaningful. Because we measure throughput in the number of bits transmitted (and not bit-meters as Gupta and Kumar), the capacity can vary widely depending on how far apart the sources and destinations are. If each source is adjacent to its destination, then we can achieve a throughput of $\Theta(n)$; if source-destination pairs are $\Theta(n)$ distance apart (as in a path graph), then the throughput drops to $O(1)$. Thus, a natural and important parameter is the *distance* between the source and destination nodes. However, even if two input instances have roughly equal average distance between s–t pairs, their throughputs can vary widely (for instance, if a small cut separates all source-destination pairs). We show, however, that there is an intermediate ground of *structured network* and *arbitrary* s–t pairs, where such a characterization is possible. The special structure we consider is a *grid network*, which is a rather natural topology.

Our Contributions

1. Suppose we have n *arbitrarily paired* s–t pairs in an $\Theta(\sqrt{n}) \times \Theta(\sqrt{n})$ size grid network. We show that if the average (hop) distance among the s–t pairs is d, then it is always possible to achieve a total throughput of $\Omega(n/d)$. There are instances where this bound is tight. The upper bound on the throughput follows easily from a packing argument; our contribution is to show that $\Omega(n/d)$ throughput is *always* achievable.

2. The $\Omega(n/d)$ throughput in a grid network can be achieved by a simple routing scheme that routes each flow along a *single* path. Both the routing and the scheduling algorithms are simple, deterministic, and distributed. Thus, for the grid topology networks, one can achieve (asymptotic) worst-case optimal throughput *without resorting to* computationally expensive LP type methods.

3. Our third result concerns an approximation bound for the throughput in a general network: arbitrary network topology and arbitrary s–t pairs. In contrast to previous work [4,1,8], we introduce two novel ideas: improved interference constraints at the *node* level, and improving the approximation ratio by imposing an *ordering* on the nodes. As a result of these two ideas, we achieve an approximation ratio of 3 for the optimal throughput, improving all previous bounds.

4. An interesting corollary of our LP formulation is that it yields *provably* optimal throughput if the network topology has a special structure, such as a tree. Tree-like topologies may be quite natural in some wireless mesh networks, especially at the peripheries.

5. We show through experimentation that our node-based LP routing delivers excellent performance. In most cases, it achieves twice the throughput of the edge-based LP, and is typically within 10% of the optimal.

6. All LP based techniques split flows across multiple paths, and an obvious question is to bound the integrality gap between the optimal multi-path and single-path

routes. Simulations studies [8] suggest that, for random inputs, natural heuristics based on the classical shortest path schemes can give acceptable results. In the full version of this paper, we show that three straightforward routing schemes can have arbitrarily small throughput. On the other hand, in the special case of grid networks, we show that one can efficiently compute a single path route whose end to end throughput is within a constant factor of the optimal single path throughput.

2   Preliminaries and Related Work

We assume a standard graph model of wireless networks. The network connectivity is described by an undirected graph $G = (V, E)$, where V denotes the set of ad-hoc wireless nodes, and E denotes the set of node-pairs that are neighbors. The communication radius of every radio node $i \in \{1, 2, \ldots, n\}$ is R; throughout the paper, we assume that the communication occurs on a single radio channel, although the extension to multiple channels is straightforward. Each communicating node causes interferences at all other nodes within distance ϱ from it, where $\varrho \geq R$, is called the *interference radius* of the node. Note that we assume that all radios have an identical communication radius R, and an identical interference radius ϱ. In order to simplify the discussion, we assume that $\varrho = R$, but all our arguments can be easily extended to the general case of $\varrho > R$.

A problem instance is a network $G = (V, E)$, and a set of k source-target pairs $(s_j, t_j), j = 1, 2, \ldots, k$, where s_j and t_j are nodes of V. We assume that each source s_j wants to transmit to its target t_j at a normalized rate of 1. For simplicity, we also assume that the channel capacity is also 1; again, these are easily generalized to different rates. Our problem is to maximize the network *throughput*, which is the total amount of traffic that can be scheduled among all the s–t pairs subject to the capacity and interference constraints.

Models of Interference. The wireless network uses a broadcast medium, which means that when one node transmits, it causes interference at the neighboring nodes, preventing them from receiving (correct) signals from other nodes. The details of which nodes cause interference at which other nodes depend on the specifics of the MAC protocol being used. In this paper, we adopt the interference model corresponding to the IEEE 802.11-like MAC protocols, which require senders to send RTS control messages and receivers to send CTS and ACK messages. Currently, this is the most widely used MAC protocol in wireless networks. Under this protocol, two edges are said to *interfere* if either endpoint (node) of one is within the interference radius ϱ of a node of the other edge. In other words, the edges ij and kl interfere if $\max\{\text{dist}(i, k), \text{dist}(i, l), \text{dist}(j, k), \text{dist}(j, l)\} \leq \varrho$. It is clear that if a set of edges pairwise interfere with each other, then only one of those edges can be active at any point of time.

There are several other models of interference in the literature. The *protocol model* introduced by Gupta and Kumar [3] assumes that the transmission from node i is received correctly at j if no other node k is transmitting within interference range ϱ of j. This model corresponds to MAC protocols that *do not require an ACK from the receiver*. The throughput of a network can be higher under the protocol model because it assumes a weaker interference condition than the 802.11-like protocols. The *transmitter model* introduced in Kumar et al. [8] assumes that two transmitting nodes are in conflict unless they are separated by *twice the interference range* (2ϱ). The interference condition assumed here is unnecessarily stronger than 802.11 MACs and leads to a lower estimate of throughput of the network. While we have chosen to work with the 802.11 model of interference, our methodology is quite general, and can be applied to these other models as well.

Related Work. Gupta and Kumar [3] provide (near) tight bounds on the throughput capacity of a *random* network, where the nodes are placed randomly in a square and sources and destinations are randomly paired. They show that the expected throughput available to each node in the network is $\Theta(1/\sqrt{n})$. Their result essentially articulates that interference leads to *geometric packing* constraint in the medium. In a follow up work, Li et al. [10] did simulations and experiments to measure the impact of interference in some realistic networks. They made the case that it might not be realistic to assume random s–t pairs. They argue that if s–t pairs are not too far from each other then the throughput improves; in fact, they observe that the throughput is bounded by $O(n/d)$ if the average s–t separation is d. They cannot tell, however, if this throughput bound can always be achieved. Kyasanur et al. [9] have recently extended the work of Gupta and Kumar [3] to study the dependence of total throughput on the number radio channels and interfaces on each network node.

While the results of Gupta-Kumar and Li et al. focused on random or grid-like networks, they did not address a very practical question: given a particular instance of a network and a set of s–t pairs, how much throughput is achievable? Jain et al. [4] formalized this problem, proved that it is NP-hard, and gave upper and lower bounds to estimate the optimal throughput. Their methods, however, do not translate to polynomial time approximation algorithms with any provable guarantees. Kodialam et al. [5] studied a variant of the throughput maximization problem for arbitrary networks, but they do not consider the effect of interference in detail. Recently Padhye et al. [11] have taken significant steps to measure interference between real radio links. Raniwala et al. [12] have designed and implemented a multichannel wireless network. Draves et al. [2] have proposed routing metrics to efficiently route in such networks. On the theoretical side, the problem of maximizing throughput in a network using multiple channels and interfaces have been studied by Alicherry et al. [1] and Kodialam et al. [6].

Kumar et al. [8,7] were the first to give a constant factor approximation algorithm for the throughput maximization problem in a network with a single radio channel. In particular, they give a 5-approximation algorithm for throughput, their algorithm assumes the *transmitter model*. As we mentioned earlier, the transmitter model is unduly restrictive compared to the 802.11-like models, and their algorithm does not give any explicit approximation bound for the 802.11 model. As mentioned above, Alicherry et al. [1] considered the problem of routing in the presence of interference with multiple radio channels and interfaces. As part of that work, they give an approximation algorithm for the throughput maximization problem with a constant factor guarantee under the 802.11-like model using interference constraints between edges. Their approximation factor is 1/8 for the case of $\varrho = R$, and it becomes progressively worse as ϱ becomes larger compared to R. By contrast, our approximation factor is 1/3, and does not depend on the ratio ϱ/R.

3   Maximum Throughput for Grid Topologies

Before we discuss our linear programming approach for computing interference-aware routes in arbitrary networks, it is worth asking to what extent one can estimate the throughput using *structural* facts, in the style of Gupta and Kumar [3]. In other words, are there simple characterizations of the network and the s–t distributions that allow us to derive good estimates of the achievable throughput *without* resorting to computationally expensive methods such as linear programming. We do not know of any result of this type for completely general setting (nor is one likely to exist), but we show below that for special network topologies, such as grids, one can obtain a bound on achievable throughput based on average separation among source-destination pairs. Furthermore, our investigation also leads to a simple and distributed routing scheme that achieves the optimal throughput using *single* paths.

Consider a grid network of size $\Theta(\sqrt{n}) \times \Theta(\sqrt{n})$, which can be thought of as a square lattice in the plane. We assume there are n source-destinations pairs, arbitrarily chosen by the user (or adversary). We assume that all sources and all destinations are distinct. We assume that $R = \varrho = 1$, each edge in the network has capacity 1, and each source wants to communicate with its destination at the rate of 1. We assume that these demands are persistent, i.e. the flow demands are constant over time and we are interested in the steady state flow. We wish to maximize the total throughput among all the s–t pairs. (For the moment, we do not worry about fairness among different pairs, but will briefly discuss that issue in Section 5.)

Manhattan Routing. We first consider the case when each s–t pair has (lattice) distance d. In the following subsection, we will generalize the result to average distances. A simple packing argument shows that the maximum possible throughput is at most $O(n/d)$; a similar observation was also made in Li et al. [10]. But it is far from obvious that $\Omega(n/d)$ throughput can *always* be realized (for adversarially chosen s–t pairs). By clustering sources on one side, and destinations on the other, it may be possible to create significant bottlenecks in routing.

In fact, one can see that a simple-minded routing scheme can lead to very low throughput. Consider, for instance, the particular choice of s–t pairs shown in Fig. 1. There are 4 source-destination pairs $\{(A, B), (C, D), (E, F), (G, H)\}$. Suppose we route each flow using the shortest paths, staying as close as possible to the straight line joining the s–t pair. These routes are shown using the dotted lines in the figure. Observe that all these paths go through a common node N, which becomes the bottleneck, and limits the total throughput to 1. Nevertheless, the following result shows that for any configuration of n source-destination pairs, one can achieve $\Theta(n/d)$ throughput.

Theorem 1. *Consider n source-destination pairs in an $\Theta(\sqrt{n}) \times \Theta(\sqrt{n})$ size grid, with all sources and all destinations distinct. Suppose that each s–t pair has (lattice) distance d. Then, one can always achieve a throughput of $\Omega(n/d)$, and this is also the best possible bound.*

Due to space limitations, the proof is given in the full version of this paper.

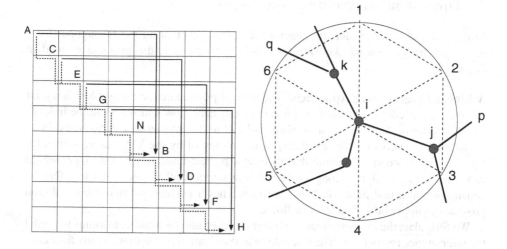

Fig. 1. Illustration of the Manhattan routing. The source destination pairs are (A,B), (C,D), (E,F) and (G,H).

Fig. 2. Interference zone for a single node i

Extension to Average or Median Distances. The strict distance requirement for all s–t pairs is clearly too restrictive. We now show that the result actually holds more broadly, for the case when d is either the *average* or the *median* distance among all pairs.

Theorem 2. *Consider n source-destination pairs in an $\Theta(\sqrt{n}) \times \Theta(\sqrt{n})$ size grid, with all sources and destinations distinct. Suppose that the* average *(lattice) distance between the s–t pairs is d. Then, one can always achieve a throughput of $\Omega(n/d)$.*

Proof. We simply observe that if the n pairs have average distance d, then at least half the pairs must be at distance less than $2d$. We set the rate for all the pairs whose separation is larger than $2d$ to zero, and route the remaining pairs using Manhattan routing. By Theorem 1, the throughput of these routes is $\Omega(n/d)$.

A very similar argument shows that a throughput of $\Omega(n/d)$ is also achievable when the *median* s–t distance is d.

These bounds characterize the throughput of an instance based on just one key parameter: the separation among the source and destination pairs. Given an instance of the problem, a network manager can now deduce the asymptotic worst-case optimal throughput of the network simply from the distances among the source-destination pairs. From a network manager's perspective, this result is an encouraging one: while the traffic matrix of a network is beyond control, the network topology is something she can control. Thus, our result suggests that in sufficiently regular network topologies, one can consistently achieve high throughput *and* do so through single path routing.

4 Throughput in Arbitrary Topologies

In this section, we consider the general problem of estimating the throughput for a given (arbitrary) network with arbitrary s–t pairs (namely, the problem defined in Section 2).

A Linear Programming Approach. The throughput maximization problem is a joint routing and scheduling problem: we need to route each flow and schedule the links so that flows can be feasibly accommodated subject to the interference constraints. In an actual wireless network, the scheduling is taken care of by the MAC layer —thus for this discussion we shall assume that there is a perfect underlying MAC layer which can schedule a solution as long as the solution respects all the flow and interference constraints. In a real network, MAC layers are never perfect and hence our solution provides an upper bound on feasible flows.

We formulate the flow problem as a linear program and then add constraints to model the interference restrictions. The throughput maximization problem with only flow constraints is just the classical max-flow problem:

$$\text{Maximize} \quad \sum_{i \in N(s)} f_{si} \text{ subject to}$$

$$\sum_{j \in N(i)} f_{ij} = \sum_{j \in N(i)} f_{ji}, \ \forall \, i \neq s, t$$

$$0 \leq \quad f_{ij} \leq 1, \ \forall \, ij \in E \tag{1}$$

where $N(i)$ denotes the set of nodes adjacent to i, and f_{ij} denotes the amount of flow in edge ij from node i to node j, for each edge $ij \in E$. The objective function maximizes the total flow out of s subject to the capacity constraint on each edge; the other constraint imposes the flow conservation condition at each intermediate node. In order to simplify the discussion, we have assumed that there is only one source-destination pair (s, t). The extension to multiple pairs is straightforward: in each term, we sum over all flows instead of just one.

We now describe how to supplement this standard multicommodity flow problem with interference constraints. The key difficulty in designing an approximation algorithm for the throughput maximization problem lies in resolving conflicts between neighbors who interfere with each other. Jain et al. [4] model this constraint using an *independent set* framework, which attempts to resolve the conflicts globally. Unfortunately, finding an independent set is NP-complete, and so it does not lead to a polynomial time approximation.

Our approach is to resolve the conflicts locally, and model the problem as a *geometric packing* problem. For example, consider an edge ij and all the edges that interfere with it. When the edge ij is active, none of the edges with which it interferes can be active. Because each edge carves out a portion of the space (its interference zone) while it is active, one can use packing arguments to derive upper and lower bounds on the throughput. Indeed, similar ideas are used in [1], where the packing constraints are formulated in the space around each edge. Unfortunately, the constant factor in their approximation is rather large, and it also depends on the ratio between the interference and the radio ranges ϱ/R. [1]

Instead of modeling the interference around edges, as has been done by others, we introduce two new ideas that lead to improved algorithms and approximation bounds. We model the interference around nodes, and introduce an ordering over nodes. These two ideas allow us to guarantee an approximation ratio of 3, which is independent of ϱ.

Modeling Interference Constraints at Nodes. Let us assume that the flow of data through the network is like fluid which is infinitely divisible. Then in a steady state, suppose an edge ij supports the flow $f_{ij} \leq 1$ (recall that each edge has unit capacity). This means that given a *unit time interval*, the edge ij is required to be active for a fraction of time f_{ij} and remains inactive for the rest of the time. Towards this end, we introduce two sets of variables τ_i and τ_{ij} as follows.

$$\tau_{ij} = f_{ij} + f_{ji} \leq 1,$$
$$\tau_i = \sum_{j \in N(i)} \tau_{ij} \leq 1, \ \forall \, i \in V. \tag{2}$$

Here τ_{ij} represents the *total* fraction of the unit time interval that an edge ij is active and similarly τ_i is the fraction of time for the node i. Using these variables, we now introduce the *node interference* constraint which enforces the interference restrictions. Consider the node i shown in Fig. 2, and the set of its neighbors (within interference range) denoted by $N(i)$. It is clear that while any node j in the set $N(i) \cup \{i\}$ is transmitting, all other nodes in this set must be inactive unless there is a single node that is communicating with j. This leads us to the following constraint:

$$\sum_{j \in N(i) \cup \{i\}} \tau_j - \sum_{j,k \in N(i) \cup \{i\}, \ jk \in E} \tau_{jk} \leq 1, \ \forall i \in V, \tag{3}$$

where E denotes the edges of the interference graph. To understand this inequality, let us consider the unit time interval and in that time interval, which nodes can be active for how long. The first term in LHS, counts the total amount of time (out of the unit time interval) that nodes are active in the neighborhood of i. The second term accounts for the fact that if two nodes j and k in the neighborhood of i are communicating with each other, the time they spend communicating to each other is counted only once.

By construction, if the nodes satisfy condition (3), then the flow is definitely free of interference. But condition (3) is actually more restrictive than necessary. For instance, consider the nodes j and k in Fig. 2, which are separated by a distance larger than the radio range. Constraint (3) implies that the edges jp and kq cannot be active at the same time, while in reality they can. Eliminating such unnecessary constraints is key to our improved analysis, and so we next introduce the idea of node ordering.

---

[1] Lemma 1 of [1] proves an approximation bound of 8 for $\varrho = 2$. They also claim an approximation bound of 4 for $\varrho = 1$, which appears to be wrong, and should be 8. Also, the approximation factor grows as the ratio ϱ/R grows. For instance, the factor is 12 for $\varrho/R = 2.5$.

Node Ordering. Consider a total order on the nodes. (We will prescribe a specific order shortly.) Observe that the interference relation is symmetric. If nodes i and j interfere with each other, then constraint (3) imposes the interference condition twice: once when we consider the neighborhood of i and once for j. Therefore, if i precedes j in the ordering, then it is enough to only consider the constraint introduced by i on j. Specifically, let $N_L(i)$ denote the set of interfering nodes *preceding* node i in the ordering, then the following relaxed constraint still ensures an interference-free schedule.

$$\sum_{j \in N_L(i) \cup \{i\}} \tau_j - \sum_{j,k \in N_L(i) \cup \{i\},\ jk \in E} \tau_{jk} \leq 1, \quad \forall i \in V. \tag{4}$$

In order to define $N_L(i)$, any arbitrary ordering over the nodes will work. To get a good approximation factor, we specify the following *lexicographical* order on the nodes: i *precedes* j if and only if, denoting the coordinates of the points by $i = (x_i, y_i)$ and $j = (x_j, y_j)$, we have either $x_i < x_j$ or $x_i = x_j$ and $y_i < y_j$.

LP-NODE. We are now ready to describe the complete linear program, LP-NODE.

$$\text{Maximize} \quad \sum_{i \in N(s)} f_{si} \quad \text{subject to}$$

$$\sum_{j \in N(i)} f_{ij} = \sum_{j \in N(i)} f_{ji}, \quad \forall\, i \neq s, t$$

$$0 \leq f_{ij} \leq 1, \quad \forall\, ij \in E$$

$$\tau_{ij} = f_{ij} + f_{ji} \leq 1,$$

$$\tau_i = \sum_{j \in N(i)} \tau_{ij} \leq 1, \quad \forall\, i \in V,$$

$$\sum_{j \in N_L(i) \cup \{i\}} \tau_j - \sum_{j,k \in N_L(i) \cup \{i\},\ jk \in E} \tau_{jk} \leq 1, \quad \forall\, i \in V. \tag{5}$$

By construction, the solution to LP-NODE leads to a feasible flow. This flow can be scheduled. One can show (the proof is available in the full paper) that f_{NODE} gives a factor 3 approximation to f_{OPT}.

Theorem 3. *The flow produced by the solution of* LP-NODE *satisfies*
$$f_{\text{NODE}} \leq f_{\text{OPT}} \leq 3 f_{\text{NODE}}.$$

Our technique can easily be extended to the case that the interference range ϱ is larger than radio range $R = 1$. Consider any $\varrho > 1$. The last constraint in LP-NODE will now include all nodes which are within interference range of i. We can see from Fig. 2, that within a semicircle of radius ϱ, we can still pack at most 3 nodes which do not interfere with each other and hence the approximation bound given above, holds for *any* $\varrho > R$. By contrast, the approximation ratio given by Alicherry et al. [1] grows monotonically with increasing ϱ; it is 8 when $\varrho = 2R$, 12 when $\varrho = 2.5R$, and so on.

Optimal Throughput for Tree-Structured Networks. If the underlying network is a tree, then we can show (the proof is available in the full paper) that a variation of our LP-NODE can solve the throughput maximization problem optimally.

Theorem 4. *If the network connectivity graph is a tree, then we can solve the through-put maximization problem optimally using a variant of* LP-NODE.

5 Experimental Results

We ran experiments on both the regular as well as random networks. The random net-works consist of n nodes spread over a square $\sqrt{n} \times \sqrt{n}$ area with radio range 3.0. Any two nodes which are within radio range can communicate. This radio range was chosen so that the network is almost always connected. We assume that we are using a bidirectional MAC protocol like 802.11 and the radio range as well as interference range are the same. We assume that each link can support 1 unit of throughput. In our evaluation, we used three algorithms:

- LP-NODE: This is our main linear program described in Section 4. This algorithm has provable worst-case approximation ratio of 3.
- LP-EDGE: This is the best previously known linear programming based scheme, as described in Alicherry et al. [1]. This algorithm has an approximation ratio of 8, under the condition that $\varrho = R$.
- OPTIMAL: Since the throughput maximization problem is NP-Complete, there is no polynomial time scheme to compute the maximum throughput. We therefore use the independent set enumeration method as described by Jain et al. [4]. We enumerate larger and larger number of independent sets and estimate the throughput until adding more independent sets do not improve the throughput any more. At this point we declare convergence and use the final throughput as optimal.

Throughput Scaling With Network Size. In this experiment, we wanted to see how well LP-NODE's performance scales with the network size. We used a random network topology where the nodes were distributed uniformly at random throughout a square area. The source and destination are located at diagonally opposite corners. We then increased the number of nodes in the network from 32 to 64 to 96. In each case, we also computed the optimal throughput f_{OPT} by running the OPTIMAL algorithm.

In Fig. 3, left, we plot the throughput of the OPTIMAL, LP-NODE and LP-EDGE algorithms. Our LP-NODE algorithm shows excellent performance and yields close to 90% of the optimal throughput. By contrast, LP-EDGE performs much worse and achieves only 50%-60% of the OPTIMAL. In fact, even with a single source-destination pair, LP-EDGE at times failed to achieve 1/3 of the optimal throughput, which one could have achieved by routing along a single path [10]! With a single s–t pair, the maximum possible throughput using multipath routing is $5/6$; by contrast, the maximum through-put using a single path is $1/3$. In these cases, the constant factors in the approximation algorithms become crucially important, and the LP-NODE algorithm does well.

Throughput Scaling with Source-Destination Pairs. In this experiment, we fixed the network and increased the number of s–t pairs in the network to evaluate the throughput that the various routing schemes achieve. We used a random network topology with 64 nodes and up to 16 source destination pairs organized in a crosshatch pattern. In Fig. 3, middle, we plot the total throughput using LP-EDGE, LP-NODE and OPTIMAL

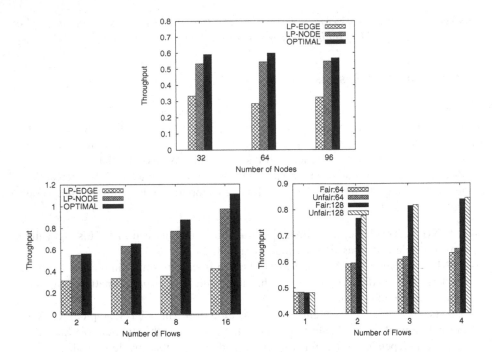

Fig. 3. Upper left: Performance of the LP-NODE, LP-EDGE and OPTIMAL algorithms compared for 32, 64 and 96 node networks. Upper right: The total throughput for different numbers of flows for a 64 node network. Down: Effect of fairness on total flow for 64 and 128 node networks. Note that the fairness constraint lowers total throughput by only a small amount.

algorithms for different number of source destination pairs. As expected we see that the throughput increases as the number of flows increases, but the dependence is not linear because interference from one set of paths reduces throughput for other pairs. Again, LP-NODE shows excellent performance, reaching near-optimal throughput in most cases, while LP-EDGE achieves less than half the throughput of LP-NODE.

Impact of Fairness on Flows. When multiple flows compete for bandwidth, the optimal flow is not necessarily fair. In practice though, fairness is an important criterion in any network protocol. To investigate the effect of fairness we again used the uniform random topology in a square area with four source destination pairs which intersect at the center of the square. For multiple flows we enforced the simplest fairness condition that each flow gets an equal amount of the total flow. We computed the total throughput using the LP-NODE algorithm and the results are shown in Fig. 3, right. As expected we see that enforcing fairness reduces total throughput, but surprisingly, *the effect is very mild.* In fact for the larger 128 node networks, the throughput for fair and unfair flows is almost identical. This is due to the fact that in larger networks the nodes have a lot of freedom in routing the flows and hence overall interference in any single node is low. Thus every pair can route equal amounts of flow without congestion.

References

1. Alicherry, M., Bhatia, R., Li, L.: Joint channel assignment and routing for throughput optimization in multiradio wireless mesh networks. In: Proc. Mobicom, ACM Press, New York (2005)
2. Draves, R.P., Padhye, J., Zill, B.: Routing in multi-radio, multi-hop wireless mesh network. In: Proc. MobiCom, ACM Press, New York (2004)
3. Gupta, P., Kumar, P.R.: The capacity of wireless networks. IEEE Trans. Inf. Theory 46, 388–404 (2000)
4. Jain, K., Padhye, J., Padmanabhan, V., Qiu, L.: Impact of interference on multi-hop wireless network performance. In: Proc. Mobicom, ACM Press, New York (2003)
5. Kodialam, M., Nandagopal, T.: Characterizing the achievable rates in multihop wireless networks. In: Proc. MobiCom, ACM Press, New York (2003)
6. Kodialam, M., Nandagopal, T.: Characterizing the capacity region in multi-radio multi-channel wireless mesh networks. In: Proc. MobiCom, ACM Press, New York (2005)
7. Kumar, V.S.A., Marathe, M.V., Parthasarathy, S., Srinivasan, A.: End-to-end packet-scheduling in wireless ad-hoc networks. In: Proc. SODA, pp. 1021–1030. ACM Press, New York (2004)
8. Kumar, V.S.A., Marathe, M.V., Parthasarathy, S., Srinivasan, A.: Algorithmic aspects of capacity in wireless networks. In: Proc. SIGMETRICS, ACM Press, New York (2005)
9. Kyasanur, P., Vaidya, N.H.: Capacity of multi-channel wireless networks: Impact of number of channels and interfaces. In: Proc. MobiCom, ACM Press, New York (2005)
10. Li, J., Blake, C., Couto, D.S.J.D., Lee, H.I., Morris, R.: Capacity of ad hoc wireless networks. In: Proc. MobiCom, ACM Press, New York (2001)
11. Padhye, J., Agarwal, S., Padmanabhan, V., Qiu, L., Rao, A., Zill, B.: Estimation of link interference in static multi-hop wireless networks. In: Proc. Internet Measurement Conf. ACM Press, New York (2005)
12. Raniwala, A., Chiueh, T.-C.: Architecture and algorithms for an ieee 802.11-based multi-channel wireless mesh network. In: Proc. INFOCOM, IEEE (2005)

Generating Minimal k-Vertex Connected Spanning Subgraphs

Endre Boros[1], Konrad Borys[1], Khaled Elbassioni[2], Vladimir Gurvich[1],
Kazuhisa Makino[3], and Gabor Rudolf[1]

[1] RUTCOR, Rutgers University, 640 Bartholomew Road, Piscataway NJ 08854-8003
{boros,kborys,gurvich,grudolf}@rutcor.rutgers.edu
[2] Max-Planck-Institut für Informatik, Saarbrücken, Germany
elbassio@mpi-sb.mpg.de
[3] Division of Mathematical Science for Social Systems, Graduate School of
Engineering Science, Osaka University, Toyonaka, Osaka, 560-8531, Japan
makino@sys.es.osaka-u.ac.jp

Abstract. We show that minimal k-vertex connected spanning sub-graphs of a given graph can be generated in incremental polynomial time for any fixed k.

1   Introduction

Vertex and edge connectivity are two of the most fundamental concepts in network reliability theory. While in the simplest case only the connectedness of an undirected graph, that is, the presence of a spanning tree, is required, in practical applications higher levels of connectivity are often desirable. Given the possibility that the edges of the network can randomly fail the reliability of the network is defined as the probability that the operating edges provide a certain level of connectivity. Most methods computing network reliability depend on the efficient generation of all (or many) minimal subsets of network edges which guarantee the required connectivity [5,14].

In this paper we consider the problems of generating minimal k-vertex connected spanning subgraphs. An undirected graph G on at least $k + 1$ vertices is *k-vertex connected* if every subgraph of G obtained by removing at most $k - 1$ vertices is connected. A subgraph of a graph G is *spanning* if it has the same vertex set as G.

For a fixed integer k we define the problem of generating minimal k-vertex connected spanning subgraphs as follows:

Input: A k-vertex connected graph G
Output: The list of all minimal k-vertex connected spanning subgraphs of G

Note that the output of the above problem may consist of exponentially many subgraphs in terms of the input size. Thus, the efficiency of generation algorithms is measured customarily in both the input and output size (see e.g., [14,10,7]). An algorithm generating all elements of a family $\mathcal{F}$ is said to run in *incremental polynomial time* if generating K elements of $\mathcal{F}$ (or all if $\mathcal{F}$ has less than K

G. Lin (Ed.): COCOON 2007, LNCS 4598, pp. 222–231, 2007.
© Springer-Verlag Berlin Heidelberg 2007

elements) can be done in time polynomial in K and the size of the input, for an arbitrary integer K.

Our problems include as a special case the problem of generating spanning trees ($k = 1$), which can be solved efficiently [12,6,11,1]. The problem of generating 2-vertex connected subgraphs and its generalization for matroids has been considered in [8].

1.1 Main Results

We show that this generation problem can be solved in incremental polynomial time.

Theorem 1. *For every K we can generate K minimal k-vertex connected spanning subgraphs of a given graph in $O(K^3 m^3 n + K^2 m^5 n^4 + K n^k m^2)$ time, where $n = |V|$, $m = |E|$.*

We remark that the running time of our algorithm depends exponentially on k. The complexity of the above problem when k is also part of the input remains an open question.

1.2 The $X - e + Y$ Method

In this section we recall a technique from [9], which is a variant of the supergraph approach introduced by [13]. Let $\mathcal{C}$ be a class of finite sets and for every $E \in \mathcal{C}$ let $\pi : 2^E \to \{0, 1\}$ be a monotone Boolean function, i.e., one for which $X \subseteq Y$ implies $\pi(X) \leq \pi(Y)$. We assume that $\pi(\emptyset) = 0$ and $\pi(E) = 1$. Let

$$\mathcal{F} = \{X \mid X \subseteq E \text{ is a minimal set satisfying } \pi(X) = 1\}.$$

Our goal is to generate all sets belonging to $\mathcal{F}$.

We remark that for every $X \subseteq E$ for which $\pi(X) = 1$ we can derive a subset $Y \subseteq X$ such that $Y \in \mathcal{F}$, by evaluating π exactly $|X|$ times. This can be accomplished by deleting one-by-one elements of X whose removal does not change the value of π. To formalize this, we can fix an arbitrary linear order $\prec$ on elements of E, without any loss of generality, and define a mapping $Project : \{X \subseteq E \mid \pi(X) = 1\} \to \mathcal{F}$ by $Project(X) = X \setminus Z$, where Z is the lexicographically first subset of X, with respect to $\prec$, such that $\pi(X \setminus Z) = 1$ and $\pi(X \setminus (Z \cup e)) = 0$ for every $e \in X \setminus Z$. Clearly, by trying to delete elements of X in their $\prec$-order, we can compute $Project(X)$, as we remarked above, by evaluating π exactly $|X|$ times.

We next introduce a directed graph $\mathcal{G} = (\mathcal{F}, \mathcal{E})$ on vertex set $\mathcal{F}$. We define the neighborhood $N(X)$ of a vertex $X \in \mathcal{F}$ as follows $N(X) = \{Project((X \setminus e) \cup Y) \mid e \in X, Y \in \mathcal{Y}_{X,e}\}$, where $\mathcal{Y}_{X,e}$ is defined by $\mathcal{Y}_{X,e} = \{Y \mid Y \text{ is a minimal subset of } E \setminus X \text{ satisfying } \pi((X \setminus e) \cup Y) = 1\}$.

In other words, for every set $X \in \mathcal{F}$ and for every element $e \in X$ we extend $X \setminus e$ in all possible minimal ways to a set $X' = (X \setminus e) \cup Y$ for which $\pi(X') = 1$ (since $X \in \mathcal{F}$, we have $\pi(X \setminus e) = 0$), and introduce each time a directed arc from X to $Project(X')$. We call the obtained directed graph $\mathcal{G}$ the *supergraph* of our generation problem.

Proposition 1 ([9]). *The supergraph $\mathcal{G} = (\mathcal{F}, \mathcal{E})$ is strongly connected.*   □

Since $\mathcal{G}$ is strongly connected by performing a breadth-first search in $\mathcal{G}$ we can generate all elements of $\mathcal{F}$. Thus, given two procedures:

- $First(X, e)$, which for every $X \in \mathcal{F}$ and $e \in X$ returns an element of $\mathcal{Y}_{X,e}$ if $\mathcal{Y}_{X,e} \neq \emptyset$ and $\emptyset$ otherwise,
- $Next(\mathcal{Y}, X, e)$, which return an element of $\mathcal{Y}_{X,e} \setminus \mathcal{Y}$ if $\mathcal{Y}_{X,e} \neq \mathcal{Y}$ and $\emptyset$ otherwise,

the procedure $Transversal(\mathcal{G})$, defined below, generates all elements of $\mathcal{F}$.

$Traversal(\mathcal{G})$

Find an initial vertex $X^0 \leftarrow Project(E)$, initialize a queue $\mathcal{Q} = \emptyset$ and a dictionary of output vertices $\mathcal{D} = \emptyset$.

Perform a breadth-first search of $\mathcal{G}$ starting from X^o:

1 **output** X^0 and insert it to $\mathcal{Q}$ and to $\mathcal{D}$
2 **while** $\mathcal{Q} \neq \emptyset$ **do**
3   take the first vertex X out of the queue $\mathcal{Q}$
4   **for** every $e \in X$ **do**
5 $\mathcal{Y} \leftarrow \emptyset, Y \leftarrow First(X, e)$
6   **while** $Y \neq \emptyset$ **do**
7   compute the neighbor $X' \leftarrow Project((X \setminus e) \cup Y)$
8   **if** $X' \notin \mathcal{D}$ **then output** X' and insert it to $\mathcal{Q}$ and to $\mathcal{D}$
9   add Y to $\mathcal{Y}, Y \leftarrow Next(\mathcal{Y}, X, e)$

Proposition 2. *Assume that the procedure $First(X, e)$ works in time $O(\phi_1(E))$, the for every K procedure $Next(\mathcal{Y}, X, e)$ outputs K elements of $\mathcal{Y}_{X,e}$ in time $\phi_2(K, E)$ and there is an algorithm evaluating π in time $O(\gamma(E))$. Then $Traversal(\mathcal{G})$ outputs K elements of $\mathcal{F}$ in time $O(K^2|E|^2\gamma(E) + K^2 log(K)|E|^2 + K|E|\phi_2(K, E) + K|E|\phi_1(E))$.*

2   Proof of Theorem 1

In this section we apply the $X - e + Y$ method to the generation of all minimal k-vertex connected spanning subgraphs.

For a given k-vertex connected graph (V, E) we define a Boolean function π as follows: for a subset $X \subseteq E$ let

$$\pi(X) = \begin{cases} 1, \text{ if (V,X) is } k\text{-vertex connected;} \\ 0, \text{ otherwise.} \end{cases}$$

Clearly π is monotone, $\pi(\emptyset) = 0$, $\pi(E) = 1$. Then $\mathcal{F} = \{X \mid X \subseteq E$ is a minimal set satisfying $\pi(X) = 1\}$ is the family of edge sets of all minimal k-vertex connected spanning subgraphs of (V, E).

2.1 $(k-1)$-Separators of $(V, X \smallsetminus e)$

Before describing procedures $First(X, e)$ and $Next(\mathcal{Y}, X, e)$ we need the additional notions and elementary results.

A *k-separator* of a graph is a set of k vertices whose removal (simultaneously removing all edges adjacent to those vertices) makes the graph no longer connected. Note that a k-vertex connected graph has no k'-separators for $k' < k$.

Let $G = (V, X)$ be a minimal k-vertex connected spanning subgraph of a k-vertex connected graph (V, E) (see Figure 1).

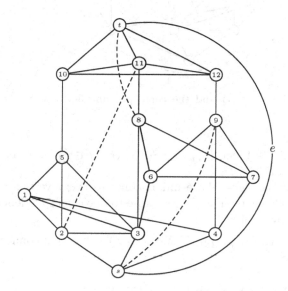

Fig. 1. 4-vertex connected graph (V, E) and its minimal 4-vertex connected subgraph $G = (V, X)$. Solid lines are edges in X.

Let $e = st$ be an arbitrary edge of G and let W be a $(k-1)$-separator of $G_e = (V, X \smallsetminus e)$. Note that W contains neither s nor t, since otherwise W would also be a $(k-1)$-separator of G. We denote by S_W and T_W the vertex sets of the components (i.e., maximal connected subgraphs) of $G_e[V \smallsetminus W]$ containing s and t, respectively.

Claim. $G_e[V \smallsetminus W]$ consists of two components, $G_e[S_W]$ and $G_e[T_W]$ (see Figure 2).

We denote by $N(\cdot)$ a neighborhood in the graph G_e. Let $\mathcal{W}$ be the set of all $(k-1)$-separators of $G_e = (V, X \smallsetminus e)$ and let $\mathcal{S} = \{S \subseteq V \mid |N(S)| = k-1, s \in S, t \notin S \cup N(S)\}$. We call an element of $\mathcal{S}$ a $(k-1)$-*source*. Note that the mapping $W \longmapsto S_W$ is a bijection between $\mathcal{W}$ and $\mathcal{S}$ whose inverse is $S \longmapsto N(S)$.

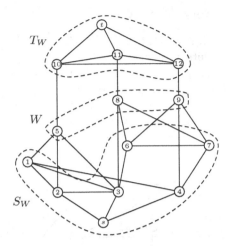

Fig. 2. 3-separator $W = \{5, 8, 9\}$ and the corresponding 3-source $S_W = \{s, 1, 2, 3, 4, 6, 7\}$

For two vertices $u, v \in V$ let $D_{u,v} = \{S_W \in \mathcal{S} \mid u \in S_W, v \in T_W\}$. For an edge $f = uv$ let $D_f = D_{u,v} \cup D_{v,u}$

We call a set of hyperedges whose union contains every vertex a *hyperedge cover*. We show that elements of $\mathcal{Y}_{X,e}$ are in one to one correspondence with the minimal hyperedge covers of $\mathcal{H}_{X,e}$.

Claim. Let $Y \subseteq E \smallsetminus X$. The graph $(V, X \smallsetminus e \cup Y)$ is k-vertex connected if and only if $\bigcup_{f \in Y} D_f = \mathcal{S}$.

2.2 Procedures $First(X, e)$ and $Next(\mathcal{Y}, X, e)$

We describe $First(X, e)$ and $Next(\mathcal{Y}, X, e)$, procedures generating all elements of $\mathcal{Y}_{X,e}$.

$First(X, e)$

  1 construct a hypergraph $\mathcal{H}_{X,e}$ on vertex set $\mathcal{S}$ with edge set $\mathcal{E} = \{D_f \mid f \in E \smallsetminus X\}$
  2 find a minimal hyperedge cover $\mathcal{C}$ of $\mathcal{H}_{X,e}$
  3 return a set $\{f \mid D_f \in \mathcal{C}\}$

$Next(\mathcal{Y}, X, e)$

  1 find a a minimal hyperedge cover $\mathcal{C}$ of $\mathcal{H}_{X,e}$ not in $\{D_f \mid f \in Y, Y \in \mathcal{Y}\}$
  2 return a set $\{f \mid D_f \in \mathcal{C}\}$

In the remainder of this section we show that we can generate minimal hyperedge covers of $\mathcal{H}_{X,e}$ efficiently.

2.3   Structure of $(k-1)$-Separators

Consider the poset $L = (\mathcal{S}, \subseteq)$ of the $(k-1)$-sources ordered by inclusion.

Proposition 3. *The poset L with operations $\cap$ and $\cup$ is a lattice.*

We show that the ordering of $(k-1)$-sources in L has a natural interpretation for the corresponding $(k-1)$-separators.

Since the graph G_e is $(k-1)$-vertex connected, by Menger's Theorem it contains $k-1$ internally vertex disjoint s-t paths. Let $P_1 = sv_1^1 \dots v_{l_1}^1 t$, $P_2 = sv_1^2 \dots v_{l_2}^2 t$, $\dots$, $P_{k-1} = sv_1^{k-1} \dots v_{l_{k-1}}^{k-1} t$ denote such a collection of paths (see Figure 3). We denote by V_P the set of all vertices belonging to the paths $P_1, \dots, P_{k-1}$. Note that not all vertices in V necessarily belong to V_P.

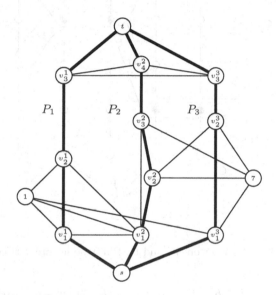

Fig. 3. Internally vertex disjoint paths P_1, P_2, P_3 of G_e represented by thick edges

Consider a $(k-1)$-separator W. Since the removal of W disconnects G_e, W contains at least one internal vertex from each path P_i, $i = 1, \dots, k-1$. As W has $k-1$ vertices, $W = \{v_{\alpha(W,1)}^1, \dots, v_{\alpha(W,k-1)}^{k-1}\}$, where $\alpha(W,i)$ is the index of the vertex of P_i belonging to W.

Claim. Let $W, U \in \mathcal{W}$. $S_W \subseteq S_U$ if and only if $\alpha(W,i) \leq \alpha(U,i)$ for all $i = 1, \dots, k-1$.

Lemma 1. *Let S_W, S_U be $(k-1)$-sources of G_e. Either $S_W \cap T_U = \emptyset$ or $T_W \cap S_U = \emptyset$.*

Proof. We partition $\{1, \dots, k-1\}$ into sets I, J and K as follows: $I = \{i \mid \alpha(W,i) > \alpha(U,i)\}$, $J = \{i \mid \alpha(W,i) = \alpha(U,i)\}$, $K = \{i \mid \alpha(W,i) < \alpha(U,i)\}$.

Let $C = \{v^i_{\alpha(U,i)} \mid i \in I\} \cup \{v^i_{\alpha(W,i)} \mid i \in I\} \cup \{v^i_{\alpha(W,i)} \mid i \in J\}$. Observe that $|C| = 2|I| + |J|$.

We show that $N(S_W \cap T_U) \subseteq C$. Note that $V \setminus ((S_W \cap T_U) \cup C) = T_W \cup S_U$ (see Figure 4). Since W and U are $(k-1)$-separators of G_e, there is no edge between $S_W \cap T_U$ and $T_W \cup S_U$, thus $N(S_W \cap T_U) \subseteq C$.

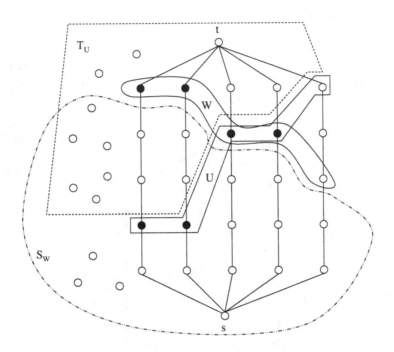

Fig. 4. $(k-1)$-separators W and U. Black nodes are vertices of C.

Let $D = \{v^i_{\alpha(U,i)} \mid i \in K\} \cup \{v^i_{\alpha(W,i)} \mid i \in K\} \cup \{v^i_{\alpha(W,i)} \mid i \in J\}$. Similarly, we obtain that $N(T_W \cap S_U) \subseteq D$.

Suppose for contradiction that $S_W \cap T_U \neq \emptyset$ and $T_W \cap S_U \neq \emptyset$. Since $S_W \cap T_U$ contains neither s nor t, the removal of $N(S_W \cap T_U)$ disconnects G. As G is k-vertex connected, we obtain $k \leq |N(S_W \cap T_U)| \leq |C|$, thus $2|I| + |J| \geq k$. Similarly, we have $2|K| + |J| \geq k$. Recall that I, J and K partition $\{1, \ldots, k-1\}$, thus $k - 1 = |I| + |J| + |K|$.

Combining this with the above inequalities we obtain $2((k-1) + |I| + |J| + |K|) \geq 2(k + |I| + |J| + |K|)$, a contradiction. $\square$

2.4   Bounding the Number of $(k-1)$-Sources

It is easy to see that the numebr of $(k-1)$-sources is at most $\binom{|V|}{k-1}$, since each one corresponds to different $(k-1)$-separators. In this section we provide a better bound on this number.

Corollary 1. *If S_W and S_U are incomparable in L then there exists some $i \in \{1, \ldots, k-1\}$ such that $|\alpha(W, i) - \alpha(U, i)| = 1$, i.e., the vertices $v^i_{\alpha(W,i)}$ and $v^i_{\alpha(U,i)}$ are adjacent on the path P_i.*

Proof. Suppose on the contrary that $|\alpha(W, i) - \alpha(U, i)| > 1$ for all $i = 1, \ldots, k-1$. Then since S_W and S_U are incomparable, by Claim 2.3 there exist $j, l \in \{1, \ldots, k-1\}$ such that $\alpha(U, j) + 1 < \alpha(W, j)$ and $\alpha(W, l) + 1 < \alpha(U, l)$. Then $v^j_{\alpha(U,j)+1} \in S_W \cap T_U$, $v^l_{\alpha(W,l)+1} \in T_W \cap S_U$ contradicting Lemma 1. $\qquad\square$

The *width* of a poset is the size of its largest antichain. We show that the width of L is bounded.

Proposition 4. *The width of L is at most 2^{k-1}.*

Proof. We associate to every $(k-1)$-separator W a 0-1 vector $\pi(W) = (\alpha(W, 1) \mod 2, \ldots, \alpha(W, k-1) \mod 2)$. By Corollary 1, if two $(k-1)$-separators W, U are incomparable, there exists some $i \in \{1, \ldots, k-1\}$ such that $|\alpha(W, i) - \alpha(U, i)| = 1$, implying $\pi(W) \neq \pi(U)$.

Since the number of different 0-1 vectors of length $k-1$ is 2^{k-1}, every antichain in P has size at most 2^{k-1}. $\qquad\square$

Corollary 2. *For every fixed k the number of $(k-1)$-sources is $O(|V|)$.*

2.5   Generating Minimal Hyperedge Covers of $\mathcal{H}_{X,e}$

In this section we reduce the problem of generating minimal hyperedge covers of $\mathcal{H}_{X,e}$ to the problem of generating minimal transversals of 2-conformal hypergraphs. For the latter problem the algorithm is provided in [3].

A *transversal* is a set of vertices intersecting every hyperedge. A hypergraph is δ-*conformal* if its transpose is δ-Helly (see [2] for other equivalent definitions).

First we show that the hypergraphs $\mathcal{H}_{X,e}$ are 2-Helly.

Claim. Either $D_{u,v} = \emptyset$ or $D_{v,u} = \emptyset$ for all $u, v \in V$.

Proof. Suppose on the contrary that we have $S_W \in D_{u,v}$ and $S_U \in D_{v,u}$. Then $u \in S_W \cap T_U$ and $v \in T_W \cap S_U$, contradicting Lemma 1. $\qquad\square$

Claim. D_f is a sublattice of L (see Figure 5).

Proof. Let $f = uv$. Without loss of generality we can assume that $D_f = D_{u,v}$. Let $S, S' \in D_{u,v}$. Then $u \in S \cap S'$ and $v \notin (S \cap S') \cup N(S \cap S')$. Similarly, $u \in S \cup S'$ and $v \notin S \cup S' \cup N(S \cup S')$. Thus $S \cap S', S \cup S' \in D_f$. $\qquad\square$

Since the edges of $\mathcal{H}_{X,e}$ are sublattices of L, the hypergraphs $\mathcal{H}_{X,e}$ are 2-Helly ([2, Example 2 on page 21]). Thus the hypergraphs $\mathcal{H}^T_{X,e}$ are 2-conformal.

Note that minimal hyperedge covers of $\mathcal{H}_{X,e}$ are minimal transversals of $\mathcal{H}^T_{X,e}$. An algorithm from [3] generates K minimal transversals of δ-conformal

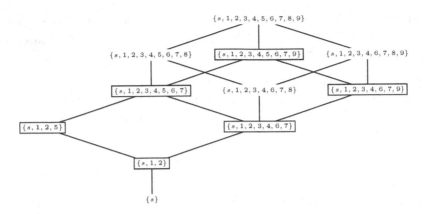

Fig. 5. Elements of $D_{2,11}$ are in black rectangles. Note that $D_{11,2} = \emptyset$.

hypergraph in $O(K^2 i^2 j + K i^{\delta+2} j^{\delta+2})$, where i and j are the numeber of vertices and number of hyperedges, respectively.

2.6   Complexity

In this section we analyze the complexity of $Traversal(\mathcal{G})$. Let $n = |V|$, $m = |E|$. Since G is k-vertex connected we have $m \geq n$.

Note that $\pi(X)$ can be evaluated in $O(k^3 |V|^2)$ time [4], thus $\gamma(E) = n^2$.

Claim. For every $X \in \mathcal{F}$ and $e \in X$ the hypergraph $\mathcal{H}_{X,e} = (\mathcal{S}, \mathcal{E})$ has $O(n)$ vertices and $O(m)$ edges and it can be constructed in $O(n^k m)$ time.

Proof of Claim 6: By Corollary 2 the number of vertices of $\mathcal{H}_{X,e} = (\mathcal{S}, \mathcal{E})$ is at most $O(n)$. The number of edges of $\mathcal{H}_{X,e} = (\mathcal{S}, \mathcal{E})$ is exactly $|E \setminus X| \leq m$ since we add an edge to $\mathcal{H}_{X,e} = (\mathcal{S}, \mathcal{E})$ for every edge of $E \setminus X$.

To construct $\mathcal{H}_{X,e} = (\mathcal{S}, \mathcal{E})$ we first need to find all $(k-1)$-sources and $(k-1)$-separators. We can check if after removing a given set of $k-1$ vertices the graph G is still connected in $O(n+m)$ time using, e.g., depth first search. Thus we can find all $(k-1)$-separators by repeating the above procedure for every $(k-1)$-element subset of V. The number of such subsets is $\binom{n}{k-1} \leq n^{k-1}$. Thus we can compute all $(k-1)$-sources and $(k-1)$-separators in $O(n^{k-1} m)$ time.

To add edges we need to check for every $f \in E \setminus X$ and every $(k-1)$-separator W if S_W belongs to D_f, which can be done in $O(n)$ time for each pair f and W. Thus the complexity of constructing edges of $\mathcal{H}$ is $O(n^2 m)$.   □

Since we can find a minimal transversal of $\mathcal{H}^T_{X,e}$ in $O(|\mathcal{E}|)$ time and by Claim 2.6 we have $\phi_1(E) = n^k m$. Recall that $\phi_2(K, E) = K^2 m^2 n + K m^4 n^4$ (see Section 2.5). Thus by Proposition 2 the complexity of $Traversal(\mathcal{G})$ is $O(K^3 m^3 n + K^2 m^5 n^4 + K n^k m^2)$.

References

1. Tamura, A., Shioura, A., Uno, T.: An optimal algorithm for scanning all spanning trees of undirected graphs. SIAM Journal on Computing 26(3), 678–692 (1997)
2. Berge, C.: Hypergraphs. Elsevier-North Holand, Amsterdam (1989)
3. Boros, E., Elbassioni, K., Gurvich, V., Khachiyan, L.: Generating maximal independent sets for hypergraphs with bounded edge-intersections. In: Farach-Colton, M. (ed.) LATIN 2004. LNCS, vol. 2976, pp. 488–498. Springer, Heidelberg (2004)
4. Cheriyan, J., Kao, M.-Y., Thurimella, R.: Algorithms for parallel k-vertex connectivity and sparse certificates. 22, 157–174 (1993)
5. Coulbourn, C.J.: The Combinatorics of Network Reliability. Oxford University Press, Oxford (1987)
6. Gabow, H.N., Myers, E.W.: Finding all spanning trees of directed and undirected trees. SIAM Journal on Computing 117, 280–287 (1978)
7. Johnson, D.S., Papadimitriou, Ch.H.: On generating all maximal independent sets. Information Processing Letters 27, 119–123 (1988)
8. Khachiyan, L., Boros, E., Borys, K., Elbassioni, K., Gurvich, V., Makino, K.: Enumerating spanning and connected subsets in graphs and matroids. Manuscript.
9. Khachiyan, L., Boros, E., Borys, K., Elbassioni, K., Gurvich, V., Makino, K.: Generating cut conjunctions and bridge avoiding extensions in graphs. In: Deng, X., Du, D.-Z. (eds.) ISAAC 2005. LNCS, vol. 3827, pp. 156–165. Springer, Heidelberg (2005)
10. Lawler, E., Lenstra, J.K., Kan, A.H.G.R.: Generating all maximal independent sets: NP-hardness and polynomial-time algorithms. SIAM Journal on Computing 9, 558–565 (1980)
11. Matsui, T.: Algorithms for finding all the spanning trees in undirected graphs. Technical report, Department of Mathematical Engineering and Information Physics, Faculty of Engineering, University of Tokyo, 1993. Report METR93-08
12. Read, R.C., Tarjan, R.E.: Bounds on backtrack algorithms for listing cycles, paths, and spanning trees. Networks 5, 237–252 (1975)
13. Schwikowski, B., Speckenmeyer, E.: On enumerating all minimal solutions of feedback problems. Discrete Applied Mathematics 117, 253–265 (2002)
14. Valiant, L.: The complexity of enumeration and reliability problems. SIAM Journal on Computing 8, 410–421 (1979)

Finding Many Optimal Paths Without Growing Any Optimal Path Trees

Danny Z. Chen[1,*] and Ewa Misiołek[2]

[1] Department of Computer Science and Engineering, University of Notre Dame,
Notre Dame, IN 46556, USA
chen@cse.nd.edu.
[2] Mathematics Department, Saint Mary's College, Notre Dame, IN 46556, USA
misiolek@saintmarys.edu.

Abstract. Many algorithms seek to compute actual optimal paths in weighted directed graphs. The standard approach for reporting an actual optimal path is based on building a single-source optimal path tree. A technique was given in [1] for a class of problems such that a single actual optimal path can be reported without maintaining any single-source optimal path tree, thus significantly reducing the space bound of those problems with no or little increase in their running time. In this paper, we extend the technique in [1] to the generalized problem of reporting many actual optimal paths with different starting and ending vertices in certain directed graphs. We show how this new technique yields improved results on several application problems, such as reconstructing a 3-D surface band bounded by two simple closed curves, finding various constrained segmentation of 2-D medical images, and circular string-to-string correction. Although the generalized many-path problem seems more difficult, our algorithms have nearly the same space and time bounds as those of the single-path cases. Our technique is likely to help improve other optimal paths or dynamic programming algorithms. We also correct an error in the time/space complexity for the circular string-to-string correction algorithm in [7] and give improved results for it.

1 Introduction

Many algorithms seek to compute actual optimal paths in a weighted directed graph $G = (V, E)$. The standard approach for finding an actual optimal path [3] builds a single-source optimal path tree using $O(|V|)$ space. Chen *et al.* [1] developed a technique for a class of problems that reports a single actual optimal path without maintaining any single-source optimal path tree, thus significantly reducing the space bound of those problems with no or little increase in their running time. Their technique is a combination of dynamic programming and divide-and-conquer methods [1]. However, some applications need to find many optimal paths with different starting and ending vertices, for which the technique in [1] does not seem immediately applicable. In this paper, we extend the

* This research was supported in part by the National Science Foundation under Grant CCF-0515203.

technique in [1] to the generalized problem of finding $k > 1$ actual optimal paths in certain directed graphs. This new technique yields improved space bounds of several problems, such as reconstructing a 3-D surface band defined by two simple closed curves [4,10], finding various constrained segmentation of 2-D medical images [2,8,11], and circular string-to-string correction [5,6,7,9]. Although this generalized many-path problem appears more difficult, our algorithms have nearly the same space and time bounds as those of the single-path cases. This new technique is likely to help improve other optimal paths or dynamic programming algorithms.

In this paper, we consider several problems where a "best" actual optimal path is sought from among $k = O(|V|)$ optimal paths in some special weighted directed graphs, namely, directed regular grid graphs. Our new space-efficient technique for these problems is based on the following observations. (1) The computation of the k optimal paths actually organizes these paths in a manner of a complete binary tree T of $O(k)$ nodes (i.e., each tree node is for one of the paths). (2) The computational process of these paths can be viewed as following a path $p(v)$ in T from the root to a node v, in the depth-first search fashion. (3) Although there are possibly many nodes on the root-to-v path $p(v)$ in T and each such node is associated with an already computed optimal path, we only need to store the two actual paths that bound the subgraph associated with the node v; the actual paths for other nodes on the path $p(v)$ can be represented in an *encoded form* that uses much less space. (4) The graph G can be represented implicitly. Consequently, the space bounds of our algorithms are basically the same as those for computing a *single* actual path, yet our time bounds are only a small increase on those of the corresponding k-path algorithms [2,4,6,7,11]. In fact, our technique can report all k actual paths with no further increase in the time and space bounds.

We illustrate our technique by applying it to the problem of reconstructing a 3-D surface band defined by two simple closed curves. The reconstructed surface consists of a sequence of triangles in 3-D with the minimum total area. This surface reconstruction problem arises in computer-aided design and manufacturing (CAD/CAM), computer graphics, biological research, and medical diagnosis and imaging. Fuchs *et al.* [4] modeled this problem as finding an actual shortest path among k shortest paths in a weighted acyclic grid graph $G = (V, E)$, where $|V| = O(|E|) = O(kl)$ and k and l are positive integer input parameters (with $k \leq l$). They gave an $O(kl \log k)$ time algorithm. If a standard shortest path algorithm [3] is used, then the space bound in [4] is $O(kl)$. Our technique yields an $O(kl(\log k)^2)$ time and $O(k + l)$ space algorithm. In fact, our technique gives a general trade-off relation between the space and time bounds (see Theorem 3).

Our technique can also solve several variations of the 3-D surface reconstruction problem in [10] and can significantly reduce the space bounds for four of the several optimization criteria considered in [10].

In the full version of the paper, we also show how to apply our technique to reduce the space needed for computing various constrained segmentation of medical images. Image segmentation is important to medical image analysis for medical diagnosis and treatment, computer vision, pattern recognition, mechanical and material study, and data mining. Its goal is to define the boundary for an object of interest in the image, separating it from its surrounding. The best known algorithms for this problem [2,11] take $O(kl \log k)$ time and $O(kl)$ space. Using our technique, the time bound is increased from that of [2,11] by only a factor of less than or equal to $\log l$, but the working space of these algorithms is decreased significantly to almost linear.

Another key application is the circular string-to-string correction problem [5,6,7,9], which seeks a minimum-cost sequence of editing operations for changing a circular string $A = (a_1, a_2, \ldots, a_k)$ to another circular string $B = (b_1, b_2, \ldots, b_l)$ (assume $k \leq l$). This problem arises in many areas such as shape recognition, pattern recognition, speech recognition, and circular DNA and protein sequence alignment, and is well studied. To our knowledge, the best known algorithm for this problem is due to Maes [7]. It is based on a divide-and-conquer scheme for computing k shortest paths in a rectangular grid graph G, and the author claimed it takes $O(kl \log k)$ time and $O((k + l) \log k)$ space. However, these time and space bounds do not appear correct. As stated in [7], it just applies Hirschberg's algorithm [5] to find each of the k paths. However, when computing a single shortest path in a subgraph G_i of the original graph G using the method in [5], unlike in the situation of finding only *one* actual path in the original grid graph G, there is no guarantee that as the recursion in [5] proceeds, the total size of the subgraphs always decreases significantly (say, by a constant fraction) from one recursion level to the next. This implies that when computing k paths in G, finding one actual path in a subgraph G_i of G using the method in [5] actually takes $O(|G_i| \log(k+l))$ time, instead of $O(|G_i|)$. Thus, to obtain an $O((k+l) \log k)$ space bound, the algorithm as stated in [7] actually takes $O(kl \log k \log(l + k))$ time (see our full paper for more detailed discussion and analysis). Applying our technique, we improve the best known circular string-to-string correction algorithm in [7], e.g., to $O(l + k(\log k)^\epsilon)$ space and $O(kl \frac{(\log k)^2}{\log \log k})$ time, for any constant ϵ with $0 < \epsilon < 1$ (the exact results are as those stated in Theorem 3).

2   Preliminaries

In this section, we sketch the approaches in [1] and [6,2,4,7,11], which will be refered to and build upon in the later sections.

Let $G = (V, E)$ be a directed graph with nonnegative edge weights. For two vertices $s, t \in V$, the standard method [3] for finding an actual optimal s-to-t path in G is to build and store a *single-source optimal path tree* T rooted at s. Then for any vertex v, an actual optimal s-to-v path in G is obtained by traversing in T the v-to-s path. It is well known that no asymptotically faster algorithm is known for reporting a single actual optimal path than for building a single-source optimal path tree. However, for a set of optimal path problems on regular grid graphs, Chen *et al.* [1] gave a technique with space bound for computing a single actual optimal s-to-t path asymptotically better than that for building a single-source tree. Their method is hinged on the *clipped tree* data structure and a variation of the divide-and-conquer called *marriage-before-conquer*.

A regular grid graph $G = (V, E)$ has vertices lying on a rectangular $k \times l$ grid and the vertices are connected by edges only to the adjacent vertices from the same row or to the vertices from the row below, thus $|V| = kl$ and $|E| = O(|V|)$. Figures 1(a),(b) show examples of such graphs. A clipped tree T_{clp} is a compressed version of a single-source tree T with the following characteristics: (1) T_{clp} contains a subset of the nodes in T, including s, t, and vertices from only τ rows of G; (2) two nodes v and w in T_{clp} form an ancestor-descendant relation in T_{clp} if and only if they form the same relation in T.

The following result in [1] is useful to us.

Lemma 1. [1] *Given a regular grid graph G of size $k \times l$, two vertices s and t in G, and a parameter τ with $1 \leq \tau \leq k$, it is possible to use $O(|T_{clp}|) = O(l\tau)$ space to report an optimal s-to-t path in G in time $O((T(\cdot)(\log k))/\log(\tau - 1))$, where $T(\cdot)$ is the running time of a standard optimal path algorithm on G.*

In some special cases the extra $(\log n)/(\log(\tau - 1))$ time factor can be avoided. For example, for the special graph in Fig. 1(a), to find a shortest path from s to t, one can first locate a vertex v_2 in the middle row of the graph, and then recursively find the shortest s-to-v_2 and v_2-to-t paths in two rectangular subgraphs of G. Observe, that the sum of the sizes of the two subgraphs in the recursive calls is half the size of the original graph, resulting in the same time bound as that of a standard shortest path algorithm.

It should be pointed out that the method in [1] may be applied to graphs even if their structures are not grid-like. In fact, the paradigm can be applied to some problems that are solved using dynamic programming, by "divide-and-conquering" the dynamic programming table instead of the graph itself.

We extend the technique in [1] to the type of problems where an optimal path is to be selected from a set of k optimal paths with k pairs of different source-destination vertices in a directed grid graph G, with $k = O(n)$.

A straightforward approach for solving such a k-path problem is to apply an optimal path algorithm to compute each of the k paths and then determine the best one. However, more efficient algorithms are possible for regular, rectangular grid graphs. A key idea, as shown in [6,2,4,7,11], is to first find one of the k paths and then recursively search for the remaining paths in increasingly smaller subgraphs of G. The description below summarizes this approach and introduces some of the notation used in the later sections.

Suppose we are given a directed acyclic graph G and a sequence of k pairs of vertices (v_i, w_i), $i = 1, 2, \ldots, k$, between which the shortest paths in G are sought. Let SP_i denote a shortest path between the pair (v_i, w_i) in G. We assume that the graph G has the following important property.

Property 2. For any i and j, $1 \leq i \leq j \leq k$, two shortest paths SP_i and SP_j can be computed in G, such that SP_i and SP_j bound a subgraph $G_{i,j}$ of G and $G_{i,j}$ contains the shortest paths SP_g in G, for all $g = i, i + 1, \ldots, j$.

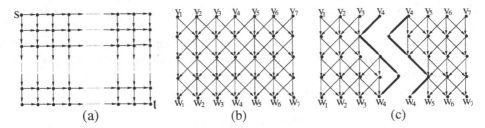

Fig. 1. (a),(b) An example of a regular rectangular grid graph G. (c) G is split along the "middle" shortest path SP_4 into two subgraphs for the recursive subproblems.

For example, the graphs in [2,4,6,7,11] all have this property since they are all regular, rectangular grid graphs embedded in the plane, the sequence of vertex pairs (v_i, w_i) are all on the outer face of the embedding, and subgraphs $G_{i,j}$ can be obtained by finding the shortest paths SP_i and SP_j. Let $SP(G_{i,j}, i, j)$ denote a procedure for computing the "middle" shortest path SP_t in the subgraph $G_{i,j}$, $t = \lfloor \frac{i+j}{2} \rfloor$. Then the k-path algorithm in [6,2,4,7,11] proceeds as follows.

1. Find SP_1 and SP_k. Let $i = 1$ and $j = k$.
2. Run $SP(G_{i,j}, i, j)$ to obtain SP_t (for the pair (v_t, w_t), with $t = \lfloor \frac{i+j}{2} \rfloor$).
3. Recursively solve the subproblems on subgraphs $G_{i,t}$ and $G_{t,j}$, respectively.

At the end of the algorithm, the "best" of these k shortest paths (i.e., the one with the overall shortest path length) is selected and and the actual path is reported. For example, to find the shortest of the $k = 7$ paths between the pairs of vertices (v_i, w_i), $i = 1, \ldots, 7$, in the graph in Fig. 1(b), we first find a shortest path between the "middle" pair (v_4, w_4), then split the graph along that path to recursively compute the remaining paths in the resulting subgraphs (Fig. 1(c)).

A straightforward implementation of this paradigm needs to construct and store both the grid graph G (say, of size $k \cdot l$) and single-source shortest path trees for each source vertex, thus using $O(|V|) = O(|E|) = O(kl)$ space. The running time of this approach is $O(T(\cdot) \log k)$, where $T(\cdot)$ is the standard time bound for computing one shortest path in G.

It should be pointed out that in the above paradigm, since the already computed shortest paths are used to define the boundaries of the subgraphs in which the remaining paths are to be searched, the computed actual paths need to be stored. A straightforward method could store $O(k)$ actual paths at the same time, thus using a lot of space. To achieve a space-efficient algorithm for the k-path problem, we must avoid storing too many actual paths simultaneously. This is made possible by exploiting the special structures of the graphs we use. We will demonstrate our algorithm by applying it to the problem of reconstructing a 3-D surface band bounded by two simple closed curves.

3   Optimally Triangulating a 3-D Surface Band

3.1   The Surface Band Triangulation and a Previous Algorithm

The desired output of the surface reconstruction problem, also called a *surface triangulation problem*, is a sequence of triangular tiles that best approximates the surface band based on a given optimization criteria. Below we define the problem and sketch its time-efficient algorithm as given in [4].

Suppose a 3-D surface $\mathcal{S}$, which is to be reconstructed, is divided into a sequence of surface "bands" by a set of mutually non-crossing simple closed curves (also called *contours*) lying on $\mathcal{S}$, such that each surface band is bounded by two of the curves. To reconstruct $\mathcal{S}$, it is sufficient to reconstruct each of the surface bands. Thus, we focus on the problem of reconstructing a single surface band bounded by two simple closed curves C_P and C_Q in 3-D [4,10].

¿From the computational view point, each of the two curves, is represented by a sequence of points along the curve. That is, a curve C_P is approximated by a closed polygonal curve $P = \{p_1, p_2, \ldots, p_k\}$ in such a way, that the portion of C_P between any two consecutive points p_i and p_{i+1} is approximated by the line segment $\overline{p_i p_{i+1}}$. The triangulation of the surface band bounded by $P = \{p_1, p_2, \ldots, p_k\}$ and $Q = \{q_1, q_2, \ldots, q_l\}$ gives a sequence of non-overlapping triangles, whose edges are the edges of P or Q, and line segments connecting a vertex of P with a vertex of Q. More precisely, each triangle has one edge that is an edge of one curve (say $\overline{p_i p_{i+1}}$ of P), called the "base" edge, and two other edges that share a vertex of the other curve (say q_j of Q), called the "bridges". Thus a triangle $\triangle p_i q_j p_{i+1}$ with base on P, has the edges $\overline{p_i p_{i+1}}$, $\overline{p_i q_j}$, and $\overline{q_j p_{i+1}}$, similarly, a triangle $\triangle q_j q_i q_{j+1}$ with base on Q, has the edges $\overline{q_j q_{j+1}}$, $\overline{q_j p_i}$, and $\overline{p_i q_{j+1}}$. The set of triangles of a feasible triangulation forms a partition of the surface band into a set of triangular faces such that the set of vertices of the faces is exactly the set of the vertices of P and Q. Also, all edges of P and Q belong to the set of the triangulation edges. See Fig. 2(a) for an example of a feasible triangulation.

An optimal triangulation is a feasible triangulation that optimizes the given criteria. Possible optimization criteria for triangulating a surface band include minimizing the total area, bending energy, and mean curvature variation, or maximizing the total twist, or optimizing a certain combination of these or other optimization criteria (see [10] for detailed descriptions of such criteria). For our discussion in this paper, we assume that the optimization criterion is to minimize the total triangulated area. However, three other optimization criteria in [10] (minimum total twist, maximum convexity, and minimum normal variation) can also be used by our technique.

Note that the number of feasible triangulations of the surface band between the two curves P and Q is prohibitively large. An efficient approach [4] models this triangulation problem as one of searching for a best path among the shortest paths between k pairs of vertices in a directed graph $H = (V_H, E_H)$, as follows. Without loss of generality, we assume $k \leq l$ throughout this section.

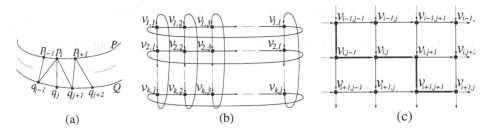

Fig. 2. (a) A feasible triangulation for a portion of the surface band between the polygonal curves P and Q. (b) A toroidal graph H defined by P and Q. (c) A path in H corresponds to a triangulation in (a).

- Every possible "bridge" segment $\overline{p_i q_j}$ corresponds to a vertex $v_{i,j}$ in V_H.
- For every $j = 1, 2, \ldots, l$, there is a directed edge $(v_{i,j}, v_{i,j+1})$ in E_H corresponding to a possible triangle $\triangle q_j p_i q_{j+1}$ with the "base" edge $\overline{q_j q_{j+1}}$, where the index $l + 1$ means 1. For every $i = 1, 2, \ldots, k$, there is a directed edge $(v_{i,j}, v_{i+1,j})$ in E_H corresponding to a possible triangle $\triangle p_i q_j p_{i+1}$ with the "base" edge $\overline{p_i p_{i+1}}$, where the index $k + 1$ means 1 (see Fig. 2(c)).
- The weight of every edge is the surface area of the triangle represented by that edge.

Note that the graph H is a toroidal graph of size $O(kl)$. See Fig. 2(b)-2(c) for an example of H; the thick path in Fig. 2(c) represents the given portion of a triangulation of the band in Fig. 2(a). Any triangulation of the surface band corresponds to a simple closed path, called a *triangulating path*, in H that visits every row and every column of the "flattened" version of H [4].

To simplify the shortest path search, a new ("flattened") graph G is created from H by appending a copy of H after the last row of H (without the wrap-around edges), and duplicating the first column (i.e., the vertices $v_{i,1}$) of the newly created graph and appending it after the last column (the appended vertices are denoted by w_i's). Fig. 3(b) shows the graph G created from the graph H in Fig. 3(a); the thick path in G corresponds to a triangulating path in H.

Now, the problem becomes that of finding the k shortest paths between the k pairs of vertices $(v_{i,1}, w_{k+i})$ in G for $i = 1, 2, \ldots, k$ ($v_{i,1}$'s are denoted as v_i's in Fig. 3), and selecting the shortest of those as the solution. Note that G thus created is a directed planar acyclic grid graph, and hence it satisfies Property 2. Therefore, this k-path problem on G can be solved in $O(kl \log k)$ time and $O(kl)$ space using the divide-and-conquer method in [4,6] and described in Sect. 2.

3.2 Our Basic Algorithm

We now present a method that significantly reduces the space bound of the solution in [4,6]. For clarity of discussion, we first present a less efficient algorithm, followed by further improvements.

Recall from Sect. 2 that in order to solve the k-path problem in a time-efficient manner, we need to recursively compute shortest paths on increasingly

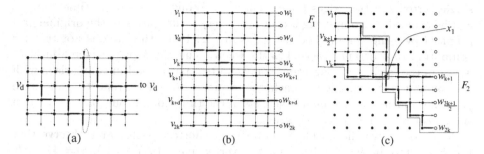

Fig. 3. (a) A "flat" version of the toroidal graph H. (b) The planar grid graph G created from H. (c) The two subgraphs F_1 and F_2 of $G_{1,k}$ for the two subsequent subproblems in the recursion. The graph $F_1 \cup F_2$ has only a single vertex and two edges (those outside the dashed boundaries) fewer than $G_{1,k}$.

smaller subgraphs bounded by two already computed paths. Therefore, since paths are needed to define subgraphs for future recursive calls, in the worst case, a straightforward algorithm may require storage of all k paths. In this section we present a method that requires storage of only a constant number of shortest paths by organizing the k shortest paths using the recursion tree and exploiting the structure of such organization. We combine our method with the technique in [1] to significantly decrease the space complexity of algorithms in [4,6] with only a small increase in time. In fact, depending on the application, one can select a trade-off between the time and space saving produced by our method. We now proceed to explain the method.

First notice, that we need not explicitly construct and store G since for any vertex of G, its two incoming and two outgoing edges as well as their weights can be determined in $O(1)$ time. Since the input consists of $k + l$ points representing the curves P and Q, the implicit representation of G uses only $O(k + l)$ space, a significant saving since G has $n = 2kl$ vertices and $m = 4kl$ edges.

Further, whenever an actual shortest path SP_i in a subgraph G_i of G from $v_{i,1}$ to w_{k+i} is computed in the k-path scheme of [4,6], we apply the space-efficient algorithm of Chen et al. [1]. This way we avoid storing a single-source tree (of size $O(kl)$) as in the standard single-source shortest path algorithm. This, however, causes a $\log k$ factor increase in the time complexity of the algorithm. To find an actual path SP_i in G_i, the method in [1] divides G_i into subgraphs and recursively reports vertices of SP_i in these subgraphs. However, G_i, although is a subgraph of the original rectangular grid graph G, has an irregular shape, see Fig. 3(c). Thus, when recursively computing SP_i on G_i, there is no guarantee that the total size of the subgraphs containing the pieces of SP_i at one recursion level decreases significantly (say, by a constant fraction) from the total size of the subgraphs in G_i at the "parent" level; i.e., from one recursion level to the next, it is not guaranteed that a constant fraction of the subgraphs is always "pruned" away. For example, in Fig. 3(c), the two subgraphs F_1 and F_2 of $G_{1,k}$ for the two subsequent subproblems are such that the graph $F_1 \cup F_2$ has only a single vertex and two edges fewer than $G_{1,k}$. While the $O(\log k)$ time increase can be avoided for computing *one* or even $O(1)$ actual paths in the original grid graph G by using the method in [1] (as in Fig. 1(a)), the same is not true for computing *each* of *many* actual paths when using the k-path scheme in [4,6]. Thus, in the setting of computing many shortest paths, and in comparison with the common tree-growing approach, the recursive process for reporting an actual path in G_i incurs a $\log k$ factor increase in the time bound, but also results in a significant reduction of needed space.

Now, to resolve the problem of storing k shortest paths, first observe that the k shortest paths SP_i in G for the vertex pairs $(v_{i,1}, w_{k+i})$, $i = 1, \ldots, k$, can be organized based on the recursion tree used by the scheme in [4,6]. This recursion tree is a complete binary tree of $O(k)$ nodes, called the *k-path tree* T_k. The root of T_k represents the "middle" path SP_t in the subgraph $G_{1,k} \subseteq G$ ($t = \lfloor \frac{1+k}{2} \rfloor$); the children of each internal node in T_k are for the subproblems on their associated subgraphs. For example, the children of the root are associated with the subgraphs $G_{1,t}$ and $G_{t,k}$, and represent their "middle" paths computed by the procedures $SP(G_{i,t}, i, t)$ and $SP(G_{t,j}, t, j)$, respectively.

If the k actual shortest paths represented by the k-path tree T_k were to be all stored at the same time, then the space bound would be $O(k(k + l))$ (since each path SP_i in G has $O(k + l)$ vertices). Fortunately, only a small subset of the paths associated with the nodes in T_k is needed at any time of the computation. We call such nodes in T_k the *active nodes*. It is sufficient to consider only the space used by the active nodes.

To compute the middle shortest path SP_t in a subgraph $G_{i,j}$, we need to store two actual shortest paths SP_i and SP_j that bound the subgraph $G_{i,j}$. Let u_t be the node in the k-path tree T_k corresponding to SP_t and $G_{i,j}$. Then an important observation is that the two paths SP_i and SP_j are always at two nodes on the root-to-u_t path in T_k. This implies that when the computation is performed at a node u of T_k, only the ancestor nodes of u (including u itself) are active. Clearly, at any time of the computation, there are only $O(\log k)$ active nodes in T_k.

In fact, the computation of the k paths in G proceeds in the order as if a depth-first search is done on the tree T_k. During the computation, we store all actual paths represented by the active nodes of T_k. In the worst case, $2 + \log k$ actual paths need to be stored (the paths for $\log k$ active nodes plus SP_1 and SP_k). In addition, we keep track of the vertex pair $(v_{i,1}, w_{k+i})$ such that the length value of SP_i is the smallest among all paths computed thus far. After all k shortest paths (and their lengths) are computed, we use the algorithm in [1] to report the actual shortest path in G whose length value is the smallest among all k paths. Given the "best" actual shortest path in G, the corresponding optimal triangulation of the 3-D surface band can be easily obtained.

Since we store $O(\log k)$ actual shortest paths, each of which has $O(k + l)$ vertices, our basic algorithm uses $O((k + l) \log k)$ space, improving the straightforward $O(kl)$ space bound. Because the k-path scheme in [4,6] takes $O(kl \log k)$ time and our space-efficient algorithm incurs a factor of $O(\log k)$ time increase, the total time of our basic algorithm is $O(kl(\log k)^2)$.

3.3   A Space Improvement Scheme

It is possible to reduce the space bound of the above basic algorithm from $O((k+l)\log k)$ to $O(k + l)$, without affecting its running time. Note that this $O(k + l)$ space bound matches the one for finding *one* actual shortest path [1]. The idea is to use a bit encoding scheme to represent most of the actual paths for the active nodes of T_k during the computation.

When computing a shortest path SP_t in a subgraph $G_{i,j}$ ($t = \lfloor\frac{i+j}{2}\rfloor$), we need to store the two actual paths SP_i and SP_j that bound $G_{i,j}$. Actually, SP_i and SP_j are the only actual paths needed for computing SP_t in $G_{i,j}$. That is, the other paths represented by the active nodes in T_k are not really needed for computing SP_t. Therefore, it is sufficient to explicitly store only SP_i and SP_j when computing SP_t. All other shortest paths of the active nodes in T_k can be temporarily stored in a "compressed" form. This "compressed" form should use less space, yet can be easily converted back and forth between the explicit form (so that these paths can be used for future computation).

Our choice of this "compressed" representation is a bit encoding scheme for the actual paths in G. Our observation is that each shortest path in the grid graph G consists of a sequence of "rightward" edges and "downward" edges, which can be represented by 0's and 1's, respectively. For example, the thick path in Fig. 3(b) can be represented by the sequence $(0, 1, 0, 0, 1, 0, 1, 1, 0, 1, 0, 0, 0)$.

Given the starting vertex of a path, it is easy to convert a bit-encoded path representation to or from an explicit actual path in $O(k + l)$ time. Such conversions need to be performed $O(k)$ times by our algorithm, which add only $O(k(k + l)) = O(kl)$ time to the algorithm (since $k \leq l$). Hence, the total running time of our algorithm is $O(kl(\log k)^2)$ as the basic algorithm.

Thus, at any time of the computation, our algorithm stores only $O(1)$ actual paths explicitly, and the other $O(\log k)$ paths using bit encoding. To analyze the space bound of our algorithm, note that storing one path using bit encoding needs $O(k + l)$ bits. Thus, storing $O(\log k)$ paths in this scheme uses $O((k + l)\log k)$ bits, which are equal to $O(k + l)$ space. Also, to explicitly store $O(1)$ actual paths with $k+l$ vertices each, $O(k+l)$ space is needed. Hence, the overall space bound of our algorithm is $O(k+l)$, and its time bound is still $O(kl(\log k)^2)$.

3.4   Our Final Algorithm

We now show how to improve both the space and time bounds of our basic algorithm. In fact, we give a trade-off relation between the space and time bounds. This is achieved by combining the space improving idea in Subsection 3.3 with a more careful application of the algorithm in [1].

Based on the idea and analysis in Subsection 3.3, it is clear that our algorithm can store all actual shortest paths needed by the computation using only $O(k+l)$ space.

Another important fact is that Chen *et al.*'s algorithm [1] can use the clipped tree T_{clp} to store $\tau \geq 3$ vertices on any s-to-t path. Our basic algorithm stores only $\tau = 3$ vertices for any sought path in the tree T_{clp} (by recording mainly the vertices in the middle column of the grid subgraph); this divides the sought path into only two subpaths for the recursive reporting. Our idea here is that a larger value of τ can be used. That is, we can store τ columns of vertices of the subgraph in T_{clp}. This means that τ vertices on the sought path are stored in T_{clp}, which cut the sought path into $\tau - 1$ subpaths for the recursive reporting. This implies that the factor of time increase incurred by our algorithm over that of the method in [4,6] becomes $\frac{\log k}{\log \tau}$ (instead of $\log k$ as for the basic algorithm). Clearly, storing τ columns of vertices of a subgraph in T_{clp} uses $O(k\tau)$ space.

We can choose τ in the following manner. Suppose we consider a function $f(k)$. If $f(k) = O(1)$, then let $\tau = O(1) \geq 3$; if $f(k) > O(1)$, then let $\tau = (f(k))^\epsilon$ for any constant ϵ with $0 < \epsilon < 1$ (e.g., τ can be k^ϵ, $(\log k)^\epsilon$, $(\log \log k)^\epsilon$, etc.). For example, by letting $\tau = (\log k)^\epsilon$, our algorithm takes $O(l + k(\log k)^\epsilon)$ space and $O(kl\frac{(\log k)^2}{\log \log k})$ time.

Theorem 3. *The problem of optimally triangulating a 3-D surface band bounded by two closed polygonal curves P and Q can be solved in $O(l + k\tau)$ space and $O(kl\frac{(\log k)^2}{\log \tau})$ time, where $k = |P|$, $l = |Q|$, $k \leq l$, and unless $\tau = O(1)$, τ is of the form $(f(k))^\epsilon$ for any chosen function $f(k) > O(1)$ and any constant ϵ with $0 < \epsilon < 1$.*

References

1. Chen, D.Z., Daescu, O., Hu, X.S., Xu, J.: Finding an optimal path without growing the tree. Journal of Algorithms 49(1), 13–41 (2003)
2. Chen, D.Z., Wang, J., Wu, X.: Image segmentation with asteroidality/tubularity and smoothness constraints. International Journal of Computational Geometry and Applications 12(5), 413–428 (2002)
3. Cormen, T.H., Leiserson, C.E., Rivest, R.L., Stein, C.: Introduction to Algorithms, 2nd edn. McGraw-Hill, New York (2001)
4. Fuchs, H., Kedem, Z.M., Uselton, S.P.: Optimal surface reconstruction from planar contours. Communications of the ACM 20(10), 693–702 (1977)
5. Hirschberg, D.S.: A linear space algorithm for computing maximal common subsequences. Communications of the ACM 18(6), 341–343 (1975)
6. Kedem, Z.M., Fuchs, H.: A fast method for finding several shortest paths in certain graphs. Proc. 18th Allerton Conf., 677–686 (1980)
7. Maes, M.: On a cyclic string-to-string correction problem. Information Processing Letters 35(2), 73–78 (1990)
8. Thedens, D.R., Skorton, D.J., Fleagle, S.R.: Methods of graph searching for border detection in image sequences with applications to cardiac magnetic resonance imaging. IEEE Trans. on Medical Imaging 14(1), 42–55 (1995)
9. Wagner, R.A., Fischer, M.J.: The string-to-string correction problem. Journal of the ACM 21(1), 168–173 (1974)
10. Wang, C., Tang, K.: Optimal boundary triangulations of an interpolating ruled surface. ASME Journal of Computing and Information Science in Engineering 5(4), 291–301 (2005)
11. Wu, X.: Segmenting doughnut-shaped objects in medical images. In: Ibaraki, T., Katoh, N., Ono, H. (eds.) ISAAC 2003. LNCS, vol. 2906, pp. 375–384. Springer, Heidelberg (2003)

Enumerating Constrained Non-crossing Geometric Spanning Trees

Naoki Katoh and Shin-ichi Tanigawa

Department of Architecture and Architectural Engineering, Kyoto University,
Kyoto 615-8450 Japan
{naoki,is.tanigawa}@archi.kyoto-u.ac.jp

Abstract. In this paper we present an algorithm for enumerating without repetitions all non-crossing geometric spanning trees on a given set of n points in the plane under edge inclusion constraints (i.e., some edges are required to be included in spanning trees). We will first prove that a set of all edge-constrained non-crossing spanning trees is connected via remove-add flips, based on the constrained smallest indexed triangulation which is obtained by extending the lexicographically ordered triangulation introduced by Bespamyatnikh. More specifically, we prove that all edge-constrained triangulations can be transformed to the smallest indexed triangulation among them by $O(n^2)$ times of greedy flips. Our enumeration algorithm generates each output graph in $O(n^2)$ time and $O(n)$ space based on reverse search technique.

1 Introduction

Given a graph $G = (V, E)$ with n vertices and m edges where $V = \{1, \ldots, n\}$, G is a *spanning tree* if and only if G is connected and does not contain any cycle. An embedding of the graph on a set of points $P = \{p_1, \cdots, p_n\} \subset \mathbf{R}^2$ is a mapping of the vertices to points in the Euclidian plane $i \mapsto p_i \in P$. The edges ij of G are mapped to straight line segments $p_i p_j$. An embedding is *non-crossing* if each pair of segments $p_i p_j$ and $p_k p_l$ have no point in common without their endpoints.

If the spanning tree embedded on the plane is non-crossing, it is called *geometric non-crossing spanning tree*, which is simply called non-crossing spanning tree (NST) in this paper. We assume in this paper that spanning trees are embedded on a fixed point set P in $\mathbf{R}^2$. Let F be a set of non-crossing line segments on P. A spanning tree containing F is called *F-constrained spanning tree*. Then in this paper we give an algorithm for enumerating all the *F-constrained non-crossing spanning trees* (F-CNST). We simply denote a vertex p_i by i and an edge $p_i p_j$ by ij throughout the paper.

The algorithm we propose requires $O(n^2)$ time per output and $O(n)$ space. For the unconstrained case (i.e. $F = \emptyset$), the algorithm by Avis and Fukuda [6] requires an $O(n^3)$ time per output and $O(n)$ space. Recently Aichholzer et al. [2] have developed an algorithm for enumerating $O(n)$ time per output based on

G. Lin (Ed.): COCOON 2007, LNCS 4598, pp. 243–253, 2007.

the Gray code enumeration, whose space complexity is not given. Although their algorithm is superior to ours in the unconstrained case, it seems that it cannot be extended to the edge-constrained case.

It is well known that the number of non-crossing spanning trees grows too rapidly to allow a complete enumeration for a significantly larger point set. In view of both theoretical and practical applications the number of objects to be enumerated or the computational cost must be reduced by imposing several reasonable constraints. For this purpose, the edge constraint would be naturally considered, and thus our algorithm has much advantage in practice.

For the edge-constrained case, in our recent paper [7], we proposed an algorithm for enumerating the F-constrained non-crossing minimally rigid frameworks embedded on a given point set in the plane. We remarked therein that based on a similar approach, we could develop an $O(n^3)$ algorithm for enumerating F-CNSTs. Although we have not given either any algorithm details or analysis of the running time, it seems difficult to improve this running time.

Let $\mathcal{O}$ be a set of objects to be enumerated. Two objects are *connected* iff they can be transformed to each other by a *local operation*, which generates one object from the other by means of a small change. Especially, it is sometimes called *(1-)flip* if they have all but one edge in common. Define a graph $\mathcal{G}_\mathcal{O}$ on $\mathcal{O}$ with a set of edges connecting between objects that can be transformed to each other by one local operation. Then the natural question is how we can design local operation so that $\mathcal{G}_\mathcal{O}$ is connected, or how we can design $\mathcal{G}_\mathcal{O}$ with small diameter. There are several known results for these questions for triangulations (e.g. [11]), pseudo-triangulations [1], geometric matchings [10], some classes of simple polygons [9] and also for NSTs [2,1,3,4,6]. Especially relevant to the historical context of our work are the results for NSTs [6,1,3,4,2]. Let $\mathcal{ST}$ be a set of all NSTs on a given set of n points. Avis and Fukuda [6] have developed 1-flip such that $\mathcal{G}_{\mathcal{ST}}$ is connected with diameter $2n - 4$. For the case of a local operation other than 1-flip, the operations with diameters of $O(\log n)$ [3] and the improved result [1] are known. Aichholzer et al. in [3,4] also tried to design 1-flip with the additional requirement, called *edge slide*, such that removed edge moves to the other one along an adjacent edge keeping one endpoint of the removed edge fixed. Aichholzer et al. in [2] showed that the graph $\mathcal{G}_{\mathcal{ST}}$ contains a Hamiltonian path by developing Gray code enumeration schemes. In this paper, we will propose 1-flip with the additional requirement such that removed and added edges are sharing one endpoint, and show that all F-CNSTs are connected by $O(n^2)$ such flips plus $O(n)$ base exchange operations. We notice that all 1-flips designed in the previous works seem to be difficult to extend to the edge-constrained case.

Main tools we use are reverse search and the *smallest indexed triangulation* (SIT). Reverse search developed by Avis and Fukuda [5,6] generates all the elements of $\mathcal{O}$ by tracing the nodes in $\mathcal{G}_\mathcal{O}$. To trace $\mathcal{G}_\mathcal{O}$ efficiently, it defines a *root* on $\mathcal{G}_\mathcal{O}$ and a *parent* for each node except the root. Define the parent-child relation satisfying the following conditions: (1) each non-root object has a unique parent, and (2) an ancestor of an object is not itself. Then, iterating going up to the parent leads to the root from any other node in $\mathcal{G}_\mathcal{O}$ if $\mathcal{G}_\mathcal{O}$ is *connected*. The set of such paths defines a spanning tree, known as a *search tree*, and the algorithm traces it by depth-first search manner. So, the necessary ingredients to use the method are an implicitly described connected graph $\mathcal{G}_\mathcal{O}$ and an implicitly defined search tree in $\mathcal{G}_\mathcal{O}$. In this paper we supply these ingredients for the problem of generating all F-CNSTs.

The idea of the SIT is derived from a lexicographically ordered triangulation developed by Bespamyatnikh [8] for enumerating triangulations efficiently. We generalize it to edge-constrained case by associating an appropriate index with each triangulation. The SIT plays a crucial role in the development of our algorithm. We conjecture that the general idea to use triangulations proposed in this paper can be extended to an efficient algorithm for enumerating non-crossing graphs other than NSTs and minimally rigid frameworks because any non-crossing graph can be augmented to a triangulation.

2   Smallest Indexed Triangulation

In this section, we define an F-constrained smallest indexed triangulation (F-CSIT). Then we show that any edge-constrained triangulation can be transformed into CSIT by $O(n^2)$ flips. We remark again that CSIT is derived from the lexicographically ordered triangulation developed by Bespamyatnikh [8] although he had not extended his results to the edge-constrained case.

2.1   Notations

Let us first define several notations. Let P be a set of n points on the plane, and for simplicity we assume that the vertices $P = \{1, \ldots, n\}$ are labeled in the increasing order of x-coordinates. We assume that x-coordinates of all points are distinct and that no three points in P are colinear. For two vertices $i, j \in P$, we use the notation, $i < j$, if $x(i) < x(j)$ holds, where $x(\cdot)$ represents a x-coordinate of a point. Considering $i \in P$, we often pay our attention only to its right point set, $\{i + 1, \ldots, n\} \subseteq P$. So, let us denote $\{i + 1, \ldots, n\}$ by P_i.

We usually denote an edge between i and j with $i < j$ by ij. For three points i, j, k, the signed area $\Delta(i, j, k)$ of a triangle Δijk tells us that k is on the left or right side of a line passing through i and j when moving along the line from i to j by $\Delta(i, j, k) > 0$ or $\Delta(i, j, k) < 0$, respectively. Then the lexicographical ordering on a set of edges is defined as follows: for $e = ij$ and $e' = kl$, e is lexicographically smaller than e' (denoted by $e \prec e'$ or $e' \succ e$) iff $i < k$, or $i = k$ and $\Delta(i, j, l) < 0^1$, and denote by $e = e'$ when they coincide. Notice that, when $i = k$, the lexicographical ordering corresponds to the clockwise ordering around i in our definition.

---

[1] In general the lexicographical ordering for edge set is defined in such a way that $e = ij$ is smaller than $e' = kl$ iff either $i < k$ or $i = k$ and $j < l$ holds. But in this paper we adopt our lexicographical ordering for efficient enumeration described in Section 4.

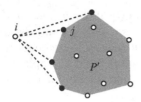

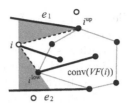

Fig. 1. Black vertices represent a set of vertices visible from i with respect to conv(P')

Fig. 2. Example of the upper and lower tangents. Bold edges represent F. Empty regions of ii^{up} and ii^{low} are shaded.

For two vertices $i, j \in P$, j is *visible* from i with respect to a constrained edge set F when an edge ij and any edge in F do not have a point in common except their endpoints. We assume that j is visible from i if $ij \in F$. And we denote a set of vertices of P_i visible from i with respect to F by $P_i(F)$.

Let conv(P') be a convex hull of a point set P'. For a vertex i with $i \notin$ conv(P'), $j \in P'$ is visible from i with respect to (the boundary of) conv(P') when $j = ij \cap$ conv(P') holds (see Fig. 1).

For an edge $e = ij$, let $l(e)$ and $r(e)$ denote the left and right endpoints of e, i.e. $l(e) = i$ and $r(e) = j$, respectively. A straight line passing through i and j splits $\mathbf{R}^2$ into two regions that are open regions of left and right sides of ij, i.e. $R^+(ij) = \{p \in \mathbf{R}^2 \mid \Delta(i,j,p) > 0\}$ and $R^-(ij) = \{p \in \mathbf{R}^2 \mid \Delta(i,j,p) < 0\}$. Similarly, the closed regions $\bar{R}^+(ij)$ and $\bar{R}^-(ij)$ are defined.

For $i \in P$, let $F(i) \subseteq F$ denote a set of constrained edges whose left endpoints coincide with i. *Upper* and *lower tangents*, ii^{up} and ii^{low}, of i with respect to (the constrained edge set) F are defined as those from i to the convex hull of $P_i(F)$, (see Fig. 2). Notice that $P_i(F) \subset \bar{R}^-(ii^{\mathrm{up}})$ and $P_i(F) \subset \bar{R}^+(ii^{\mathrm{low}})$ hold. Then they define *empty region* in which no point of P exists as we describe below. Let l be a line perpendicular to x-axis passing through i, and let e_1 and e_2 be the closest edges from i among F intersecting with l in the upper and lower side of i (if such edge exists). Then there exists no point of P inside the region bounded by l, e_1 (resp. e_2), and the line through i and i^{up} (resp. i^{low}). When e_1 (resp. e_2) does not exist, the empty region is defined by the one bounded by l and the line through i and i^{up} (resp. i^{low}). We call this fact *empty region property* of the upper and lower tangents.

2.2 Constrained Smallest Indexed Triangulation

Although we have not explain our *index* for an triangulation yet, let us first define the triangulation called F-constrained smallest indexed triangulation in this paper.

Definition 1. *Let ii^{up} and ii^{low} be the upper and lower tangents of $i \in P$ with respect to F, and denote a set of edges of $F(i) \cup \{ii^{\mathrm{up}}, ii^{\mathrm{low}}\}$ by $ii_0, ii_1, \ldots, ii_k$ arranged in clockwise order around i, (where $i_0 = i^{\mathrm{up}}$ and $i_k = i^{\mathrm{low}}$ hold). Let*

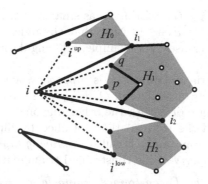

Fig. 3. The part of the CSIT around i, where bold edges represent F. The CSIT has edges between i and black vertices.

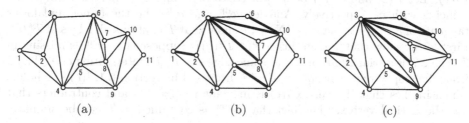

Fig. 4. (a)SIT. (b)CSIT. (c)non-CSIT.

$H_l = \operatorname{conv}(P_i(F) \cap \bar{R}^-(ii_l) \cap \bar{R}^+(ii_{l+1}))$ *for l with $0 \leq l \leq k-1$. Then for every l with $0 \leq l \leq k-1$ the F-constrained smallest indexed triangulation (F-CSIT) has an edge ij with $j \in H_l$ if and only if j is visible from i with respect to the convex hull H_l (see Fig. 3).*

We call H_l of Definition 1 *the convex hull of $P_i(F)$ bounded by ii_l and ii_{l+1}*. We give an example of CSIT in Fig.4, and also give an example when $F = \emptyset$. Notice that CSIT always has the edges of $F(i) \cup \{ii^{\mathrm{up}}, ii^{\mathrm{low}}\}$ for all $i \in P$.

Lemma 1. *The CSIT is a triangulation of the point set P.*

2.3 Greedy Flipping in Constrained Triangulations

Let T^* denote CSIT on a given point set. For any F-constrained triangulation, an *index* of T is defined as a pair of integers $n - c$ and d, and denoted by $\operatorname{index}(T) = (n - c, d)$, where $c \in \{1, \ldots, n-1\}$ and $d \in \{1, \ldots, n-3\}$ are a label of a *critical vertex* of T and a *critical degree* of the critical vertex, respectively. The *critical vertex* is the smallest label of a vertex whose incident edges differ from the corresponding set of incident edges in T^*. The *critical degree* is the number of edges incident to the critical vertex not contained in T^*. The index of T^* is defined to be $(0, 0)$. Then, for two triangulations T and T' with $\operatorname{index}(T) = $

$(n - c, d)$ and $\text{index}(T') = (n - c', d')$, T has smaller index than that of T' when $c > c'$, or $c = c'$ and $d < d'$ holds. Note that the index decreases as the label of the critical vertex increases. For example, a triangulation in Fig. 4 (c) has an index $(8, 2)$.

For an edge e in a triangulation T, e is *flippable* when two triangles incident to e in T form a *convex* quadrilateral Q. *Flipping* e in T generates a new triangulation by replacing e of T with the other diagonal of Q. Such operation is called *improving flip* if the triangulation obtained by flipping e has a smaller index than the previous one, and e is called *improving flippable*. Now let us show the greedy flipping property of the constrained triangulations.

Lemma 2. *Let T be the F-constrained triangulation with $T \neq T^*$ and c be the critical vertex of T. Then there exists at least one improving flippable edge incident to c in $T \setminus T^*$.*

Proof. Let $F(c)$ and $T^*(c)$ be sets of edges of F and T^* whose left endpoints coincide with c, respectively. And let cc^{up} and cc^{low} be the upper and lower tangents of c with respect to F. Now we show that T contains all edges of $T^*(c)$. First let us show that T contains $cc^{\text{up}}(\in T^*(c))$. Suppose that cc^{up} is missing in T. Since T is a triangulation, T has some edge $e \in T \setminus T^*$ intersecting cc^{up}. Then, from the empty region property of cc^{up} discussed in Section 2.1, $l(e) < c$ holds, which implies that the vertex $l(e)$ is incident to $e(\neq T^*)$ and contradicts that c is the critical vertex. (The fact that cc^{low} is contained in T can be similarly proved.)

Next let us show that each edge cv of $T^*(c)$ other than $F(c) \cup \{cc^{\text{up}}, cc^{\text{low}}\}$ is contained in T. Suppose that cv is missing in T. Then there exists some edge $e \in T \setminus T^*$ intersecting cv. Let $cc_0, \ldots, cc_k$ be the edges of $F(c) \cup \{cc^{\text{up}}, cc^{\text{low}}\}$ arranged in clockwise ordering around c. Since $cv \in T^*(c)$, there exists a unique l with $0 \leq l \leq k - 1$ of the convex hull of $P_c(F)$ bounded by cc_l and cc_{l+1} such that v is on the boundary of the convex hull. Then the edges cc_l, cc_{l+1} and the part of the boundary edges of such convex hull (convex chain) from c_l to c_{l+1} forms a pseudo-triangle with three corners c, c_l and c_{l+1}. Since there exists no point of P inside of such pseudo-triangle, the fact that e intersects cv implies that e also intersects at least one of cc_l and cc_{l+1}, which contradicts that T contains all edges of $F(c) \cup \{cc^{\text{up}}, cc^{\text{low}}\}$. Hence T contains $T^*(c)$.

Now let us show that there exists at least one improving flippable edge $e^* \notin T^*$ incident to c. Since c is a critical vertex, T has at least one edge $e' \notin T^*$. Let cp_1 and cp_2 be a pair of edges of $T^*(c)(\subset T)$ such that e' exists between cp_1 and cp_2 and an angle $\angle p_1 c p_2$ is minimum for all pairs of $T^*(c)$ (see Fig. 5). Consider a set of edges in T incident to c between cp_1 and cp_2, and denote them by $cq_1, cq_2, \ldots, cq_{\bar{j}}$ in clockwise order around c. Note that $cq_j \in T \setminus T^*$ holds for all $j = 1, \ldots, \bar{j}$, and then no vertex of q_j is inside of the triangle $\triangle cp_1p_2$, since T^* contains empty triangle face $\triangle cp_1p_2$. Therefore, all edges $cq_1, \ldots, cq_{\bar{j}}$ intersect p_1p_2. Let q_{j^*} be a vertex furthest from the line through p_1 and p_2 among q_j. Then a quadrilateral $cq_{j^*-1}q_{j^*}q_{j^*+1}$ is convex because q_{j^*-1}, q_{j^*} and q_{j^*+1} are not colinear, and flipping cq_{j^*} produces a triangulation with a smaller index than the previous one because $c < q_{j^*-1}$ and $c < q_{j^*+1}$ hold now. $\qquad\square$

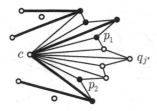

Fig. 5. Existence of an improving flippable edge cq_{j^*}. Bold edges represent the edges of F, and black vertices represent the vertices incident to c in T^*.

Theorem 1. *Every F-constrained triangulation T can be transformed into F-CSIT by $O(n^2)$ flips.*

Proof. From Lemma 2, $T(\neq T^*)$ always has an improving flippable edge, and flipping such edge reduces index(T). Since the number of distinct indices is $O(n^2)$, T can be transformed into F-CIST by $O(n^2)$ improving flips. □

3 Constrained Non-crossing Spanning Trees

Let F be a non-crossing edge set on P, and we assume that F is a forest. In this section we show that a set of F-constrained non-crossing spanning trees on P, denoted by $\mathcal{ST}$, is connected by $O(n^2)$ flips.

Let $E = \{e_1 \prec \ldots \prec e_m\}$ and $E' = \{e'_1 \prec \ldots \prec e'_m\}$ be lexicographically ordered edge lists. Then E is lexicographically smaller than E' if $e_i \prec e'_i$ for the smallest i such that $e_i \neq e'_i$. Consider the F-CSIT, which is denoted by $T^*(F)$ in what follows. *F-constrained smallest indexed spanning tree (F-CSIST)* is a F-constrained non-crossing spanning tree that is a subset of $T^*(F)$, and we denote a set of all F-CSISTs by $\mathcal{CSIST}$. Define ST^* as a spanning tree consisting of the lexicographically smallest edge list among $\mathcal{CSIST}$. The following lemma holds from the known fact about matroid (see e.g. [12]). Namely each $ST \in \mathcal{CSIST}$ is a base of graphic matroid restricted to the edge set of $T^*(F)$ (see the proof of Theorem 3 of [7] for more details).

Lemma 3. *Every non-crossing spanning tree of $\mathcal{CSIST}$ can be transformed into ST^* by at most $n-1$ flips.*

Now we will define an index for each spanning tree $ST \notin \mathcal{CSIST}$ to represent how far it is from one of $\mathcal{CSIST}$. For each F-constrained triangulation T we have defined its index with respect to F by index$_F(T) = (n - c, d)$, which represents how far T is from $T^*(F)$ by means of the critical vertex c and the critical degree d, (see the definitions in Section 2.3). We associate ST-constrained smallest indexed triangulation $T^*(ST)$ with each spanning tree ST, and define an index of ST denoted by index(ST) = $(n - c_{ST}, d_{ST})$ as index$_F(T^*(ST))$, We also call c_{ST} the critical vertex of ST. Fig. 6 shows an example of ST whose critical vertex is 1 and index(ST) is $(7, 2)$.

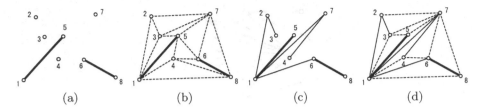

Fig. 6. (a) F, (b) $T^*(F)$, (c) ST, and (d) $T^*(ST)$, where bold edges represent F and dotted edges represent added edges for triangulations

For $i \in P$, let $ST(i)$ and $T^*(ST; i)$ be the edge sets of ST and $T^*(ST)$ whose left endpoints coincide with i, respectively. It is clear from the definition of CSIT (Definition 1) that the newly added edges for obtaining $T^*(ST)$ from ST are not flippable in $T^*(ST)$ except for the upper and lower tangents. Thus the next observation follows:

Observation 1. *For $i \in P$ any edge of $T^*(ST; i) \setminus (ST(i) \cup \{ii^{\text{up}}, ii^{\text{low}}\})$ is not flippable in $T^*(ST)$, where ii^{up} and ii^{low} are upper and lower tangents of i with respect to ST.*

Then we derive the followings from Lemma 2 and Observation 1:

Lemma 4. *Let $ST \notin \mathcal{CSIST}$ and c be the critical vertex of ST. Then (i) there exists at least one improving flippable edge in $T^*(ST) \setminus T^*(F)$, and (ii) an edge $e \in T^*(ST)$ is improving flippable if and only if e is flippable and $e \in ST(c) \setminus (F(c) \cup \{cc^{\text{up}}, cc^{\text{low}}\})$, where cc^{up} and cc^{low} are upper and lower tangents of c with respect to ST.*

Proof. Since $T^*(ST) \neq T^*(F)$ holds, (i) immediately follows from Lemma 2. Let us show (ii). From the definition of the index of $T^*(ST)$, we notice that flipping an improving flippable edge decreases the number of edges incident to c. Hence all improving flippable edges must be incident to c with their left endpoints. From Observation 1 any edge of $T^*(ST; c) \setminus (ST(c) \cup \{cc^{\text{up}}, cc^{\text{low}}\})$ is not flippable. Then the proof is completed by showing that neither cc^{up} nor cc^{low} are improving flippable edges. Suppose that cc^{up} is improving flippable. (The other case is similarly proved.) Then there exists a triangle face $\triangle cc^{\text{up}}v$ incident to cc^{up} in $T^*(ST)$ with $v \in R^+(cc^{\text{up}})$. Since cc^{up} is improving flippable edge, $c < v$ holds, and then $v \in P_c(ST)$. This contradicts the empty region property of cc^{up} with respect to ST.   □

Lemma 5. *Every F-constrained non-crossing spanning tree $ST \notin \mathcal{CSIST}$ can be transformed into a spanning tree in $\mathcal{CSIST}$ by at most $O(n^2)$ flips.*

Proof. Let c be a critical vertex of ST. From Lemma 4 there exists an edge $cc^* \in ST \setminus F$ that is improving flippable in $T^*(ST)$. There exist two vertices c_1^* and c_2^* incident to both c^* and c in $T^*(ST)$ such that $cc^* c_1^*$ and $cc^* c_2^*$ are triangle faces of $T^*(ST)$. When removing cc^* from ST, the set of vertices of $ST - cc^*$ is partitioned into two components, where c^* and c belong to different components, and c_1^* can belong to only one of them. Therefore adding one of cc_1^* or $c_1^*c^*$ to $ST - cc^*$, we obtain a new non-crossing spanning tree ST'. Note that index of $T^*(ST')$ is smaller than that of $T^*(ST)$ because $T^*(ST')$ does not have cc^* but has $c_1^*c_2^*$ instead and the critical degree decreases, i.e. the edge-constraint of cc^* is released and improving flip occurs in the underlying triangulation. Repeating this procedure, the underlying triangulation eventually reaches the F-CSIT. $\square$

From Lemmas 3 and 5 we have the following theorem:

Theorem 2. *Every F-constrained non-crossing spanning tree is connected by $O(n^2)$ flips.*

4 Enumerating Constrained Non-crossing Spanning Trees

Let ST^* be an F-CSIST with the lexicographically smallest edge list as defined in Section 3. Let I_{ST} be a set of improving flippable edges in ST. We define the following *parent function* $f : \mathcal{ST} \setminus \{ST^*\} \to \mathcal{ST}$ based on the results of the previous section:

Definition 2. *(Parent function) Let $ST \in \mathcal{ST}$ with $ST \neq ST^*$, and c be a critical vertex of ST. $ST' = ST - e_1 + e_2$ is the parent of ST, where*
Case 1: $ST \in \mathcal{CSIST}$,
- $e_1 = \max\{e \mid e \in ST \setminus ST^*\}$, *and* $e_2 = \min\{e \in ST^* \setminus ST \mid ST - e_1 + e \in ST\}$,
Case 2: $ST \notin \mathcal{CSIST}$,
- $e_1 = cc^* = \min\{e \in I_{ST}\}$, *and* e_2 *is either* cc_1^* *or* $c_1^*c^*$ *such that* $ST - e_1 + e_2 \in \mathcal{ST}$, *where* c_1^* *is a vertex of triangle face* $\Delta cc^*c_1^*$ *in* $T^*(ST)$ *with* $\Delta(c, c^*, c_1^*) > 0$.

Note that $I_{ST} \neq \emptyset$ and I_{ST} is the subset of $ST(c) \setminus F$ from Lemma 4. Therefore e_1 of Case 2 always exists. There exist two triangle faces, $\Delta cc^*c_1^*$ and $\Delta cc^*c_2^*$, incident to both cc^* in $T^*(ST)$. Here, we adopt c_1^* in order to define the unique parent. In Fig. 7 we show how the parent function works in Case 2. From Lemmas 3 and 5 these parent-child relationships form the search tree of $\mathcal{ST}$ explained in Section 1. To simplify the notations, we denote the parent function depending on Cases 1 and 2 by $f_1 : \mathcal{CSIST} \setminus \{ST^*\} \to \mathcal{CSIST}$ and $f_2 : \mathcal{ST} \setminus \mathcal{CSIST} \to \mathcal{ST}$, respectively.

Let $\mathsf{elist}_{ST'}$ and elist_{K_n} be the edge lists of ST' and K_n ordered lexicographically, and let $\mathsf{elist}_{ST'}(i)$ and $\mathsf{elist}_{K_n}(i)$ be their i-th element. Then, based on the algorithm in [5,6], we describe our algorithm in Fig. 8. The parent function needs $O(n + T_{\mathrm{CSIT}})$ time for each execution, where T_{CSIT} denotes the time to calculate $T^*(ST)$. Then, the while-loop from lines 4 to 15 has $|ST'| \cdot |K_n|$ iterations which require $O(n^3(n + T_{\mathrm{CSIT}}))$ time if simply checking lines 8 and 9. We can improve it to $O(n^2)$ time. Because of the lack of space, let us explain the outline and its rigorous description is omitted in this proceeding version.

Remember that the removing edge e_1 and the adding edge e_2 must share one endpoint in Case 2 of the parent function, so the inner while-loop from lines 6 to 14 can be reduced to $O(n)$ iterations. Then, to achieve $O(n^2)$ time algorithm,

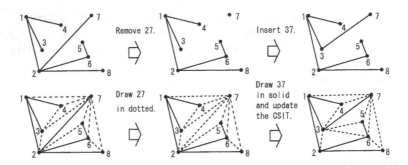

Fig. 7. An example of the parent function for $ST \notin CSIST$, where $index(ST) = (6,2)$ and $F = \emptyset$. Removing 27 and adding 37 we obtain a new spanning tree with index $(6,1)$.

Algorithm Enumerating F-constrained non-crossing spanning trees.

1: $ST^* := F\text{-CSIST}$ with lexicographically smallest edge list;
2: $ST' := ST^*$; $i,j := 0$; Output(ST');
3: **repeat**
4:     **while** $i \leq |ST'|$ **do**
5:         **do** $\{i := i + 1; e_{\text{rem}} := \text{elist}_{ST'}(i); \}$ **while**($e_{\text{rem}} \in F$);
6:         **while** $j \leq |K_n|$ **do**
7:             **do** $\{j := j + 1; e_{\text{add}} := \text{elist}_{K_n}(j); \}$ **while**($ea \in ST'$));
8:             **if** $ST' - e_{\text{rem}} + e_{\text{add}} \in ST$ **then**
9:                 **if** $f_1(ST' - e_{\text{rem}} + e_{\text{add}}) = ST'$ **or** $f_2(ST' - e_{\text{rem}} + e_{\text{add}}) = ST'$ **then**
10:                     $ST' := ST' - e_{\text{rem}} + e_{\text{add}}$; $i,j := 0$; Output(ST');
11:                     **go to** line 4;
12:                 **end if**
13:             **end if**
14:         **end while**
15:     **end while**
16:     **if** $ST' \neq ST^*$ **then**
17:         $ST := ST'$;
18:         **if** $ST \in CSIST$ **then** $ST' := f_1(ST)$; **else** $ST' := f_2(ST)$;
19:         determine integers pair (i,j) such that $ST' - \text{elist}_{ST'}(i) + \text{elist}_{K_n}(j) = ST$;
20:         $i := i - 1$;
21:     **end if**
22: **until** $ST' = ST^*$ **and** $i = |ST'|$ **and** $j = |K_n|$;

Fig. 8. Algorithm for enumerating F-constrained non-crossing spanning trees

we need to show that the inner-while loop can be implemented in $O(n)$ time. Let ST and ST' be two distinct F-CNSTs for which $ST = ST' - e_{\text{rem}} + e_{\text{add}}$ for $e_{\text{rem}} \in ST' \setminus F$ and $e_{\text{add}} \in K_n \setminus ST'$. Our goal is to enumerate all the edge pairs of $(e_{\text{rem}}, e_{\text{add}})$ such that ST is a child of ST' in linear time for each e_{rem}. More precisely we will describe necessary and sufficient conditions for which e_{rem} and e_{add} satisfy either $f_1(ST) = ST'$ or $f_2(ST) = ST'$. And then for each

$e_{\text{rem}} \in ST' \setminus F$ we can enumerate, in $O(n)$ time with $O(n)$ space, a set of edge pairs satisfying these conditions. Since the number of candidates for e_{rem} is $O(n)$, we have the following theorem:

Theorem 3. *The set of all F-constrained non-crossing spanning trees on a given point set can be reported in $O(n^2)$ time per output using $O(n)$ space.*

Acknowledgment

We would like to thank Professor David Avis and Professor Ileana Streinu for their contributions of crucial ideas and many research discussions. This work is supported by the project *New Horizons in Computing*, Grant-in-Aid for Scientific Research on Priority Areas, NEXT Japan.

References

1. Aichholzer, O., Aurenhammer, F., Huemer, C., Krasser, H.: Transforming spanning trees and pseudo-triangulations. Inf. Process. Lett. 97(1), 19–22 (2006)
2. Aichholzer, O., Aurenhammer, F., Huemer, C., Vogtenhuber, B.: Gray code enumeration of plane straight-line graphs. In: Proc. 22th European Workshop on Computational Geometry (EuroCG '06), pp. 71–74. Greece (2006)
3. Aichholzer, O., Aurenhammer, F., Hurtado, F.: Sequences of spanning trees and a fixed tree theorem. Comput. Geom. 21(1-2), 3–20 (2002)
4. Aichholzer, O., Reinhardt, K.: A quadratic distance bound on sliding between crossing-free spanning trees. In: Proc. 20th European Workshop on Computational Geometry (EWCG04), pp. 13–16 (2004)
5. Avis, D., Fukuda, K.: A pivoting algorithm for convex hulls and vertex enumeration of arrangements and polyhedra. Discrete and Computational Geometry 8, 295–313 (1992)
6. Avis, D., Fukuda, K.: Reverse search for enumeration. Discrete Applied Mathematics 65(1-3), 21–46 (1996)
7. Avis, D., Katoh, N., Ohsaki, M., Streinu, I., Tanigawa, S.: Enumerating constrained non-crossing minimally rigid frameworks,
   http://arxiv.org/PS_cache/math/pdf/0608/0608102.pdf
8. Bespamyatnikh, S.: An efficient algorithm for enumeration of triangulations. Comput. Geom. Theory Appl. 23(3), 271–279 (2002)
9. Hernando, M.C., Houle, M.E., Hurtado, F.: On local transformation of polygons with visibility properties. In: Du, D.-Z., Eades, P., Sharma, A.K., Lin, X., Estivill-Castro, V. (eds.) COCOON 2000. LNCS, vol. 1858, pp. 54–63. Springer, Heidelberg (2000)
10. Hernando, C., Hurtado, F., Noy, M.: Graphs of non-crossing perfect matchings. Graphs and Combinatorics 18(3), 517–532 (2002)
11. Hurtado, F., Noy, M., Urrutia, J.: Flipping edges in triangulations. Discrete & Computational Geometry 22(3), 333–346 (1999)
12. Welsh, D.J.A.: Matroids: fundamental concepts. In: Graham, R.L., Grötschel, M., Lovász, L. (eds.) Handbook of Combinatorics, vol. I, pp. 481–526. North-Holland, Amsterdam (1995)

Colored Simultaneous Geometric Embeddings*

U. Brandes[1], C. Erten[2], J. Fowler[3], F. Frati[4], M. Geyer[5], C. Gutwenger[6],
S. Hong[7], M. Kaufmann[5], S.G. Kobourov[3], G. Liotta[8],
P. Mutzel[6], and A. Symvonis[9]

[1] Department of Computer & Information Science, University of Konstanz
ulrik.brandes@uni-konstanz.de
[2] Department of Computer Science, Isik University
cesim@isikun.edu.tr
[3] Department of Computer Science, University of Arizona
{jfowler,kobourov}@cs.arizona.edu
[4] Department of Computer Science, University of Roma Tre
frati@dia.uniroma3.it
[5] Wilhelm-Schickard-Institute of Computer Science, University of Tübingen
{geyer,mk}@informatik.uni-tuebingen.de
[6] Department of Computer Science, University of Dortmund
{petra.mutzel,carsten.gutwenger}@cs.uni-dortmund.de
[7] NICTA Ltd. and School of Information Technologies, University of Sydney
seokhee.hong@nicta.com.au
[8] School of Computing, University of Perugia
liotta@diei.unipg.it
[9] School of Applied Math. & Phys. Sciences, National Technical University of Athens
symvonis@math.ntua.gr

Abstract. We introduce the concept of *colored simultaneous geometric embeddings* as a generalization of simultaneous graph embeddings with and without mapping. We show that there exists a universal pointset of size n for paths colored with two or three colors. We use these results to show that colored simultaneous geometric embeddings exist for: (1) a 2-colored tree together with any number of 2-colored paths and (2) a 2-colored outerplanar graph together with any number of 2-colored paths. We also show that there does not exist a universal pointset of size n for paths colored with five colors. We finally show that the following simultaneous embeddings are not possible: (1) three 6-colored cycles, (2) four 6-colored paths, and (3) three 9-colored paths.

1 Introduction

Visualizing multiple related graphs is useful in many applications, such as software engineering, telecommunications, and computational biology. Consider the case where a pair of related graphs is given and the goal is to visualize them so as to compare the two, e.g., evolutionary trees obtained by different algorithms.

---

* Work on this paper began at the BICI Workshop on Graph Drawing, held in Bertinoro, Italy in March 2006.

G. Lin (Ed.): COCOON 2007, LNCS 4598, pp. 254–263, 2007.
© Springer-Verlag Berlin Heidelberg 2007

When visually examining relational information, such as a graph structure, viewers construct an internal model called the mental map, for example, using the positions of the vertices relative to each other. When viewing multiple graphs the viewer has to reconstruct this mental map after examining each graph and a common goal is to aid the viewer in this reconstruction while providing a readable drawing for each graph individually. Simultaneous embeddings [4] aid in visualizing multiple relationships between the same set of objects by keeping common vertices and edges of these graphs in the same positions.

A simultaneous geometric embedding is a generalization of the traditional planar graph embedding problem, where we look for a common embedding of multiple graphs defined on the same vertex set. We omit the "geometric" clarification in the rest of the paper as we only consider straight-line drawings. There are two main variations of the problem. In *simultaneous embedding with mapping* the embedding consists of plane drawings for each of the given graphs on the same set of points, with corresponding vertices in the different graphs placed at the same point. In *simultaneous embedding without mapping* the embedding consists of plane drawings for each of the given graphs on the same set of points, where any vertex can be placed at any of the points in the point set.

Restricted subclasses of planar graphs, such as pairs of paths, pairs of cycles, and pairs of caterpillars, admit a simultaneous embedding with mapping, while there exist pairs of outerplanar graphs and triples of paths that do not [4]. Recently, it was shown that pairs of trees do not always have such embeddings [9]. Fewer results are known for the less restricted version of the problem where the mapping is not predefined. While it is possible to simultaneously embed without mapping any planar graph with any number of outerplanar graphs, it is not known whether any pair of planar graphs can be simultaneously embedded without mapping [4].

Simultaneous embedding is related to universal pointsets, graph thickness, and geometric thickness. While de Fraysseix *et al.* [6] showed that there does not exist a universal pointset of size n in the plane for n-vertex planar graphs, Bose [3] showed that a set of n points in general position is a universal pointset for trees and outerplanar graphs. Using simultaneous embedding techniques, Duncan *et al.* [8] showed that degree-four graphs have geometric thickness two.

As we show, colored simultaneous embeddings allow us to generalize the problems above so that the versions with and without mappings become special cases. Formally, the problem of *colored simultaneous embedding* is defined as follows. The input is a set of planar graphs $G_1 = (V, E_1)$, $G_2 = (V, E_2)$, ..., $G_r = (V, E_r)$ on the same vertex set V and a partition of V into k classes, which we refer to as *colors*. The goal is to find plane straight-line drawings D_i of G_i using the same $|V|$ points in the plane for all $i = 1, \ldots, r$, where vertices mapped to the same point are required to be of the same color.

We call such graphs k-colored graphs. Given the above definition, simultaneous embeddings with and without mapping correspond to colored simultaneous embeddings with $k = |V|$ and $k = 1$, respectively. Thus, when a set of input graphs allows for a simultaneous embedding without mapping but does not allow for a simultaneous embedding with mapping, there must be a threshold for the number of colors beyond which the graphs can no longer be embedded simultaneously.

In this paper we present the first results about colored simultaneous embeddings. We study different values of k and show that any line-separated set of points of size n is a universal pointset for n-vertex 2-colored paths. Moreover, there exists a universal pointset of size n for n-vertex 3-colored paths while there is no such universal pointset n-vertex 5-colored paths. We also show how to simultaneously embed a 2-colored outerplanar graph and any number of 2-colored paths. Finally we show the existence of three 6-colored cycles (or four 6-colored paths, or three 9-colored paths) that cannot be simultaneously embedded.

2   Two-Colored Simultaneous Embeddings

We begin by showing the existence of a universal pointset for 2-colored paths. The following lemma extends a result of Abellanas et al. [1] on proper 2-colorings of paths.

Lemma 1. *Given a 2-colored path P of r red and b blue vertices and a set S of r red and b blue points separated by a line and in general position, there exists a planar straight-line embedding of P into S.*

Proof. Without loss of generality we can assume that S is separated by a vertical line, and that the red points are on the left of that line. Let $P = v_0, v_1, \ldots v_n$ and let P_i be the drawing of the path after the first i vertices of P have been embedded. Let H_i be the lower convex envelope of the points of S not used by P_i. We maintain the following invariants for all $i = 0, \ldots, n-1$ for which the colors of v_i and v_{i+1} are different:

1. The drawing of P_i does not intersect H_i.
2. The point p_i into which the most recent vertex v_i has been embedded can see a point of H_i of the other color and P_i does not intersect the area bounded by this line of sight and the vertical line from p_i upward.

Assume that v_i is of different color than v_{i+1} and let h, $1 \le h \le n-i$, be maximal such that $v_{i+1}, v_{i+2}, \ldots v_{i+h}$ all have the same color. To maintain the above invariants, we find a line that cuts off the required number h of points of color different from v_i from H_i (identified with the area on and above it). Assume v_i is red (which implies that it has been placed at a point p_i in the left half-plane) and v_{i+1} is blue; see Fig. 1.

Consider now the red end-point r_i of the unique edge of H_i that crosses the vertical separation line. We rotate a ray emanating from r_i counterclockwise until either h unused blue points are encountered, or a red point r_i' lies on the ray. In the latter case, we continue by rotating counterclockwise the ray around r_i'. We repeat this process until h blue points are found, and let B_i be the set of identified blue points. Let C_{B_i} be the convex hull of B_i. These points can be added to the path, as follows: Let a be the first blue point of H_i that is hit by a ray emanating

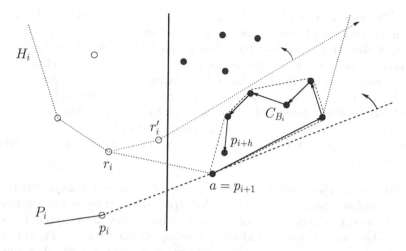

Fig. 1. Embedding a 2-colored path

from p_i and rotated counterclockwise. Point a also belongs to C_{B_i}. We can then connect p_i to point a. From point a we move counterclockwise along C_{B_i} until the right-most point of C_{B_i} is reached, while adding each encountered point to the drawing of the path. The remaining points of B_i are taken in decreasing value of their x-coordinates until the final point, p_{i+h}.

The resulting path ending at p_{i+h} satisfies the invariants: P_{i+h} does not intersect H_{i+h} and since p_{i+h} is the leftmost point of B_i the second invariant is also satisfied.   □

Using Lemma 1 we can embed k 2-colored paths for any $k > 0$ on a set of 2-colored points in general position in the plane that are separated by a straight-line, provided we have sufficient number of points of each color. The resulting set of points is a universal one for these k 2-colored paths, which yields the following theorem:

Theorem 1. *Any number of 2-colored paths can be simultaneously embedded.*

2.1   A Tree and Paths on Two Colors

We first show that it is always possible to draw a 2-colored tree in such a way that the two colors are separated by a line.

Lemma 2. *Any 2-colored tree can be embedded so that the colors are separated by a straight line.*

Proof. We use a divide-and-conquer approach and recursively process the tree from an arbitrary root node. We begin by drawing a vertical line l and assigning the left side to color 1 and the right side to color 2. Next we sort the children of the root by their colors. Let j of the children have color 1 and k children have color 2.

We can assume without loss of generality that the root is of color 1 and can place it on the left side of line l. The j children of color 1 are placed consecutively, such that the first is strictly beneath and to the left of the root, the second is strictly beneath and to the left of the first, and so on. We place the k children of color 2 to the right of line l in a similar fashion. We place the first child strictly beneath and to the right of the root, the second strictly beneath and to the right of the first, and so on. Note that every child has unobstructed line of sight to an horizontal sliver of the plane on both sides of line l. Thus, we can recursively place the children of the $j + k$ vertices until the entire tree has been processed.   □

Now using the result from Lemma 2 we can embed a 2-colored tree on a set of 2-colored points in the plane that are separated by a straight-line. Then we can perturb the positions of the vertices until they are in general position. This can be done without introducing crossings as shown in [4]. From Lemma 1, the resulting set of points is a universal one for 2-colored paths. Together these two results yield the next theorem:

Theorem 2. *A 2-colored tree and any number of 2-colored paths can be simultaneously embedded.*

2.2   Planar Graph and Paths on Two Colors

We have seen that in order to simultaneously embed a 2-colored planar graph G with any number of 2-colored paths it suffices to find a plane drawing of G in which the vertex sets of the same color, V_1 and V_2, can be separated by a line. Let G_1 and G_2 be the two subgraphs induced by the vertex sets V_1 and V_2 respectively. We call such a partition a *bipartition*, and the edges with vertices from both graphs are called *bipartition edges*.

Next we present a characterization of the class of 2-colored planar graphs that can be separated by a line. We make extensive use of the characterization and the embedding algorithm for HH layouts by Biedl *et al.* [2]. An HH layout is a drawing of a planar bipartition without crossings (but not necessarily using straight-line edges), in which the two vertex sets are separated by a horizontal line. We begin with the characterization of planar bipartitions that can be drawn as HH layouts.

Lemma 3. *[2] Planar bipartitions can be realized as HH layouts only if the subgraph D of the dual graph induced by the dual edges of the bipartition edges is connected.*

Moreover, it is shown in [2] that D is Eulerian and that it is possible to construct y-monotone HH layouts with few bends in linear time. The construction is roughly as follows. Find an Eulerian circuit of D that separates the sets V_1 and V_2. Then dummy vertices, that will become bends later, are introduced along the bipartition edges. Next the chain of dummy vertices is processed in the order of the Eulerian circuit and the straight-line drawing algorithm of Chrobak and Kant [5] is applied to the two subgraphs separately by placing one of them below (without loss of generality, say, G_1) and the other above the chain. The final result is straight-line planar drawing with the exception of the bipartition edges which have exactly one bend each; see Fig. 2(a).

This approach does not produce exactly the result that we need. We now show how to obtain a drawing with no bends, while not introducing any crossings, after applying the above technique to the planar bipartition and obtaining the HH layout (which may have some bends).

Lemma 4. *From each HH layout with some bends on the separation line, we can derive a straight-line drawing, while keeping the two partitions separated by a line.*

Proof. We begin by directing all the edges upward with respect to the basic HH layout L in order to obtain an upward planar embedding E of G. A theorem of Di Battista and Tamassia [7] states that the upward planar embedding E can be realized as a straight-line upward drawing. The resulting drawing, however, may not separate the two sets by a straight horizontal line. Below we show how to obtain the needed straight-line drawing in which the two sets are indeed separable by a line.

Let Γ_1 be the upward embedding of the graph G_1 with an upper boundary B_1 made of vertices adjacent to the bipartition edges. We extend Γ_1 by adding a top vertex t which we connect to all the boundary vertices by edges (v, t), where $v \in B_1$. Now we can apply the straight line drawing algorithm of Di Battista and Tamassia to the extended embedding and obtain an upward straight-line drawing, with the vertices on the boundary B_1 drawn with increasing x-coordinates; see Fig 2(b). After removing vertex t, B_1 is once again the upper boundary. Similarly, we can extend the embedding Γ_2 of G_2 in order to obtain a drawing with x-monotone lower boundary B_2.

Next we stretch the two layouts in the x-direction so that the slopes of the boundary edges become smaller. In particular, we stretch the layouts until all slopes are less than $40°$. Note that stretching preserves both planarity and upwardness of the layouts.

Finally we place the two layouts of Γ_1 and Γ_2 above each other and at vertical distance twice the larger of their widths. Now we can safely insert the bipartition edges which connect the two boundaries B_1 and B_2. By the choice of separation distance, the slopes of the bipartition edges are larger than $60°$. Thus the bipartition edges cannot introduce any crossings and now the two parts can be separated by an horizontal line as desired; see Fig. 2(c).   □

Lemma 1 and the algorithm above yield the following lemma:

Lemma 5. *Let G be a planar bipartition graph in which the dual graph of the subgraph induced by the bipartition edges is connected. (a) Then a straight-line drawing for G can be constructed where the two parts are separated by a horizontal line. (b) Since the bipartition includes a 2-coloring, G plus any number of 2-colored paths can be simultaneously embedded.*

As 2-colored outerplanar graphs fulfill the conditions of Lemma 5, we have the following theorem:

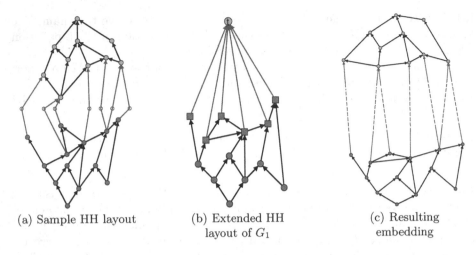

(a) Sample HH layout   (b) Extended HH
                             layout of G_1

(c) Resulting
    embedding

Fig. 2. HH Layouts

Theorem 3. *A 2-colored outerplanar graph and any number of 2-colored paths can be simultaneously embedded.*

3 k-Colored Simultaneous Embeddings

In this section we extend the investigation to more than two colors. We recall that there exist three paths which do not admit a simultaneous embedding with mapping [4], whereas it is easy to see that any number of paths have a simultaneous embedding without mapping. Now we consider k-colored paths and/or k-colored k-cycles for $3 \leq k \leq 9$.

3.1 Three Colors

As in the case of 2-colored embeddings we are looking for a universal pointset for paths. A slight modifications of the original universal pointset for 2-colored paths allows us to extend its utility to the 3-colored case.

Theorem 4. *Any number of 3-colored paths can be simultaneously embedded.*

Proof. Let P be any 3-colored path with c_1 vertices of color 1, c_2 vertices of color 2 and c_3 vertices of color 3, where $c_1 + c_2 + c_3 = n$. Let l_1, l_2 and l_3 be three line-segments with a common endpoint O and meeting at 120° angle. Place c_1 points along l_1, c_2 points along l_2, and c_3 points along l_3, ensuring that the origin O is not used.

Next map every vertex of the path, in order, to the point of the corresponding color that is closest to the origin and is not already taken. Since every point has line of sight to any other point and for a given p_i of P the previous path only blocks line of sight to the points already taken, the result is a plane drawing.   □

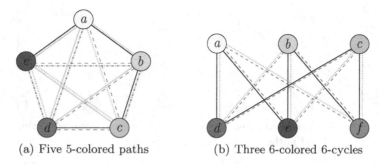

(a) Five 5-colored paths   (b) Three 6-colored 6-cycles

Fig. 3. Sets of k-colored graphs for $k \in \{5, 6\}$ on distinctly colored points whose unions form a K_5 and a $K_{3,3}$

3.2   Four and Five Colors

While universal pointsets exist for 1-colored paths, 2-colored paths and 3-colored paths, we have not been able to find one for 4-colored paths. However, we can show that for $k > 4$ universal pointsets for k-colored paths do not exist.

Theorem 5. *There does not exist a universal pointset for 5-colored paths.*

Proof. Consider the following five 5-colored paths on 5 points given in Fig. 3(a) whose union is K_5 where each edge in the K_5 belongs to exactly two paths:

1. $a-c-d-b-e$ (thin red dashed edges),
2. $a-d-e-b-c$ (thick light purple alternating dash and dot edges),
3. $b-a-c-e-d$ (thick green dotted edges),
4. $b-d-a-e-c$ (thick yellow solid edges), and
5. $e-a-b-c-d$ (thin blue solid edges).

   In any drawing of K_5 there must be at least one crossing. If this crossing is formed by a pair of edges from different paths then a simultaneous embedding might be possible. However, the paths above were chosen in such a way that every pair of edges either belongs to the same path or is incident. As straight-line incident edges cannot form the crossing pair it suffices to examine all pairs of non-adjacent edges in order to verify that they occur in at least one of the paths.

3.3   Six and Nine Colors

Here we consider sets of graphs on pointsets of six or more colors, in which the sets of graphs to simultaneously embed have cardinality less than five.

Lemma 6. *There exist three 6-colored cycles that cannot be simultaneously embedded.*

Proof. Consider the following three cycles, also shown in Fig. 3(b):
1. $e-a-d-c-f-b-e$ (thin blue solid edges),
2. $e-a-f-b-d-c-e$ (thin red dashed edges), and
3. $a-f-c-e-b-d-a$ (thick green dotted edges).

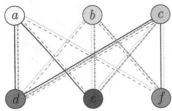 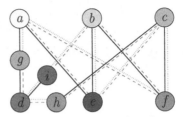

(a) One 5-colored and three   (b) Three 9-colored paths
6-colored paths

Fig. 4. Sets of k-colored graphs for $k \in \{6, 9\}$ on distinctly colored points whose unions form a $K_{3,3}$ or a subdivision thereof

A visual examination of Fig. 3(b) shows that the union of these cycles forms a $K_{3,3}$. Moreover, every edge in the $K_{3,3}$ belongs to two of the three cycles. In any drawing of $K_{3,3}$ there must be at least one crossing. Since there are only three paths altogether, every pair of edges in the $K_{3,3}$ must share a common 6-cycle, which forces a self-intersecting cycle.   □

Lemma 7. *There exist four 6-colored paths that cannot be simultaneously embedded.*

Proof. Fig. 4(a) depicts the following set of one 5-colored path and three 6-colored paths whose union forms $K_{3,3}$:

1. $e-a-d-c-f$ (thin blue solid edges),
2. $e-a-f-b-d-c$ (thin red dashed edges),
3. $a-f-c-e-b-d$ (thick green dotted edges), and
4. $a-d-c-e-b-f$ (thick brown dash-and-dots edges).

Every edge in $K_{3,3}$ belongs to at least two of the four paths. As a result, since there are more than three paths, it is easy to manually inspect all 18 pairs of non-adjacent edges to verify that each pair shares a common path. Thus at least one of the paths must be self-intersecting.   □

Lemma 8. *There exist three 9-colored paths that cannot be simultaneously embedded.*

Proof. Fig. 4(b) shows that every edge in the subdivided $K_{3,3}$ union belongs to exactly two of the following three paths:

1. $h-c-f-b-e-a-g-d-i$ (thin blue solid edges),
2. $g-d-h-c-e-a-f-b-i$ (thin red dashed edges), and
3. $g-a-f-c-e-b-i-d-h$ (thick green dotted edges).

Since there are only three 9-colored paths altogether, every pair of edges in the subdivided $K_{3,3}$ must share a common path forcing a self-intersecting path. Note that this result is a simplified version of Theorem 2 of Brass *et al.* [4].   □

4 Conclusions and Open Problems

Table 1 summarizes the current status of the newly formulated problem of colored simultaneous embedding. A "✓" indicates that it is always possible to simultaneously embed the type of graphs, a "✗" indicates that it is not always possible, and a "?" indicates an open problem.

Table 1. k-colored simultaneous embeddings: results and open problems

	$k=1$	$k=2$	$k=3$	$k=4$	$k=5$	$k=6$	$k=9$	$k=n$
Paths $P_1 \ldots P_3$	✓	✓	✓	?	?	?	✗	✗
Paths $P_1 \ldots P_4$	✓	✓	✓	?	?	✗	✗	✗
Any number of paths	✓	✓	✓	?	✗	✗	✗	✗
Planar Graph G and Path P	✓	✓	?	?	?	?	✗	✗
Outerplanar Graph G and Path P	✓	✓	?	?	?	?	?	?
Tree T and Path P	✓	✓	?	?	?	?	?	?
Two trees T_1, T_2	✓	?	?	?	?	?	?	✗
Two planar graphs G_1, G_2	?	?	?	?	?	?	✗	✗

References

1. Welsh, D.J.A.: Matroids: fundamental concepts. In: Graham, R.L., Grötschel, M., Lovász, L. (eds.) Handbook of Combinatorics, vol. I, pp. 481–526. North-Holland, Amsterdam (1999)
2. Biedl, T., Kaufmann, M., Mutzel, P.: Drawing planar partitions: HH-drawings. In: 24th Workshop on Graph-Theoretic Concepts in Computer Science (WG), pp. 124–136 (1998)
3. Bose, P.: On embedding an outer-planar graph in a point set. Computational Geometry: Theory and Applications 23(3), 303–312 (2002)
4. Brass, P., Cenek, E., Duncan, C.A., Efrat, A., Erten, C., Ismailescu, D., Kobourov, S.G., Lubiw, A., Mitchell, J.S.B.: On simultaneous graph embedding. In: 8th Workshop on Algorithms and Data Structures (WADS), pp. 243–255 (2003)
5. Chrobak, M., Kant, G.: Convex grid drawings of 3-connected planar graphs. Intl. Journal of Computational Geometry and Applications 7(3), 211–223 (1997)
6. de Fraysseix, H., Pach, J., Pollack, R.: How to draw a planar graph on a grid. Combinatorica 10(1), 41–51 (1990)
7. Di Battista, G., Tamassia, R.: Algorithms for plane representation of acyclic digraphs. Theoretical Computer Science 61(2-3), 175–198 (1988)
8. Duncan, C.A., Eppstein, D., Kobourov, S.G.: The geometric thickness of low degree graphs. In: 20th Annual ACM-SIAM Symposium on Computational Geometry (SCG), pp. 340–346 (2004)
9. Kaufmann, M., Vrťo, I., Geyer, M.: Two trees which are self-intersecting when drawn simultaneously. In: 13th Symposium on Graph Drawing (GD), pp. 201–210 (2005)

Properties of Symmetric Incentive Compatible Auctions*

Xiaotie Deng[1], Kazuo Iwama[2], Qi Qi[1], Aries Wei Sun[1], and Toyotaka Tasaka[2]

[1] Department of Computer Science, City University of Hong Kong
csdeng@cityu.edu.hk, qi.qi@student.cityu.edu.hk, sunwei@cs.cityu.edu.hk
[2] School of Informatics, Kyoto University, Japan
iwama@kuis.kyoto-u.ac.jp, TasakaToyotaka@t13.mbox.media.kyoto-u.ac.jp

Abstract. We formalize the definition of symmetric auctions to study fundamental properties of incentive compatible auction protocols. We characterize such auction protocols for those with a fixed number of items to sell and study some important properties of those with an indefinite number of sales.

1 Introduction

In recent years, auction based protocols have become a popular solution to E-commerce trading systems. Their adoption in real commercial processes have been at a scale never seen previously. Originally, auctions are used to sell normal items that are in limited supply[8,4,7]. Largely because of digital goods of zero marginal cost, auction models with unlimited supply [6,3,2,1,5] have become an important research topic in recent year. The study of digital goods has also invented new variations of auctions and provoked new thinkings in the principles governing auction protocol designs.

In particular, a new concept of "competitive auction" introduced in [6] made it the first priority that the total sales of the auctioneer be maximized, approximately within a constant factor of a benchmark optimum. To achieve that, several important properties are examined in details at a technical depth that has not been necessary for many previously well known auction protocols. At the same time, a nature question has arisen to demand an extensive careful study is that what are the basic principles we have to stick to in economic transactions at this micro level. Our study is motivated to study this fundamental question to understand the limitation and the possibility in auction protocols.

In designing auction protocols, a number of common properties are usually desired, though sometimes not explicitly stated. We should focus on incentive compatible protocols, for which bidding the trues value would be an optimal strategy of every participating agent. In addition, we will restrict our discussion to those with the *symmetric* property. We should use a specific definition of symmetric auction protocols. We comment that there may be other formalization

---

* This research is supported by SRG grant (7001989) of City University of Hong Kong.

G. Lin (Ed.): COCOON 2007, LNCS 4598, pp. 264–273, 2007.

of the concept symmetry for auction designs. Our choice is one of the simplest forms.

We study several other intersting properties to analyse their effects on incentive compatible auction protocol designs as well as their relationships with each other. We introduce those properties and present some preliminary results in §2.

As we shall expect, the *symmetric* property will *not* prevent lower bidders to win over higher bidders. Even though it might be possible for some special types of discriminative auctions, it is not *"reasonable"* and unfair in most situations. In §3 we define the property a bidder wins if any lower one does as "Winner Monotone". We also define two other equally plaussible monotone properties, "Price Function Monotone" and "Revenue Monotone". Our investigations into the relationships of the three properties give very interesting results and show that we should not take for granted that an arbitrarily designed auction is "reasonable".

Then an interesting question arises, when will an auction be "reasonable"? A most common benchmark auction is the celebrated Vickrey auction [8]. We study Vickrey auction in depth in §4. Our analysis show two *sufficient* and *necessary* conditions for a symmetric incentive compatible auction to be equivalent to Vickrey auction. One is that the number of winners is always fixed (See §4.1). Another is that the auction is homogeneous, revenue monotone, price function monotone, and single winning price (See §4.2).

We conclude our work in §5 with a discussion on future works.

2 Preliminaries

We present the models of auction protocols, the mathematical notations for agents' parameters, as well as acronyms for important properties to be discussed.

2.1 Basic Model and Notations

We restrict our attentions to auctions satisfying the following properties:

1. Every bidder wants at most one item.
2. The items sold by the auction are the same.
3. The auction is deterministic.
4. The auction is carried out in a one-round sealed-bid manner.
5. The bidders know the auction protocol.

The auctions we study in this paper can be viewed as algorithms that take the bidders' bids $(b^{(1)}, b^{(2)}, \ldots, b^{(n)})$ as input, and, upon termination, give the result as output. The output consists of two parts. The first part is allocation $w^{(i)}$, $i = 1, 2, \ldots, n$. $w^{(i)}$ indicates the number of items agent $a^{(i)}$ has won. In our model $w^{(i)}$ is either 0 or 1. We call an agent a **winner** if it has won at least one item, or a **loser** otherwise. The second part is price $p^{(i)}$, $i = 1, 2, \ldots, n$. $p^{(i)}$ indicates the price that agent $a^{(i)}$ needs to pay for each unit. Notice that if agent $a^{(i)}$ is a loser that has won no item, it pays a price 0, while $p^{(i)}$ may not be equal to 0. The auctioneer's revenue is defined as $\sum_{i=1}^{n} w^{(i)} \times p^{(i)}$.

For convenience, we use the following notations and terminologies throughout the paper:

n the total number of bidders
k the total number of winners
$a^{(i)}$ the i-th agent (or bidder) in the auction
$b^{(i)}$ the bid submitted by $a^{(i)}$. $b^{(i)} \geq 0$.
$v^{(i)}$ the private valuation on the product of $a^{(i)}$. We restrict our discussion on
    normal goods characterized by $v^{(i)} \geq 0$.
$\mathbf{b}^{(\sim i)} = (b^{(1)}, b^{(2)}, \ldots, b^{(i-1)}, b^{(i+1)}, \ldots, b^{(n)})$
$w^{(i)}$ The number of items agent $a^{(i)}$ has won from the auction.
$p^{(i)}$ The price agent $a^{(i)}$ should pay for *each unit*. $p^{(i)} \geq 0$.
$u^{(i)}$ The utility of agent $a^{(i)}$. $u^{(i)} = (v^{(i)} - p^{(i)}) \times w^{(i)}$.

The one shot auction protocol takes the sealed-bids of the participants, determines an allocation policy of the selling item, as well as the prices charged to all the winners. For the incentive compatible property, the most fundamental property in the tradition of auction protocol design since Vickery [8], the description of allocation policies can be much simplified by a set of pricing functions, $f_i(\cdot)$, one each for the participating agents. $a^{(i)}$ wins the item if its bid $b^{(i)}$ is larger than $f_i(\cdot)$, loses otherwise. In the event bid $b^{(i)}$ is equal to $f_i(\cdot)$, the agent is called a zero winner or a zero loser depending on whether it is allocated with the selling item by the auction protocol.

2.2   Basic Properties and Their Notations

We introduce important properties and their acronyms so that they can be referred to conveniently.

Definition 1 (Symmetric Auction). *An auction is symmetric if and only if for any input:*

1. *$b^{(i)} = b^{(j)} \Rightarrow p^{(i)} = p^{(j)}$.*
2. *$p^{(i)}$ remains the same if two other agents exchange their bids.*

Definition 2 (Homogeneous Auction). *An auction is homogeneous if after multiply every bidder's bid by a factor $m > 0$, the result of the auction satisfies:*

1. *A bidder wins if and only if it previously wins.*
2. *The return value of the price function of every bidder is equal to m times its previous value.*

IR Individual rational. An auction is said to be individual rationality if $b^{(i)} < p^{(i)} \rightarrow w^{(i)} = 0, \forall i = 1, 2, \ldots, n$.
IC Incentive compatible. Also called *truthful*. An auction is incentive compatible if and only if reporting truth is each agent (or bidder)'s dominant strategy.
SYM Symmetric. An auction is said to be symmetric if it satisfies Definition 1.
SEL(n,k) The auction is participated by n bidders and selects exactly k winners. $0 < k < n$ unless otherwise stated.

SWP Single Winning Price. An auction is said to be single winning price if and only if all winners' prices are the same, i.e. $w^{(i)} > 0$ and $w^{(j)} > 0 \Rightarrow$ $p^{(i)} = p^{(j)}, \forall 1 \leq i < j \leq n$.

HOMO Homogeneous. An auction is homogeneous if it satisfies Definition 2.

2.3 Useful Results

Lemma 1. *In an auction satisfying* **SEL(n,k)**, *there are at most k bidders with positive utilities.*

Lemma 2. *In an auction satisfying* **IR** *and* **IC**,

$$w^{(i)} \times (b^{(i)} - p^{(i)}) \geq 0, \forall i = 1, 2, \ldots, n$$

where $w^{(i)}$ is the number of items agent $a^{(i)}$ has won, $b^{(i)}$ is the bid of agent $a^{(i)}$, $p^{(i)}$ is the price agent $a^{(i)}$ need to pay for each unit of items it has won.

We restate Theorem 2.5 on page 5 in [6] here:

Lemma 3 (Bid Independent Pricing. Folklore, see e.g. [6]). *In an auction satisfying* **IR** *and* **IC**, *the pricing function of $a^{(i)}$ does not depend on $b^{(i)}$.*

In other words, $a^{(i)}$'s price function does not take its bid $b^{(i)}$ as a variable. i.e.

$$p^{(i)} = f_i(b^{(\sim i)})$$

And for any bidder $a^{(i)}$, if $b^{(i)} > p^{(i)}$, $a^{(i)}$ is a winner.

From the above lemma and the definition of **SYM**, if $b^{(i)} = b^{(j)}$ and $a^{(i)}$ wins while $a^{(j)}$ loses, then we must have $p^{(i)} = b^{(i)}$.

Lemma 4. *In an auction satisfying* **IR**, **IC** *and* **SYM**, *every bidder $a^{(i)}$'s pricing function is independent of its index i.*

Given Lemma 4, we can remove the subscript from every bidder $a^{(i)}$'s price function $f_i(\cdot)$, and denote it by $f(\cdot)$. Thus, the output of the auction is indepent of the order of the bidders. In the following parts, we always assume the bid vector is sorted in decreasing order.

3 Monotone Properties

In general, the symmetric property would not prevent lower bidders to win over higher bidders. Even though this may sometimes be acceptable for special types of discriminative auctions, it is not reasonable for symmetric auction where all bidders are regarded equal. To handle this abnormality, auction protocols are often required to be monotone.

However, careful examination into monotone properties shows that there may be different properties of monotoneness. We should distinguish three types of monotone properties:

1. **Winner Monotone (WM).** High bidder wins if any lower one does.
2. **Price Function Monotone (PFM).** The price function f is a non-decreasing function in each of its variables.
3. **Revenue Monotone (RM).** The total value of the sales is a non-decreasing function in each of its variables.

All the three monotone properties cover some aspects of the generic concept of monotonicity. However, they are not the same and yet they are not all independent. In this section we examine the relationships among them.

3.1   Some Exotic Auctions

We introduce some exotic auctions to help disprove some of the implication relations. Each auction is carefully selected to perform as much disproof as possible. It is straight forward to prove that all the auctions presented in this subsection satisfies **IR**, **IC** and **SYM**. The analysis of monotonicities of these auctions can be found in §**??**.

Auction 1 (k-Winner Vickrey Auction). $p^{(i)}$ *is set to the k-th highest bid of all other bidders. A bidder who bids higher than its price will* definitely *win. A bidder who bids lower than its price will* definitely *lose. A bidder who bids equal to its price may win or lose. The auction always selects k winners.*
    The auction is **WM**, **PFM** *and* **RM**.

Auction 2. $p^{(i)}$ *is set as the minimum bid times the number of minimum bids in the bid vector containing all other bidders' bids. Anyone who bids higher than its price is a winner.*
    The auction is **WM** *but it is not* **PFM**.

Auction 3. $p^{(i)}$ *is set as the minimum bid of all other bidders. Any bidder bid higher than its price is winner.*
    The auction is **PFM** *and* **WM**. *But it is not* **RM**.

Auction 4. $p^{(i)}$ *is set as follows: 1) 98, If at least one other bidder bids no less than 100; 2) 101, otherwise. Any $a^{(i)}$ bidding higher than its price is a winner.*
    The auction is **RM**. *But it is neither* **PFM** *nor* **WM**.

Auction 5. $p^{(i)}$ *is set as follows: 1) 0, if no other bidder bids no less than 100; 2) 101, if one and only one other bidder bids no less than 100; 3) 100, if more than one bidders bid no less than 100.*
    Everyone who bids no less than its price is a winner.
    The auction is **WM** *and* **RM**, *but not* **PFM**.

3.2   Implication Relationships

In this subsection, we derive the implication relationships among the three monotone properties, with the help of the examples introduced in the previous subsection.

Lemma 5. *In an auction satisfying **IR**, **IC** and **SYM**, PFM ⇒ WM.*

However, this is the only implication we know of the monotone properties. As we should show next, none of the other possible implications holds.

Theorem 1. *In an auction satisfying **IR**, **IC** and **SYM**, the implication relationships between any two of the three monotone properties are as follows:*
PFM ⇒ WM, WM ⇏ PFM; RM ⇏ WM, WM ⇏ RM; PFM ⇏ RM, RM ⇏ PFM.

Moreover, two of the monotone properties cannot strengthen the implication relationships either, as we should see in the following theorem.

Theorem 2. *In an auction satisfying **IR**, **IC** and **SYM**, the implication relationships from any two to the other one of the three monotone properties are as follows:*
PFM and RM ⇒ WM; WM and RM ⇏ PFM; WM and PFM ⇏ RM.

Theorem 3. *It is possible that the three monotone properties, **WM**, **PFM** and **RM**, simultaneously exist in an auction satisfying **IR**, **IC** and **SYM**.*
In other words, the existence of any one or two of the three monotone properties does not imply the non-existence of the other monotone properties (or property).

4   Vickrey Auction in Depth

In this section, we study Vickrey auction in depth. Our analysis show two sufficient and necessary conditions for a symmetric incentive compatible auction to be equivalent to Vickrey auction. One is that the number of winners is always fixed. Amazingly, symmetry is all we need for this class to be a Vickery auction. When the number of winners is determined by the auction protocol, we consider a variation of Vickery auction protocols. We show its necessary and sufficient condition is that the auction is homogeneous, revenue monotone, price function monotone, and single winning price.

4.1   Fixed Number of Winners

In this subsection, we prove that for auctions with a fixed number of winners, a symmetric incentive compatible auction is all we need for it to be the Vickrey auction. Though it seems to be a classical type result which may have already known, we have not been able to locate a statement of such a theorem in the literatures. To be on the safe side, we would not claim the result as completely new but to include it only for the sake of completeness.

Since it is trivial that Vickrey auction must select a fixed number of winners, we only need to prove the sufficient part.

We first prove the 1-winner case, with an extension to the k-winner case later.

1 Winner

Theorem 4. *An auction satisfying* **IR**, **IC**, **SYM** *and* **SEL(n,1)** *must be equivalent to 1-item n-bidder Vickrey auction. i.e.:*

1. *The total number of winners equals to 1.*
2. *(One of) the highest bidder wins*
3. *The winner pays the price of the highest bid of all other bidders*

k Winners. The result can be extended to protocols with exactly k winners with $n \geq k + 1$ bidders. However the proof is neither simple nor obvious.

We put the bids in a set $\mathbb{S}$ without duplicate values. Then we construct an n-dimensional counter vector $c = (c_1, c_2, \ldots, c_n)$, where c_i is the number of bidders whose bids equals to the $i - th$ highest bid in set $\mathbb{S}$. If there is no such bidder, $c_i = 0$. Obviously, for any auction with n bidders and $k < n$ winners, the set of c's possible values is limited.

We define the order on c's possible value set as follows:

1. If there are more non-zero elements in c than c', then $c > c'$; else
2. If there are more non-zero elements in c' than c, then $c < c'$; else
3. If $c_i = c'_i, \forall i$, then $c = c'$; else
4. Let $i = \min\{k | c_k \neq c'_k\}$. If $c_i > c'_i$ then $c > c'$, else $c < c'$.

Lemma 6. *An auction satisfying* **IR**, **IC**, **SYM** *and* **SEL(n,k)** *must satisfy:*

1. $b^{(i)} > b^{(k+1)} \Rightarrow b^{(i)} > p^{(i)}$
2. $b^{(i)} < b^{(k+1)} \Rightarrow b^{(i)} < p^{(i)}$

where $b^{(k+1)}$ is the $(k + 1)$-th highest bid.

Proof. We prove the lemma by mathematical induction.

1. **(Initial Step).** The smallest possible value of counter vector c is $(n, 0, \ldots, 0)$. At this time, all the bidders have the same bids, and the lemma is trivially true.
2. **(Induction Step).** In induction step, we want to show that the lemma is true for counter vector c if the induction assumption (the lemma is true for all smaller counter vectors) is true.
   (a) We first prove $b^{(i)} > b^{(k+1)} \Rightarrow b^{(i)} > p^{(i)}$. The argument is by contradiction. Suppose $\exists i, b^{(i)} > b^{(k+1)}$, but $b^{(i)} \leq p^{(i)}$. Now we decrease $a^{(i)}$'s bid to $\widetilde{b^{(i)}} = b^{(k+1)}$. Let c' denote the new counter vector after this change. Obviously $c' < c$ and the $(k + 1)$-th highest bid now is still $b^{(k+1)}$.

   By Lemma 3, $a^{(i)}$'s new price $\widetilde{p^{(i)}} = p^{(i)} \geq b^{(i)} > \widetilde{b^{(i)}}$. By **SYM**, $\forall j, b^{(j)} = \widetilde{b^{(i)}} \Rightarrow p^{(j)} = \widetilde{p^{(i)}} > b^{(j)}$.

   Thus all bidders bidding $b^{(k+1)}$ will *definitely* lose and there are at most $k - 1$ bidders bidding higher than $b^{(k+1)}$ in the current situation or we say in c'. By **SEL(n,k)**, $\exists L, b^{(L)} < b^{(k+1)}$ and $b^{(L)} \geq p^{(L)}$ and $a^{(L)}$ wins. *This contradicts the induction assumption that the lemma stands for c.*

(b) We then prove $b^{(i)} < b^{(k+1)} \Rightarrow b^{(i)} < p^{(i)}$. Similarily, suppose $\exists i, b^{(i)} < b^{(k+1)}$ but $b^{(i)} \geq p^{(i)}$.

Now we increase $a^{(i)}$'s bid to $\widetilde{b^{(i)}} = b^{(k+1)}$. Let c' denote the new counter vector after this first change.

By Lemma 3, $a^{(i)}$'s new price $\widetilde{p^{(i)}} = p^{(i)} \leq b^{(i)} < \widetilde{b^{(i)}}$. By **SYM**, $\forall j, b^{(j)} = \widetilde{b^{(i)}} \Rightarrow p^{(j)} = \widetilde{p^{(i)}} < b^{(j)}$.

Thus, all bidders bidding $b^{(k+1)}$ now will *definitely* win and there are at least $k + 2$ bidders bidding no lower than $b^{(k+1)}$ in the current situation or we say in c'. By **SEL(n,k)**, $\exists H, b^{(H)} > b^{(k+1)}$ and $b^{(H)} \leq \widetilde{p^{(H)}}$ and $a^{(H)}$ loses.

Now we decrease $a^{(H)}$'s bid to $\widetilde{b^{(H)}} = b^{(k+1)}$. Let c'' denote the new counter vector after this *second* change. Obviously $c'' < c$ and the $(k+1)$-th highest bid is still $b^{(k+1)}$.

By Lemma 3, $a^{(H)}$'s new price remains $\widetilde{p^{(H)}} \geq b^{(H)} > \widetilde{b^{(H)}}$. By **SYM**, $\forall j, b^{(j)} = \widetilde{b^{(H)}} \Rightarrow p^{(j)} = \widetilde{p^{(H)}} > b^{(j)}$.

Thus all bidders bidding $b^{(k+1)}$ will *definitely* lose under current situation.

Now there are at most $k - 1$ bidders bidding higher than $b^{(k+1)}$. By **SEL(n,k)**, $\exists L, b^{(L)} < b^{(k+1)}$ and $b^{(L)} \geq p^{(L)}$ and $a^{(L)}$ wins. *This contradicts the induction assumption that the lemma stands for c''.*

3. **(Conclusion Step).** From the above reasonings, we conclude that the lemma is true for all possible values of the counter vector.

The above reasonings complete the proof of the lemma.

Lemma 7. *An auction satisfying* **IR, IC, SYM** *and* **SEL(n,k)** *must satisfy:*

1. $b^{(i)} = b^{(k+1)}$ *and* $a^{(i)}$ *wins* $\Rightarrow p^{(i)} = b^{(k+1)}$
2. $b^{(i)} > b^{(k+1)} \Rightarrow p^{(i)} = b^{(k+1)}$

where $b^{(k+1)}$ *is the* $(k + 1)$-*th highest bid.*

Theorem 5. *An auction satisfying* **IR, IC, SYM** *and* **SEL(n,k)** *must be equivalent to k-item n-bidder Vickrey auction. i.e.:*

1. *Each bid higher than the $(k + 1)$-th highest bid will **definitely** win.*
2. *Each bid lower than the $(k + 1)$-th highest bid will **definitely** lose.*
3. *Each winner's price **must** be equal to the $(k + 1)$-th highest bid.*
4. *The total numbder of winners is equal to k.*

4.2   Homogeneous Monotone Auction and Vickrey Auction

In this subsection, we prove that **SWP**, **PFM**, **RM** and **HOMO** is a sufficient and necessary condition that a symmetric incentive compatible auction is equivalent to the Vickrey auction. Again it is trivial that Vickrey auction satisfies those properties, we only need to prove the sufficient part.

Lemma 8. *In an auction satisfying* **IR**, **IC**, **SYM**, **SWP**, **PFM**, **RM** *and* **HOMO**, *if a bid vector* **b** *of size* n *results in exactly* k *winners, then after increasing any bidder* $a^{(c)}$*'s bid by* δ *and* $0 < \delta < \frac{b^{(c)}}{n}$ *the auction will still result in exactly* k *winners.*

Proof. Let the original winning price be p, the auctioneer's revenue be $\widetilde{R}$ and the number of winners be $\widetilde{k}$ after $a^{(c)}$'s bid increasing by δ.

By **HOMO**, if we increase every bidder's bid to $\frac{b^{(c)}+\delta}{b^{(c)}}$ times larger, the auction will still have exactly k winners and the winning price is $\frac{b^{(c)}+\delta}{b^{(c)}} \times p$.

By **RM**, after only increasing $b^{(c)}$ by δ, auctioneer's revenue should be no less than before but no more than increasing every bid to $\frac{b^{(c)}+\delta}{b^{(c)}}$ times larger. Or:

$$p \times k \le \widetilde{R} \le \frac{b^{(c)}+\delta}{b^{(c)}} \times p \times k$$

By **PFM**, after changing $b^{(c)}$, every bidder's price should be no less than before but no more than changing every bid $\frac{b^{(c)}+\delta}{b^{(c)}}$ times larger. Or:

$$p \times \widetilde{k} \le \widetilde{R} \le \frac{b^{(c)}+\delta}{b^{(c)}} \times p \times \widetilde{k}$$

From the above two inequations we get:

$$p \times k \le \frac{b^{(c)}+\delta}{b^{(c)}} \times p \times \widetilde{k}$$

$$p \times \widetilde{k} \le \frac{b^{(c)}+\delta}{b^{(c)}} \times p \times k$$

Which further gives the following inequation:

$$\frac{b^{(i)}}{b^{(i)}+\delta} \times k \le \widetilde{k} \le \frac{b^{(i)}+\delta}{b^{(i)}} \times k$$

Since $0 < \delta < \frac{b^{(i)}}{n}$, we must have:

$$\frac{b^{(i)}}{b^{(i)}+\delta} \times k > k-1$$

$$\frac{b^{(i)}+\delta}{b^{(i)}} \times k < k+1$$

Hence:

$$k-1 < \widetilde{k} < k+1$$

And we must have $\widetilde{k} = k$. This completes the proof. □

Lemma 9. *In an auction satisfying* **IR**, **IC**, **SYM**, **SWP**, **PFM**, **RM** *and* **HOMO**, *if a bid vector* **b** *of size* n *results in exactly* k *winners, then after increasing any bidder* $a^{(c)}$*'s bid by* δ *and* $\delta > 0$ *the auction will still result in exactly* k *winners.*

Lemma 10. *In an auction satisfying* $\underline{IR}$, $\underline{IC}$, $\underline{SYM}$, $\underline{SWP}$, $\mathbf{PFM}$, $\mathbf{RM}$ *and* $\underline{HOMO}$, *if a bid vector* $\mathbf{b}$ *of size* n *results in exactly* k *winners, then any bid vector* $\mathbf{b}'$ *of size* n *satisfying* $b'^{(i)} \geq b^{(i)}, \forall i$, *will also results in* k *winners.*

Theorem 6. *An auction satisfying* $\underline{IR}$, $\underline{IC}$, $\underline{SYM}$, $\underline{SWP}$, $\mathbf{PFM}$, $\mathbf{RM}$ *and* $\underline{HOMO}$ *must have a fixed number* k *of winners as long as the total number of bidders* n *is being held fixed and every bid is a positive real number.*

Theorem 7. *In an auction satisfying* $\underline{IR}$, $\underline{IC}$, $\underline{SYM}$, $\underline{SWP}$, $\mathbf{PFM}$, $\mathbf{RM}$ *and* $\underline{HOMO}$, *when every bid is a positive real number, there must exist a function* f *such that for any bidder number* n, *the auction must be equivalent to* $(k = f(n))$-*winner Vickrey auction. i.e.:*

1. *Bidding higher than the* $(k + 1)$-*th highest bid will* **definitely** *win.*
2. *Bidding lower than the* $(k + 1)$-*th highest bid will* **definitely** *lose.*
3. *Each winner's price* **must** *be equal to the* $(k + 1)$-*th highest bid.*
4. *The total number of winners is equal to* $k = f(n)$.

5 Conclusions

In this paper, we have formally defined some basic properties of symmetric incentive compatible single-item single-unit auctions. We have shown that some of the relationships among them are quite complex. In §3 we have studied the implication relationships among three monotone properties. In §4 we have studied two sufficient and necessary conditions that an auction is equivalent to the Vickrey auction. Our results are of substaintial value to auction research.

In the end we provide a non-restricting list of some interesting open problems as our future works. Can we extends our results to multi-unit auctions where a bidder demands multiple units of a single item? Can we extend the implications relationships among the monotone properties to randomized auctions? What are the foundational properties of the more generalized auctions that need not to be symmetric, where the order of the bids does matter?

References

1. Aggarwal, G., Goel, A., Motwani, R.: Truthful auctions for pricing search keywords. In: ACM Conference on Electronic Commerce (EC), ACM Press, New York (2006)
2. Aggarwal, G., Hartline, J.: Knapsack auctions. In: SODA (2006)
3. Bu, T., Qi, Q., Sun, A.W.: Unconditional competitive auctions with copy and budget constraints. In: Spirakis, P.G., Mavronicolas, M., Kontogiannis, S.C. (eds.) WINE 2006. LNCS, vol. 4286, Springer, Heidelberg (2006)
4. Clarke, E.H.: Multipart pricing of public goods. Public Choice 11, 17–33 (1971)
5. Deng, X., Huang, L., Li, M.: On walrasian price of cpu time. In: Wang, L. (ed.) COCOON 2005. LNCS, vol. 3595, Springer, Heidelberg (2005)
6. Goldberg, A., Hartline, J., Karlin, A., Saks, M., Wright, A.: Competitive auctions. Games and Economic Behavior 55(2), 242–269 (2006)
7. Groves, T.: Incentives in teams. Econometrica 41(4), 617–631 (1973)
8. Vickrey, W.: Counterspeculation, auctions, and competitive sealed tenders. Journal of Finance 16, 8–37 (1961)

Finding Equilibria in Games of No Chance

Kristoffer Arnsfelt Hansen, Peter Bro Miltersen, and Troels Bjerre Sørensen

Department of Computer Science, University of Aarhus, Denmark
{arnsfelt,bromille,trold}@daimi.au.dk

Abstract. We consider finding maximin strategies and equilibria of explicitly given extensive form games with imperfect information but with no moves of chance. We show that a maximin pure strategy for a two-player game with perfect recall and no moves of chance can be found in time linear in the size of the game tree and that all pure Nash equilibrium outcomes of a two-player general-sum game with perfect recall and no moves of chance can be enumerated in time linear in the size of the game tree. We also show that finding an optimal behavior strategy for a one-player game of no chance without perfect recall and determining whether an equilibrium in behavior strategies exists in a two-player zero-sum game of no chance without perfect recall are both **NP**-hard.

1 Introduction

In a seminal paper, Koller and Megiddo [3] considered the complexity of finding maximin strategies in *two-player zero-sum imperfect-information extensive form games*. An extensive form game is an explicitly given game tree with *information sets* modeling hidden information (for details, see [3] or any text book on game theory). A main result of Koller and Megiddo was the existence of a polynomial time algorithm for finding an equilibrium in behavior strategies (or equivalently, a pair of maximin behavior strategies) of such a game when the game has *perfect recall*. Informally speaking, a game has perfect recall when a player never forgets what he once knew (for a formal definition, see below). In contrast, for the case of imperfect recall, the problem of finding a maximin strategy was shown to be **NP**-hard.

Pure equilibria (i.e, equilibria avoiding the use of randomization) play an important role in game theory and it is of special interest to know if a game possesses such an equilibrium. For the case of a zero-sum games, one may determine if a game has a pure equilibrium by computing a maximin pure strategy for each of the two players and checking that these strategies are best responses to one another. Unfortunately, Blair *et al.* [1] established that the problems of finding a maximin pure strategy of a two-player extensive form game or determining whether a pure equilibrium exists are both **NP**-hard, even for the case of zero-sum games of perfect recall. Their proof is an elegant reduction from the EXACT PARTITION (or BINPACKING) problem and relies heavily on the fact that the extensive form game is allowed to contain *chance* nodes, i.e., random events not controlled by either of the two players.

G. Lin (Ed.): COCOON 2007, LNCS 4598, pp. 274–284, 2007.
© Springer-Verlag Berlin Heidelberg 2007

Extensive form games *without* chance nodes is a very natural special case to consider (natural non-trivial examples include such popular parlor games as variants of *Spoof*). In this paper we consider the equilibrium computation problems considered by Koller and Megiddo and by Blair *et al.* for this special case. Our main results are the following:

First, we show that *a maximin* pure *strategy for a two-player extensive form game of no chance with imperfect information but perfect recall can be found in time linear in the size of the game tree.* As stated above, Blair *et al.* show that *with* chance moves, the problem is **NP**-hard. Apart from the obvious practical interest, the example is also interesting in light of the recent work of von Stengel and Forges [6]. They introduced the notion of *extensive form correlated equilibria* (EFCEs) of two-player extensive form games. They showed that finding such equilibria in games *without* chance moves can be done in polynomial time while finding them in games *with* chance moves may be **NP**-hard. They remark that EFCE seems to be the first example of a game-theoretic solution concept where the introduction of chance moves marks the transition from polynomial-time solvability to **NP**-hardness. Our result combined with the result of Blair *et al.* provides a second and much more elementary such example.

Second, we extend the above result from maximin pure strategies to pure Nash equilibria. We show that *all pure Nash equilibrium outcomes of a two-player general-sum extensive form game of no chance with imperfect information but perfect recall can be enumerated in time linear in the size of the game tree.* Here, an outcome is a leaf of the tree defining the extensive form. Also, given one such pure Nash equilibrium outcome, we can in linear time construct a pure equilibrium (in the form of a strategy profile) with that particular outcome. In contrast, the recent breakthrough result of Chen and Deng [2] implies that finding a behavior Nash equilibrium for a game of this kind is **PPAD**-hard.

The results of Blair *et al.* and those of Koller and Megiddo give a setting where finding a pure equilibrium is **NP**-hard while finding an equilibrium in behavior strategies can be done in polynomial time. Considering games without perfect recall, we give an example of the opposite. We show that *determining whether a one-player game in extensive form with imperfect information, imperfect recall and no moves of chance has a behavior strategy that yields a given expected payoff is* **NP**-*hard.* In contrast, it is easy to see that finding an optimal *pure* strategy for such a game can be done in linear time. Our result strengthens a result of Koller and Megiddo [3, Proposition 2.5] who showed **NP**-hardness of finding a maximin behavior strategy in a *two-player* game with imperfect recall and no moves of chance. Koller and Megiddo [3, Example 2.12] also showed that a maximin behavior strategy in such a two-player game may require irrational behavior probabilities. We give a one-player example with the same property.

Finally, we show that determining whether a Nash equilibrium in behavior strategies exists in a two-player extensive form zero-sum game with no moves of chance but without perfect recall is **NP**-hard.

The rest of the paper is organized as follows. In section 2, we formally define the objects of interest and introduce the associated terminology (for a less concise introduction, see the paper by Koller and Megiddo, or any textbook on game theory). In sections 3,4,5 and 6, we prove each of the four results mentioned above.

2 Preliminaries

A two-player *extensive form game* is given by a finite rooted tree with pairs of payoffs (one payoff for each of the two players) at the leaves, and information sets partitioning nodes of the tree. In a *zero-sum* game, the sum of each payoff pair is zero. A *general-sum* game is a game without this requirement. In this paper, we do not consider games with nodes of chance, so every node in the tree is owned by either Player 1 or to Player 2. All nodes in an information set belong to the same player. Intuitively, the nodes in an information set are indistinguishable for the player they belong to. In a one-player game, all nodes belong to Player 1. Actions of a player are denoted by labels on edges of the tree. Given a node u and an action c that can be taken in u, we let apply(u, c) be the unique successor node v of u with the edge (u, v) being labeled c. Each node in an information set has the same set of outgoing actions. The set of possible actions in information set h we denote C_h. The actions belong to the player owning the nodes of the information set. *Perfect recall* means that all nodes in an information set belonging to a player share the sequence of actions and information sets belonging to that player that are visited on the path from the root to each of the nodes.

A *pure strategy* for a player assigns to each information set belonging to that player a chosen action. A *behavior strategy* assigns to each action at each information set belonging to that player a probability. A pure strategy can also be seen as a behavior strategy that only uses the probabilities 0 and 1. Thus, concepts defined below for behavior strategies also apply to pure strategies. A (pure or behavior) *strategy profile* is a pair of (pure or behavior) strategies, one for each player. Given a pure strategy profile for a game without chance nodes, there is a unique path in the tree from the root to a leaf formed by the chosen actions of the two players. The leaf is called the *outcome* of the profile. A behavior strategy profile defines in the natural way a probability distribution on the leaves of the tree and hence a probability distribution on payoffs for each of the two players. So given a behavior strategy profile we can talk about the expected payoff for each of the two players.

A *maximin pure strategy* for a player is a pure strategy that yields the maximum possible payoff for that player assuming a worst case opponent, i.e., the maximum possible guaranteed payoff. A *maximin behavior strategy* for a player is a behavior strategy that yields the maximum possible expected payoff for that player assuming a worst case opponent, i.e., the maximum possible guaranteed expected payoff. A *Nash equilibrium* is a strategy profile (s_1, s_2) so that no strategy s_1' yields strictly better payoff for Player 1 than s_1 when Player 2 plays s_2 and no strategy s_2' yields strictly better payoff for Player 2 than s_2 when Player 1 plays s_1.

Kuhn [5] showed that for an extensive form two-player zero-sum game with perfect recall, a pair of maximin behavior strategies is a Nash equilibrium. The expected payoff for Player 1 is the same in any such equilibrium and is called the *value* of the game. Any extensive form general-sum game with perfect recall in fact possesses a Nash equilibrium in behavior strategies.

3   Maximin Pure Strategies in Games with Perfect Recall

Consider a two-player extensive form game G with perfect recall and without chance nodes. We shall consider computing a maximin pure strategy for one of the players, say, Player 1. For the purpose of computing such a strategy, we can consider G to be a zero-sum game where Player 1 (henceforth the max-player) attempts to maximize his payoff and Player 2 (henceforth the min-player) attempts to minimize the payoff of Player 1. Let G' be the zero-sum game obtained from G by dissolving all information sets of the min-player into singletons.

Note that the set of strategies for the max-player is the same in G and G'. For the min-player, however, the set of strategies is larger in G' thereby making G' a better game that G for the min-player, so its *value* as a zero-sum game is at most the value of G. However we have the following key lemma. Note that the lemma fails badly for games containing chance nodes.

Lemma 1. *A pure strategy π for the max-player has the same payoff against an optimal counter strategy in G as it has against an optimal counter strategy in G' (note that the statement makes sense as the max-player has the same set of strategies in the two games).*

Proof. Let σ be a pure best counter strategy against π in G'. As there are no chance nodes, σ and π defines a single path in the tree of G' from the root to a leaf. Due to perfect recall, none of the choices made by the min-player along the path are choices of the same information set. Thus, the same sequence of choices can also be made by a strategy in G. Thus, there is a counter strategy in G that achieves the same payoff against π as σ does in G', and since the set of possible counter strategies is bigger in G', the best in each game each achieves exactly the same payoff.

To compute the best payoff that can be obtained by a pure strategy in G', we define for information set h of G' a value $\mathrm{pval}(h)$ ("pure value") inductively in the game-tree as follows.

- If h belongs to the min-player, and therefore consists of a single node u, define
$$\mathrm{pval}(h) = \min_{c \in C_h} \mathrm{pval}(\mathrm{apply}(u, c))$$

- If h belongs to max-player, define
$$\mathrm{pval}(h) = \max_{c \in C_h} \min_{u \in h} \mathrm{pval}(\mathrm{apply}(u, c))$$

The induction is well-founded due to perfect recall and the fact that there are no chance nodes, see [6, Lemma 3.2].

Lemma 2. *For every pure strategy π for the max-player, there exists a pure strategy σ for the min-player with the following property. For every information set h of the max-player there is some node $u \in h$ such that play from u using the pair of strategies (π,σ) yields payoff at most* pval(h). *Similarly, for every information set h of the min-player, play from the single node u of h using the pair of strategies (π,σ) yields payoff at most* pval(h).

Proof. Given a pure strategy π for the max-player, we construct the strategy σ inductively in the game tree. Let h be a given information set of the max-player. Then, by definition of pval(h) there must be a path from some node $u \in h$ using the action chosen by π out of u (say, L), then going through min-nodes to an information set g of the max-player with pval(g) $\leq$ pval(h), or to a leaf l with payoff less than or equal to pval(h).

In the latter case we simply let σ take the choices defining the path to the leaf l. In the former case, by induction, we know we have constructed a pure strategy σ for min from g onwards so that for some node $v \in g$, play from g using π and σ leads to payoff at most pval(g). Note that we have a path from u to some (possibly) other node $v' \in g$ using min-nodes. We claim that there is a path from some node $\bar{u} \in h$ to v using min-nodes and also choosing the action L in $\bar{u}$ (see Fig. 1).

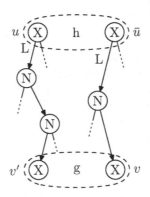

Fig. 1. Finding $\bar{u}$

Indeed, assume that this is not the case. Then the sequence of information sets and own actions encountered by max on the way to v differs from the corresponding sequence in some of other node (namely v') in the information set of v, contradicting perfect recall. But then, the node $\bar{u}$ establishes the induction claim, with the desired strategy σ taking the choices defining the path from $\bar{u}$ to v.

It remains to provide the first actions for the min-player in case the root node belongs to the min-player. In this case there is a path from the root r, going through min-nodes to an information set h of max with $\mathrm{pval}(h) \leq \mathrm{pval}(r)$, or to a leaf l with payoff equal to $\mathrm{pval}(r)$. As before we let σ take the choices defining this path.

With this we can now obtain the following result.

Theorem 1. *Given a two-player extensive form game with perfect recall G without chance nodes, we can compute a maximin pure strategy for a player in linear time in the size of the game tree.*

Proof. We describe how to compute a maximin strategy for one of the players, say Player 1. By Lemma 1 we can compute this by computing a pure maximin strategy in the game G'. We compute the pval function of the information sets in G' and let the strategy of the max-player be the choices that obtains the maximum in the definition of pval for every information set, i.e., the choice in information set h is $\mathrm{argmax}_{c \in C_h} \min_{u \in h} \mathrm{pval}(\mathrm{apply}(u, c))$. We claim that the value $\mathrm{pval}(r)$ assigned to the root is the best guaranteed payoff the max-player can get in G' using some pure strategy. Indeed the max-player is guaranteed payoff $\mathrm{pval}(r)$, where r is the root of G', playing this strategy, and Lemma 2 establishes this is the best he can be guaranteed.

Note also that having computed the maximin pure strategy, we can determine whether it is also maximin as a behavior strategy by computing the value v of the game in polynomial time using, e.g., the algorithm of Koller and Megiddo [3] or the more practical one by Koller, Megiddo and von Stengel [4] and checking if the computed pure value $\mathrm{pval}(r)$ of the root equals v.

4   Enumerating All Pure Equilibria of Games with Perfect Recall

Let G be a 2-player *general sum* extensive form game with perfect recall and without chance nodes. Let (π, σ) be a pair of pure strategies. For (π, σ) to be a pure equilibrium we must have that π is a best response to σ and vice versa. Play using the pair (π, σ) will lead to a unique leaf of G, since there are no chance nodes. Consider now a leaf l of G, as a potential outcome of a pure equilibrium. Clearly the actions along the path from the root r of G to the leaf must be such that they follow the path. Hence what remains are to find the actions of the remaining information sets. Player 1 must find pure actions in his remaining information sets such that Player 2 can not obtain greater payoff than she receives at l. Similarly Player 2 must find pure actions in her information sets such that Player 1 can not obtain greater payoff than he receives at l. Given l, we can define zero-sum games G_1 and G_2 by modifying G such that such actions, if they exist, can be found in linear time using Theorem 1.

We can simply construct G_1 from G as follows (the construction of G_2 being the same with Player 1 and Player 2 exchanged). Player 1 will be the max-player of G_1 and Player 2 will be the min-player. For every information set of Player

1 along the path from the root to l we remove all choices (and the subgames below) except the ones agreeing with the path. The payoff at a leaf in G_1 is the negative of the payoff that Player 2 receives in the corresponding leaf in G. The following lemma is immediate.

Lemma 3. *There is a pure strategy for Player 1 in G leading towards l ensuring that Player 2 can obtain at most payoff p if and only if there is a pure strategy for the max-player of G_1 ensuring payoff at least $-p$.*

Using this lemma, is is easy to check in linear time if a given leaf l with payoffs (p_1, p_2) is a pure equilibrium outcome: We check that the maximin pure strategy for Player 1 in G_1 ensures payoff at least $-p_2$ and we check that the maximin pure strategy for Player 2 in G_2 ensures payoff at least $-p_1$. Also, given such an outcome, we can in linear time construct a pure strategy equilibrium with this outcome: The equilibrium is the profile consisting of transferring in the obvious way to G the maximin pure strategies for Player 1 in G_1 and for Player 2 in G_2.

Since we can check in linear time if a given leaf is an outcome, we can enumerate the set of outcomes in quadratic time. To get a linear time algorithm, we will go one step further and work with a derived game that is independent of the leaf l.

Let G_1' be the zero-sum game obtained from G by dissolving the information sets of Player 2 and letting payoff at a leaf in G_1' be the negative of the payoff that Player 2 receives in the corresponding leaf in G. We define the pval function on G_1' as in section 3.

Let T_1 be a tree on the information sets of Player 1 and the leaves together with a root, such that the parent of an information set or leaf is the first information set on the path to the root in G_1' or the root itself.

Define a *point of deviation* with respect to a given leaf l, to be a node in T not on the path from the root to l, but sharing the sequence of actions leading to the node with a node on the path from the root to l. Thus only nodes that have their parents on the path can be a points of deviation. See Fig. 2 for an example. Intuitively, a point of deviation is an information set where Player 1 first observes that Player 2 has deviated from the strategy leading to l.

The following lemma is easy to establish.

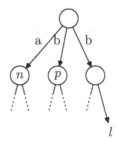

Fig. 2. Node p is a point of deviation, node n is not

Lemma 4. *There is a pure strategy for Player 1 in G leading towards l ensuring that Player 2 can obtain at most the payoff p if and only if for every point of deviation h with respect to l we have* $\mathrm{pval}(h) \geq -p$.

Theorem 2. *Given a 2-player general-sum extensive form game with perfect recall G without chance nodes, we can in linear time in the size of the game tree enumerate the set of leaves that are outcomes of pure equilibria.*

Proof. Using Lemma 4, we compute the leaves l such that Player 1 has a pure strategy leading towards l ensuring that Player 2 can obtain at most the payoff received at l and conversely Player 2 has a pure strategy leading towards l ensuring that Player 1 can obtain at most the payoff received at l. These sets can be computed separately; we describe how to compute the former.

We construct the game G_1' and compute the pval function on G_1' in linear time. In linear time we then construct the tree T_1 and record the computed pval values in the nodes. Finally we traverse the tree T. During this traversal we maintain the minimum pval value that is on any sibling to the nodes on the path to the root, corresponding to the points of deviations relevant for the leaves in the subtree of the current node. Once we visit a leaf we can then directly decide the criteria of Lemma 4 by comparing with the payoff of the leaf.

5   Optimal Behavior Strategies in One-Player Games Without Perfect Recall

In this section we consider *one-player* games without perfect recall and no moves of chance and show **NP**-hardness of the problem of determining whether a behavior strategy yielding an expected payoff of at least a given rational number exists. In contrast, it is straightforward to see that the corresponding problem for pure strategies is in **P**: For each leaf of the game, one checks if this leaf can be reached by a sequence of actions so that the same action is taken in all nodes in a given information set. This results strengthens the result of Koller and Megiddo [3, Proposition 2.6] who showed **NP**-hardness of the problem of determining whether some behavior strategy in a *two-player* game without perfect recall guarantees a certain expected payoff (against any strategy of the opponent). Also, our reduction is heavily based on their reduction but uses imperfect recall to eliminate one of the players. Before giving the proof, we give a simple example showing that an optimal strategy may require irrational behavior probabilities (therefore, strictly speaking, "finding" an optimal strategy is not a well-defined computational problem which leads to considering the stated decision problem instead). A corresponding two-player example was given by Koller and Megiddo [3, Example 2.12]. Our one-player game of Fig. 3 is in fact somewhat simpler than their example. All nodes in the game are included in the same information set. The player can choose either L or R. Thus, a behavior strategy is given by a single probability p_L with $p_R = 1 - p_L$. By construction, the expected payoff is $-2p_L^3 - (1 - p_L)^3$. This is maximized for $p_L = \sqrt{2} - 1$.

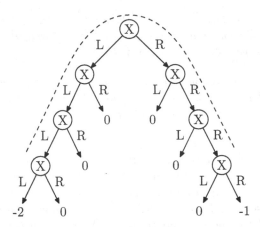

Fig. 3. A one-player game where the rational behavior is irrational

Theorem 3. *The following problem is* **NP***-hard: Given a rational number v and a one-player extensive form game without chance nodes and a rational number v, does some behavior strategy ensure expected payoff at least v?*

Proof. The proof is by reduction from 3SAT. Given a 3-CNF formula F with m clauses we construct a game G as follows.

Assume without loss of generality that m is a power of 2, $m = 2^k$. First G will consist of a complete binary tree of depth $2k$, whose nodes are contained in a single information set. If on the path from the root to a node, the same choice is made in step $2(i-1)+1$ and $2i$ for some $i \in \{1, \ldots, k\}$, the game is terminated and the player receives payoff 0. Otherwise, we will associate a clause to the node in the following way: For $i = 1, \ldots, k$ we interpret the choices made at step $2(i-1)+1$ and $2i$ as defining a binary choice. With the choices (left,right) we associate the bit 0, and with choices (right,left) we associate the bit 1. Having defined in this way k bits, we may associate a uniquely determined clause with the node.

From this node we let the player, for each of the three variables in the clause, select a truth value. If one of these choices satisfies the clause, the player receives payoff 1, and 0 otherwise. We place the nodes corresponding to the same variable in a single information set. In particular, the player does not know the clause.

The proof is now concluded by the following claim: The player can obtain expected payoff $\frac{1}{m}$ if and only if F is satisfiable.

Assume first that F is satisfiable. The player will make the first $2k$ choices by choosing left with probability $\frac{1}{2}$. The rest of the choices are made according to a satisfying assignment to F. With probability $(\frac{1}{2})^k = \frac{1}{m}$, the player gets to a node corresponding to a clause, and will obtain payoff 1. The expected payoff is therefore $\frac{1}{m}$.

Assume on the other hand that the player can obtain expected payoff $\frac{1}{m}$. Suppose that the player chooses left with probability p in the first $2k$ choices. The probability that the player reaches a node associated with a given clause is $(p(1-p))^k \leq \frac{1}{m^2}$, independently of the given node. Since the player can in fact obtain expected payoff $\frac{1}{m}$, we have that at every node associated with a clause the player must obtain payoff 1, and thus his strategy gives a satisfying assignment to F.

6   Determining Whether a Two-Player Game Without Perfect Recall has an Equilibrium

Our final hardness result again uses a reduction very similar to Koller and Megiddo [3, Proposition 2.6]. In this case, we use the imperfect recall to force Player 1 to use an almost pure strategy.

Theorem 4. *The following problem is* **NP**-*hard: Given a two-player zero-sum extensive form game without chance nodes, does the game possess a Nash equilibrium in behavior strategies?*

Proof. The proof is by reduction from 3SAT. Given a 3-CNF formula F with m clauses we construct a zero-sum two-player game G as follows.

Player 1 (the max-player) starts the game by making two actions, each time choosing a clause of F. We put all corresponding $m + 1$ nodes (the root plus m nodes in the next layer) of the game in one information set. If he fails to choose the same clause twice, he receives a payoff of $-m^3$ and the game stops. Otherwise, Player 2 (the min-player) then selects a truth value for each of the three variables in the clause. We place all nodes of Player 2 corresponding to the same variable in a single information set. If one of the choices of Player 2 satisfies the clause, Player 1 receives payoff 0. If none of them do, Player 1 receives payoff 1.

The proof is now concluded by the following claim: G has an equilibrium in behavior strategies if and only if F is satisfiable.

Assume first that F is satisfiable. G then has the following equilibrium (which happens to be pure): Player 2 plays according to a satisfying assignment while Player 1 uses an arbitrary pure strategy. The payoff is 0 for both players and no player can modify their behavior to improve this so we have an equilibrium.

Next assume that G has an equilibrium. We shall argue that F has a satisfying assignment. We first observe that Player 1 in equilibrium must have expected payoff at least 0. If not, he could switch to an arbitrary pure strategy and would be guaranteed a payoff of at least 0. Now look at the two actions (i.e., clauses) that Player 1 is most likely to choose. Let clause i be the most likely and let clause j be the second-most likely. If Player 1 chooses i and then j he gets a payoff of $-m^3$. His maximum possible payoff is 1 and his expected payoff is at least 0. Hence, we must have that $-m^3 p_i p_j + 1 \geq 0$. Since $p_i \geq 1/m$, we have that $p_j \leq 1/m^2$. Since clause j was the second most likely choice, we in fact have that $p_i \geq 1 - (m-1)(1/m^2) > 1 - 1/m$. Thus, there is one clause that Player 1 plays with probability above $1 - 1/m$. Player 2 could then guarantee an expected payoff of less than $1/m$ for Player 1 by playing any assignment satisfying this clause. Since we are actually playing an equilibrium, this would not decrease the payoff of Player 1 so Player 1 currently has an expected payoff less than $1/m$. Now look at the assignment defined by the most likely choices of Player 2 (i.e, the choices he makes with probability at least $\frac{1}{2}$, breaking ties in an arbitrary way). We claim that this assignment satisfies F. Suppose not. Then there is some clause not satisfied by F. If Player 1 changes his current strategy to the pure strategy choosing this clause, he obtains an expected payoff of at least $(1/2)^3 \geq 1/m$ (supposing, wlog, that $m \geq 8$). This contradicts the equilibrium property and we conclude that the assignment in fact does satisfy F.

Acknowledgments

We would like to thank Daniel Andersson, Lance Fortnow, and Bernhard von Stengel for helpful comments and discussions.

References

1. Blair, J.R.S., Mutchler, D., van Lent, M.: Perfect recall and pruning in games with imperfect information. Computational Intelligence 12, 131–154 (1996)
2. Chen, X., Deng, X.: Settling the complexity of two-player Nash equilibrium. In: 47th Annual Symposium on Foundations of Computer Science, pp. 261–272 (2006)
3. Koller, D., Megiddo, N.: The complexity of two-person zero-sum games in extensive form. Games and Economic Behavior 4, 528–552 (1992)
4. Koller, D., Megiddo, N., von Stengel, B.: Fast algorithms for finding randomized strategies in game trees. In: Proceedings of the 26th Annual ACM Symposium on the Theory of Computing, pp. 750–759 (1994)
5. Kuhn, H.W.: Extensive games and the problem of information. Annals of Mathematical Studies 28, 193–216 (1953)
6. von Stengel, B., Forges, F.: Extensive form correlated equilibrium: Definition and computational complexity. Technical Report LSE-CDAM-2006-04, London School of Economics, Centre for Discrete and Applicable Mathematics (2006)

Efficient Testing of Forecasts

Ching-Lueh Chang[1] and Yuh-Dauh Lyuu[2]

[1] Department of Computer Science and Information Engineering, National Taiwan
University, Taipei, Taiwan
d95007@csie.ntu.edu.tw
[2] Department of Computer Science and Information Engineering, National Taiwan
University, Taipei, Taiwan
lyuu@csie.ntu.edu.tw

Abstract. Each day a weather forecaster predicts a probability of each
type of weather for the next day. After n days, all the predicted probabilities and the real weather data are sent to a test which decides whether
to accept the forecaster as possessing predicting power. Consider tests
such that forecasters who know the distribution of nature are passed
with high probability. Sandroni shows that any such test can be passed
by a forecaster who has no prior knowledge of nature [San03], provided
that the duration n is known to the forecaster in advance. On the other
hand, Fortnow and Vohra [FV06] show that Sandroni's result may require forecasters with high computational complexity and is thus of little
practical relevance in some cases. We consider forecasters who select a
deterministic Turing-machine forecaster according to an arbitrary distribution and then use that machine for all future forecasts. We show
that forecasters even more powerful than the above ones are required for
Sandroni's result. Also, we show that Sandroni's result does not apply
when the duration n is not known to the forecaster in advance.

1 Introduction

The weather metaphor motivates the investigations in this paper. A weather
forecaster predicts the probability of rain on each day. How do we measure
the forecaster's predicting power? Dawid [Daw82] proposes testing whether the
announced forecasts are *well calibrated* in the following sense:

> Suppose that, in a long (conceptually infinite) sequence of weather
> forecasts, we look at all those days for which the forecast probability of
> precipitation was, say, close to some given value ω and (assuming these
> form an infinite sequence) determine the long run proportion p of such
> days on which the forecast event (rain) in fact occurred. The plot of p
> against ω is termed the forecaster's empirical calibration curve. If the
> curve is the diagonal $p = \omega$, the forecaster may be termed (empirically)
> *well calibrated*.

A forecaster who knows the distribution of nature can produce well calibrated
forecasts [Daw82].

G. Lin (Ed.): COCOON 2007, LNCS 4598, pp. 285–295, 2007.

Foster and Vohra [FV98] show that a forecaster can produce forecasts that will be well calibrated on any distribution adopted by nature. Fudenberg and Levine [DF99], Lehrer [Leh01], Sandroni, Smorodinsky and Vohra [SSV03] consider stronger forms of calibration tests. They show that a forecaster can pass these stronger tests on any distribution of nature. Thus, to pass these calibration tests, a forecaster needs no prior knowledge about the distribution of nature in that he does not need to assume anything about the distribution of nature [San03]. In this sense, if a forecaster passes a test on any distribution of nature, it means that the forecaster needs no prior knowledge about the distribution of nature to pass that test.

Ideally, we want a test to reject forecasters who have no prior knowledge about the distribution of nature. As a corollary, it is desirable that a test rejects each forecaster on some distribution of nature, for otherwise some forecaster passes the test on all distributions of nature, meaning it needs no prior knowledge about nature to pass the test. A second desirable property is that a test passes forecasters who know the distribution of nature. These two desirable properties are not simultaneously satisfied by calibration tests (see [FV98, DF99, Leh01, SSV03]). Sandroni [San03] shows that no test satisfies both desirable properties as long as the duration is made known beforehand to the forecaster. In this known-duration case, if a test is such that forecasters who know the distribution of nature are accepted with high probability, then there exists a forecaster that can pass that test on any distribution of nature.

Fortnow and Vohra [FV06] consider computational bounds on the forecaster and the test. They construct efficient tests that pass forecasters who know the distribution of nature, whereas only computationally powerful forecasters have a chance to pass these tests on all possible distributions of nature. This result implies that Sandroni's result requires the absence of computational constraints on the forecaster.

There is also closely related literature on theory testing [DF06, OS06b, OS06a]. In these models, an expert proposes a distribution of a stochastic process, which is tested for truthfulness. Some papers [ANW06, FS07] discuss identifying which of several experts knows the true distribution of the stochastic process.

In this paper, we present two results. The first proceeds along the computational perspective of Fortnow and Vohra [FV06], who show that Sandroni's [San03] result may require forecasters with high computational complexity. We consider the family of forecasters who select a deterministic Turing-machine forecaster of any time complexity according to an arbitrary distribution and then use that machine for all future forecasts. We show that Sandroni's result requires forecasters even more powerful than the above. We achieve it by exhibiting a test such that forecasters who know the distribution of nature are passed with high probability, whereas each of the above-mentioned forecasters is rejected with high probability on some n-day weather sequence for all sufficiently large n. Unlike the results in [FV06] where Turing-machine forecasters with a certain computational complexity are rejected on some distributions of nature, the above-mentioned forecasters may adopt uncomputable distributions over Turing-machines of arbitrary time complexity.

Our second result shows that Sandroni's [San03] result does not apply when the duration n of forecasting is unknown beforehand to the forecaster. We do so by exhibiting a test which passes forecasters who know the distribution of nature, whereas any forecaster is rejected on some n-day weather sequence for infinitely many durations n.

Our paper is organized as follows. Section 2 gives formal definitions. Section 3–4 present our results. Section 5 concludes the paper.

2 Definitions

The following setup extends that proposed by Sandroni [San03]. A finite state space $S \equiv \{1, \ldots, K\}$ categorizes the outcome on each period into one of $1, \ldots, K$ where $K \geq 2$. For $\ell \in \mathbb{N}$, denote by S^ℓ the ℓ-dimensional Cartesian product of S and $S^* \equiv \bigcup_{\ell \geq 0} S^\ell$. Let $n \in \mathbb{N}$ be the duration of forecasting (in periods), which may or may not be known to the forecaster and the test. Given a finite outcome sequence $s \in S^*$, the data generating process assigns a probability distribution over the outcomes for the next period. A null outcome sequence is denoted λ.

A distribution over the outcomes of S is called a forecast. The set of distributions over S is denoted ΔS [San03]. A (possibly randomized) forecaster F announces a forecast for the $(t+1)$th period given the outcomes of the first t periods and the previous forecasts by F for these t periods. Formally, the input to a forecaster F is an outcome sequence s and a forecast sequence f where $(s, f) \in \bigcup_{t \geq 0} S^t \times (\Delta S)^t$. F's output is a forecast for the following period's outcome. Let $s = (s_1, \ldots, s_{n-1}) \in S^{n-1}$. If the data generating process adopts s as the outcome sequence for the first $n-1$ periods, the n forecasts announced by a forecaster F are denoted

$$F(\lambda; 1), F(s_1; 2), \ldots, F(s_1, \ldots, s_{n-1}; n).$$

Here $F(s_1, \ldots, s_{i-1}; i)$ is F's forecast for the ith period. It depends only on the first $i-1$ outcomes in s and the forecasts

$$F(\lambda; 1), \ldots, F(s_1, \ldots, s_{i-2}; i-1)$$

for them, for $i \geq 1$. For convenience, we write

$$F(s) \equiv (F(\lambda; 1), F(s_1; 2), \ldots, F(s_1, \ldots, s_{n-1}; n)) \tag{1}$$

for the full forecast sequence (we may also write $F(s_1, \ldots, s_{n-1})$ as $s = (s_1, \ldots, s_{n-1})$). When the forecaster is a Turing machine, we assume a reasonable encoding for its output format. For example, we may let it output a forecast as a vector consisting of its predicted probabilities of the outcomes in S.

Let $\mathcal{P}$ be the distribution adopted by the data generating process. Given any outcome sequence, $\mathcal{P}$ determines the probability distribution over S for the next period. Given $\mathcal{P}$ and an outcome sequence $s = (s_1, \ldots, s_n)$, the correct forecast sequence $\hat{f}(s) = (\hat{f}_1, \ldots, \hat{f}_n)$ is one such that $\hat{f}_j$ equals the probability distribution that the data generating process assigns for the jth period conditioned on $(s_1, \ldots, s_{j-1})$ being the outcome sequence for the first $j - 1$ periods, where $1 \le j \le n$. Sensibly, $\hat{f}_j$ depends only on $\mathcal{P}$ and $(s_1, \ldots, s_{j-1})$, not on future outcomes $(s_j, \ldots, s_n)$. We may write $\hat{f}_j$ as $\hat{f}_j(s_1, \ldots, s_{j-1})$ to make the dependency explicit. According to this notation,

$$\hat{f}(s) \equiv \left(\hat{f}_1(\lambda), \hat{f}_2(s_1), \ldots, \hat{f}_n(s_1, \ldots, s_{n-1}) \right). \tag{2}$$

A test is deterministic. It receives an outcome sequence s and a corresponding forecast sequence f where $(s, f) \in \bigcup_{t \ge 0} S^t \times (\Delta S)^t$ and decides whether to accept the forecaster as possessing predicting power. A natural criterion is that, whatever the distribution $\mathcal{P}$ adopted by the data generating process, an outcome sequence $s = (s_1, \ldots, s_n)$ together with the correct forecast sequence $\hat{f}(s)$ must be accepted with high probability. Here the probability is taken over the random variables $s_1, \ldots, s_n$ being the outcome sequence whose distribution is determined by the data generating process. A test satisfying this criterion is said to pass the truth with high probability [FV06].

A few comparisons with Sandroni's [San03] definitions can be made. In Sandroni's definition, the duration n is known to the forecaster and the test; the data after the nth period are simply ignored. The forecaster receives its input from $\bigcup_{t=0}^{n-1} S^t \times (\Delta S)^t$, and the test receives its input from $S^n \times (\Delta S)^n$. This paper, however, considers the consequences on the same forecaster (and test) over arbitrarily large duration n. Thus, in our definition the forecaster receives its input from $\bigcup_{t \ge 0} S^t \times (\Delta S)^t$, and the test receives its input from $\bigcup_{n \ge 0} S^n \times (\Delta S)^n$.

3   Forecasters with Arbitrary Time Complexity

We first review previous theorems. The following theorem is due to Sandroni [San03] (for the interpretations of this theorem please refer to their paper).

Theorem 1. *([San03]) Let $S \equiv \{1, \ldots, K\}$ be the finite state space, $0 < \epsilon < 1$, and $n \in \mathbb{N}$ be the duration of forecasting. Consider forecasters receiving input from $\bigcup_{t=0}^{n-1} S^t \times (\Delta S)^t$ and tests receiving input from $S^n \times (\Delta S)^n$. If a test T_n passes the truth with probability at least $1 - \epsilon$, there is a forecaster F_n (which may be randomized) that is accepted by T_n with probability at least $1 - \epsilon$ on any data generating process.*

The existence of the forecaster implied in Theorem 1 is shown via Fan's minimax theorem [Fan53]. The proof is nonconstructive, and there seem no reasons for the implied forecaster to be efficient. Indeed, Fortnow and Vohra [FV06] proceed along this computational perspective and show the following.

Theorem 2. *([FV06]) Let $S \equiv \{1, \ldots, K\}$ be the finite state space. For any time-constructible $t(n)$, there is a test T of time complexity at most $poly(n) \, t(n)$ and an infinite sequence $s^* = (s_1^*, s_2^*, \ldots)$ over S with the following properties:*

1. *For each duration n, the test T passes the truth with high probability.*
2. *For any deterministic Turing-machine forecaster F with time complexity $t(n)$, T rejects $((s_1^*, \ldots, s_n^*), F(s_1^*, \ldots, s_{n-1}^*))$ for sufficiently large n.*

Theorem 2 complements Theorem 1 by giving a test which passes the truth with high probability, whereas any deterministic forecaster of time complexity $t(n)$ is rejected on some outcome sequence if the duration is sufficiently long. Fortnow and Vohra [FV06] also show that Theorem 2 carries over to randomized forecasters of time complexity $t(n)$.

Theorem 3. *([FV06]) Let $S \equiv \{1, \ldots, K\}$ be the finite state space. For any time-constructible $t(n)$, there is a test T of time complexity at most $poly(n) \, t(n)$ and an infinite sequence $s^* = (s_1^*, s_2^*, \ldots)$ over S with the following properties:*

1. *For each duration n, the test T passes the truth with high probability.*
2. *For any randomized Turing-machine forecaster F with time complexity $t(n)$, T rejects $((s_1^*, \ldots, s_n^*), F(s_1^*, \ldots, s_{n-1}^*))$ with high probability for sufficiently large n.*

Clearly, for the test in Theorem 3, the forecaster implied in Theorem 1 must be highly complicated—it cannot simply be the same randomized $t(n)$-time Turing machine for all durations n. Fortnow and Vohra [FV06] also show tests such that the forecasters implied in Theorem 1 can be used to do factorization or even solve **PSPACE**-complete problems, a strong indication of the high complexity of the forecasters in Theorem 1. Based on these results, Fortnow and Vohra [FV06] conclude that Theorem 1 may be of little practical relevance in some cases.

We are now ready to describe our results. In Theorem 2, the test T has a running time of $poly(n) \, t(n)$ and is guaranteed to reject any $t(n)$-time forecasters when the data generating process adopts s^* as the outcome sequence. The test, therefore, is more complex in terms of running time than the forecasters it is to reject. The $poly(n) \, t(n)$ running time of T is inherent in the proof of [FV06] in that T simulates deterministic Turing machines of time complexity $t(n)$. We improve upon Theorem 2 by allowing T to run in $poly(n)$ time and reject every Turing-computable deterministic forecaster of any time complexity—not just $t(n)$. For our test T, the forecasters implied in Theorem 1 can not be the same deterministic Turing machine (of any time complexity) for all durations n.

Lemma 4. *Let $S \equiv \{1, \ldots, K\}$ be the finite state space. There is a polynomial-time test T and an infinite sequence $s^* = (s_1^*, s_2^*, \ldots)$ over S with the following properties:*

1. *For each duration n, the test T passes the truth with high probability.*
2. *For any Turing-computable deterministic forecaster F, T rejects*

$$((s_1^*, \ldots, s_n^*), F(s_1^*, \ldots, s_{n-1}^*))$$

for sufficiently large n.

Proof. Consider the doubly fractal sequence [AW]

$$1, 1, 2, 1, 2, 3, 1, 2, 3, 4, 1, 2, 3, 4, 5, \ldots$$

Let $h(j)$ be the jth element of the sequence. Note that $h(j) \leq j$. We begin by describing the test T. The input to T is some outcome sequence $(s_1, \ldots, s_n)$ and a forecast sequence $(f_1, \ldots, f_n)$ where f_i is the forecast probability distribution over S for the ith period. For each $1 \leq j \leq n$ the test T marks j if s_j equals the smallest outcome among those with the smallest forecast probability in f_j. Now T rejects if for some $1 \leq k \leq \log n$, every $j \leq n$ with $h(j) = k$ gets marked; otherwise, T accepts. It is clear that T runs in polynomial time.

We proceed to construct s^*. Let $F_1, F_2, \ldots$ be an enumeration of deterministic Turing machines that compute recursive functions. Such machines must halt on all inputs. Pick s_1^* arbitrarily from S. Inductively, once $(s_1^*, \ldots, s_{j-1}^*)$ is determined, set $s_j^* \in S$ to be the outcome with the smallest forecast probability in $F_{h(j)}(s_1^*, \ldots, s_{j-1}^*; j)$. If there are ties, s_j^* should be the one with the smallest value. In this way s^* is constructed. The intuition is that s_j^* is set to foil the forecast of $F_{h(j)}$ as much as possible.

We now argue item 1. Let the data generating process adopt an arbitrary distribution $\mathcal{P}$. Let $s = (s_1, \ldots, s_n)$ be the outcome sequence. Here s and thus $\hat{f}(s)$ (defined in Eq. (2)) consist of random variables whose distributions are determined by the data generating process. On input s and $\hat{f}(s)$, what is the probability that T marks every $1 \leq j \leq n$ with $h(j) = k$, for a particular $k \leq \log n$? Conditioned on any particular realization $(w_1, \ldots, w_{j-1})$ of the outcome sequence for the first $j - 1$ periods where $h(j) = k$, let $u \in S$ be the smallest outcome among those with the smallest probability assigned by $\hat{f}_j(w_1, \ldots, w_{j-1})$. Then j gets marked if and only if s_j turns out to be u, which happens with probability at most $1/K$ since u is least probable among all K possible outcomes. Since the conditional realization $(w_1, \ldots, w_{j-1})$ is arbitrary, the probability that T marks every $1 \leq j \leq n$ with $h(j) = k$ is at most

$$(1/K)^{|\{1 \leq j \leq n | h(j) = k\}|}.$$

For $k \leq \log n$, it is not hard to see that

$$|\{1 \leq j \leq n \mid h(j) = k\}| > \sqrt{n}.$$

Therefore, the probability that T marks every $1 \leq j \leq n$ with $h(j) = k$ for some $k \leq \log n$ is at most

$$\sum_{k=1}^{\log n} (1/K)^{\sqrt{n}} = o(1).$$

That is, the correct forecast sequence is rejected only with $o(1)$ probability. This completes item 1.

We now move on to item 2. Fix a k and consider F_k. From the construction of s^*, the forecast sequence $F_k(s_1^*, \ldots, s_{n-1}^*)$ made by F_k is such that for every j

with $h(j) = k$, s_j^* is the smallest among the outcomes to which $F_k(s_1^*, \ldots, s_{j-1}^*; j)$ assigns the smallest probability. Therefore, on input $(s_1^*, \ldots, s_n^*)$ and the forecast sequence $F_k(s_1^*, \ldots, s_{n-1}^*)$ announced by F_k, each j with $h(j) = k$ gets marked by T and thus T rejects if n satisfies $k \leq \log n$. □

The next theorem generalizes Lemma 4 by including a class of randomized forecasters.

Theorem 5. *Let $S \equiv \{1, \ldots, K\}$ be the finite state space. There is a polynomial-time test T and an infinite sequence $s^* = (s_1^*, s_2^*, \ldots)$ over S with the following properties:*

1. *For each duration n, the test T passes the truth with high probability.*
2. *Consider any forecaster F that adopts an arbitrary distribution over deterministic Turing-machine forecasters to select one machine for use, and then uses that same machine for all future forecasts. The test T rejects*

$$\left((s_1^*, \ldots, s_n^*), F(s_1^*, \ldots, s_{n-1}^*) \right)$$

with high probability for sufficiently large n.

Proof. T and s^* are as in Lemma 4, which already shows that T passes the truth with high probability for each duration n. Let F adopt a distribution $\mathcal{Q}$ over deterministic Turing-machine forecasters to select one machine for use, and then uses that same machine for all future forecasts. Let $F_1, F_2, \ldots$ be an enumeration of deterministic Turing-machine forecasters. Clearly,

$$\sum_{i=1}^{\infty} \Pr[F \text{ selects } F_i] = 1.$$

For any $\epsilon > 0$, there is an $m \in \mathbb{N}$ such that

$$\sum_{i=1}^{m} \Pr[F \text{ selects } F_i] > 1 - \epsilon. \tag{3}$$

Assume the data generating process adopts s^* as the outcome sequence. Lemma 4 guarantees that T rejects all $F_1, \ldots, F_m$ for sufficiently long duration n. Once F selects a machine for forecasting, it stays with that machine thereafter. Hence inequality (3) shows that, with probability more than $1 - \epsilon$, T rejects F. □

4   Impossibility of Working for All Durations

A natural generalization of Theorem 1 to the case where the duration n is unknown to the forecaster is to have the forecaster implied in Theorem 1 pass the test with high probability for all durations n and data generating processes. This is because a forecaster who does not know the duration n beforehand is guaranteed to pass a test with high probability if and only if it is so for every duration n (note that $n \in \mathbb{N}$ may be arbitrary). Formally, given an arbitrary test T that passes the truth with high probability, is it possible to find a forecaster F such that for each duration n and each data generating process, F is accepted by T with high probability? However, this is impossible as the following theorem shows.

Theorem 6. *Let $S \equiv \{1, \ldots, K\}$ be the finite state space and $0 < \epsilon < 1$. There is a polynomial-time test T with the following properties:*

1. *For each duration n, the test T passes the truth with probability greater than $1 - \epsilon$.*
2. *For every randomized forecaster F and $\delta > 0$, there are infinitely many $n \in \mathbb{N}$ and outcome sequences $s^* = (s_1^*, \ldots, s_n^*)$ such that T accepts $\left(s^*, F(s_1^*, \ldots, s_{n-1}^*)\right)$ with probability at most $1/2 + \delta$.*

Proof. We first describe the test T. Consider an arbitrary outcome sequence $s = (s_1, \ldots, s_n)$ and a forecast sequence $f = (f_1, \ldots, f_n)$. Denote by $f_i^{[1]}$ the probability f_i assigns to the outcome $1 \in S$. On input (s, f), the test T rejects only in the following two cases:

case 1. $s_1 = \cdots = s_n = 1$ and $\prod_{i=1}^{n} f_i^{[1]} < \epsilon/2$.
case 2. $s_1 = \cdots = s_{n-1} = 1, s_n \neq 1$, and $f_n^{[1]} > 1 - \epsilon/2$.

The test T clearly runs in polynomial time.

To argue item 1, fix $n \in \mathbb{N}$ and a distribution $\mathcal{P}$ adopted by the data generating process. Let $s = (s_1, \ldots, s_n)$ be the outcome sequence for the first n periods and $\hat{f}(s)$ be the corresponding correct forecast sequence. Here s and thus $\hat{f}(s)$ are random variables whose distributions are determined by the data generating process. By the definition of the correct forecast sequence $\hat{f}(s)$, the probability that $s_1 = \cdots = s_n = 1$ is

$$\prod_{i=1}^{n} \Pr\left[s_i = 1 \mid s_1 = \cdots = s_{i-1} = 1\right] = \prod_{i=1}^{n} \hat{f}_i^{[1]}(1^{i-1}),$$

where $\hat{f}_i^{[1]}(1^{i-1})$ is the probability assigned to the outcome $1 \in S$ in $\hat{f}_i(1^{i-1})$. Note that $\hat{f}_i^{[1]}(1^{i-1})$ is not a random variable but a fixed number determined by $\mathcal{P}$ for $1 \leq i \leq n$. Now, feed $(s, \hat{f}(s))$ to T. The test T rejects due to case 1 only if $s_1, \ldots, s_n$ all turn out to be 1 and $\prod_{i=1}^{n} \hat{f}_i^{[1]}(1^{i-1}) < \epsilon/2$. But

$$\prod_{i=1}^{n} \hat{f}_i^{[1]}(1^{i-1})$$

is precisely the probability that $s_1 = \cdots = s_n = 1$. Hence, if $\prod_{i=1}^{n} \hat{f}_i^{[1]}(1^{i-1}) < \epsilon/2$ holds, T rejects due to case 1 with probability less than $\epsilon/2$. Similarly, T rejects due to case 2 only if $s_1, \ldots, s_{n-1}$ turn out to be 1 but s_n turns out otherwise, and $\hat{f}_n^{[1]}(1^{n-1}) > 1 - \epsilon/2$. Assume $\hat{f}_n^{[1]}(1^{n-1}) > 1 - \epsilon/2$. When conditioned on

$s_1 = \cdots = s_{n-1} = 1$, the probability that s_n turns out 1 is $\hat{f}_n^{[1]}(1^{n-1}) > 1 - \epsilon/2$, so that s_n turns out otherwise with probability less than $\epsilon/2$. Therefore, the probability that T rejects due to case 2 is less than $\epsilon/2$. To sum up, the probability that T rejects $(s, \hat{f}(s))$ on either case is less than $\epsilon/2 + \epsilon/2 = \epsilon$. This completes item 1.

We now move on to item 2. Let F be an arbitrary (possibly randomized) forecaster and $\delta > 0$. Let $m > d_{\epsilon,\delta}$ for some $d_{\epsilon,\delta}$ to be determined later. Assume the data generating process picks the outcome sequence $1^m \in S^m$ for the first m periods. T rejects $(1^m, F(1^{m-1}))$ because of case 1 if $\prod_{i=1}^m F^{[1]}(1^{i-1}; i) < \epsilon/2$. Thus, either

$$\Pr\left[T \text{ accepts } (1^m, F(1^{m-1}))\right] \leq \frac{1}{2} + \delta, \tag{4}$$

or

$$\Pr\left[\prod_{i=1}^m F^{[1]}(1^{i-1}; i) \geq \epsilon/2\right] > \frac{1}{2} + \delta, \tag{5}$$

where the probability is taken over the random variables in $F(1^{m-1})$. Assume inequality (5) holds. The event $\prod_{i=1}^m F^{[1]}(1^{i-1}; i) \geq \epsilon/2$ implies

$$\left|\{i \mid F^{[1]}(1^{i-1}; i) \leq 1 - \epsilon/2, 1 \leq i \leq m\}\right| \leq c_\epsilon$$

where

$$c_\epsilon \equiv \frac{\log(\epsilon/2)}{\log(1 - \epsilon/2)}.$$

The above and inequality (5) imply

$$\Pr\left[\left|\{i \mid F^{[1]}(1^{i-1}; i) \leq 1 - \epsilon/2, 1 \leq i \leq m\}\right| \leq c_\epsilon\right] > \frac{1}{2} + \delta$$

where the probability is taken over the random variables in $F(1^{m-1})$. Now set $d_{\epsilon,\delta} \equiv c_\epsilon(1 + 2\delta)/\delta$. The above inequality trivially implies

$$\Pr\left[\left|\{i \mid F^{[1]}(1^{i-1}; i) \leq 1 - \epsilon/2, m - d_{\epsilon,\delta} < i \leq m\}\right| \leq c_\epsilon\right] > \frac{1}{2} + \delta,$$

which is equivalent to

$$\Pr\left[\left|\{i \mid F^{[1]}(1^{i-1}; i) > 1 - \epsilon/2, m - d_{\epsilon,\delta} < i \leq m\}\right| \geq d_{\epsilon,\delta} - c_\epsilon\right] > \frac{1}{2} + \delta \tag{6}$$

where the probability is taken over the random variables in $F(1^{m-1})$.

Now add the random variable r uniformly distributed over $\{m - d_{\epsilon,\delta} + 1, m - d_{\epsilon,\delta} + 2, \ldots, m\}$ and independent of the random variables in $F(1^{m-1})$. Inequality (6) and the independence of r from the random variables in $F(1^{m-1})$ imply

$$\Pr\left[F^{[1]}(1^{r-1}; r) > 1 - \epsilon/2\right] > \left(\frac{1}{2} + \delta\right)\frac{d_{\epsilon,\delta} - c_\epsilon}{d_{\epsilon,\delta}} \tag{7}$$

where the probability is taken over r and the random variables in $F(1^{m-1})$. Due to the independence of r from the random variables in $F(1^{m-1})$, we have

$$\Pr\left[F^{[1]}(1^{r-1};r) > 1 - \epsilon/2\right] = \sum_{i=m-d_{\epsilon,\delta}+1}^{m} \frac{\Pr\left[F^{[1]}(1^{i-1};i) > 1 - \epsilon/2\right]}{d_{\epsilon,\delta}} \tag{8}$$

where the probability on the left-hand side is taken over r and the random variables in $F(1^{m-1})$, and those within the summation are over the random variables in $F(1^{m-1})$. Inequalities (7) and (8) imply the existence of a number $i(m) \in \{m - d_{\epsilon,\delta} + 1, m - d_{\epsilon,\delta} + 2, \ldots, m\}$ with

$$\Pr\left[F^{[1]}(1^{i(m)-1};i(m)) > 1 - \epsilon/2\right] > (\frac{1}{2} + \delta) \cdot \frac{d_{\epsilon,\delta} - c_\epsilon}{d_{\epsilon,\delta}} = \frac{1}{2} + \frac{\delta}{2} \tag{9}$$

where the probability is taken over the random variables in $F(1^{m-1})$.

Consider the sequence $1^{i(m)-1}2 \in S^*$ standing for $\overbrace{1 \cdots 1}^{i(m)-1} 2$. Observe that inequality (9) holds also when the probability is taken over the random variables in $F(1^{i(m)-1})$ because 1^m and $1^{i(m)-1}$ share a common prefix, $1^{i(m)-1}$. Inequality (9) therefore says that, if the data generating process adopts the outcome sequence $1^{i(m)-1}2$ for the first $i(m)$ periods, then with probability more than $1/2 + \delta/2$ the forecast that F makes for the $i(m)$th period attaches more than $1 - \epsilon/2$ to the probability on the outcome 1. But then T rejects $(1^{i(m)-1}2, F(1^{i(m)-1}))$ with probability more than $1/2 + \delta/2$ (see case 2). In summary, for each $m > d_{\epsilon,\delta}$, either inequality (4) holds, or T rejects $(1^{i(m)-1}2, F(1^{i(m)-1}))$ with probability more than $1/2 + \delta/2$.   ☐

Our proof of Theorem 6 actually shows the following stronger result.

Theorem 7. *Let $S \equiv \{1, \ldots, K\}$ be the finite state space and $0 < \epsilon < 1$. There is a polynomial-time test T with the following properties:*

1. *For each duration n, the test T passes the truth with probability greater than $1 - \epsilon$.*

2. *Consider an arbitrary, possibly randomized forecaster F and $\delta > 0$. Let $d_{\epsilon,\delta} \equiv (1 + 2\delta)\log(\epsilon/2)/(\delta\log(1 - \epsilon/2))$. For sufficiently large $m \in \mathbb{N}$, there is an $n \in \{m - d_{\epsilon,\delta} + 1, m - d_{\epsilon,\delta} + 2, \ldots, m\}$ and an outcome sequence $s^* = (s_1^*, \ldots, s_n^*) \in S^n$ such that T accepts $(s^*, F(s_1^*, \ldots, s_{n-1}^*))$ with probability at most $1/2 + \delta$.*

Thus, for the test T in Theorem 7, the outcome sequences on which a forecaster (computable or not) cannot perform well are not rare. One difference of Theorem 7 from [OS06a] is that T may accept sequence pairs with prefixes rejected by T.

5   Conclusion

We have built computationally efficient tests with various desirable properties. The tests that we construct pass the truth with high probability, and it is hard in various ways to pass these tests for all data generating processes. Thus, it is hard for a forecaster to pass our tests without prior knowledge on the underlying data generating process.

Unlike most previous works except that by Fortnow and Vohra [FV06], our results take a computational perspective on forecast testing. Our tests run in polynomial time, and our first result requires the forecaster to have enormous computational power to pass the test on all data generating processes. As suggested by Fortnow and Vohra [FV06], we believe that taking a computational perspective may shed some light on many other problems previously studied without computational considerations.

References

[ANW06] Al-Najjar, N.I., Weinstein, J.: Comparative testing of experts, Levine's Working Paper Archive 321307000000000590, Department of Economics, UCLA (2006)

[AW]   Adams-Watters, F.T.: http://www.research.att.com/~njas/sequences/a002260.txt.

[Daw82] Dawid, A.P.: The well calibrated Bayesian. Journal of the American Statistical Association 77(379), 605–613 (1982)

[DF99]   Levine, D., Fudenberg, D.: Conditional universal consistency. Games and Economic Behavior 29, 104–130 (1999)

[DF06]   Dekel, E., Feinberg, Y.: Non-Bayesian testing of a stochastic prediction. Review of Economic Studies 73(4), 893–906 (2006)

[Fan53] Fan, K.: Minimax theorems. Proceedings of the National Academy of Science USA. 39, 42–47 (1953)

[FS07]   Feinberg, Y., Stewart, C.: Testing multiple forecasters. Research Paper Series 1957, Graduate School of Business, Stanford University (2007)

[FV98]   Foster, D.P., Vohra, R.V.: Asymptotic calibration. Biometrika 85(2), 379–390 (1998)

[FV06]   Fortnow, L., Vohra, R.V.: The complexity of forecast testing, Tech. Report TR06-149, Electronic Colloquium on Computational Complexity (2006)

[Leh01] Lehrer, E.: Any inspection rule is manipulable. Econometrica 69(5), 1333–1347 (2001)

[OS06a] Olszewski, W., Sandroni, A.: Counterfactual predictions, Tech. report, Northwestern University, Department of Economics (2006)

[OS06b] Olszewski, W., Sandroni, A.: Strategic manipulation of empirical tests, Tech. report, Northwestern University, Department of Economics (2006)

[San03] Sandroni, A.: The reproducible properties of correct forecasts. International Journal of Game Theory 32(1), 151–159 (2003)

[SSV03] Sandroni, A., Smorodinsky, R., Vohra, R.V.: Calibration with many checking rules. Mathematics of Operations Research 28(1), 141–153 (2003)

When Does Greedy Learning of Relevant Attributes Succeed?
— A Fourier-Based Characterization —

Jan Arpe* and Rüdiger Reischuk

Institut für Theoretische Informatik, Universität zu Lübeck
Ratzeburger Allee 160, 23538 Lübeck, Germany
{arpe,reischuk}@tcs.uni-luebeck.de

Abstract. We introduce a new notion called *Fourier-accessibility* that allows us to precisely characterize the class of Boolean functions for which a standard greedy learning algorithm successfully learns all relevant attributes. If the target function is Fourier-accessible, then the success probability of the greedy algorithm can be made arbitrarily close to one. On the other hand, if the target function is *not* Fourier-accessible, then the error probability tends to one. Finally, we extend these results to the situation where the input data are corrupted by random attribute and classification noise and prove that greedy learning is quite robust against such errors.

1 Introduction

For many application areas, greedy strategies are natural and efficient heuristics. In some cases, such as simple scheduling problems, greedy strategies find a global optimum (see, e.g., [21, Chap. 4]). For the vast majority of optimization problems, however, greedy heuristics do not achieve optimal solutions for all inputs. In such a case, one can sometimes show that the greedy algorithm at least achieves a nontrivial approximation to an optimal solution. A prominent example is the greedy algorithm for the SET COVER problem [20, 16, 27, 17].

A different approach is to ask "What is the subset of the input space for which a greedy algorithm outputs an optimal solution?". This question has rarely been answered. One notable exception is the characterization of transportation problems using the Monge property by Shamir and Dietrich [26].

We investigate the performance of greedy algorithms for the problem of *relevant feature selection*. Confronted with an unknown *target function* $f : \{0,1\}^n \to \{0,1\}$ that is only specified by randomly drawn examples $(x^k, f(x^k))$, $x^k \in \{0,1\}^n$, $k = 1, \ldots, m$, the task is to detect which variables x_i (also referred to as *attributes* or *features*) are relevant to f. This problem is central to many data mining applications; specifically, if f is a so-called *d-junta*, which means that it only depends on a small number d of all n attributes. A survey of this topic has been provided by Blum and Langley [13].

To infer relevant attributes from randomly drawn examples, the key task is to find a minimal set of attributes R admitting a consistent hypothesis h (i.e., $h(x^k) = f(x^k)$

---

* Supported by DFG research grant Re 672/4.

G. Lin (Ed.): COCOON 2007, LNCS 4598, pp. 296–306, 2007.
© Springer-Verlag Berlin Heidelberg 2007

for all k) that depends only on the variables in R. By standard arguments [14], once the sample size m exceeds $\text{poly}(2^d, \log n)$, with high probability there remains only one such hypothesis—the target function itself. Finding such a set R is equivalent to solving the following SET COVER instance. The ground set is the set of all pairs $\{k, \ell\}$ such that $f(x^k) \neq f(x^\ell)$. A pair $\{k, \ell\}$ may be covered by an attribute x_i if $x_i^k \neq x_i^\ell$. The goal is to cover the ground set by as few attributes as possible. Using this reduction, we can apply the greedy heuristic for SET COVER: the algorithm, which we call GREEDY, successively selects the attribute that covers most of the remaining edges and deletes them [20, 16].

For relevant feature selection, this approach has been proposed by Almuallim and Dieterich [4] and Akutsu and Bao [1]. Experimental results have been obtained in various areas [4, 2, 3, 15]. Akutsu et al. [3] have shown how to implement GREEDY such that its running time is only $O(d \cdot m \cdot n)$.

In this paper, we are mainly concerned with uniformly distributed attributes. For this case, Akutsu et al. [3] have proven that with high probability, GREEDY successfully infers the relevant variables for the class of Boolean monomials and that a small sample of size $\text{poly}(2^d, \log n)$ already suffices. Fukagawa and Akutsu [18] have extended this result to functions f that are unbalanced with respect to all of their relevant variables (i.e., for x uniformly chosen at random, $\Pr[f(x) = 1 | x_i = 0] \neq \Pr[f(x) = 1 | x_i = 1]$ for each relevant x_i).

Our first major result is a concise characterization of the class of target functions for which GREEDY is able to infer the relevant variables. This class properly contains the functions mentioned above. The new characterization is based on a property of the Fourier spectrum of the target function, which we call *Fourier-accessibility*. Recall that for $I \subseteq [n]$, the *Fourier coefficient* $\hat{f}(I)$ is equal to the correlation between $f(x)$ and the parity $\bigoplus_{i \in I} x_i$ (see Sect. 2 for a precise definition). A function $f : \{0, 1\}^n \to \{0, 1\}$ is

Table 1. Examples of Boolean functions and their Fourier spectra

$f(x_1, x_2, x_3)$	$\hat{f}(\emptyset)$	$\hat{f}(1)$	$\hat{f}(2)$	$\hat{f}(3)$	$\hat{f}(\{1,2\})$	$\hat{f}(\{1,3\})$	$\hat{f}(\{2,3\})$	$\hat{f}(\{1,2,3\})$
$f_1 = x_1 \oplus (x_2 \wedge x_3)$	$1/2$	$-1/4$	0	0	$-1/4$	$-1/4$	0	$1/4$
$f_2 = (x_1 \oplus x_2) \wedge x_3$	$1/4$	0	0	$-1/4$	$-1/4$	0	0	$1/4$

Fourier-accessible if for each relevant variable, one can find a sequence $\emptyset \subsetneq I_1 \subsetneq \ldots \subsetneq I_s \subseteq [n]$ such that $i \in I_s$ and for all $j \in \{1, \ldots, s\}$, $|I_j \setminus I_{j-1}| = 1$ and $\hat{f}(I_j) \neq 0$.

We prove that GREEDY correctly infers all relevant variables of Fourier-accessible d-juntas from $m = \text{poly}(2^d, \log n, \log(1/\delta))$ uniformly distributed examples with probability at least $1 - \delta$. On the other hand, it is shown that if a function f is *not* Fourier-accessible, then the error probability of GREEDY is at least $1 - d^2/(n - d)$. In particular, this probability tends to 1 if d is fixed and $n \to \infty$, or if $d \to \infty$ and $n \in \omega(d^2)$. Thus, the average-case analysis of the greedy algorithm results in a *dichotomy:* for a given function, *either* the relevant variables are inferred correctly with high probability, *or* with high probability at least some relevant variables are not detected at all.

Coping with errors in the input data has been well studied in numerical analysis, but hardly in discrete algorithms. We have shown in [7, 8] that the relevant attributes of a

d-junta can still be learned efficiently if the input data are corrupted by random noise. In Sect. 6, we describe how to extend our analysis to noisy situations and show that greedy learning is highly fault-tolerant.

There is a long tradition of relating algorithmic learning problems to spectral properties of Boolean functions, see, e.g., [22, 23, 12]. Specifically, Mossel et al. [24] have combined spectral and algebraic methods to reduce the worst-case running time for learning the class of all n-ary d-juntas to roughly $n^{0.7 \cdot d}$ (a trivial approach is to test all $\Theta(n^d)$ sets of potentially relevant variables). The novelty of our analysis lies in the following. While the greedy algorithm investigated in this paper does *not* exploit any properties of the Fourier spectrum explicitly, we show that Fourier-accessibility is necessary and sufficient for this algorithm to work successfully.

There is a simple Fourier-based algorithm that learns the relevant attributes of Fourier-accessible functions: It estimates all first-level coefficients $\hat{f}(i)$ until it finds a nonzero coefficient, which implies that x_i is relevant. Subsequently, it recurses for the subfunctions $f_{x_i=0}$ and $f_{x_i=1}$ (see also [24]). The point is that this simple greedy algorithm coincidentally also succeeds for exactly those functions. While one may argue in favour of the Fourier-based algorithm that it can easily be generalized to cope with functions with vanishing Fourier coefficients at low levels, it is explained in [6] how the greedy algorithm can also be extended naturally to cope with such functions as well.

This paper is organized as follows. The terminology and the learning model are introduced in Sect. 2. The reduction to SET COVER and the GREEDY algorithm are presented in Sect. 3. Sect. 4 provides three major lemmata used in the proof of our main results for GREEDY, which are presented in Sect. 5. In Sect. 6, we study the robustness of GREEDY against corruption of the input data. In Sect. 7, issues of further research are discussed. Due to space constraints, most proofs have been omitted. A technical report for the noise-free case is available [9]. A detailed exposition of the topics presented in this paper including all proofs can be found in the first author's PhD thesis [6].

2 Preliminaries

For $n \in \mathbb{N}$, let $[n] = \{1, \ldots, n\}$. We consider the problem of inferring the relevant variables of an unknown function $f : \{0,1\}^n \to \{0,1\}$ from randomly drawn examples. Variable x_i is *relevant* to f if $f_{x_i=0} \neq f_{x_i=1}$, where $f_{x_i=a}$ denotes the restriction of f with variable x_i set to a. The set of variables that are relevant to f is denoted by $\mathrm{rel}(f)$, whereas $\mathrm{irrel}(f)$ denotes the set of variables that are *irrelevant* (i.e., not relevant) to f. A function with $|\mathrm{rel}(f)| \leq d$ is called a *d-junta*. The restriction of f to its relevant variables is called its *base function*.

We assume that an algorithm for inferring the relevant variables receives a sequence S of randomly generated *examples* (x^k, y^k), $k \in [m]$, where $x^k \in \{0,1\}^n$ is drawn according to the uniform distribution and $y^k = f(x^k) \in \{0,1\}$. Such a sequence is called a *sample* for f of size m. If for another function h, $y^k = h(x^k)$ for all $k \in [m]$, h is said to be *consistent* with S.

If x is randomly generated, then $f : \{0,1\}^n \to \{0,1\}$ is a Bernoulli random variable, thus we will use the notation $\Pr[f = b] = \Pr_x[f(x) = b]$ for $b \in \{0,1\}$ and $\mathrm{Var}(f) =$

$\Pr[f = 0] \Pr[f = 1]$. The following well-known Chernoff bound can, e.g., be found in [5].

Lemma 1 (Chernoff Bound). *Let X be a random variable that is binomially distributed with parameters n and p, and let $\mu = pn$ be the expectation of X. Then for all ϵ with $0 \le \epsilon \le 1$, $\Pr[|X - \mu| > \epsilon n] < 2e^{-2\epsilon^2 n}$.*

The space $\mathbb{R}^{\{0,1\}^n}$ of real-valued functions on the hypercube with inner product $\langle f, g \rangle = 2^{-n} \sum_{x \in \{0,1\}^n} f(x)g(x)$ is a Hilbert space of dimension 2^n. It has an orthonormal basis $(\chi_I \mid I \subseteq [n])$, where $\chi_I(x) = (-1)^{\sum_{i \in I} x_i}$ for $x \in \{0,1\}^n$, see, e.g., [11]. Let $f : \{0,1\}^n \to \mathbb{R}$ and $I \subseteq [n]$. The *Fourier coefficient of f at I* is $\hat{f}(I) = 2^{-n} \sum_{x \in \{0,1\}^n} f(x) \cdot \chi_I(x)$. Thus, $\hat{f}(I)$ measures the correlation between $f(x)$ and $\chi_I(x)$. If $I = \{i\}$, we write $\hat{f}(i)$ instead of $\hat{f}(\{i\})$. We have the *Fourier expansion formula* $f(x) = \sum_{I \subseteq [n]} \hat{f}(I) \cdot \chi_I(x)$ for all $x \in \{0,1\}^n$.

Definition 1 (Fourier support). *Let $f : \{0,1\}^n \to \{0,1\}$. The* Fourier support *of f is* $\mathrm{supp}(\hat{f}) = \{I \subseteq [n] \mid \hat{f}(I) \ne 0\}$. *The* Fourier support graph $\mathrm{FSG}(f)$ *of f is the subgraph of the n-dimensional Hamming cube induced by $\mathrm{supp}(\hat{f})$.*

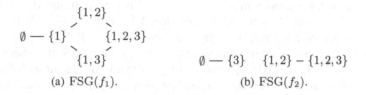

(a) $\mathrm{FSG}(f_1)$.   (b) $\mathrm{FSG}(f_2)$.

Fig. 1. Fourier support graphs of functions f_1 and f_2 presented in Table 1.

Fourier coefficients are connected to relevant variables as follows (cf. [24, 8]):

Lemma 2. *Let $f : \{0,1\}^n \to \{0,1\}$. Then for all $i \in [n]$, x_i is relevant to f if and only if there exists $I \subseteq [n]$ such that $i \in I$ and $\hat{f}(I) \ne 0$.*

Hence, whenever we find a nonzero Fourier coefficient $\hat{f}(I)$, we know that all variables $x_i, i \in I$, are relevant to f. Moreover, all relevant variables can be detected in this way, and we only have to check out subsets of size at most $d = |\mathrm{rel}(f)|$. However, there are $\Theta(n^d)$ such subsets, an amount that one would generally like to reduce. The best known learning algorithm to date that for *all* d-juntas is guaranteed to find the relevant attributes runs in time roughly $n^{0.7 \cdot d}$ [24]. In contrast, greedy heuristics require only time polynomial in n with an exponent independent of d. For our characterization of the functions to which GREEDY is applicable, we introduce the concept of *Fourier-accessibility*.

Definition 2 (Fourier-accessible). *Let $f : \{0,1\}^n \to \{0,1\}$ and $i \in [n]$. Variable x_i is* accessible *(w.r.t. f) if there exists a sequence $\emptyset = I_0 \subsetneq I_1 \subsetneq \ldots \subsetneq I_s \subseteq [n]$ such that (1) $i \in I_s$, (2) for all $j \in [s]$, $|I_j \setminus I_{j-1}| = 1$, and (3) for all $j \in [s]$, $\hat{f}(I_j) \ne 0$. The set of variables that are accessible with respect to f is denoted by $\mathrm{acc}(f)$, whereas the set of inaccessible variables with respect to f is denoted by $\mathrm{inacc}(f)$. The function f is called* Fourier-accessible *if and only if every variable that is relevant to f is also accessible, i.e., $\mathrm{acc}(f) = \mathrm{rel}(f)$.*

Equivalently, x_i is accessible if and only if there exists $I \in \text{supp}(\hat{f})$ with $i \in I$ such that there is a path in $\text{FSG}(f)$ from $\emptyset$ to I. Since $\hat{f}(\emptyset) = \Pr[f(x) = 1]$, $\emptyset \in \text{supp}(\hat{f})$ whenever $f \not\equiv 0$. Hence f is Fourier-accessible if and only if the union of all subsets $I \in \text{supp}(\hat{f})$ that belong to the connected component of $\emptyset$ in $\text{FSG}(f)$ equals $\text{rel}(f)$.

Throughout the paper, if f is clear from the context, we call a variable that is relevant to f simply *relevant*. Similarly, a variable that is accessible with respect to f is simply called *accessible*. Simple examples of a Fourier-accessible function f_1 and a non-Fourier-accessible function f_2 are given in Table 1. The corresponding Fourier support graphs are presented in Fig. 1.

In our algorithm analyses, we will consider the *expanded attribute space* of attributes $x_I = \bigoplus_{i \in I} x_i$ for $I \subseteq [n]$. The connection between these expanded attributes and the functions χ_I used to define the Fourier transform is given by $\chi_I(x) = (-1)^{x_I}$.

3   The Reduction to Set Cover and the Greedy Algorithm

With a sample $S = (x^k, y^k)_{k \in [m]} \in (\{0,1\}^n \times \{0,1\})^m$, we associate the *functional relations graph* $G_S = (V, E)$ which is defined as follows (see also [3, 7]). Its vertices correspond to the examples of S, i.e., $V = [m]$. They are partitioned into the subset of examples $V^{(0)}$ with $y^k = 0$, and the examples $V^{(1)}$ with $y^k = 1$. G_S is the complete bipartite graph with the vertex set partition $[m] = V^{(0)} \cup V^{(1)}$. Given S, our primary goal is to determine a set of variables $R \subseteq \{x_1, \ldots, x_n\}$ such that there exists *some* function $g : \{0,1\}^n \to \{0,1\}$ with $\text{rel}(g) \subseteq R$ that is consistent with the sample. In this case, R is said to *explain the sample*. Note that g may not be identical to the original function f, nor may the set R contain all relevant variables of f.

In order to find an explaining set of variables, we have to specify, for each edge $\{k, \ell\} \in E$, a relevant variable that differs in x^k and x^ℓ. Such a variable is said to *explain the edge*. Formally, an edge $\{k, \ell\} \in E$ may be *covered* by attribute x_i if and only if $x_i^k \neq x_i^\ell$. The set of edges that may be covered by x_i is denoted by E_i. A set R of variables thus explains the sample S if and only if these variables explain all edges. The previous discussion is formally summarized by the following lemma:

Lemma 3. *Let $S \in (\{0,1\}^n \times \{0,1\})^m$ be a sample and $R \subseteq \{x_1, \ldots, x_n\}$. Then R explains S if and only if $E = \cup_{x_i \in R} E_i$, where E is the edge set of the functional relations graph G_S.*

The lemma provides a reduction from the problem of inferring small sets of explaining variables to the problem of finding a small cover of E by sets from $E_1, \ldots, E_n$. This allows us to use algorithms for the set cover problem to find explaining variables. The best known and most generic algorithm for this problem is a greedy algorithm that successively picks a set that covers the largest amount of elements not covered so far. This algorithm, which we call GREEDY, is defined as follows.

If there are several sets of maximum cardinality in step 7, GREEDY picks one of them at random. The notion of success for GREEDY is captured as follows.

Definition 3 (λ-**success**). *Let $f : \{0,1\}^n \to \{0,1\}$, S be a sample for f, and $\lambda \geq 1$. GREEDY is λ-successful on input S if and only if $|\text{GREEDY}(S)| \leq \lambda \cdot |\text{rel}(f)|$ and*

Algorithm 1.. GREEDY

```
1: input S = ((x₁ᵏ,...,xₙᵏ), yᵏ)ₖ∈[m]
2: E ← {{k,ℓ} | k,ℓ ∈ [m], yᵏ ≠ yℓ}
3: R ← ∅
4: while E ≠ ∅ do
5:    for i = 1 to n do
6:       Eᵢ ← {{k,ℓ} ∈ E | xᵢᵏ ≠ xᵢℓ}
7:    select xᵢ ∉ R with maximum |Eᵢ|
8:    E ← E \ Eᵢ
9:    R ← R ∪ {xᵢ}
10: output GREEDY(S) = R
```

GREEDY$(S) \supseteq$ rel(f). GREEDY *is* successful *(or* succeeds*) if and only if it is 1-* successful, *i.e.,* GREEDY$(S) = $ rel(f), *otherwise we say that it* fails. GREEDY λ*-fails if and only if it is not λ-successful.*

4 Key Lemmata for the Algorithm Analysis

In this section, we provide three key lemmata that will be used in the proofs of our main results in Sect. 5. For technical reasons, it is useful to consider the edge sets $E_I = \{\{k,\ell\} \in E \mid x_I^k \neq x_I^\ell\}$ corresponding to the attributes from the expanded attribute space. Since x_I^k and x_I^ℓ differ if and only if the number of $i \in I$ with $x_i^k \neq x_i^\ell$ is odd, we obtain that $E_I = \triangle_{i \in I} E_i$, where $\triangle$ denotes the symmetric difference.

Suppose that GREEDY has put the variables $x_{i_1}, \ldots, x_{i_s}$ into R after s rounds. Hence, all edges in $E' = E_{i_1} \cup \cdots \cup E_{i_s}$ have been covered. The number of remaining edges that can be covered by variable x_i in the next round is $|E_i \setminus E'|$. Provided that $x_{i_1}, \ldots, x_{i_s}$ are all relevant, we would like to estimate the set size $|E_i \setminus E'|$ in dependence of properties of f. As we do not see any direct way of doing so, we take a detour via the cardinalities of the sets E_I. These turn out to be quite efficiently approximable, as we will show in Lemma 5. But let us first show how to express the cardinality of $E_i \setminus E'$ in terms of the cardinalities of the sets E_I, $I \subseteq \{i_1, \ldots, i_s\}$:

Lemma 4. *Let $S \in (\{0,1\}^n \times \{0,1\})^m$ be a sample and $G_S = (V, E)$ be the corresponding functional relations graph. Let $R \subsetneq [n]$ and $i^* \in [n] \setminus R$ and define $E' = \bigcup_{i \in R} E_i$. Then $|E_{i^*} \setminus E'| = 2^{-|R|} \sum_{I \subseteq R}(|E_{I \cup \{i^*\}}| - |E_I|)$.*

Now we are concerned with the estimation of the cardinalities $|E_I|$, $I \subseteq [n]$. For $a, b \in \{0,1\}$, let $\alpha_I^{ab} = \Pr[x_I = a \wedge f(x) = b]$, where $x \in \{0,1\}^n$ is drawn according to the uniform distribution. It follows that $\alpha_I^{a0} + \alpha_I^{a1} = \Pr[x_I = a] = 1/2$ for $I \neq \emptyset$ and $\alpha_I^{0b} + \alpha_I^{1b} = \Pr[f(x) = b]$ for all $I \subseteq [n]$.

A (noise-free) sample of size m consists of the outcomes of m independent draws of $x^k \in \{0,1\}^n$ and the corresponding classifications $y^k = f(x^k) \in \{0,1\}$. In the following, all probabilities and expectations are taken with respect to the random experiment of "drawing a sample of size m" for an arbitrary but fixed m. For all $I \subseteq [n]$ and all pairs of example indices $k, \ell \in [m]$ with $k \neq \ell$, the probability that $\{k,\ell\} \in E_I$ is

$\Pr[x_I^k \neq x_I^\ell \wedge y^k \neq y^\ell] = 2(\alpha_I^{00}\alpha_I^{11} + \alpha_I^{10}\alpha_I^{01})$. Since there are $\frac{1}{2}(m-1)m$ such pairs, the expectation of $|E_I|$ is $\alpha_I(m-1)m$ with $\alpha_I = \alpha_I^{00}\alpha_I^{11} + \alpha_I^{10}\alpha_I^{01}$.

We prove a Chernoff style mass concentration for the cardinalities $|E_I|$. It shows that for a sufficiently large sample size, $|E_I|$ is likely to be close to $\alpha_I \cdot m^2$.

Lemma 5. *There exist $c_1, c_2 > 0$ such that for every $f : \{0,1\}^n \to \{0,1\}$, given a uniformly distributed sample S of size m for f, for all $I \subseteq [n]$ and all $\epsilon \in [0,1]$,*
$$\Pr\left[\left||E_I| - \alpha_I m^2\right| > \epsilon m^2\right] < c_1 e^{-c_2 \epsilon^2 m}.$$

Before stating the third lemma, let us briefly take a closer look at the cardinalities $|E_i|$ for irrelevant variables x_i. Since for these, the value of x_i is independent of the classification $f(x)$, $\alpha_i^{ab} = \frac{1}{2}\Pr[f(x) = b]$. Consequently, $\alpha_i = \frac{1}{2}\Pr[f(x) = 0]\Pr[f(x) = 1] = \frac{1}{2}\text{Var}[f]$. Hence, the expectation of $|E_i|$ is $\frac{1}{2}\text{Var}[f]m(m-1) \approx \frac{1}{2}\text{Var}[f]m^2$. The following lemma generalizes this result to *arbitrary* $I \subseteq [n]$, revealing an unexpected relationship between the cardinalities $|E_I|$ and the Fourier coefficients $\hat{f}(I)$. Recall that for $I \subseteq [n]$ with $I \not\subseteq \text{rel}(f)$, $\hat{f}(I) = 0$ by Lemma 2.

Lemma 6. *Let $I \subseteq [n]$ with $I \neq \emptyset$. Then $\alpha_I = (\text{Var}[f] + \hat{f}(I)^2)/2$.*

5   Analysis of Greedy

In this section, we state and prove our main results. Let us start with the positive result, the class of functions for which GREEDY is successful.

Theorem 1. *There is a polynomial p such that the following holds. Let $f : \{0,1\}^n \to \{0,1\}$ be a Fourier-accessible concept, $d = |\text{rel}(f)|$, and $\delta > 0$. Let S be a uniformly distributed sample for f of size $m \geq p(2^d, \log n, \log(1/\delta))$. Then $\text{GREEDY}(S) = \text{rel}(f)$ with probability at least $1 - \delta$.*

Proof. We only provide a brief sketch here, the full proof can be found in [6, 9].

First we can show that with probability at least $1 - \delta/2$, GREEDY outputs *at least* d variables, provided that m is sufficiently large. Once this has been shown, let the sequence of variables output by GREEDY start with $x_{i_1}, \dots, x_{i_d}$. By Lemma 5, it happens with probability at least $\rho = 1 - n^d \cdot c_1 \cdot e^{-c_2 \epsilon^2 m}$ that for all $I \subseteq [n]$ such that $1 \leq |I| \leq d$, we have $\left||E_I| - \alpha_I m^2\right| \leq \epsilon m^2$. For this case, we show by induction that the variables $x_{i_1}, \dots, x_{i_s}$ are all relevant for $s \in [d]$. This implies that GREEDY halts exactly after d steps since E can always be covered by the sets E_i with $x_i \in \text{rel}(f)$. For $s = 0$, $R_0 = \emptyset \subseteq \text{rel}(f)$. For the induction step, we pick an $i^* \in \text{rel}(f) \setminus R_s$ and an $I^* \subseteq R_s$ such that $\hat{f}(I^* \cup \{i^*\}) \neq 0$ and hence $|\hat{f}(I^* \cup \{i^*\})| \geq 2^{-d}$ (since the Fourier coefficients of d-juntas are always multiples of 2^{-d}). Such an i^* exists since f is Fourier-accessible. Now we use Lemmata 4, 5, and 6 and obtain that for a suitable choice of ϵ in Lemma 5, every $x_j \in \text{irrel}(f)$ satisfies $|E_{i^*}^{(s)}| > |E_j^{(s)}|$. Consequently, in step $s + 1$, GREEDY prefers the relevant variable x_{i^*} to all irrelevant variables. Finally, we have to choose m such that $\rho \geq 1 - \delta/2$. $\square$

Example 1. The function $f_1 : \{0,1\}^3 \to \{0,1\}$ in Table 1 is Fourier-accessible. By Theorem 1, for any function $f : \{0,1\}^n \to \{0,1\}$ that has f_1 as its base function,

GREEDY succeeds with probability at least $1 - \delta$ for sample size polynomial in 2^d, $\log n$, and $\log(1/\delta)$.

If a function is not Fourier-accessible, then one of its relevant variables is not accessible. The proof of Theorem 1 shows that GREEDY first outputs all accessible variables with high probability. Once all of these have been output, the intuition is that the non-accessibility of the other relevant variables makes them statistically indistinguishable from the irrelevant variables. In particular, each inaccessible but relevant variable will be selected by GREEDY with the same probability as each irrelevant variable. Assuming that the number of irrelevant variables is much larger than the number of relevant ones, it becomes very likely that GREEDY picks an irrelevant variable and thus fails. The following result describes the class of functions for which GREEDY fails.

Theorem 2. *Let* $f : \{0,1\}^n \to \{0,1\}$ *be a function that is not Fourier-accessible and* $\lambda \geq 1$. *Given a sample* S *for* f *of arbitrary size,* GREEDY λ-*fails on input* S *with probability at least* $1 - \frac{\lambda d^2}{n - \lambda d}$, *where* $d = |\operatorname{rel}(f)|$.

Corollary 1. *Let* $p_\lambda(n,d)$ *denote the probability that for any given function* f *with* $|\operatorname{rel}(f)| = d$ *that is not Fourier-accessible and for any uniformly distributed sample* S *for* f, GREEDY λ-*fails. Then for fixed* $\lambda \geq 1$,
(a) for fixed d, $\lim_{n \to \infty} p_\lambda(n,d) = 1$ *and*
(b) for $d \to \infty$ *and* $n = n(d) \in \omega(d^2)$, $\lim_{d \to \infty} p_\lambda(n,d) = 1$.

Example 2. The function $f_2 : \{0,1\}^3 \to \{0,1\}$ in Table 1 is not Fourier-accessible. By Theorem 2, for any function $f : \{0,1\}^n \to \{0,1\}$ that has f_2 as its base function, GREEDY fails with probability at least $1 - \frac{9}{n-3}$.

Note that Theorem 2 not only says that GREEDY (with high probability) fails for functions that are not Fourier-accessible, but that GREEDY even fails to find all relevant variables of the target function in $\lambda \cdot |\operatorname{rel}(f)|$ rounds for any $\lambda \geq 1$. In addition, note that the claim in Theorem 2 is independent of the sample size.

In the literature, it has often been emphasized that GREEDY has a "logarithmic approximation guarantee" (see [1, 3, 13, 18]), i.e., given a sample S for f of size m, GREEDY finds a set of at most $(2 \ln m + 1) \cdot |\operatorname{rel}(f)|$ variables that explain S. Theorem 2 shows that if f is not Fourier-accessible, then with probability at least $\frac{(2 \ln m + 1)d^2}{n - (2 \ln m + 1)d}$, these variables *do not contain all relevant variables* (where $d = |\operatorname{rel}(f)|$). Thus, GREEDY *misses* some relevant variable with high probability, provided that $m \in 2^{o(n)}$. Hence the positive approximability properties of the greedy strategy for the SET COVER problem do not translate to the learning situation. The fact that GREEDY outputs at most $(2 \ln m + 1) \cdot |\operatorname{rel}(f)|$ variables only guarantees that any sample of size m can be explained by this amount of more or less arbitrary variables.

6 Robustness Against Noise

The technical analysis of GREEDY and the Fourier spectrum for d-juntas in the previous section can be extended to the situation where the input data contain errors.

We can show that GREEDY is *extremely robust* against noise, which will be modelled as follows. Instead of receiving suitable examples $(x, f(x))$, the learning algorithm now obtains *noisy examples* of the form $(x \oplus \xi, f(x) \oplus \zeta)$, where in the *noise vector* $\xi = (\xi_1, \ldots, \xi_n) \in \{0, 1\}^n$ each ξ_i is set to 1 independently with probability p_i, and the *classification noise bit* $\zeta \in \{0, 1\}$ is set to 1 with probability η. To avoid that some attribute or the classification is turned into a purely random bit, the noise has to be bounded away from $1/2$: we require that there exist $\gamma_a, \gamma_b > 0$ such that $p_i \leq (1 - \gamma_a)/2$ for all $i \in [n]$ and $\eta \leq (1 - \gamma_b)/2$. Let P denote the product distribution on $\{0, 1\}^n$ induced by the probabilities $p_1, \ldots, p_n$. A sample S for a function f that is corrupted by such a noise process is called a (P, η)-*noisy sample*.

We can show that given a (P, η)-noisy sample S for a Fourier-accessible function f of size polynomial in 2^d, $\log(n/\delta)$, γ_a^{-d}, and γ_b, GREEDY still outputs all relevant variables of f. It may be the case that the sets E_i that correspond to the relevant variables do not suffice to explain all edges in E due to noise. Even worse, it may happen that some edges cannot be explained at all: the sample may contain contradictive examples. For this reason, we have to employ a variant of GREEDY that gets the number d of relevant attributes as a parameter and outputs a set of d variables that is supposed to contain the relevant ones. In other words, the `while`-loop in line 4 of Algorithm 1 is replaced with the statement "`while` $|R| \leq d$ `do`". We denote this algorithm by GREEDY$_d$.

We adjust the definition of the probabilities α_I^{ab} and define, for $a, b \in \{0, 1\}$, $\beta_I^{ab} = \Pr[x_I \oplus \xi_I = a \wedge f(x) \oplus \zeta = b]$, where $\Pr[\xi_i = 1] = p_i$, $\Pr[\zeta = 1] = \eta$, and $\xi_I = \bigoplus_{i \in I} \xi_i$. While in the noise-free scenario, the expectation of $|E_I|$ is $\alpha_I(m-1)m$, the expectation of $|E_I|$ is now equal to $\beta_I(m-1)m$ with $\beta_I = \beta_I^{00}\beta_I^{11} + \beta_I^{10}\beta_I^{01}$. The next lemma is completely analogous to Lemma 5:

Lemma 7. *There exist $c_1, c_2 > 0$ such that for every $f : \{0, 1\}^n \to \{0, 1\}$, given a uniformly distributed (P, η)-noisy sample S of size m for f, for all $I \subseteq [n]$ and $\epsilon \in [0, 1]$, $\Pr\left[\left||E_I| - \beta_I m^2\right| > \epsilon m^2\right] < c_1 e^{-c_2 \epsilon^2 m}$.*

Proving an analog of Lemma 6 requires some more computation:

Lemma 8. *Let $f : \{0, 1\}^n \to \{0, 1\}$, $I \subseteq [n]$ with $I \neq \emptyset$, and $\lambda_I = \prod_{i \in I}(1 - 2p_i)$. Then $\beta_I = \frac{1}{2}\left((1 - 2\eta)^2 \cdot \mathrm{Var}[f] + \eta \cdot (1 - \eta) + (1 - 2\eta)^2 \cdot \lambda_I^2 \cdot \hat{f}(I)^2\right)$*

Theorem 1 generalizes to the scenario of noisy data as follows:

Theorem 3. *There is a polynomial p such that the following holds. Let $f : \{0, 1\}^n \to \{0, 1\}$ be a Fourier-accessible function, $d = |\mathrm{rel}(f)|$, and $\delta > 0$. Let S be a uniformly distributed (P, η)-noisy sample S for f of size $m \geq p\left(2^d, \log n, \log(1/\delta), \gamma_a^d, \gamma_b\right)$. Then GREEDY$_d(S) = \mathrm{rel}(f)$ with probability at least $1 - \delta$.*

So far, we cannot exclude that there may be (fixed) noise distributions for which *strictly more* functions may be learned than can be learned without noise (compare to the situation of noisy circuits as discussed in [25]). However, we can show that Theorem 2 and Corollary 1 also hold for uniformly distributed (P, η)-noisy samples. The proofs are similar to those for the noise-free case.

The high fault tolerance of GREEDY can be further generalized to a situation where we do not have to assume statistical independence for the corruption of individual attributes. Instead of flipping each attribute value independently with probability p_i, let the noise vectors ξ be drawn according to an arbitrary distribution $P : \{0,1\}^n \to [0,1]$ that satisfies $\Pr[\xi_I = 1] \leq \frac{1}{2}(1 - \gamma_a^{|I|})$ for all $I \subseteq [n]$ with $1 \leq |I| \leq d$. All results of this section are still valid in this setting (in Lemma 8, the definition of λ_I has to be replaced with $1 - 2\Pr[\xi_I = 1]$). In fact, product distributions with $p_i \leq \frac{1}{2}(1 - \gamma_a)$ are a special case of this scenario. Further details of this most general result have to be omitted due to space contraints, but can be found in [6].

7 Concluding Remarks

The first issue left for future research is the investigation of the performance of the greedy algorithm in variations of the learning scenario considered in this paper: attributes and classifications may take more than two values, attributes may be non-uniformly distributed, etc.

For non-uniform attribute distributions—although a generic notion of Fourier coefficients can be given [10, 19]—Lemma 5 with a similar definition of α_I does not hold any more. It is easy to find examples such that (a) there are $x_i, x_j \in \text{irrel}(f)$ such that the expected sizes of E_i and E_j differ or (b) there are $x_i \in \text{rel}(f)$ and $x_j \in \text{irrel}(f)$ such that the expected sizes of E_i and E_j are equal, although $\hat{f}(i) = \hat{f}(j) = 0$. Thus, a completely different analysis is needed for such a setting. Again, we refer to [6, 9] for more details.

The second issue is to stick to the learning scenario and investigate variants of the greedy heuristic. If an edge is labeled by exactly one variable, then this variable has to be selected in order to explain the sample. For this reason, Almuallim and Dietterich [4] proposed to assign the weight $\sum_{e \in E_i} \frac{1}{c(e)-1}$ to x_i (where $c(e)$ is equal to the number of variables that can cover e) and then find a set cover by selecting variables of maximum weight. Since for $n \gg |\text{rel}(f)|$, each edge is labeled by roughly $n/2$ irrelevant variables, such a weighting is unlikely to help much during the first rounds of the algorithm. Thus, it seems unlikely that there are functions for which this heuristic outperforms the algorithm analyzed in this paper.

References

[1] Akutsu, T., Bao, F.: Approximating Minimum Keys and Optimal Substructure Screens. In: Cai, J.-Y., Wong, C.K. (eds.) COCOON 1996. LNCS, vol. 1090, pp. 290–299. Springer, Heidelberg (1996)

[2] Akutsu, T., Miyano, S., Kuhara, S.: Algorithms for Identifying Boolean Networks and Related Biological Networks Based on Matrix Multiplication and Fingerprint Function. J. Comput. Biology 7(3-4), 331–343 (2000)

[3] Akutsu, T., Miyano, S., Kuhara, S., Simple, A.: A Simple Greedy Algorithm for Finding Functional Relations: Efficient Implementation and Average Case Analysis. Theoret. Comput. Sci. 292(2), 481–495 (2003)

[4] Almuallim, H., Dietterich, T.G.: Learning Boolean Concepts in the Presence of Many Irrelevant Features. Artificial Intelligence 69(1-2), 279–305 (1994)

[5] Alon, N., Spencer, J.: The Probabilistic Method. Wiley-Intersci. Ser. Discrete Math. Optim. John Wiley and Sons, Chichester (1992)

[6] Arpe, J.: Learning Concepts with Few Unknown Relevant Attributes from Noisy Data. PhD thesis, Institut für Theoretische Informatik, Universität zu Lübeck (2006)

[7] Arpe, J., Reischuk, R.: Robust Inference of Relevant Attributes. In: Gavaldá, R., Jantke, K.P., Takimoto, E. (eds.) ALT 2003. LNCS (LNAI), vol. 2842, pp. 99–113. Springer, Heidelberg (2003)

[8] Arpe, J., Reischuk, R.: Learning Juntas in the Presence of Noise. In: Cai, J.-Y., Cooper, S.B., Li, A. (eds.) TAMC 2006. LNCS, vol. 3959, pp. 387–398. Springer, Heidelberg, Invited to appear in special issue of TAMC 2006 in Theoret. Comput. Sci., Series A (2006)

[9] Arpe, J., Reischuk, R.: When Does Greedy Learning of Relevant Attributes Succeed?— A Fourier-based Characterization. Technical Report ECCC TR06-065, Electronic Colloquium on Computational Complexity (2006)

[10] Bahadur, R.R.: A Representation of the Joint Distribution of Responses to n Dichotomous Items. In: Solomon, H. (ed.) Studies in Item Analysis and Prediction, pp. 158–168. Stanford University Press, Stanford (1961)

[11] Bernasconi, A.: Mathematical Techniques for the Analysis of Boolean Functions. PhD thesis, Università degli Studi di Pisa, Dipartimento di Ricerca in Informatica (1998)

[12] Blum, A., Furst, M., Jackson, J.C., Kearns, M., Mansour, Y., Rudich, S.: Weakly Learning DNF and Characterizing Statistical Query Learning Using Fourier Analysis. In: Proc. 26th STOC 1994, pp. 253–262 (1994)

[13] Blum, A., Langley, P.: Selection of Relevant Features and Examples in Machine Learning. Artificial Intelligence 97(1-2), 245–271 (1997)

[14] Blumer, A., Ehrenfeucht, A., Haussler, D., Warmuth, M.K.: Occam's Razor. Inform. Process. Lett. 24(6), 377–380 (1987)

[15] Boros, E., Horiyama, T., Ibaraki, T., Makino, K., Yagiura, M.: Finding Essential Attributes from Binary Data. Ann. Math. Artif. Intell. 39(3), 223–257 (2003)

[16] Chvátal, V.: A Greedy Heuristic for the Set Covering Problem. Math. Oper. Res. 4(3), 233–235 (1979)

[17] Feige, U.: A Threshold of ln n for Approximating Set Cover. J. ACM 45(4), 634–652 (1998)

[18] Fukagawa, D., Akutsu, T.: Performance Analysis of a Greedy Algorithm for Inferring Boolean Functions. Inform. Process. Lett. 93(1), 7–12 (2005)

[19] Furst, M.L., Jackson, J.C., Smith, S.W.: Improved Learning of AC^0 Functions. In: Proc. 4th COLT 1991, pp. 317–325

[20] Johnson, D.S.: Approximation Algorithms for Combinatorial Problems. J. Comput. System Sci. 9(3), 256–278 (1974)

[21] Kleinberg, J., Tardos, É.: Algorithm Design. Addison-Wesley, Reading (2005)

[22] Linial, N., Mansour, Y., Nisan, N.: Constant Depth Circuits, Fourier Transform, and Learnability. J. ACM 40(3), 607–620 (1993)

[23] Mansour, Y.: Learning Boolean Functions via the Fourier Transform. In: Roychodhury, V., Siu, K.-Y., Orlitsky, A. (eds.) Theoretical Advances in Neural Computation and Learning, pp. 391–424. Kluwer Academic Publishers, Dordrecht (1994)

[24] Mossel, E., O'Donnell, R.W., Servedio, R.A.: Learning functions of k relevant variables. J. Comput. System Sci. 69(3), 421–434 (2004)

[25] Reischuk, R.: Can Large Fanin Circuits Perform Reliable Computations in the Presence of Noise? Theoretical Comput. Sci. 240(4), 319–335 (2000)

[26] Shamir, R., Dietrich, B.: Characterization and Algorithms for Greedily Solvable Transportation Problems. In: Proc. 1st SODA 1990, pp. 358–366

[27] Slavík, P.: A Tight Analysis of the Greedy Algorithm for Set Cover. In: Proc. 28th STOC 1996, pp. 435–441.

The Informational Content of Canonical Disjoint NP-Pairs

Christian Glaßer[1], Alan L. Selman[2,*], and Liyu Zhang[2]

[1] Lehrstuhl für Informatik IV, Universität Würzburg, Am Hubland,
97074 Würzburg, Germany
glasser@informatik.uni-wuerzburg.de
[2] Department of Computer Science and Engineering, University at Buffalo,
Buffalo, NY 14260
{selman,lzhang7}@cse.buffalo.edu

Abstract. We investigate the connection between propositional proof systems and their canonical pairs. It is known that simulations between proof systems translate to reductions between their canonical pairs. We focus on the opposite direction and study the following questions.

Q1: Where does the implication $[can(f) \leq^{pp}_m can(g) \Rightarrow f \leq_s g]$ hold, and where does it fail?

Q2: Where can we find proof systems of different strengths, but equivalent canonical pairs?

Q3: What do (non-)equivalent canonical pairs tell about the corresponding proof systems?

Q4: Is every NP-pair (A, B), where A is NP-complete, strongly many-one equivalent to the canonical pair of some proof system?

In short, we show that both parts of Q1 and Q2 can be answered with 'everywhere', which generalize previous results by Pudlák and Beyersdorff. Regarding Q3, inequivalent canonical pairs tell that the proof systems are not "very similar", while equivalent, P-inseparable canonical pairs tell that they are not "very different". We can relate Q4 to the open problem in structural complexity that asks whether unions of disjoint NP-complete sets are NP-complete. This demonstrates a new connection between proof systems, disjoint NP-pairs, and unions of disjoint NP-complete sets.

1 Introduction

One reason it is important to study canonical pairs of propositional proof systems (proof systems) is their role in connecting proof systems with disjoint NP-pairs (NP-pairs) [7]. Razborov [13] first defined the canonical pair, $can(f) = (SAT^*, REF(f))$, for every proof system f. He showed that if there exists an optimal proof system f, then its canonical pair is a complete pair for DisjNP. In a recent paper [8], we show that every NP-pair is polynomial-time many-one equivalent to the canonical pair of some proof system. So the degree structure of the class of NP-pairs and of all canonical pairs is identical.

* Research partially supported by NSF grant CCR-0307077.

G. Lin (Ed.): COCOON 2007, LNCS 4598, pp. 307–317, 2007.
© Springer-Verlag Berlin Heidelberg 2007

Beyersdorff [1] studies proof systems and their canonical pairs from a proof theoretic point of view. He defines the subclasses DNPP(P) of NP-pairs that are representable in some proof system P and shows that the canonical pairs of P are complete for DNPP(P). This interesting result tells us that for certain meaningful subclasses of NP-pairs, complete pairs do exist. Beyersdorff also compares the simulation order of proof systems with the hardness of their canonical pairs, which we will address in this paper too.

Encouraged by these exciting results on proof systems and their canonical pairs, we continue this line of research and concentrate on the following correspondence between proof systems and NP-pairs. For proof systems f and g,

$$f \leq_s g \quad \Rightarrow \quad can(f) \leq_m^{pp} can(g). \tag{1}$$

Pudlák [12] and Beyersdorff [1] give counter examples for the converse. This raises the following questions which we investigate in this paper.

Q1: Where does the following implication hold, and where does it fail?

$$can(f) \leq_m^{pp} can(g) \quad \Rightarrow \quad f \leq_s g \tag{2}$$

Q2: Where can we find proof systems of different strengths whose canonical pairs are equivalent?

Q3: What do (non-)equivalent canonical pairs tell about the corresponding proof systems?

Moreover, it is known that every NP-pair is many-one equivalent to the canonical pair of some proof system [8]. Here we investigate the same question for strongly many-one reductions. It is easy to see that this question must be restricted to pairs whose first component is NP-complete.

Q4: Is every NP-pair (A, B), where A is NP-complete, strongly many-one equivalent to the canonical pair of some proof system?

Theorem 3 addresses the first part of Q1: The theorem asserts that, for any two disjoint NP-pairs (A, B) and (C, D), there are proof systems f and g such that $can(f) \equiv_m^{pp} (A, B)$, $can(g) \equiv_m^{pp} (C, D)$, and implication (2) holds nontrivially.

Corollary 2 addresses the second part of Q1: The following assertion is equivalent to the reasonable assumption that optimal proof systems do not exist. For every proof system f there is a proof system g such that f and g is a counter example to implication (2). More strongly, there is an infinite chain of proof systems $g_0, g_1, \cdots$, such that $f <_s g_0 <_s g_1 <_s \cdots$, but the canonical pairs of all of these proof systems are many-one equivalent. In this way, we address Q2.

In section 4 we answer Q3 in different ways. Equivalent canonical pairs do not tell much about the mere simulation order of two proof systems (Theorem 5). However, inequivalent canonical pairs tell us that the corresponding proof systems do not simulate each other except on a P-subset of TAUT (Proposition 3). Hence these proof systems are not "very similar". In contrast, equivalent,

P-inseparable canonical pairs tell us that none of the corresponding proof systems is almost everywhere super-polynomially stronger than the other one (Theorem 6). So these proof systems are not "very different".

In section 5 we can relate Q4 to the open problem in structural complexity [3,6] that asks whether unions of disjoint NP-complete sets are NP-complete. We show under the hypothesis NP $\neq$ coNP that if Q4 has an affirmative answer, then unions of disjoint NP-complete sets are NP-complete. This demonstrates a new connection between proof systems, NP-pairs, and problems in structural complexity. Finally, in section 6 we obtain connections between proof systems and the Turing-degrees of their canonical pairs.

2   Preliminaries

A disjoint NP-pair is a pair (A, B) of nonempty sets A and B such that $A, B \in$ NP and $A \cap B = \emptyset$. Let DisjNP denote the class of all disjoint NP-pairs.

Given a disjoint NP-pair (A, B), a *separator* is a set S such that $A \subseteq S$ and $B \subseteq \overline{S}$; we say that S *separates* (A, B). Let $Sep(A, B)$ denote the set of all separators of (A, B). For disjoint NP-pairs (A, B), the fundamental question is whether $Sep(A, B)$ contains a set belonging to P. In that case the pair is P-*separable*; otherwise, the pair is P-*inseparable*. There is evidence [4,5] that P-inseparable disjoint NP-pairs exist, and this will be our main hypothesis in the paper. The following proposition summarizes known results.

Proposition 1

1. P $\neq$ NP $\cap$ coNP *implies that* DisjNP *contains a P-inseparable pair.*
2. P $\neq$ UP *implies that* DisjNP *contains a P-inseparable pair. [4].*
3. *If* DisjNP *contains P-inseparable pairs, then it contains a P-inseparable pair whose components are NP-complete. [4].*

While it is probably the case that DisjNP contains P-inseparable pairs, there is an oracle relative to which P $\neq$ NP and P-inseparable pairs in DisjNP do not exist [9]. So P $\neq$ NP probably is not a sufficiently strong hypothesis to show the existence of P-inseparable pairs in DisjNP. On the other hand, if there exist secure public-key cryptosystems (for example, if RSA cannot be cracked in polynomial-time), then there exist P-inseparable disjoint NP-pairs [4].

All reducibilities in the paper are polynomial time computable. We review the notions of reducibilities between disjoint pairs. The original notions are nonuniform [4], here we state the equivalent uniform versions [4,5].

Definition 1. *Let* (A, B) *and* (C, D) *be disjoint pairs.*

1. (A, B) *is* many-one reducible in polynomial-time *to* (C, D), $(A, B) \leq_m^{pp} (C, D)$, *if there exists a polynomial-time computable function* f *such that* $f(A) \subseteq C$ *and* $f(B) \subseteq D$.
2. (A, B) *is* Turing reducible in polynomial-time *to* (C, D), $(A, B) \leq_T^{pp} (C, D)$, *if there exists a polynomial-time oracle Turing machine* M *such that for every separator* S *of* (C, D), $L(M, S)$ *is a separator of* (A, B).

Köbler, Meßner, and Torán [10] define the following stronger version of many-one reductions between disjoint NP-pairs:

Definition 2. *Let (A, B) and (C, D) be disjoint pairs. (A, B) is strongly many-one reducible in polynomial-time to (C, D), $(A, B) \leq_{sm}^{pp} (C, D)$, if there exists a polynomial-time computable function f such that $f(A) \subseteq C$, $f(B) \subseteq D$, and $f(\overline{A \cup B}) \subseteq \overline{C \cup D}$.*

Definition 3. *A disjoint pair (A, B) is $\leq_m^{pp}$-hard for NP if for every separator L of (A, B), SAT $\leq_m^p L$.*

Definition 4. *For any disjoint pair (A, B), the polynomial-time Turing-degree (Turing-degree for short) of (A, B) is defined as*

$$\mathbf{d}(A, B) = \{(C, D) \mid (C, D) \text{ is a disjoint pair and } (A, B) \equiv_T^{pp} (C, D)\}.$$

In an earlier paper [8] we investigated the restriction of Turing-degrees of disjoint pairs on DisjNP and showed that every countable distributive lattice can be embedded into the interval between any two comparable but inequivalent restricted Turing-degrees of disjoint NP-pairs. It follows trivially that every countable distributive lattice can be embedded into the interval between any two comparable but inequivalent Turing-degrees of disjoint pairs if both degrees contain some disjoint NP-pair.

Let **SAT** denote the set of satisfiable formulas and let UNSAT $\stackrel{df}{=} \overline{SAT}$. Moreover, let TAUT denote the set of tautologies. Cook and Reckhow [2] defined a *propositional proof system* (proof system for short) to be a function $f : \Sigma^* \to$ TAUT such that f is onto and f is polynomial-time computable. For every tautology α, if $f(w) = \alpha$, then we say w is an f-*proof* of α.

The *canonical* NP-*pair* (canonical pair for short) of f [13,12] is the disjoint NP-pair (SAT*, REF(f)), denoted by $can(f)$, where

$$\text{SAT}^* = \{(x, 0^n) \mid x \in \text{SAT}\} \quad \text{and}$$

$$\text{REF}(f) = \{(x, 0^n) \mid \neg x \in \text{TAUT and } \exists y[|y| \leq n \text{ and } f(y) = \neg x]\}.$$

Conversely, for every disjoint NP-pair (A, B), we can define a proof system $f_{A,B}$ as follows. Let $\langle \cdot, \cdot \rangle$ be a polynomial-time computable, polynomial-time invertible pairing function such that $|\langle v, w \rangle| = 2|vw|$. Choose a g that is polynomial-time computable and polynomial-time invertible such that $A \leq_m^p \text{SAT}$ via g. Let N be an NP-machine that accepts B in time p.

$$f_{A,B}(z) \stackrel{df}{=} \begin{cases} \neg g(x) & : \quad \text{if } z = \langle x, w \rangle, |w| = p(|x|), N(x) \text{ accepts along path } w \\ x & : \quad \text{if } z = \langle x, w \rangle, |w| \neq p(|x|), |z| \geq 2^{|x|}, x \in \text{TAUT} \\ \text{true} & : \quad \text{otherwise} \end{cases}$$

Clearly, $f_{A,B}$ is a propositional proof system for every disjoint NP-pair (A, B).

Theorem 1 ([8]). *For every $(A, B) \in$ DisjNP, $(A, B) \equiv_m^{pp} can(f_{A,B})$.*

Let f and f' be two propositional proof systems. We say that f *simulates* f' ($f' \leq_s f$) if there is a polynomial p and a function $h : \Sigma^* \to \Sigma^*$ such that for every $w \in \Sigma^*$, $f(h(w)) = f'(w)$ and $|h(w)| \leq p(|w|)$. Furthermore, if the function h can be computed in polynomial-time, f *p-simulates* f' ($f' \leq_p f$). A proof system is *(p-)optimal* if it $(p\text{-})$simulates every other proof system.

In Section 4, we will need the following generalization of the concept "simulation". We say that f *simulates* f' *on a subset* S of TAUT, if there is a polynomial p and a function $h : \Sigma^* \to \Sigma^*$ such that for every $w \in \Sigma^*$, $f'(w) \in S$ implies that $f(h(w)) = f'(w)$ and $|h(w)| \leq p(|w|)$. Moreover, f *simulates* f' *except on a subset* S of TAUT, if f simulates f' on TAUT $- S$. Obviously, a proof system f simulates a proof system f' if and only if f simulates f' on TAUT.

We use $f <_s g$ to denote that $f \leq_s g$ and $g \not\leq_s f$. We use $(A, B) <_m^{pp} (C, D)$ to denote that $(A, B) \leq_m^{pp} (C, D)$ and $(C, D) \not\leq_m^{pp} (A, B)$.

3 Proof Systems and Many-One Degrees of Canonical Pairs

We recall the fundamental relation between proof systems and canonical pairs.

Proposition 2 ([12,8]). *Let f and g be proof systems.*

$$f \leq_s g \;\Rightarrow\; can(f) \leq_m^{pp} can(g)$$

In this section, we investigate the converse of the above proposition. We show results that address both parts of Q1 and Q2.

We start our investigations with the observation that refuting an implication that is slightly weaker than (2) is equivalent to proving the existence of P-inseparable disjoint NP-pairs. This is done by a purely complexity theoretic proof that does not rely on specific properties of concrete proof systems.

Theorem 2. *The following statements are equivalent.*

1. *P-inseparable disjoint NP-pairs exist.*
2. *There exist proof systems f and g such that $can(f) <_m^{pp} can(g) \not\Rightarrow f \leq_s g$.*

Proof. If P-inseparable disjoint NP-pairs do not exist, then all canonical pairs of proof systems are P-separable and hence are equivalent. This shows $2 \Rightarrow 1$.

For the other direction, assume that P-inseparable disjoint NP-pairs exist and define the following set of propositional formulas.

EASY $\overset{df}{=} \{x \mid x$ is a propositional formula such that $x = (b \vee \overline{b} \vee y)$ for a suitable variable b and a suitable formula $y\}$

EASY is a subset of TAUT. Also, EASY $\in$ P. Let true $\overset{df}{=} (b \vee \overline{b} \vee b)$ and define a proof system as follows.

$$f(z) \overset{df}{=} \begin{cases} x & : & \text{if } z = \langle x, \varepsilon \rangle \text{ and } x \in \text{EASY} \\ x & : & \text{if } z = \langle x, y \rangle \text{ and } |y| > 2^{|x|} \text{ and } x \in \text{TAUT} \\ \text{true} & : & \text{otherwise.} \end{cases}$$

Note that f is a proof system. Observe that the elements in EASY are the only tautologies that have polynomial-size f-proofs. All other tautologies do not have polynomial-size f-proofs. This makes $can(f)$ P-separable which is witnessed by the following separator:

$$S = \{(x, 0^n) \mid [n \leq 2^{|x|} \text{ and } \neg x \notin \text{EASY}] \text{ or } [n > 2^{|x|} \text{ and } x \in \text{SAT}]\}$$

By assumption there exists a P-inseparable disjoint NP-pair (A, B). Hence, by Theorem 1 there exists a proof system g' such that $can(g')$ and (A, B) are many-one equivalent. Now define another proof system.

$$g(z) \stackrel{df}{=} \begin{cases} g'(w) & : \quad \text{if } z = 0w \text{ and } g'(w) \notin \text{EASY} \\ \text{true} & : \quad \text{if } z = 0w \text{ and } g'(w) \in \text{EASY} \\ x & : \quad \text{if } z = 1w,\ w = \langle x, y \rangle,\ |y| = 2^{|x|}, \text{ and } x \in \text{EASY} \\ \text{true} & : \quad \text{otherwise.} \end{cases}$$

Note that g is a proof system. Observe that formulas in $\text{EASY} - \{\text{true}\}$ do not have polynomial-size g-proofs. It follows that g does not simulate f, since f provides polynomial-size proofs for elements in EASY.

Now we verify that $can(g') \leq_m^{pp} can(g)$ via the reduction that maps $(x, 0^n)$ to $(x, 0^{n+1})$. If $(x, 0^n) \in \text{SAT}^*$, then $(x, 0^{n+1}) \in \text{SAT}^*$ and we are done. Let $(x, 0^n) \in \text{REF}(g')$. So there exists some w such that $|w| \leq n$ and $g'(w) = (\neg x)$. Note that $(\neg x) \notin \text{EASY}$, since formulas in EASY do not start with a negation. From the definition of g it follows that $g(0w) = g'(w) = (\neg x)$. So $(x, 0^{n+1}) \in \text{REF}(g)$.

So $can(g') \leq_m^{pp} can(g)$ and therefore, $(A, B) \leq_m^{pp} can(g)$. Hence $can(g)$ is P-inseparable. This shows $can(f) <_m^{pp} can(g)$.   □

The examples given by Pudlák [12] and Beyersdorff [1] show that the simulation order of proof systems is not necessarily reflected by the reducibility of their canonical pairs. However, as the next theorem shows, the canonical pairs of proof systems that satisfy implication (2) in a non-trivial way, vary over all degrees of disjoint NP-pairs. More precisely, for each pair of many-one degrees of disjoint NP-pairs, there do exist proof systems whose canonical pairs lie in the respective degrees such that their simulation order is consistent with the reducibility of the canonical pairs. This answers the first part of Q1 in the sense that implication (2) can be satisfied non-trivially for arbitrary canonical pairs.

Theorem 3. Let $(A, B), (C, D) \in \text{DisjNP}$ such that $(A, B) \leq_m^{pp} (C, D)$. Then there exist proof systems f_1 and f_2 such that $f_1 \leq_p f_2$, $can(f_1) \equiv_m^{pp} (A, B)$, and $can(f_2) \equiv_m^{pp} (C, D)$.

Proof. Let $\langle \cdot, \cdot \rangle$ be a polynomial-time computable, polynomial-time invertible pairing function such that $|\langle v, w \rangle| = 2|vw|$. Choose g_1 that is polynomial-time computable and polynomial-time invertible such that $A \leq_m^p \text{SAT}$ via g_1. Let N_1 be an NP-machine that accepts B in time p_1. Define the following function f_1.

$$f_1(z) \stackrel{df}{=} \begin{cases} \neg g_1(x) & : \quad \text{if } z = \langle x, w \rangle,\ |w| = p_1(|x|),\ N_1(x) \text{ accepts along path } w \\ x & : \quad \text{if } z = \langle x, w \rangle,\ |w| \neq p_1(|x|),\ |z| \geq 2^{|x|},\ x \in \text{TAUT} \\ \text{true} & : \quad \text{otherwise} \end{cases}$$

The proof of Theorem 1 shows that f_1 is a proof system and $can(f_1) \equiv_m^{pp} (A, B)$. Now choose g_2 that is polynomial-time computable and polynomial-time invertible such that $C \leq_m^p$ SAT via g_2. Let N_2 be an NP-machine that accepts D in time p_2. Without loss of generality, we assume for every $n \geq 0$, $p_1(n) \neq p_2(n)$ and $range(g_1) \cap range(g_2) = \emptyset$. Define the following function f_2.

$$f_2(z) \stackrel{df}{=} \begin{cases} \neg g_1(x) & : \text{if } z = \langle x, w \rangle, |w| = p_1(|x|), N_1(x) \text{ accepts along path } w \\ \neg g_2(x) & : \text{if } z = \langle x, w \rangle, |w| = p_2(|x|), N_2(x) \text{ accepts along path } w \\ x & : \text{if } z = \langle x, w \rangle, |w| \neq p_i(|x|) \text{ for } i = 1, 2, |z| \geq 2^{|x|}, x \in \text{TAUT} \\ true & : \text{otherwise} \end{cases}$$

Clearly f_2 is also a proof system, since for every tautology y, $f_2(\langle y, 0^{2^{|y|}} \rangle) = y$. Also, we notice that each f_1-proof z is also an f_2-proof for the same tautology except for $z \in \{\langle x, w \rangle \mid |w| = p_2(|x|) \wedge |\langle x, w \rangle| \geq 2^{|x|} \wedge x \in \text{TAUT}\}$, which is a finite set. So, $f_1 \leq_p f_2$.

It remains to show $can(f_2) \equiv_m^{pp} (C, D)$. We only show $can(f_2) \leq_m^{pp} (C, D)$. The proof for $(C, D) \leq_m^{pp} can(f_2)$ is the same as that for $(A, B) \leq_m^{pp} can(f_1)$, for which we refer the reader to [8].

Let g many-one reduce (A, B) to (C, D). Choose elements $c \in C$ and $d \in D$. Define a reduction function h as follows.

```
1    input (y, 0ⁿ)
2    if n ≥ 2^|y|+1 then
3        if y ∈ SAT then output c else output d
4    endif
5    if g₁⁻¹(y) exists then output g(g₁⁻¹(y))
6    if g₂⁻¹(y) exists then output g₂⁻¹(y)
7    output c
```

Line 3 needs quadratic time in n. So h is polynomial-time computable.

Assume $(y, 0^n) \in$ SAT*. Then $y \in$ SAT. If we reach line 3, then we output $c \in C$. Otherwise we reach line 5. If $g_1^{-1}(y)$ exists (hence, $g_2^{-1}(y)$ does not exist, since the ranges of g_1 and g_2 are disjoint), then $g_1^{-1}(y) \in A$ and so, $g(g_1^{-1}(y)) \in C$. Otherwise we reach line 6. If $g_2^{-1}(y)$ exists, then $g_2^{-1}(y) \in C$ as $y \in$ SAT. So in all cases (output in line 5, 6 or 7), we output an element in C.

Assume $(y, 0^n) \in$ REF(f_2) (in particular $y \in$ UNSAT). So there exists z such that $|z| \leq n$ and $f(z) = \neg y$. If we reach line 3, then we output $d \in D$. Otherwise we reach line 5. So far we have $\neg y \neq true$ and $|z| \leq n < 2^{|y|+1}$. Therefore, $f(z) = \neg y$ must be due to line 1 or line 2 in the definition of f_2. It follows that either $g_1^{-1}(y)$ exists or $g_2^{-1}(y)$ exists (but not both). If $g_1^{-1}(y)$ exists, then $g_1^{-1}(y) \in B$ (by line 1 of f_2's definition) and we output $g(g_1^{-1}(y))$, which belongs to D. Otherwise, $g_2^{-1}(y)$ exists and we output $g_2^{-1}(y)$, which belongs to D as well (by line 2 of f_2's definition). This shows $can(f_2) \leq_m^{pp} (C, D)$ via h. $\square$

The proof system g constructed in Theorem 2 might seem "pathological", since tautologies from an easy subset of TAUT have proofs of super-polynomial length. One might wonder whether Theorem 2 can be proved without such pathology. The corresponding proof systems are formalized as follows.

Definition 5. *A proof system f is* well-behaved *if for every polynomial-time decidable $S \subseteq$ TAUT there exists a polynomial p such that for all $x \in S$,*

$$\min\{|w| \mid f(w) = x\} \leq p(|x|).$$

However, well-behaved proof systems probably do not exist. Meßner [11] shows that the existence of well-behaved proof systems implies the existence of optimal proof systems which we believe not to exist. So it is probably the case that no proof system is well-behaved and therefore, every proof system has long proofs on some polynomial-time decidable subset of TAUT. This shows that the proof system constructed in Theorem 2 is not uncommon. Even more, we can apply the arguments used in Theorem 2 to every non-well-behaved proof system.

Theorem 4. *Let f be a proof system that is not well-behaved. For every $(A, B) \in$ DisjNP, there exists a proof system g such that $can(g) \equiv_m^{pp} (A, B)$ and $g \not\leq_s f$.*

With help of Theorem 4 we can now give an answer to Q2: All non-well-behaved proof systems provide examples for proof systems that have equivalent canonical pairs, but that differ with respect to their strengths. Moreover, we can answer the second part of Q1 in the sense that all non-well-behaved proof systems provide counter examples for implication (2).

Corollary 1. *For every proof system f that is not well-behaved, there exists a proof system g such that $can(f) \equiv_m^{pp} can(g)$ and $f <_s g$. In particular,*

$$can(g) \leq_m^{pp} can(f) \not\Rightarrow g \leq_s f.$$

If we assume that optimal proof systems do not exist, then Corollary 1 provides even stronger answers: With regard to Q1, *all* proof systems provide counter examples for the implication (2). With regard to Q2, *all* proof systems provide examples that have equivalent canonical pairs, but that differ with respect to their strengths. Even more, each proof system is the origin of an infinite, strictly ascending chain of proof systems whose canonical pairs are equivalent.

Corollary 2. *The following statements are equivalent.*

1. *Optimal proof systems do not exist.*
2. *For every proof system f there exists a proof system g such that $can(f) \equiv_m^{pp} can(g)$ and $f <_s g$.*
3. *For every proof system f there exists an infinite chain of proof systems $g_0, g_1, \ldots$ such that $f <_s g_0 <_s g_1 <_s \cdots$ and $can(f) \equiv_m^{pp} can(g_0) \equiv_m^{pp} can(g_1) \cdots$.*
4. *For every proof system f there exists a proof system g such that*

$$can(g) \leq_m^{pp} can(f) \not\Rightarrow g \leq_s f.$$

4   Proof Systems with Equivalent Canonical Pairs

We have seen in the last section that the degree structure of canonical pairs does not necessarily reflect the simulation order of the corresponding proof systems. In this section we study the related question Q3.

We first show that equivalent canonical pairs do not tell much about the simulation order of two proof systems.

Theorem 5. *For every disjoint NP-pair (A, B), there exist proof systems f, g, and h such that*

- $can(f) \equiv_m^{pp} can(g) \equiv_m^{pp} can(h) \equiv_m^{pp} (A, B)$,
- $f <_s g$ *and* $f <_s h$,
- $g \not\leq_s h$ *and* $h \not\leq_s g$.

However, from another point of view, the proof systems defined in the proof of Theorem 5 are actually quite "similar" to each other. They differ only super-polynomially on an easy subset of TAUT. More precisely, the construction of proof systems with equivalent canonical pairs but arbitrary simulation order hinges on the following fact.

Proposition 3. *If proof systems f and g simulate each other except on a P-subset of TAUT, then $can(f) \equiv_m^{pp} can(g)$.*

So here the question is whether we can construct proof systems f and g with equivalent canonical pairs such that the proof systems are "very different". For example, do there exist proof systems f and g such that $can(f) \equiv_m^{pp} can(g)$ and f is almost everywhere super-polynomially stronger than g? The following theorem shows that such an extreme difference is only possible for proof systems whose canonical pairs are P-separable.

Theorem 6. *Let f and g be proof systems such that $can(g) \leq_m^{pp} can(f)$. If for almost all tautologies x and for every polynomial p, the length of the shortest f-proof of x is not bounded by p in the length of the shortest g-proof of x, then $can(f)$ and $can(g)$ are P-separable.*

Let us summarize what we have seen: Proposition 3 says that if two proof systems are "very similar", then they have equivalent canonical pairs. Theorem 6 tells us that if two proof systems are "very different" from each other, then either they have P-separable canonical pairs or their canonical pairs are inequivalent.

We continue to follow the question to what extent proof systems can differ, while still having equivalent canonical pairs. Under the hypothesis that P-inseparable disjoint NP-pairs exist, we show that Proposition 3 does not hold when the P-subset is replaced with an NP-subset (Corollary 3). So altering f-proofs on a P-subset of TAUT does not change the many-one degree of $can(f)$, but altering f-proofs on an NP-subset of TAUT can do so.

Theorem 7. *Let f be a proof system such that $can(f)$ is not $\leq_m^{pp}$-complete for DisjNP. Then there exists a proof system f' such that $can(f) <_m^{pp} can(f')$ and f and f' simulate each other except on an NP-subset of TAUT.*

Corollary 3. *The following statements are equivalent.*

1. *P-inseparable NP-pairs exist.*
2. *There exist proof systems f and g whose canonical pairs are not many-one equivalent, but that simulate each other except on an NP-subset of TAUT.*

Corollary 4. *If* $P \neq NP \cap coNP$, *then there exist proof systems* f *and* g *whose canonical pairs are not many-one equivalent, but that simulate each other except on an* NP-*subset of* TAUT.

Under the hypothesis that P-inseparable disjoint NP-pairs exist, we can show that proof systems whose difference cannot be "covered" by any P-subset of TAUT may still have equivalent canonical pairs. Hence, the converse of Proposition 3 does not hold, unless P-inseparable disjoint NP-pairs do not exist.

Theorem 8. *Let* (A,B) *be a* P-*inseparable* NP-*pair. Then there exist proof systems* f *and* f' *such that* $can(f) \equiv_m^{pp} can(f') \equiv_m^{pp} (A, B)$ *and for every* P-*subset* S *of* TAUT *it holds that* f *and* f' *do not simulate each other on* TAUT $- S$.

5   Strongly Many-One Degrees of Canonical Pairs

Every disjoint NP-pair is many-one equivalent to the canonical pair of some proof system [8]. We ask the same question for strongly many-one reductions. Note that if a disjoint NP-pair (A, B) is strongly many-one equivalent to the canonical pair of some proof system, then A must be NP-complete. So we arrive at question Q4 which is closely related to the following open problem [3].

Q5: Is the union of two disjoint NP-complete sets NP-complete?

For this, we first translate Q4 into the question whether certain NP-pairs are many-one hard for NP (Corollary 5). From this we show under the hypothesis NP $\neq$ coNP that if Q4 has an affirmative answer, then Q5 has an affirmative answer. It suffices to demand that Q4 has answer 'yes' only for $A = $ SAT.

Theorem 9. *Let* (A, B) *be a disjoint* NP-*pair. If* $(A, \overline{A \cup B})$ *is* $\leq_m^{pp}$-*hard for* NP, *then there exists a proof system* f *such that* $(SAT^*, REF(f)) \equiv_{sm}^{pp} (A, B)$.

Proposition 4. *Let* (A, B) *be a disjoint* NP-*pair such that* $A \cup B \neq \Sigma^*$. *If there exists a proof system* f *such that* $(SAT^*, REF(f)) \equiv_{sm}^{pp} (A, B)$, *then* $(A, \overline{A \cup B})$ *is* $\leq_m^{pp}$-*hard for* NP.

Corollary 5. *The following are equivalent for a disjoint* NP-*pair* (A, B) *where* $A \cup B \neq \Sigma^*$.

1. $(A, \overline{A \cup B})$ *is* $\leq_m^{pp}$-*hard for* NP.
2. *There exists a proof system* f *such that* $(SAT^*, REF(f)) \equiv_{sm}^{pp} (A, B)$.

Corollary 6. *Assume* NP $\neq$ coNP. *If for all disjoint* NP-*pairs* (SAT, B) *there exists a proof system* f *such that* $(SAT^*, REF(f)) \equiv_{sm}^{pp} (SAT, B)$, *then unions of disjoint* NP-*complete sets are* NP-*complete.*

6   Proof Systems and Turing-Degrees of Canonical Pairs

We consider the connection between proof systems and the more general Turing-degrees of their canonical pairs.

Proposition 5. *Let f and g be proof systems such that $can(f) \leq_T^{pp} can(g)$. Then there exists a proof system g' such that $can(g') \equiv_T^{pp} can(g)$ and $f \leq_p g'$.*

Corollary 7. *Let $\mathbf{d}_1 < \mathbf{d}_2$ be two Turing-degrees of disjoint NP-pairs. Then for every proof system f such that $can(f) \in \mathbf{d}_1$, there exists a proof system g such that $can(g) \in \mathbf{d}_2$ and $f <_s g$.*

Proposition 6. *For all $(A, B), (C, D) \in \mathrm{DisjNP}$ such that $(A, B) <_T^{pp} (C, D)$, there exist proof systems f and g such that $can(f) \equiv_m^{pp} (A, B)$, $can(g) \equiv_m^{pp} (C, D)$, $f \nleq_s g$, and $g \nleq_s f$.*

References

1. Beyersdorff, O.: Disjoint NP-pairs from propositional proof systems. In: Cai, J.-Y., Cooper, S.B., Li, A. (eds.) TAMC 2006. LNCS, vol. 3959, pp. 236–247. Springer, Heidelberg (2006)
2. Cook, S.A., Reckhow, R.A.: The relative efficiency of propositional proof systems. The Journal of Symbolic Logic 44(1), 36–50 (1979)
3. Glaßer, C., Pavan, A., Selman, A.L., Sengupta, S.: Properties of NP-complete sets. SIAM Journal on Computing 36(2), 516–542 (2006)
4. Grollmann, J., Selman, A.L.: Complexity measures for public-key cryptosystems. SIAM Journal on Computing 17(2), 309–335 (1988)
5. Glaßer, C., Selman, A.L., Sengupta, S., Zhang, L.: Disjoint NP-pairs. SIAM Journal on Computing 33(6), 1369–1416 (2004)
6. Glaßer, C., Selman, A.L., Travers, S., Wagner, K.W.: The complexity of unions of disjoint sets. In: Thomas, W., Weil, P. (eds.) STACS 2007. LNCS, vol. 4393, Springer, Heidelberg (2007)
7. Glaßer, C., Selman, A.L., Zhang, L.: Survey of disjoint NP-pairs and relations to propositional proof systems. In: Goldreich, O., Rosenberg, A.L., Selman, A.L. (eds.) Theoretical Computer Science. LNCS, vol. 3895, Springer, Heidelberg (2006)
8. Glaßer, C., Selman, A.L., Zhang, L.: Canonical disjoint NP-pairs of propositional proof systems. Theoretical Computer Science 370, 60–73 (2007)
9. Homer, S., Selman, A.L.: Oracles for structural properties: The isomorphism problem and public-key cryptography. Journal of Computer and System Sciences 44(2), 287–301 (1992)
10. Köbler, J., Messner, J., Torán, J.: Optimal proof systems imply complete sets for promise classes. Information and Computation 184(1), 71–92 (2003)
11. Meßner, J.: On the Simulation order of proof systems. PhD thesis, Universität Ulm, Abteilung Theoretische Informatik (December 2000)
12. Pudlák, P.: On reducibility and symmetry of disjoint NP-pairs. Theoretical Computer Science 295, 323–339 (2003)
13. Razborov, A.A.: On provably disjoint NP-pairs. Technical Report TR94-006, Electronic Computational Complexity Colloquium (1994)

On the Representations of NC and Log-Space Real Numbers*

Fuxiang Yu

Computer Science Department
Stony Brook University
Stony Brook, New York, U.S.A.
fuxiang@cs.sunysb.edu

Abstract. We study the representations of **NC** and Log-space real numbers in this paper. We show that the classes of the **NC** and Log-space real numbers under the general left cut representation are among the most expressive representations. [1] On the other hand, although the general left cut representation and the Cauchy function representation have the same expressive power in **P**, the expressive power of the Cauchy function representation is weaker than that of the general left cut representation in **NC** if $\mathbf{P}_1 \neq \mathbf{NC}_1$. In addition, although the expressive power of the standard left cut representation is weaker than that of the Cauchy function representation in **P**, the expressive powers of these two representations are incomparable in **NC** if $\mathbf{P}_1 \neq \mathbf{NC}_1$. Similar results hold in Log-space.

Keywords: Complexity, representations of real numbers, Cauchy function, left cut, **P**, **NC**, Log-space, expressive power.

1 Introduction

The computability and complexity of real numbers have been widely studied, since real numbers are the main objects of continuous computation. The first issue is how to represent a real number, and the second issue is how to compute the representations of a real number effectively or efficiently.

For a given real number $x \in [0, 1]$, there are many representations. To name a few, x can be represented by (1) a Cauchy sequence $\{x_n\}$ of rational numbers such that for all n, $|x_n - x| < 2^{-n}$; (2)the standard left cut, which is the set of rational numbers that are no more than x; and (3) the binary expansion $x = (0.b_1 b_2 \cdots)_2$. These representations and many others are, mathematically, equivalent to each other.

---

* This material is based upon work supported by National Science Foundation under grant No. 0430124.
[1] We say a class A of languages is more expressive than another class B of languages, if $B \subsetneq A$. Now we also say A has more expressive power. For example, **EXP** has more expressive power than **P**.

However, when computability and complexity are considered, it becomes more complicated and interesting. For example, a recent work of Chen et al. [3] showed that the primitive recursive versions of these representations can lead to different notions of primitive recursive real numbers.

When studying the complexity of real numbers, *dyadic numbers* $0.b_1 b_2 \cdots b_n$ (i.e., finite binary fractions) are the basic computational objects. Now a *Cauchy function representation* of a real number $x \in [0, 1]$ is a function ϕ from natural numbers to dyadic numbers such that for all n, $\phi(n)$ has n bits and $|\phi(n) - x| <$ 2^{-n}. A number x is *Cauchy polynomial-time computable* if a Cauchy function representation ϕ of x is polynomial-time computable. Similarly, we can define the *standard left cut representation* of x and the *binary expansion representation* of x using dyadic numbers as the basic objects. Ko [7] has shown that the Cauchy function representation is more expressive, unconditionally, than the standard left cut representation and the binary expansion as long as polynomial-time computability is concerned. On the other hand, if for every Cauchy function representation ϕ of a real number x, we define a set $LC_\phi = \{d \leq \phi(n) : n \in$ $\mathbb{N}, d$ is a dyadic number of n bits.$\}$, called the *general left cut* of x associated with ϕ, then the general left cut representation is as expressive as the Cauchy function representation in **P**. There are also some other representations, for example, Ko [6] presented some interesting results on the continuous fraction representation.

In this paper, we study the parallel-time complexity issues of real numbers. More precisely, we investigate which representation is the most expressive in parallel-time complexity classes such as **NC** and **L** (short for *Log-Space*). We consider four representations: standard left cut, general left cut, binary expansion, and the Cauchy function representation. As expected, the general left cut representation is among the most expressive representations; however, we still obtain some interesting results, which are summarized below.

(1) The general left cut representation is the most expressive representation in **NC** and **L**.
(2) There exists a real number x that has a **NC** (or **L**) computable Cauchy function representation but the standard left cut of x is not **NC** (or respectively, **L**) computable.
(3) If $\mathbf{P}_1 \neq \mathbf{NC}_1$ (or $\mathbf{P}_1 \neq \mathbf{L}_1$), then there exists a real number x whose standard left cut is **NC** (or respectively, **L**) computable, but no Cauchy function representation of x is **NC** (or respectively, **L**) computable.

In other words, If $\mathbf{P}_1 \neq \mathbf{NC}_1$ (or $\mathbf{P}_1 \neq \mathbf{L}_1$), then the expressive powers of the Cauchy function representation and the standard left cut representation are incomparable under **NC** (or respectively, **L**). As it seems that the Cauchy function representation has been used as one main representation in parallel computation of real numbers (see, e.g., Ko [7], Hoover [5], and Yu [8]), this result should not be ignored.

2   Notations

2.1   Representations

This paper involves notions used in both discrete computation and continuous computation. The basic computational objects in discrete computation are integers and strings in $\{0,1\}^*$. The length of a string w is denoted $\ell(w)$.

The basic computational objects in continuous computation are dyadic rationals $\mathbb{D} = \{m/2^n : m \in \mathbb{Z}, n \in \mathbb{N}\}$. Each dyadic rational d has infinitely many binary representations with arbitrarily many trailing zeros. For each such representation s, we write $\ell(s)$ to denote its length. If the specific representation of a dyadic rational d is understood (often the shortest binary representation), then we write $\ell(d)$ to denote the length of this representation. We let $\mathbb{D}_n$ denote the class of dyadic rationals with at most n bits in the fractional part of its binary representation.

We use $\mathbb{R}$ to denoted the set of real numbers. We say a function $\phi : \mathbb{N} \to \mathbb{D}$ is a *Cauchy function representation* of a real number x, if (i) for all $n \geq 0$, $\phi(n) \in \mathbb{D}_n$, and (ii) for all $n \geq 0$, $|\phi(n) - x| \leq 2^{-n}$. For any $x \in \mathbb{R}$, there is a unique function $b_x : \mathbb{N} \to \mathbb{D}$ that binary converges to x and satisfies the condition $x - 2^{-n} < b_x(n) \leq x$ for all $n \geq 0$. We call this function b_x the *standard Cauchy function* for x. The binary expansion of x is represented by b_x. For any Cauchy function representation ϕ of x, there exists a general left cut representation $LC_\phi = \{d \in \mathbb{D}_n : n \in \mathbb{N}, d \leq \phi(n)\}$ associated with ϕ, and LC_{b_x}, also denoted LC_x, is called the standard left cut.

2.2   Complexity Classes

In this paper we consider mainly the circuit complexity class **NC** as well as complexity classes defined based on Turing machines listed as follows (see, e.g., Du and Ko [4]).

P : the class of sets accepted by deterministic polynomial-time (Turing) machines.

L : the class of sets accepted by deterministic Turing machines restricted to use an amount of memory logarithmic in the size of the input.

Recall that $\mathbf{NC} = \cup_{i \geq 0} \mathbf{NC}^i$, where $\mathbf{NC}^i$ is the class of languages $A \subset \{0,1\}^*$ such that there exists a circuit family $\{C_n\}$ with the following properties (see, e.g., Du and Ko [4]).

1. there exists a Turing machine M that constructs (the encoding of) each C_n in log-space.
2. for all n, C_n has n input nodes with each node fan-in 2 and fan-out 1 and accepts $A_n = A \cap \{0,1\}^n$.
3. there exist a polynomial function p and a constant $k > 0$, such that for all n, the size $size(C_n)$ of C_n is no more than $p(n)$ and the depth $depth(C_n)$ of C_n is no more than $k \log^i n$.

We call $\{C_n\}$ an $\mathbf{NC}^i$ circuit family.

These complexity classes have properties such as $\mathbf{NC}^1 \subseteq \mathbf{L} \subseteq \mathbf{NC} \subseteq \mathbf{P}$. The interested readers are referred to Du and Ko [4].

NC functions can be viewed as an extension of **NC** languages. Namely, now for $i \geq 0$ and an $\mathbf{NC}^i$ circuit family $\{C_n\}$, each node in C_n is allowed to have multiple fan-outs, and a function $f : \{0,1\}^* \to \{0,1\}^*$ is computable by $\{C_n\}$ if for each n and each $x \in \{0,1\}^n$, C_n computes $f(x)$. In other words, f is $\mathbf{NC}^i$ computable if the language $A_f = \{\langle x, i \rangle : \text{the } i\text{-th bit of } f(x) \text{ is } 1\}$ is in $\mathbf{NC}^i$. For more details, see, for example, Allender et al. [2].

If $\mathcal{C}$ is a complexity class of sets, we use $\mathcal{C}_1$ to denote the complexity class $\{A \in \mathcal{C} : A \subseteq \{0\}^*\}$, called the unary version of $\mathcal{C}$. Similarly, if FC is a class of functions, FC_1 is the corresponding class of functions whose domain is $\{0\}^*$. As shown in Ko [7], the unary classes are closely related to complexity classes of real numbers, because Cauchy functions are unary functions.

We define **NC** computable real numbers under different representations:

$\mathbf{NC}_{SLC}$: the class of real numbers whose standard left cuts are **NC** computable.

$\mathbf{NC}_{BE}$: the class of real numbers whose standard Cauchy functions are **NC** computable.

$\mathbf{NC}_{GLC}$: the class of real numbers x such that there exists a general left cut of x that is **NC** computable.

$\mathbf{NC}_{CF}$: the class of real numbers x such that there exists a Cauchy function of x that is **NC** computable.

Similarly for **L**, we define $\mathbf{L}_{SLC}$, $\mathbf{L}_{BE}$, $\mathbf{L}_{GLC}$, $\mathbf{L}_{CF}$, and for **P**, we define $\mathbf{P}_{SLC}$, $\mathbf{P}_{BE}$, $\mathbf{P}_{GLC}$, $\mathbf{P}_{CF}$.

3   Main Results

In this section, we present the main results. The first part contains results that do not rely on assumptions that $\mathbf{P}_1 \neq \mathbf{L}_1$ and $\mathbf{P}_1 \neq \mathbf{NC}_1$, and the second part contains results that rely on these assumptions. The main result is that, if $\mathbf{P}_1 \neq \mathbf{L}_1$ (or $\mathbf{P}_1 \neq \mathbf{NC}_1$), then the four classes of Log-space (or respectively, **NC**) computable real numbers under four different representations are all distinct classes (see Figure 1), while $\mathbf{L}_{GLC}$ is the most powerful one.

These results are more complicated than the complexity of these representations in **P**, as shown in Ko [7]:

$$\mathbf{P}_{SLC} = \mathbf{P}_{BE} \subsetneq \mathbf{P}_{GLC} = \mathbf{P}_{CF}.$$

3.1   *Absolute* Results

We first present some *absolute* results, which do not rely on assumptions on the complexity classes such as $\mathbf{P}_1 \neq \mathbf{L}_1$. Note that we can obtain trivial results $\mathbf{L}_{BE} \subseteq \mathbf{NC}_{BE} \subseteq \mathbf{P}_{BE}$ and so on, since $\mathbf{L} \subseteq \mathbf{NC} \subseteq \mathbf{P}$. We omit such results.

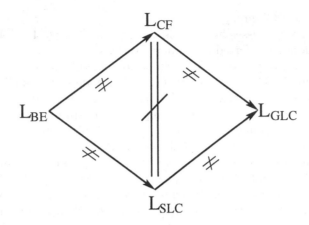

Fig. 1. Assuming $\mathbf{P}_1 \neq \mathbf{L}_1$

Theorem 3.1. $\mathbf{NC}_{BE} \subseteq \mathbf{NC}_{CF}$, $\mathbf{L}_{BE} \subseteq \mathbf{L}_{CF}$.

Proof. This is obvious since the standard Cauchy function b_x of a real number x is a Cauchy function of x. □

Theorem 3.2. $\mathbf{NC}_{BE} \subseteq \mathbf{NC}_{SLC}$, $\mathbf{L}_{BE} \subseteq \mathbf{L}_{SLC}$.

Proof. For any number $x \in \mathbf{NC}_{BE}$, we have that b_x is in $\mathbf{NC}$. If $x \in \mathbb{D}$, then $x \in \mathbf{NC}_{SLC}$. Assume that $x \notin \mathbb{D}$. For any $n \in \mathbb{N}$ and $d \in \mathbb{D}_n$, it is in $\mathbf{NC}$ to compare $b_x(n)$ and d. Because $x \notin \mathbb{D}$, $d \leq b_x(n)$ implies that $d < x$, and $d > b_x(n)$ implies that $d > x$. This completes the proof for $\mathbf{NC}_{BE} \subseteq \mathbf{NC}_{SLC}$. Similarly, we can prove that $\mathbf{L}_{BE} \subseteq \mathbf{L}_{SLC}$. □

The following theorem extends Theorem 2.8 of Ko [7].

Theorem 3.3. *There exists a real number* $x \in \mathbf{L}_{CF} - \mathbf{P}_{SLC}$. *Thus,* $\mathbf{L}_{CF} - \mathbf{L}_{SLC} \neq \emptyset$ *and* $\mathbf{NC}_{CF} - \mathbf{NC}_{SLC} \neq \emptyset$.

Proof. We first define a function $T(n)$ inductively: $T(1) = 1$, and $T(n+1) = 2^{2^{2^{2^{T(n)}}}}$. By a simple diagonalization, we can find a set $A \subseteq \{0\}^*$ such that A is computable in space $T(n)$ but not in space $\log T(n)$ (see, for example, Aho et al. [1]). Without loss of generality, let $0 \in A$. Define

$$x = \sum_{i=1}^{\infty} (2 \cdot \chi_A(0^i) - 1) 2^{-2^{T(i)}}.$$

First we show that $x \in \mathbf{L}_{CF}$. We can compute, for each n, a dyadic rational $\phi(n)$ as follows:

(1) find the integer k such that $T(k) \leq \log n < T(k+1)$.
(2) compute $\phi(n) = \sum_{i=1}^{k} (2 \cdot \chi_A(0^i) - 1) 2^{-2^{T(i)}}$.

It is clear that the value $\phi(n)$ computed above differs from x by at most $2^{-(2^{T(k+1)}-1)} \leq 2^{-n}$. So the above procedure computes a function $\phi \in CF_x$. Furthermore, since $T(k) \leq \log n$, the computation of $\phi(n)$ can be done in space $O(\log n)$. Thus, $x \in \mathbf{L}_{CF}$.

Next we show that $x \notin \mathbf{P}_{SLC}$. Assume, by way of contradiction, that SLC_x is computable in polynomial time. We will find a Turing machine M computing $\chi_A(0^n)$ in space $\log T(n)$.

Let M_a be a Turing machine that computes χ_A in space $T(n)$. The new machine for $\chi_A(0^n)$ works as follows. First, it simulates M_A on inputs $0, 0^2, \cdots,$ 0^{n-1} and computes $d = \sum_{i=1}^{n-1}(2\chi_A(0^i)-1)2^{-2^{T(i)}}$. Then it determines whether $d \in SLC_x$ and concludes that $\chi_A(0^n) = 1$ iff $d \in SLC_x$.

It is clear that the above machine M indeed computes χ_A. Assume that LC_x is computable in time $p(n)$ and hence in space $2^{p(n)}$. Then the space usage of the machine M is bounded by $O(\sum_{i=1}^{n-1}T(i)) + 2^{p(2^{T(n-1)})} < 2^{2^{2^{T(n-1)}}} = \log T(n)$ for almost all n, which contradicts to the fact that A is not computable in space $\log T(n)$.

In the above we have constructed a number $x \in \mathbf{L}_{CF} - \mathbf{P}_{SLC}$. As $\mathbf{L}_{CF} \subseteq \mathbf{NC}_{CF}$ and $\mathbf{L}_{SLC} \subseteq \mathbf{NC}_{SLC} \subseteq \mathbf{P}_{SLC}$, we have $\mathbf{L}_{CF} - \mathbf{L}_{SLC} \neq \emptyset$ and $\mathbf{NC}_{CF} - \mathbf{NC}_{SLC} \neq \emptyset$. $\square$

Next we consider general left cuts.

Theorem 3.4. $\mathbf{NC}_{CF} \subseteq \mathbf{NC}_{GLC}$, $\mathbf{L}_{CF} \subseteq \mathbf{L}_{GLC}$.

Proof. For a number $x \in \mathbf{NC}_{CF}$, let $\phi \in CF_x$ be in **NC**. Note that the general left cut L_ϕ of x associated with ϕ is $\{d \in \mathbb{D}_n : n \in \mathbb{N}, d \leq \phi(n)\}$. Similar to the proof of Theorem 3.2, it is in **NC** to compare a dyadic number d and $\phi(n)$. $\square$

3.2 Results Under Assumptions $\mathbf{P}_1 \neq \mathbf{L}_1$ and $\mathbf{P}_1 \neq \mathbf{NC}_1$

We use a specific coding system for the instantaneous descriptions (IDs) of the computation of a time-bounded Turing machine so that we can discuss the simulation of the machine (see Ko [7]). We assume that our machine M works on two tape symbols: 0 and 1 (and a special blank symbol), has k states $q_1, \cdots,$ q_k, uses a single tape, and has a time bound ψ. For each string $s \in \{0,1\}^*$ of length $\ell(s) = n$, each ID of the computation of $M(s)$ is encoded by a string in $\{0,1\}^*$ of length $2\psi(n) + 2k + 4$: we encode type symbols 0 and 1 by 01 and 10, respectively, and the blank symbol by 00, and the state symbol q_i by $11(01)^i(10)^{k-i}11$, where the state symbol appears just to the left of the tape symbol that is currently scanned by the tape head. (To see this in another way, at any moment, there are at most $\psi(n)$ symbols (of 0, 1 and the blank) in the computation, whose length is at most $2\psi(n)$ in our encoding, and the state is of length $2k + 4$.) Thus, for any input s of length n, the computation of $M(s)$ is encoded by a $(\psi(n) + 1) \cdot (2\psi(n) + 2k + 4)$-bit string $\alpha_0\alpha_1 \cdots \alpha_{\psi(n)}$, where α_i is the code of the i-th ID in the computation of $M(s)$. Note that we fix the length of the codes for IDs to be exactly $2\psi(n) + 2k + 4$, and fix the number of IDs in the computation to be exactly $\psi(n) + 1$.

We say a function $\phi : \mathbb{N} \to \mathbb{N}$ is *log-space computable* if it is log-space computable when both inputs and outputs are written in the binary form. Note that if ϕ is log-space computable, then it is also log-space computable when both inputs and outputs are written in the unary form.

We use the following lemma (Lemma 4.15 of Ko [7]).

Lemma 3.5. *Let M be a Turing machine having k states and having a time bound ψ, which is log-space computable. Then for any input st of length $\ell(s) = \ell(t) = \psi(n) + 2k + 4$ such that s is an ID of $M(u)$ for some string u of length n, we can determine, in log space, whether t is the successor of s in the computation of $M(u)$, or t is less than the successor ID, or t is larger than the successor ID. Furthermore, we can compute, in log space, the maximum m such that the first m bits of t agree with those of the successor ID of s.*

Theorem 3.6. *If $\mathbf{P}_1 \neq \mathbf{L}_1$, then $\mathbf{L}_{SLC} - \mathbf{L}_{CF} \neq \emptyset$.*

Proof. Let $T \in \mathbf{P}_1 - \mathbf{L}_1$ be computed by a deterministic Turing machine M in time $p(n)$. Assume that p is log-space computable. We use the encoding system described above. Assume that M has k states and so on input 0^n, an ID of $M(0^n)$ is of length $2\psi(n) + 2k + 4$. We let $s_{n,i}$, $0 \leq i \leq p(n)$, be the i-th ID of the computation of $M(0^n)$. Define a real number x whose binary expansion is

$$x = 0.01\tau(s_{1,0})\tau(s_{1,1}) \cdots \tau(s_{1,p(1)})\tau(s_{2,0}) \cdots \tau(s_{2,p(2)}) \cdots,$$

where τ is the local translation function defined by $\tau(0) = 01$, $\tau(1) = 10$, and $\tau(ab) = \tau(a)\tau(b)$ for all $a, b \in \{0,1\}^+$.

We first show that $x \in \mathbf{P}_{CF} - \mathbf{L}_{CF}$. Let $q(n) = 2p(n) + 2k + 4$ and $r(n) = \sum_{i=1}^{n} 2q(i) \cdot (p(i)+1)$. From $T \in \mathbf{P}_1$, it is easy to see that $x \in \mathbf{P}_{CF}$. Furthermore, note that for each 0^n, we need only $\tau(s_{n,p(n)})$, or, from the $(r(n)-2q(n)+3)$-rd bit to the $r(n) + 2$ bit of the binary expansion of x, to determine whether $0^n \in T$. Therefore, if $x \in \mathbf{L}_{CF}$, then we can compute, in log space, an approximation value d to x such that $|d - x| \leq 2^{r(n)+4}$, and by our coding system, the first $r(n)+2$ bits of d must be identical to those of x and so we can determine whether $0^n \in T$ in log space, which contradicts to the assumption of $T \in \mathbf{P}_1 - \mathbf{L}_1$.

Next we show that $x \in \mathbf{L}_{SLC}$, that is, $SLC_x = \{d \in \mathbb{D} : d < x\}$ is in $\mathbf{L}$. We only need to consider dyadic numbers of even lengths, because for a dyadic number d of an odd length, we can add a trailing zero to d. For a dyadic number $d \in \mathbb{D}_{2n}$, we can tell whether $d < x$ or $d > x$ in log space (note that since $x \notin \mathbf{L}_{CF}$, $x \notin \mathbb{D}$ and $d \neq x$). We achieve this by generating the initial IDs $s_{i,0}$ and applying Lemma 3.5 successively to each substring uv of d, where u is already been verified to be equal to $\tau(s_{i,j})$ for some j and $\ell(v) = \ell(u)$; furthermore, we check whether $v > \tau(w)$, $v = \tau(w)$ or $v < \tau(w)$, where w is the next ID of u. If $v > \tau(w)$, $d > x$; if $v < \tau(w)$, $d < x$; if $v = \tau(w)$, we continue with the process. If all bits of d agree with the first $2n$ bits of x, $d < x$. $\square$

Corollary 3.7. *If $\mathbf{P}_1 \neq \mathbf{NC}_1$, then $\mathbf{NC}_{SLC} - \mathbf{NC}_{CF} \neq \emptyset$.*

Proof. The proof is the same as that of Theorem 3.6, except that we let $T \in \mathbf{P}_1 - \mathbf{NC}_1$ and the constructed number $x \in \mathbf{L}_{SLC} \subseteq \mathbf{NC}_{SLC}$. $\square$

Corollary 3.8. *If* $\mathbf{P}_1 \neq \mathbf{L}_1$, *then* $\mathbf{L}_{BE} \subsetneq \mathbf{L}_{SLC}$.

Proof. Because we have $\mathbf{L}_{BE} \subseteq \mathbf{L}_{CF}$, $\mathbf{L}_{BE} \subseteq \mathbf{L}_{SLC}$ and $\mathbf{L}_{SLC} - \mathbf{L}_{CF} \neq \emptyset$.   □

Corollary 3.9. (a) *If* $\mathbf{P}_1 \neq \mathbf{L}_1$, *then* $\mathbf{L}_{CF}$ *and* $\mathbf{L}_{SLC}$ *are incomparable.*
  (b) *If* $\mathbf{P}_1 \neq \mathbf{NC}_1$, *then* $\mathbf{NC}_{CF}$ *and* $\mathbf{NC}_{SLC}$ *are incomparable.*

Proof. Statement (a) follows Theorems 3.6 and 3.3. Statement (b) follows Corollary 3.7 and Theorem 3.3.   □

Corollary 3.10. (a) *If* $\mathbf{P}_1 \neq \mathbf{L}_1$, *then* $\mathbf{L}_{CF} \subsetneq \mathbf{L}_{GLC}$.
  (b) *If* $\mathbf{P}_1 \neq \mathbf{NC}_1$, *then* $\mathbf{NC}_{CF} \subsetneq \mathbf{NC}_{GLC}$.

In summary, if $\mathbf{P}_1 \neq \mathbf{L}_1$, then $\mathbf{L}_{SLC}$, $\mathbf{L}_{BE}$, $\mathbf{L}_{GLC}$ and $\mathbf{L}_{CF}$ are four distinct classes (see Figure 1). If $\mathbf{P}_1 \neq \mathbf{NC}_1$, then $\mathbf{NC}_{SLC}$, $\mathbf{NC}_{BE}$, $\mathbf{NC}_{GLC}$ and $\mathbf{NC}_{CF}$ are four distinct classes. The converse also holds, since Ko [7] has shown that $\mathbf{P}_{CF} = \mathbf{L}_{CF} \Leftrightarrow \mathbf{P}_1 = \mathbf{L}_1$ and $\mathbf{P}_{CF} = \mathbf{NC}_{CF} \Leftrightarrow \mathbf{P}_1 = \mathbf{NC}_1$. We state it as a theorem.

Theorem 3.11. *The following are equivalent:*

(a) $\mathbf{P}_1 \neq \mathbf{L}_1$.
(b) $\mathbf{L}_{SLC}$, $\mathbf{L}_{BE}$, $\mathbf{L}_{GLC}$ *and* $\mathbf{L}_{CF}$ *are four distinct classes.*

  Similar results hold for **NC** *real numbers.*

The following corollary is an *absolute* result.

Corollary 3.12. $\mathbf{L}_{CF} \neq \mathbf{L}_{SLC}$, $\mathbf{NC}_{CF} \neq \mathbf{NC}_{SLC}$.

Proof. If $\mathbf{L}_{CF} = \mathbf{L}_{SLC}$, then from Theorem 3.11, $\mathbf{P}_1 = \mathbf{L}_1$, and from Ko's results, $\mathbf{L}_{CF} = \mathbf{P}_{CF}$, which implies $\mathbf{L}_{SLC} = \mathbf{P}_{CF}$ and furthermore $\mathbf{P}_{SLC} = \mathbf{P}_{CF}$. However, from Theorem 2.8 of Ko [7], $\mathbf{P}_{SLC} \subsetneq \mathbf{P}_{CF}$. Therefore, $\mathbf{L}_{CF} \neq \mathbf{L}_{SLC}$. We can prove in a similar way that $\mathbf{NC}_{CF} \neq \mathbf{NC}_{SLC}$.   □

4 Conclusion

In this paper we have shown that the four representations of real numbers, standard left cut, general left cut, binary expansion and Cauchy function representation, have distinct expressive powers in **L** and **NC**, unless $\mathbf{L}_1 = \mathbf{P}_1$ or $\mathbf{NC}_1 = \mathbf{P}_1$, respectively. As the expressive powers of these four representations only fall into two categories in **P**, our results show that the expressive powers of these representations are more complicated in the parallel-time complexity world unless some sequential-time complexity class collapses to some parallel-time complexity class. Note that whether $\mathbf{L} = \mathbf{P}$ and whether $\mathbf{NC} = \mathbf{P}$ are important open questions, maybe just second to the **P** versus **NP** question, in the complexity theory.

References

1. Aho, A.V., Hopcroft, J.E., Ullman, J.D.: The design and analysis of computer algorithms. Addison-Wesley, Reading (1974)
2. Allender, E., Loui, M.C., Regan, K.W.: Other Complexity Classes and Measures. Chapter 29. In: Atallah, M.J. (ed.) Algorithms and Theory of Computation Handbook, CRC Press, Boca Raton (1999)
3. Chen, Q., Su, K., Zheng, X.: Primitive recursiveness of real numbers under different representation. In: CCA. Electronic Notes in Theoretical Computer Science, vol. 167, Elsevier, Amsterdam (2007)
4. Du, D.-Z., Ko, K.-I.: Theory of Computational Complexity. John Wiley & Sons, Chichester (2000)
5. Hoover, H.J.: Feasible real functions and arithmetic circuits. SIAM J. Comput. 19(1), 182–204 (1990)
6. Ko, K.-I.: On the continued fraction representation of computable real numbers. Theor. Comput. Sci. 47(3), 299–313 (1986)
7. Ko, K.-I.: Complexity Theory of Real Functions. Birkhäuser, Boston (1991)
8. Yu, F.: On some complexity issues of NC analytic functions. In TAMC. In: Cai, J.-Y., Cooper, S.B., Li, A. (eds.) TAMC 2006. LNCS, vol. 3959, pp. 375–386. Springer, Heidelberg (2006)

Bounded Computable Enumerability and Hierarchy of Computably Enumerable Reals[*]

Xizhong Zheng

[1] Department of Computer Science, Jiangsu University, Zhenjiang 212013, China
[2] Theoretische Informatik, BTU Cottbus, D-03044 Cottbus, Germany
zheng@informatik.tu-cottbus.de

Abstract. The computable enumerability (c.e., for short) is one of the most important notion in computability theory and is regarded as the first weakening of the computability. In this paper, we explore further possible weakening of computable enumerability. By restricting numbers of possible big jumps in an increasing computable sequence of rational numbers which converges to a c.e. real number we introduce the notion of h-bounded c.e. reals and then shown that it leads naturally to an Ershov-style hierarchy of c.e. reals. However, the similar idea does not work for c.e. sets. We show that there is a computability gap between computable reals and the reals of c.e. binary expansions.

Keywords: c.e. sets, c.e. reals, bounded c.e. reals, Ershov's Hierarchy.

1 Introduction

A set A is *computably enumerable* (*c.e.*, for short) if it can be enumerated by an effective procedure. That is, the elements of A can be enumerated effectively one after another, if A is not empty. The computable enumerability is one of the most important notion in the computability theory. This is not only because a lot of problems in mathematics and computer science correspond to the c.e. sets, but also the c.e. sets are regarded as the first natural weakening of the computable sets.

The definition of c.e. sets is generalized by Putnam [6], Gold [5] and Ershov [4] to k-c.e. for any constants k and h-c.e. for any functions h. The main idea behind these definitions is to allow the *mind-changes* or, as Putnam called, *trial and error* in the effective enumerations. For example, in an effective procedure to enumerate a set A, some number n may be enumerated into A at a stage, be removed from A at a later stage, and possibly be put into A afterwards again, and so on. If the number of such kind of *mind-changes* is bounded by a constant k for each n, then A is called k-c.e. A is called h-c.e. for a function h if the number of mind-changes about n is bounded by $h(n)$ for all n. An h-c.e. set is also called ω-c.e. if h is a computable function. Ershov [4] shows that, if $f(n) < g(n)$ for

---

[*] This work is supported by DFG (446 CHV 113/240/0-1) and NSFC (10420130638).

infinitely many n, then there is a g-c.e. set which is not f-c.e. This leads to a nice *Ershov's hierarchy* of the Δ_2^0-sets:

$$\mathbf{EC} = 0\text{-}\mathbf{CE} \subseteq k\text{-}\mathbf{CE} \subsetneqq (k+1)\text{-}\mathbf{CE} \subsetneqq \omega\text{-}\mathbf{CE} \subsetneqq \Delta_2^0,$$

for all constant k, where $\mathbf{EC}$ (for effectively computable), k-$\mathbf{CE}$ and ω-$\mathbf{CE}$ denote the classes of computable, k-c.e. and ω-c.e. sets, respectively. Especially, the class 1-$\mathbf{CE}$ is just the class of c.e. sets. The 2-c.e. sets are usually called *d-c.e.* because they are the differences of c.e. sets. This hierarchy is even valid for the corresponding classes of Turing degrees as shown by Cooper [1]. Namely, there is a $(k+1)$-c.e. degree which is not k-c.e.; There is an ω-c.e. degree which is not k-c.e. for any constant k, and there is a Δ_2^0-degree which is not ω-c.e.

The Ershov's hierarchy classifies Δ_2^0-sets nicely to different classes according to the different levels of enumerability. However, from the computability point of view, it has also some unnatural properties. For example, the classes k-$\mathbf{CE}$ for $k > 0$ are not symmetrical. That is, a set and its complement do not necessarily belong to the same class, although they have exactly the same computability. The classes of Ershov's hierarchy, as binary expansions, do not correspond to the classes of reals of weaker computability defined by computable sequences of rational numbers (see e.g., [13,15]). These problems disappear if we consider the *h-bounded computability* (*h*-b.c.) introduced in section 2. In this case, we are only interested in how many mind-changes are necessary to approximate a set and do not demand an empty set to begin with.

To explore the notion weaker than c.e. we have to look at other properties which are "stronger" than simply counting the changes of membership of n in an approximation to the set A. A natural candidate here is considering the changes of the initial segments $A_s \upharpoonright n$ instead of $A_s(n)$ in an approximation (A_s) to A. Accordingly we define the notion of *h-initially-bounded computable* (*h*-i.b.c., for short) sets. Unfortunately, the classes of *h*-i.b.c. sets collapse to the set of computable sets for any constant functions h. Therefore, *h*-i.b.c. sets do not lead to an Ershov hierarchy of c.e. sets and we do not achieve a reasonable notion weaker than c.e. by this approach.

As a new approach we consider the computable enumerability of reals. By definition c.e. reals are limits of computable increasing sequences of rational numbers. By restricting the numbers of possible big jumps in an increasing sequence of rational numbers converges to a c.e. real we introduce the notion of *h-bounded c.e.* (*h*-b.c.e.) reals. We can see that the k-b.c.e. reals form an Ershov-style hierarchy of c.e. reals for constants k. This hierarchy is valid even in the sense of Turing degrees. A very interesting result here is that non-computable k-b.c.e. reals do not have c.e. binary expansions. This means that there are reals which have some weak computability between the computable reals and strongly c.e. reals (the reals of c.e. binary expansions).

The paper is organized as follows. Section 2 introduces the new notions of *h*-bounded computable and *h*-initially bounded computable sets and discusses their basic properties. In section 3 we consider *h*-bounded c.e. real numbers and show an Ershov hierarchy of k-b.c.e. reals. In the last section 4 we extend the

Ershov hierarchy of classes of k-b.c.e. reals to the classes of Turing degrees which contain k-b.c.e. real numbers.

2   Bounded Computable Sets

This section discusses the bounded computability of sets. Firstly the notion of h-c.e. sets is symmetrized to the notion of h-bounded computable sets. Then we introduce the notion of initially-bounded computable sets to explore the possible hierarchy of c.e. sets. We will see that an Ershov style hierarchy for initially bounded computable sets fails.

By definition, a set A is c.e. (computably enumerable) means that there is a computable sequence (A_s) of finite sets which converges to A such that $A_0 = \emptyset$ and $A_s \subseteq A_{s+1}$ for all s. This has been generalized by Putnam [6], Gold [5], and Ershov [4] to the following: A is called h-*computably enumerable (h-c.e.)* if A has an h-*enumeration* (A_s) which is a computable sequence of finite sets converging to A such that $A = \emptyset$ and

$$(\forall n)(|\{s : A_s(n) \neq A_{s+1}(n)\}| \leq h(n)). \tag{1}$$

We identify in this paper a set with its characteristic function. Thus we have $A(n) = 1$ if $n \in A$ and $A(n) = 0$ if $n \notin A$. Alternatively, a set A is h-c.e. iff there is a computable function f such that $f(0, n) = 0$, $\lim_s f(s, n) = A(n)$ and

$$(\forall n) \left(|\{s : f(s, n) \neq f(s+1, n)\}| \leq h(n)\right). \tag{2}$$

Obviously, c.e. sets are just the h-c.e. sets for the constant function $h(n) \equiv 1$. It is worth noting that the condition $f(0, n) = 0$ in the second definition of c.e. sets is necessary because otherwise the 1-c.e. sets were not necessarily c.e. This corresponds to requiring $A_0 = \emptyset$ in the first definition which is however superfluous because A_0 is a finite set and the h-c.e. is invariable under the change on a finite set. Here the computable sequence (A_s) is regarded as an enumeration instead of an approximation of A. If we are more interested in how effectively a set can be approximated, we should consider the following variation.

Definition 1. *A set A is called h-bounded computable (h-b.c., for short) if there is a computable sequence (A_s) of finite sets which converges to A such that*

$$(\forall n)(|\{s \geq n : A_{s+1}(n) \neq A_s(n)\}| \leq h(n)). \tag{3}$$

The additional condition $s \geq n$ in (3) ignores inessential changes of $A_s(n)$ before the stage $s := n$. This can be replaced equivalently by $s \geq g(n)$ for an increasing computable function g. We summarize some important properties of h-b.c. sets without proof as follows.

Theorem 1. *Let h be a function and let A be a set.*

1. *If A is h-c.e. or co-h-c.e., then A is h-b.c.*
2. *If A is h-b.c., the A is h'-c.e. for $h'(n) := h(n) + 1$.*

3. *If A is h-b.c., then so is the complement $\overline{A}$.*
4. *A is h-b.c. iff there is a computable function f such that $\lim_s f(s,n) = A(n)$ and $|\{s : f(s,n) \neq f(s+1,n)\}| \leq h(n)$ for all n.*
5. *A is h-b.c. iff there is a computable function g and a computable sequence (A_s) of finite sets such that $(\forall n)(|\{s \geq g(n) : A_{s+1}(n) \neq A_s(n)\}| \leq h(n))$.*

Analogous to the well known fact that a set A is computable if and only if A as well as its complement $\overline{A}$ are c.e., we have the following result.

Theorem 2. *For any functions h and $h_1(n) := h(n) + 1$, a set A is h-b.c. if and only if both A and its complement $\overline{A}$ are h_1-c.e.*

Proof. We need only to prove the direction "$\Leftarrow$". Suppose that (A_s) and (B_s) are h_1-enumerations of A and $\overline{A}$, respectively. Assume w.l.o.g. that $A_0 = B_0 = \emptyset$. Since $\lim_s A_s(n) = A(n) \neq \overline{A}(n) = \lim_s B_s(n)$ for all n, we can define a total computable increasing function $v : \mathbb{N} \to \mathbb{N}$ by

$$\begin{cases} v(0) := \min\{s : A_s(0) \neq B_s(0)\}; \\ v(n+1) := \min\{s > v(n) : A_s(n+1) \neq B_s(n+1)\}. \end{cases}$$

Then we define a computable function f by

$$f(s,n) := \begin{cases} A_{v(n)}(n) & \text{if } A_{v(n)}(n) = 1 \ \& \ s \leq v(n); \\ A_s(n) & \text{if } A_{v(n)}(n) = 1 \ \& \ s > v(n); \\ 1 \dotminus B_{v(n)}(n) & \text{if } A_{v(n)}(n) = 0 \ \& \ s \leq v(n); \\ 1 \dotminus B_s(n) & \text{if } A_{v(n)}(n) = 0 \ \& \ s > v(n). \end{cases}$$

Obviously, we have $\lim_s f(s,n) = A(n)$ for all n. Given an n, suppose that $A_{v(n)}(n) = 1$. If $s \geq v(n)$ such that $f(s,n) \neq f(s+1,n)$, then $A_s(n) \neq A_{s+1}(n)$. Since $A_{v(n)}(n) = 1 \neq 0 = A_0(n)$ and (A_s) is an h_1-enumeration, there are at most $h_1(n) - 1$ such stages $s \geq v(n)$. The same holds if $A_{v(n)}(n) = 0$. This implies that

$$(\forall n)(|\{s \geq v(n) : f(s,n) \neq f(s+1,n)\}| \leq h(n)).$$

Because $f(s,n) = f(s+1,n)$ hold for all $s < v(n)$, we have actually

$$(\forall n)(|\{s : f(s,n) \neq f(s+1,n)\}| \leq h(n)),$$

and hence, by Theorem 1.4, A is h-b.c. $\qquad\square$

From Theorem 1 we have seen that h-bounded computability is not very far from the h-computably enumerability. The notion introduced in the next definition is essentially different from h-computable enumerability. Here we count the changes of the initial segment $A_s \upharpoonright n$ instead of the changes of the single membership $A_s(n)$ for different indices s.

Definition 2. *Let $h : \mathbb{N} \to \mathbb{N}$ be a function and let $A \subseteq \mathbb{N}$ be a set.*

1. *A sequence (A_s) of finite sets converges h-initially-bounded effectively to A if $\lim_{s \to \infty} A_s = A$ and*

$$(\forall n)\,(|\{s \geq n : A_s \upharpoonright n \neq A_{s+1} \upharpoonright n\}| \leq h(n))\,. \tag{4}$$

2. *A set A is called h-initially-bounded computable (h-i.b.c., for short) if there is a computable sequence of finite sets which converges h-initially-bounded effectively to A.*

For constant function $h \equiv k$, we call h-i.b.c. sets simply k-i.b.c. Thus A is computable iff it is 0-i.b.c. Analogous to h-b.c. sets, the condition $s \geq n$ in (4) can be replaced by $s \geq g(n)$ for an increasing computable function g.

The following proposition can be proved straightforwardly.

Proposition 1. *Let $h : \mathbb{N} \to \mathbb{N}$ be a function and let A be a set.*

1. *If A is h-c.e., then it is g-i.b.c. for the function $g(n) := \sum_{i<n} h(i)$;*
2. *If A is h-i.b.c., then it is g-c.e. for the function $g(n) := h(n+1) + 1$;*
3. *If A is h-i.b.c., then so is its complement $\overline{A}$.*
4. *If A and B are h-i.b.c. and g-i.b.c., respectively, then $A \cap B$, $A \cup B$ and $A \backslash B$ are $(g + h)$-i.b.c.*

Given a constant k, any k-c.e. set is h-i.b.c. for $h(n) := kn$ by Proposition 1.1. Actually, even for the function g whose distance to the function $h(n) := kn$ is bounded by a constant, any k-c.e. set is also g-i.b.c. as shown in the next theorem. This condition is even necessary.

Theorem 3. *Let k be a constant.*

1. *If A is k-c.e., then it is h_c-i.b.c. for any constant c, where $h_c(n) := kn \doteq c$.*
2. *If h is a nondecreasing computable function such that*

$$(\forall c)(\exists n)(h(n) + c < kn), \tag{5}$$

then there is a k-c.e. set which is not h-i.b.c.

Proof. 1. Let (A_s) be a k-enumeration of A. That is, (A_s) is a computable sequence of finite sets satisfying (1) for $h(n) \equiv k$ such that $A_0 = \emptyset$, $\lim_s A_s = A$. Let N be a natural number such that $A_s \upharpoonright c = A \upharpoonright c$ hold for all $s \geq N$. Define a computable sequence (B_s) of finite sets by $B_s := A_{N+s}$. The initial segment $B_s \upharpoonright n$ can change only at the positions $m \in [c, n)$ and at most $k(n \doteq c) \leq h_c(n)$ times. Therefore (B_s) converges h_c-initial-bounded effectively to A and hence A is h_c-i.b.c.

2. Let h be a nondecreasing computable function which satisfies condition (5). Therefore, we can define an increasing computable sequence (v_s) inductively as follows:

$$\begin{cases} v_0 := 0 \\ v_{n+1} := (\mu t)(kt > kv_n + h(t)). \end{cases}$$

Let $I_n := [v_n; v_{n+1})$. Then the (I_s) is a computable sequence of disjoint intervals of natural numbers such that $k \cdot l(I_n) > h(v_{n+1})$ for all n, where $l(I_n)$ is the length of the interval I_n.

Let (V_e) be a computable enumeration of all approximable sets and $(V_{e,s})_s$ be a computable sequence of finite sets which approximates the set V_e. Now we can construct a k-c.e. set A which is not h-i.b.c., i.e., A satisfies, for all e, the following requirements:

$$R_e \quad : \quad (V_{e,s}) \text{ converges } h\text{-initially-bounded effectively to } V_e \implies A \neq V_e.$$

The set A is constructed in stages such that, for any e, there is an $n_e \in I_e$ which witnesses the inequality $A(n_e) \neq V_e(n_e)$. The witness n_e can be chosen in the following way. We reserve the interval I_e exclusively for the requirement R_e. Suppose that $A \cap I_e = \emptyset$ and let $n_e := v_e$ at the stage $s := v_{e+1}$. We put n_e into A if it is not yet in V_e. If n_e enters V_e, then delete n_e from A and it can enter A again if n_e leaves V_e at a later stage. In order to guarantee that A is k-c.e., we do such kind of actions at most k times for the same n_e. After that, if $V_e(n_e)$ changes again, then we define $n_e := n_e + 1$ and let this new witness be outside of A or put it into A at the beginning depending on if n_e is already in V_e or not. Repeat the above procedure again for the new n_e, and so on. We continue this process until $n_e = v_{e+1}$. If this process stops for some $n_e < v_{e+1}$, then n_e is a right witness for R_e. Otherwise, if n_e arrives v_{e+1}, then this means that the initial segment $V_{e,s} \upharpoonright v_{e+1}$ changes at least $k \cdot l(I_e) > h(v_{e+1})$ times and the sequence $(V_{e,s})$ does not converge h-initially bounded effectively. In this case, the requirement R_e is satisfied trivially. More precisely, the set A can be constructed by a priority methods without injury.   □

As an immediately corollary of Proposition 1, any c.e. set is h_c-i.b.c. for the function $h_c(n) := n \dotminus c$. However, if $h \in o(\text{id})$, then there is a c.e. set which is not h-i.b.c., where id is the identity function. This implies especially that there is a c.e. set which is not k-i.b.c. for any constant k. Actually, the next theorem shows that any non-computable c.e. sets are not k-i.b.c. for any constant k.

Theorem 4. *Let k be a constant. Any k-i.b.c. set is computable.*

Proof. Let A be a k-i.b.c. set and let (A_s) be a computable sequence of finite sets which converges h-initially-bounded effectively to A. Suppose w.l.o.g. that k is the least constant such that

$$|\{s \geq n : A_s \upharpoonright n \neq A_{s+1} \upharpoonright n\}| = k$$

for infinitely many n. Thus, for any $m \in \mathbb{N}$, we can always effectively find an $n > m$ and an index $s_0 > n$ such that $A_s \upharpoonright n$ changes already k times between the stages n and s_0. Therefore, $A_s(m)$ does not change any more after stage s_0 and we have $A(m) = A_{s_0}(m)$. This implies that A is computable.   □

Theorem 4 implies that the classes of k-i.b.c. sets collapse to the the first level—the class of computable sets. Therefore the k-i.b.c. sets do not lead to an Ershov-style hierarchy of c.e. sets.

3   Bounded C.E. Real Numbers

In this section we turn to the computability of real numbers. Motivated from the notions of h-i.b.c. sets and h-effectively computable real numbers of [12] we introduce the notion of h-bounded c.e. reals and show that this leads naturally to an Ershov-style hierarchy of c.e. reals.

A real number x is called *computably enumerable* (c.e., for short) if there is an increasing computable sequence (x_s) of rational numbers which converges to x. Equivalently x is c.e. iff its (left) Dedekind cut $L_x := \{r \in \mathbb{Q} : r < x\}$ is a c.e. set of rational numbers. As it is pointed out in Soare [8,9], the c.e. real numbers correspond naturally to c.e. sets and can be regarded as the first weakening of computable real numbers (see [3,11,14]). We explore now the weakening of c.e. reals by bounding the number of big jumps in the approaching sequences.

Definition 3. *Let h be a function.*

1. *An increasing sequence (x_s) converges to x h-effectively if, for all n, the length of the index-chain $n = s_0 < s_1 < s_2 < \cdots < s_k$ which satisfies*

$$(\forall i < k)\left(x_{s_{i+1}} - x_{s_i} > 2^{-n}\right)$$

   is bounded by $k \leq h(n)$.
2. *A real number x is called h-bounded c.e. (h-b.c.e., for short) if there is an increasing computable sequence (x_s) of rational numbers which converges to x h-effectively.*

The class of all h-b.c.e. real numbers is denoted by h-**BCE**. Especially, if $h(n) \equiv k$ is a constant function, then k-**BCE** denotes the class of all k-b.c.e. real numbers. We call a real *bounded computably enumerable* (b.c.e., for short) if it is k-b.c.e. for a constant k. The class of all b.c.e. real numbers is denoted by **BCE**. The next lemma follows immediately from the definition.

Lemma 1.   *1. A real number x is computable if and only if it is 0-b.c.e.*
2. Any c.e. real number x is h-b.c.e. for the function $h(n) := 2^n$.

Now we show that k-b.c.e. reals form an Ershov-style hierarchy.

Theorem 5. *For any constant k, there is a $(k+1)$-b.c.e. real number which is not k-b.c.e.*

Proof. For any constant k, we are going to construct an increasing computable sequence (x_s) of rational numbers which converges $(k+1)$-effectively to a non-k-b.c.e. real number x. To this end, the limit x has to satisfy all the following requirements

R_e : $(\varphi_e(s))_s$ is increasing and converges k-effectively to $y_e \implies x \neq y_e$

where (φ_e) is a computable enumeration of all partial computable functions $\varphi_e : \subseteq \mathbb{N} \to \mathbb{Q}$.

The sequence (x_s) is constructed in stages. To satisfy a single requirement R_e, we choose a rational interval I^{e-1} of the length $2^{-n_{e-1}}$ for some natural number n_{e-1} and divide it equidistantly into at least $2k+5$ subintervals $I_0, I_1, \cdots, I_{m_e}$ of the same length 2^{-n_e} for some natural number n_e. Our goal is to find a *witness interval* I^e for R_e from $I_0, I_1, \cdots, I_{m_e}$ such that any element of I^e satisfy the requirement R_e. As default, let $I^e := I_1$ be the first candidate of I^e and define x_s to be the middle point of I_1 at the beginning. If at some stage s_0 we find that there is a $t_0 \geq n_e$ such that $\varphi_e(t_0) \in I_1$, then $I^e := I_3$ is the new candidate of the witness interval and redefine x_{s_0+1} to be the middle point of the interval I_3. And if at a later stage $s_1 > s_0$, we find a $t_1 > t_0$ such that $\varphi_e(t_1) \in I_3$, then change x_{s_1+1} to the middle point of $I^e := I_5$ which is the new current candidate of witness interval and so on. Obviously, at most $k+1$ redefinitions guarantee that the limit $x := \lim_{s\to\infty} x_s$ is different from the possible limit $\lim_{s\to\infty} \varphi_e(s)$ if it converges k-effectively and we achieve finally a correct witness interval I^e. On the other hand, the sequence (x_s) converges $(k+1)$-effectively.

In order to satisfy all requirements simultaneously, a finite injury priority construction suffices. Here we mention only two important points. Firstly, the candidates of witness interval for R_e are chosen from the subintervals of the current candidate of witness interval for R_{e-1}. Secondly, the default candidate of witness interval for R_e (the interval I_0 explained above) should be chosen in such a way that the current x_s which is defined from the current candidate interval for R_{e-1} should be exactly its middle point. This avoids the extra jumps of the sequence (x_s) and guarantees that the constructed sequence converges $(k+1)$-effectively indeed.   □

It is natural to ask whether b.c.e. reals exhaust all c.e. real numbers. We will give a negative answer to this question. Actually we achieve a more stronger result. Recall that a real x is called *strongly c.e.* (see [2]) if it has a c.e. binary expansion A, i.e., $x = x_A := \sum_{i \in A} 2^{-(i+1)}$. As pointed out by Jockusch (see [8]) the class of strongly c.e. real numbers form a proper subset of c.e. real numbers. The next theorem implies that there is a c.e. real number which is not b.e.c.

Theorem 6. *If a real number is both b.c.e. and strongly c.e., then it must be computable.*

Proof. Let x be a b.c.e. real number for some constant k with a c.e. binary expansion A (i.e., $x = x_A$), and suppose that x is irrational. Then we have a strictly increasing computable sequence (x_s) of rational numbers which converges to x k-effectively for some constant k as well as a computable sequence (A_s) of finite sets such that $A_s \subset A_{s+1}$ and $A = \lim_{s\to\infty} A_s$. Since both sequences (x_s) and (x_{A_s}) are strictly increasing, we can define inductively two computable increasing functions $u, v : \mathbb{N} \to \mathbb{N}$ as follows.

$$\begin{cases} u(0) := 0 \\ v(0) := \min\{s : x_{A_s} > x_0\} \\ u(n+1) := \min\{s > u(n) : x_s > x_{A_{v(n)}}\} \\ v(n+1) := \min\{s > v(n) : x_{A_s} > x_{u(n)}\}. \end{cases}$$

Let $y_s := x_{u(s)}$ and $B_s := A_{v(s)}$ for all s. Then (y_s) and (x_{B_s}) are computable subsequences of (x_s) and (x_{A_s}), respectively which satisfy the following

$$y_0 < x_{B_0} < y_1 < x_{B_1} < y_2 < x_{B_2} < \cdots$$

As a subsequence of (x_s), the sequence (y_s) converges k-effectively to x too. That is, for any n, the length of any index-chain $s_0 < s_1 < s_2 < \cdots < s_t$ such that $y_{s_{i+1}} - y_{s_i} > 2^{-n}$ for all $i < t$ is bounded by $t \leq k$.

Given a natural number n. If at a stage $s + 1 > n$ some element less than n is enumerated into B_{s+1}, i.e. $B_s \upharpoonright n \neq B_{s+1} \upharpoonright n$, then we have $x_{B_{s+1}} - x_{B_s} > 2^{-n}$, and hence $y_{s+2} - y_s > 2^{-n}$. This means that the initial segment $B_s \upharpoonright n$ can be changed after the stage n at most $2k$ times. That is, we have

$$|\{s > n : B_s \upharpoonright n \neq B_{s+1} \upharpoonright n\}| \leq 2k, \tag{6}$$

for all natural numbers n. Therefore, the computable sequence (B_s) converges $2k$-initially bounded effectively to A and the set A is $2k$-i.b.c. By theorem 4, A is computable and hence the real x is computable. $\square$

Although the proof of Theorem 6 is very simple, the result seems quite surprising. As we have mentioned before, the computable enumerability of a set is the first weakening of the computability. Therefore, the strongly computable enumerability of a real number can be regarded as the first weakening of the computability of real numbers with respective to binary expansion. However, Theorem 6 shows that there is a gap between computability and strong computability of real numbers if we consider how effectively a real number can be approximated by computable sequence of rational numbers.

4   Hierarchy of Turing Degrees

In this section we discuss the hierarchy Turing degrees which contain k-b.c.e. real numbers for different constant k. This strengthens the hierarchy of Theorem 5. To simplify the notation we identify a real number x with its characteristic binary sequence in this section.

Let (Φ_e) be an effective enumeration of computable partial functionals. By definition, a real number x is *Turing reducible* to another real number y (denoted by $x \leq_T y$) if there is an index e such that $x(n) = \Phi_e^y(n)$ for all n. Two real numbers x and y are *Turing equivalent* (notation $x \equiv_T y$) if $x \leq_T y$ and $y \leq_T x$ hold. In other words, x is Turing equivalent to y if there are indices i and j such that $x(n) = \Phi_i^y(n)$ and $y(n) = \Phi_j^x(n)$ for all n. From these it is not difficult to find (possibly different) indices i and j such that

$$(\forall n) \left(x \upharpoonright n = \Phi_i^y(n) \ \& \ y \upharpoonright n = \Phi_j^x(n) \right). \tag{7}$$

We say that x is (i, j)-*Turing equivalent* to y (denote by $x \equiv_T^{(i,j)} y$) if they satisfy condition (7). For the (i, j)-Turing equivalence relates more closely to the topological property of real numbers. The following important technical lemma will be used later.

Lemma 2 (Rettinger and Zheng [7]). *For any rational interval I_0 and any natural numbers i, j, t there are two open rational intervals $I \subseteq I_0$ and J such that*

$$(\forall x, y) \left(x \equiv_T^{(i,j)} y \implies (x \in I \implies y \in J) \ \& \ (y \in J \implies x \in I_0) \right). \tag{8}$$

We say that an intervals I is (i,j)-*reducible* to another interval J (denoted by $I \preceq^{(i,j)} J$) if they satisfy the following condition

$$(\forall x, y) \left(x \in I \ \& \ x \equiv_T^{(i,j)} y \implies y \in J \right).$$

By Lemma 2, there are $I \subseteq I_0$ and J such that $I \preceq^{(i,j)} J$ for any given interval I_0. If all elements of I_0 are not (i,j)-Turing equivalent to some element, then this holds trivially. Actually, the Lemma 2 holds even in an more effective sense. Namely, if there exists $x \in I_0$ which is (i,j)-Turing equivalent to some y, then the intervals I and J which satisfy condition (7) can be effectively found (see Lemma 2.2 of [7] for details). This fact will be used in the proof of the following theorem.

Theorem 7. *For any constant k, there is a $(k+1)$-b.c.e. real number which is not Turing equivalent to any k-b.c.e. real numbers.*

Proof. We construct an increasing computable sequence (x_s) of rational numbers which converges $(k+1)$-effectively to a non-k-b.c.e. real number x. The limit x has to satisfy, for all i, j, k, all the following requirements

$$R_{\langle i,j,k \rangle} : \quad (\varphi_k(s)) \text{ converges } k\text{-effectively to } y_k \implies x \not\equiv_T^{(i,j)} y_k.$$

We explain firstly the idea how a single requirement R_e for $e := \langle i, j, k \rangle$ can be satisfied. Choose arbitrarily a rational interval I_{e-1}. We want to find a rational subinterval $I_e \subseteq I_{e-1}$ such that all $x \in I_e$ satisfy R_e. This interval I_e is called a *witness interval* of R_e.

Divide the interval I_{e-1} into at least $2k+5$ rational subintervals $I^0, I^1, I^2, \ldots$ with the same length 2^{-n_e} for some natural number n_e. According to (the effective version of) Lemma 2, either we can find an interval I_{2t+1} (for some $t \leq k$) whose elements do not (i,j)-Turing equivalent to any real number, or we can find rational subintervals $I_e^t \subseteq I^t$ and rational intervals J^t such that $I_e^t \preceq^{(i,j)} J^t$ for $t = 1, 3, \cdots, 2k+3$. W.l.o.g. we can assume that the distances between any pair of J-intervals is larger than 2^{-m_e} for some natural number m_e.

The sequence (x_s) can be constructed in stages. At the beginning we choose I_e^1 as default candidate of the witness interval of R_e and let x_s be its middle point. We do not change x_s as long as the sequence $(\varphi_k(t))$ does not enters the interval J^1 after the stage $s := m_e$. Otherwise, if at some stage s_0 we find a $t_0 > m_e$ such that $\varphi_k(t_0) \in J^1$, then change the candidate of witness interval of R_e to the interval I_e^3 and redefine x_{s_0+1} as the middle point of I_e^3. If at a late stage $s_1 > s_0$, we find a new $t_1 > t_0$ such that $\varphi_k(t_1) \in J^3$, then change the candidate interval to I_e^5 and redefine x_{s_1+1} as its middle point, and so on. We allow at most $k + 1$ times such kind of redefinition which suffices to guarantee that the limit of the sequence (x_s) is different to the possible limit of $\lim_{s\to\infty} \varphi_k(s)$ if it converges k-effectively.

We need only a standard finite injury priority construction to construct the computable sequence (x_s) and x_s is chosen from the actually smallest witness intervals defined at the stage s. The details are omitted here. $\qquad\square$

References

1. Cooper, B.S.: Degrees of Unsolvability. Ph.D thesis, Leicester University, Leicester, England (1971)
2. Downey, R.G.: Some computability-theoretic aspects of reals and randomness. In: The Notre Dame lectures. Assoc. Symbol. Logic. Lect. Notes Log., vol. 18, pp. 97–147. Urbana, IL (2005)
3. Downey, R.G., Hirschfeldt, D.R.: Algorithmic Randomness and Complexity. Springer, Heidelberg, Monograph to be published
4. Ershov, Y.L.: A certain hierarchy of sets. i, ii, iii. (Russian). Algebra i Logika. 7(1), 47–73 (1968), 7(4), 15–47 (1968), 9, 34–51 (1970)
5. Gold, E.M.: Limiting recursion. J. Symbolic Logic 30, 28–48 (1965)
6. Putnam, H.: Trial and error predicates and the solution to a problem of Mostowski. J. Symbolic Logic 30, 49–57 (1965)
7. Rettinger, R., Zheng, X.: A hierarchy of Turing degrees of divergence bounded computable real numbers. J. Complexity 22(6), 818–826 (2006)
8. Soare, R.I.: Cohesive sets and recursively enumerable Dedekind cuts. Pacific J. Math. 31, 215–231 (1969)
9. Soare, R.I.: Recursion theory and Dedekind cuts. Trans. Amer. Math. Soc. 140, 271–294 (1969)
10. Soare, R.I.: Recursively enumerable sets and degrees. A study of computable functions and computably generated sets. Perspectives in Mathematical Logic. Springer, Heidelberg (1987)
11. Weihrauch, K.: Computable Analysis, An Introduction. Springer, Heidelberg (2000)
12. Zheng, X.: Classification of the computably approximable real numbers. Theory of Computing Systems (to appear)
13. Zheng, X.: Recursive approximability of real numbers. Mathematical Logic Quarterly 48(Suppl. 1), 131–156 (2002)
14. Zheng, X.: Computability Theory of Real Numbers. Habilitation's thesis, BTU Cottbus, Germany (February 2005)
15. Zheng, X., Rettinger, R.: Weak computability and representation of reals. Mathematical Logic Quarterly 50(4/5), 431–442 (2004)

Streaming Algorithms Measured in Terms of the Computed Quantity*

Shengyu Zhang

California Institute of Technology
Computer Science Department and Institute for Quantum Information,
1200 E California Bl, MC 107-81, Pasadena, CA 91125, USA
shengyu@caltech.edu

Abstract. The last decade witnessed the extensive studies of algorithms for data streams. In this model, the input is given as a sequence of items passing only once or a few times, and we are required to compute (often approximately) some statistical quantity using a small amount of space. While many lower bounds on the space complexity have been proved for various tasks, almost all of them were done by reducing the problems to the cases where the desired statistical quantity is at one extreme end. For example, the lower bound of triangle-approximating was showed by reducing the problem to distinguishing between graphs without triangle and graphs with only one triangle.

However, data in many practical applications are not in the extreme, and/or usually we are interested in computing the statistical quantity only if it is in some range (and otherwise reporting "too large" or "too small"). This paper takes this practical relaxation into account by putting the computed quantity itself into the measure of space complexity. It turns out that all three possible types of dependence of the space complexity on the computed quantity exist: as the quantity goes from one end to the other, the space complexity can goes from max to min, remains at max, or goes to somewhere between.

1 Introduction

Data stream is a very natural and important model for massive data sets in many applications, where the input data is given as a stream of items with only one or a few passes, and usually we want to determine or approximate some statistical quantity of the input stream. See [16] for an excellent and comprehensive survey.

Many algorithms are designed and many lower bounds on the space complexities are proven for various types of problems such as, just to name a few, frequency moments [1, 7, 14, 3, 6], vector distance [11, 17], and some graph problems [2, 4, 9, 13]. Almost all lower bounds were proved by reducing the problem to the cases where the desired statistical quantity is at an extreme end (and this

---

* This work was mostly done when the author was a graduate student in Computer Science Department at Princeton University, supported in part by NSF grants CCR-0310466 and CCF-0426582.

is further reduced to the communication complexity of some related problems). For example, the lower bound for approximating the number of triangles was proved by a reduction to distinguishing between the graph containing 0 and 1 triangle; the lower bound for the infinity frequency moment F_∞^* was proved by reducing the problem to distinguishing between $F_\infty^* = 1$ and $F_\infty^* = 2$.

Despite its theoretical correctness, this reduction to extreme cases can be misleading for many practical applications for at least the following two reasons. First, the extreme case may not happen at all in practice. For example, the number of triangles in a graph has many implications in various applications. In a social network, the number of triangles characterizes the average strength of the ties in the community [8, 18]. But note that in most (if not all) practical communities, there are a large number of triangles, and those extreme cases (0 and 1 triangle) are never the case. As another example, in many applications such as data mining, the number of common neighbors of two vertices in a graph shows the amount of common interest. A canonical example is that if two commodities have a large number of common buyers, then putting these two commodities close to each other in a supermarket will make more sales for both of them. Similar to the triangle example, the maximal number of common neighbors from data in practice is always large.

The second reason is from the user side. Even if some data happen to have the quantity at extreme, we are not interested in it in this case. For example, if two commodities have a very small number (such as one) of common buyers, then it barely means anything because that buyer may just happen to buy them. Therefore, we have a *threshold range* in mind within which we care about the quantity; if the quantity is outside the range, we will be satisfied if the algorithm can report "too low" or "too high".

Due to these two reasons, it is natural to ask the following question: is the hardness of a problem essentially due to the extreme cases? To answer this question, we study the space complexity in terms of both input size and the threshold range. In particular, for any input size n and any possible quantity value $q(n)$, the stream space complexity $s(n, q(n))$ is, roughly speaking, the minimal space used to compute $f(x)$ for all inputs in $\{x : f(x) = \Theta(q(n))\}$.

This question has been occasionally studied implicitly. In [2], Bar-Yosseff, Kumar, and Sivakumar initialized the study of graph problems in the adjacency stream model, where the graph is given by a sequence of edges (i, j) in an arbitrary order[1]. In particular they studied the problem of approximating the number of triangles, giving a one-pass algorithm using $O(\frac{1}{\epsilon^2} \log \frac{1}{\delta}(1 + \frac{T_1 + T_2}{T_3})^3 \log n)$ space, where T_i is the number of unordered triples containing i edges. Unfortunately, they could not show when it is better than the naive sampling algorithm (which uses $O(\frac{1}{\epsilon^2} \log \frac{1}{\delta}(1 + \frac{T_0 + T_1 + T_2}{T_3}))$) space), and they asked this as an open problem. They also gave an $\Omega(n^2)$ lower bound for general graph, by reducing the problem at one extreme end (distinguishing between graphs with no triangle *vs.*

---

[1] They also proposed the incidence stream model, where each item in the stream is a vertex with all its neighbors. In this paper we only study the adjacency stream model.

with one triangle) to the communication complexity of some Boolean function. They then ask as another open problem for a lower bound in terms of T_1, T_2, T_3 [2]. Jowhari and Ghodsi [13] later proved a lower bound of $\Omega(n/T_3)$. In this paper we will show that their algorithm is always asymptotically worse than the naive sampling algorithm (for any graph) by proving $\left(1 + \frac{T_1+T_2}{T_3}\right)^3 \geq \Omega\left(1 + \frac{T_0+T_1+T_2}{T_3}\right)$ using algebraic graph theoretical arguments. Also, we prove a lower bound of $\min\{\Omega(n^3/T_3), \Omega(n^2)\}$, which matches the naive sampling algorithms, and the proof is much simpler than the previous (weaker) one [2]. It should be noted that subsequent papers [13, 5] improve the upper bound and finally [5] achieve $O(\frac{1}{\epsilon^2} \log \frac{1}{\delta} \cdot (1 + \frac{T_1+T_2}{T_3}))$. So our lower bound does not mean that the naive sampling is always the best, but that it is the best if the algorithm aims at dealing with all the graphs with T_3 in the known range.

For the problem of computing the maximal number of common neighbors, previously Buchsbaum, Giancarlo and Westbrook [4] gave a lower bound of $\Omega(n^{3/2}\sqrt{c})$ to compute the exact value, where c is the max number of the common neighbors. In this paper, after observing a matching upper bound, we consider the approximation version of the problem, showing that approximating the number needs $\tilde{\Theta}(n^{3/2}/\sqrt{c})$ space. Compared to the triangle counting example where the space complexity $\Theta(\min\{n^3/T_3, n^2\})$ drops from the maximum possible value ($\Theta(n^2)$) to the minimum possible value (constant), in this common neighbor example, the space complexity drops from some large value $\Theta(n^{3/2})$ to some small value $\Theta(n)$. Note that the $\tilde{\Theta}(n)$ space capacity is a well-studied model (called the semi-stream model) for graph problems, which is interesting [9] partly because $\tilde{\Theta}(n)$ is affordable in some Internet applications but higher space is not.

Not surprisingly, there are also many other problems whose space complexity, though first proved by considering extreme inputs, remains hard even if the computed quantity is not at extreme. We will give a simple example in this category too.

2   Preliminaries and Definitions

We say that an algorithm $\mathcal{A}$ (ϵ, δ)-approximates the function f on input x if $\mathbf{Pr}[|\mathcal{A}(x) - f(x)| \leq \epsilon f(x)] \geq 1 - \delta$. In this paper, we will think of ϵ and δ as small constants. A graph $G = (V, E)$ is given in the adjacency streaming model if the input is a sequence of edges $(i, j) \in E$ in an arbitrary order.

2.1   Formulation of the Notion

The most naive way to formulate the notion in Section 1 is to define the space complexity $s(n, q)$ to be the minimum space needed to compute the quantity

---

[2] A lower bound in terms of all T_1, T_2 and T_3 does not seem quite justified: after all, for an unknown given graph, we do not know what T_i's are. But a lower bound in terms of mere T_3 is well-justified for the two reasons mentioned earlier.

$f(x)$ for all those x satisfying $f(x) = q$. However, this is obviously a useless definition because if we already know that $f(x) = q$ then we do not need any computation. Thus we need to be a little more careful about the definition.

Definition 1. *A function f has stream space complexity $\Theta(s(n, q(n)))$ if for any constants $c_2 > c_1 > 0$, the best algorithm (ϵ, δ)-approximating f on any input in $\{x : c_1 q(n) \le f(x) \le c_2 q(n)\}$ uses space $\Theta(s(n, q(n)))$.*

Several comments are in order. First, as mentioned in Section 1, we may desire that for those inputs that are not in the range, the algorithm outputs "too high" or "too low". Actually, in most (if not all) cases, the algorithm working for Definition 1 can be easily modified (with a multiplicative constant factor cost added) such that for any constants d_1 and d_2 with $c_1 < d_1 < d_2 < c_2$ and $d_2 - d_1 \ge 2\epsilon$, the algorithm has the following additional property: it outputs "too low" if $f(x) < d_1 q(n)$ and "too high" if $f(x) > d_2 q(n)$; for those inputs x with $c_1 q(n) \le f(x) \le d_1 f(x)$ or $d_2 q(n) \le f(x) \le c_2 f(x)$, either an ϵ-approximation or a "too low/high" is considered correct. Second, distinguishing between $f(x) \le (1-\epsilon)cq$ and $f(x) \ge (1+\epsilon)cq$ with success probability $1 - \delta$ is clearly a relaxation of the above task for any constant c (since we can let $d_1 = (1 - \epsilon)c$ and $d_2 = (1 + \epsilon)c$). The lower bounds showed in this paper apply to this easier task.

A basic fact that will be used in the proofs is as follows. The problem Index is a streaming problem where the input is an n-bit string x followed by an index $i \in [n]$, and the task is to output x_i with success probability at least $1 - \delta$.

Fact 1. *The Index problem needs $(1 - 2\delta)n$ bits of memory.*

A generalization of the fact is to consider k bits instead of just one bit. In the problem k-Index, the input is an n-bit string x followed by k indices $i_1, ..., i_k \in [n]$. The task is to distinguish between "$x_i = 1, \forall i = i_1, ..., i_k$" and "$x_i = 0, \forall i = i_1, ..., i_k$" with success probability at least $1 - \delta$.

Fact 2. *The k-Index problem needs $(1 - 2\delta)n/k$ bits of memory.*

This is easy to see by repeating each bit in Fact 1 k times. Also, a simple random sampling argument shows an $O(\frac{n}{k} \log \frac{1}{2\delta})$ upper bound for the number of memory cells.

3 Three Types of Dependence of the Space Complexity on the Computed Quantity

In this section, we will show three types of dependence of space complexity $s(n, q(n))$ on $q(n)$. In Section 3.1, we show a dependence which is the strongest possible: as $q(n)$ goes from one end (constant) to the other ($\Theta(n^3)$), the space complexity $s(n, q(n))$ drops from the maximal possible value ($\Theta(n^2)$) to the minimal possible value (constant). In Section 3.2, we show a weaker dependence: $s(n, q(n))$ drops from $\Theta(n^{3/2})$ to $\Theta(n)$. In Section 3.3, we show one example in which $s(n, q(n))$ is independent of $q(n)$.

3.1   Strong Dependence

The first problem that we study is triangle counting: Given a graph in the adjacency streaming model, (ϵ, δ)-approximate the number of triangles in the graph. Recall that T_i is the number of unordered triples of vertices with i edges, and thus T_3 is the number of triangles. The following theorem gives lower bounds that match the naive upper bounds: $O(n^3/T_3)$ for $T_3 \geq \frac{n}{3(1+\epsilon)}$ (by random sampling) and $O(n^2)$ space for $T_3 < \frac{n}{3(1+\epsilon)}$ (by storing all the edges).

Theorem 1. *Any streaming algorithm distinguishing between $T_3 \leq (1 - \epsilon)t$ and $T_3 \geq (1 + \epsilon)t$ with error probability $\delta = 1/3$ needs $\Omega(n^3/t)$ space for $t \geq \frac{n}{3(1+\epsilon)}$ and $\Omega(n^2)$ space for $t < \frac{n}{3(1+\epsilon)}$.*

Proof. Consider the case $t \geq \frac{n}{3(1+\epsilon)}$ first. Let the input graph G consist of 2 parts. One part is an $(n/3, n/3)$-bipartite graph $H = (L, R, E_H)$ (where L and R are left and right side vertex sets), and another part $J = (V_J, E_J)$ contains $n/3$ vertices. Partition L into $n/3k$ blocks $L_1, ..., L_{n/3k}$, each of size $k = \sqrt{3(1+\epsilon)t/n}$; similarly partition $R = R_1 \cup ... \cup R_{n/3k}$. (See Figure 1.) Denote by $H_{i,j}$ the subgraph $(L_i, R_j, E_H|_{L_i \times R_j})$. Now let the stream first give the graph H, with the promise that each subgraph $H_{i,j}$ is either empty or complete. Clearly it needs $(n/3k)^2$ bits of information to specify H. We claim that the streaming algorithm needs to basically keep all these $(n/3k)^2$ bits of information in order to approximate the number of triangles in the whole graph.

Actually, we claim that for any (i, j), by choosing the rest of the graph in an appropriate way, we can know whether $H_{i,j}$ is empty or complete with probability $1 - \delta$. Suppose we want to know whether $H_{i,j}$ is empty or complete, we let the remaining stream contain all edges in $\{(a, b) : a \in L_i \cup R_j, b \in V_J\}$. If $H_{i,j}$ is

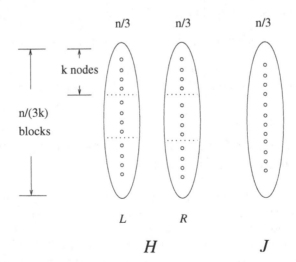

Fig. 1. A graph for illustration of the proof of triangle counting problem

complete, then G contains $k^2n/3 = (1 + \epsilon)t$ triangles; if H_i is empty, then G contains no triangle. Since the algorithm can distinguish between $T_3 \leq (1 - \epsilon)t$ and $T_3 \geq (1 + \epsilon)t$ with probability $1 - \delta$, it follows that after the first half of the stream (that specifies H) has passed, we can extract, for any (i, j), the one bit information about whether $H_{i,j}$ is empty or complete with probability $1 - \delta$. Therefore by Fact 1, we need

$$(1 - 2\delta)\left(\frac{n}{3k}\right)^2 = \frac{(1 - 2\delta)n^3}{27(1 + \epsilon)t} = \Omega\left(n^3/t\right) \tag{1}$$

bits of memory.

Note that in the above analysis, we implicitly require that block size $k \geq 1$ and the number of blocks $n/(3k) \geq 1$, for which we need $\frac{n}{3(1+\epsilon)} \leq t \leq \frac{(1/2-\delta)n^3}{27(1+\epsilon)}$. For $t > \frac{(1/2-\delta)n^3}{27(1+\epsilon)}$, the lower bound is trivially true. For $t < \frac{n}{3(1+\epsilon)}$, let $k = 1$ and then the graph has $n/3$ triangles if $H_{i,j}$ is complete. Similar arguments give the lower bound of $(1/2 - \delta)n^2/9 = \Omega(n^2)$, which completes our proof.

Another open question asked in [2] is about the comparison of their algorithm and the naive random sampling one. Their algorithm uses $O(\frac{1}{\epsilon^3}\log\frac{1}{\delta}(1 + \frac{T_1+T_2}{T_3})^3\log n)$ space, and they asked when the algorithm is better than the naive sampling algorithm which uses $O(\frac{1}{\epsilon^2}\log\frac{1}{\delta}(1 + \frac{T_0+T_1+T_2}{T_3}))$ space. We now show by some simple algebraic graph theory arguments that the algorithm in [2] is always no better than the naive random sampling one.

Proposition 1. *For any graph we have*

$$\left(1 + \frac{T_1 + T_2}{T_3}\right)^3 \geq \Omega\left(1 + \frac{T_0 + T_1 + T_2}{T_3}\right), \tag{2}$$

Proof. First observe that $T_0 + T_1 + T_2 + T_3 = \binom{n}{3}$ and that $T_1 + 2T_2 + 3T_3 = m(n - 2)$ where m is the number of edges in the graph. The latter implies that $T_1 + T_2 + T_3 = \Theta(mn)$. Thus it is enough to prove that $T_3^2 = O(m^3)$, which can be easily done using algebraic arguments as follows. Suppose A is the adjacency matrix of the graph, then $Tr(A^3) = 6T_3$. Now notice that $Tr(A^3) = \sum_i \sum_k a_{ik}b_{ki}$ where $B = [b_{ij}]_{ij} = A^2$. It is easy to see that B is also symmetric, so $Tr(A^3) = \sum_{ik} a_{ik}b_{ik} \leq \sqrt{(\sum_{ik} a_{ik}^2)(\sum_{ik} b_{ik}^2)}$. Note that $\sum_{ik} a_{ik}^2 = \sum_{ik} a_{ik} = 2m$, and $\sqrt{\sum_{ik} b_{ik}^2} = \|B\|_2 = \|A \cdot A\|_2 \leq \|A\|_2^2 = 2m$, we thus have $T_3 \leq (2m)^{3/2}/6$, as desired.

As mentioned in Section 1, new algorithms are known ([13, 5]) which are better than the naive sampling algorithm for some graphs. So the above proposition is mainly of discrete math interest.

3.2   Weak Dependence

The second problem that we study is max common neighbor counting: Given a graph in the adjacency stream model, (ϵ, δ)-approximate the maximum number of common neighbors, *i.e.* $mcn(G) = \max_{u,v} |\{w : (u, w) \in E, (v, w) \in E\}|$. In [4], it is showed that computing the *exact* value of $mcn(G)$ needs $\Omega(n^{3/2}\sqrt{c})$ space, where c is the max number of common neighbors. In this paper, after observing a matching upper bound for this exact counting problem, we consider the approximate version of the problem and show that the space complexity of approximating $mcn(G)$ is $\tilde{\Theta}(n^{3/2}/\sqrt{c})$.

In both the upper and lower bounds, we will use the following theorem in extremal graph theory. Denote by $ex(n, H)$ is the maximal number of edges that an n-vertex graph can have without containing H as a subgraph. The upper bound is by Kovari, Sos and Turan [15], and the lower bound is by Furedi [12].

Theorem 2. $\frac{1}{2}\sqrt{t}n^{3/2} - O(n^{4/3}) \leq ex(n, K_{2,t+1}) \leq \frac{1}{2}\sqrt{t}n^{3/2} + n/4$.

Now there is a very easy algorithm: keep all the edge information until the number of edges exceeds $\frac{1}{2}\sqrt{t}n^{3/2} + n/4$, in which case the graph contains $K_{2,t+1}$ for sure; otherwise, use the kept edge information to decide whether $mcn(G) \geq t$.

Now we give the algorithm to approximate $mcn(G)$ as in the Algorithm **Approx-mcn(G)** box. Its analysis is given by the theorem below.

Algorithm **Approx-mcn(G)**
Input: a data stream of edges $(i, j) \in E$ of a graph G in arbitrary order, two constants $a \in (0, 1)$ and $b > 1$.
Output: an (ϵ, δ)-approximate of $mcn(G)$ if $mcn(G) \in [ac, bc]$.

1. Use a counter to count the total number m of edges. Stop and output "$mcn(G) > bc$" if $m > M \equiv \frac{1}{2}n^{3/2}\sqrt{bc} + n/4$.
2. Randomly pick (with replacement) $t = \frac{1}{a\epsilon^2}(\log\frac{3n^2}{\delta})\frac{2n}{c}$ vertices $v_1, ..., v_t$. Denote this multi-set by T.
3. Keep all edges incident to T. (If the number exceeds $\frac{6Mt}{\delta n}$, output FAIL.)
4. Use the kept edge information to get

$$c' = \max_{u,v \in V-T} \sum_{i=1}^{t} \mathbf{1}[v_i \text{ is a common neighbor of } u \text{ and } v]$$

   where $\mathbf{1}[\phi]$ is the indicator variable for the event ϕ.
5. Output $c'n/t$ as estimate to $mcn(G)$. If $c' = 0$, output $mcn(G) < ac$.

Theorem 3. *For any c and any constants $a \in (0, 1)$ and $b > 1$, Algorithm **Approx**-mcn(G) (ϵ, δ)-approximates $mcn(G)$ for those G with $mcn(G) \in [ac, bc]$, and the algorithm uses space $O(n^{3/2}\log^2 n/\sqrt{c})$.*

Proof. First it is obvious that if the number of edge exceeds M which is larger than $ex(n, K_{2,bc})$, then $mcn(G) > bc$ for sure. Now consider $m = |E| < M$. By Markov's Inequality, the total degree of vertices in T is at most

$$\frac{3}{\delta}\frac{2mt}{n} \le \frac{6Mt}{\delta n} = O\left(\frac{n^{3/2}(\log n + \log \frac{1}{\delta})}{\epsilon^2 \delta \sqrt{c}}\right) \qquad (3)$$

with probability $1 - \delta/3$. Now assume $mcn(G) = s \in [ac, bc]$, then $\exists\ u_0, v_0$ sharing a set S_0 of s common neighbors. Fix u_0, v_0 and S_0. Since

$$\mathbf{Pr}[u_0 \in T \text{ or } v_0 \in T] \le 2t/n \le \delta/3 \qquad (4)$$

if $c \ge \frac{12}{a\delta\epsilon^2}\log\frac{3n^2}{\delta}$. Let $X_i(u,v)$ be the indicator random variable for the event "the i-th vertex picked is a common neighbor of u and v", and let $X(u,v) = \sum_{i=1}^{t} X_i(u,v)$. Then under the condition that $u_0, v_0 \notin T$, we have $X(u_0, v_0) \le c'$. Now by Chernoff's bound,

$$\mathbf{Pr}\left[X(u_0, v_0) < \frac{(1-\epsilon)ts}{n}\right] < e^{-\frac{\epsilon^2 ts}{2n}} \le e^{-\frac{\epsilon^2 tac}{2n}} = \delta/(3n^2). \qquad (5)$$

Therefore, $\mathbf{Pr}[c'n/t < (1 - \epsilon)s] \le \delta/(3n^2) < \delta/3$. On the other hand, by definition of c', we have

$$\mathbf{Pr}[c'n/t > (1 + \epsilon)s] = \mathbf{Pr}[c' > (1 + \epsilon)st/n] \qquad (6)$$
$$= \mathbf{Pr}[\exists u, v \in V - T, s.t.\ X(u,v) > (1 + \epsilon)st/n] \qquad (7)$$
$$\le n^2 \cdot \mathbf{Pr}[X(u,v) > (1 + \epsilon)st/n \mid u, v \notin T] \qquad (8)$$
$$\le n^2 \cdot \delta/(3n^2) = \delta/3. \qquad (9)$$

Putting all things together, the algorithm outputs an ϵ-approximation with probability at least $1 - \delta$.

The analysis of space that the algorithm uses is as follows. It needs $O(\log n)$ to store an vertex v or an edge (u,v), and the algorithm needs to store t vertices and $6MT/(\delta n)$ edges. Step 4 is space efficient since we can reuse space to check each pair (u,v). Thus the total number of bits used in the algorithm is $O(\frac{Mt}{\delta n}\log n) = O\left(\frac{n^{3/2}(\log n + \log\frac{1}{\delta})}{\epsilon^2 \delta \sqrt{c}}\log n\right)$, which is $O\left(\frac{n^{3/2}\log^2 n}{\sqrt{c}}\right)$ if ϵ and δ are constants.

We can also prove a matching lower bound as follows.

Theorem 4. *Distinguishing between $mcn(G) \le (1 - \epsilon)c$ and $mcn(G) \ge (1 + \epsilon)c$ with small constant error probability δ needs $\Omega(n^{3/2}/\sqrt{c})$ space for small constant ϵ (say $\epsilon = 0.1$).*

Proof. Consider the extremal graph H with $n - (1 - \epsilon)c - 1$ vertices and no $K_{2,(1-\epsilon)c+1}$ as a subgraph. For each vertex, partition its neighbors into subsets of size $2\epsilon c$, with possibly one subset of smaller size. The total number of edges in the regular (*i.e.* not smaller) subsets is at least

$$\sqrt{(1-\epsilon)c}(n-(1-\epsilon)c-1)^{3/2}/2-(n-(1-\epsilon)c-1)2\epsilon c-O(n^{4/3}) = \Omega(\sqrt{c}n^{3/2}) \quad (10)$$

if $\epsilon \le 0.1$. Now consider the graph G with n vertices, $n - (1 - \epsilon)c - 1$ of which are for the graph H, and the rest $(1 - \epsilon)c + 1$ vertices are denoted by u and S with $|S| = (1 - \epsilon)c$. (See Figure 2.)

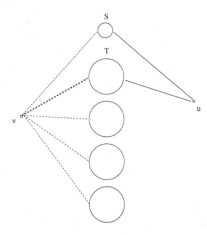

Fig. 2. A graph for illustration of the proof of max common neighbor counting problem

We will use Fact 2 to show the lower bound. Let the streaming first provide (a subgraph of) H, where for any $v \in H$ and each of its $2\epsilon c$-subsets T, the edges from v to T may all exist or all not. Fix the content of the memory of the algorithm. Then for any fixed $v \in H$ and any of its $2\epsilon c$-subsets T, we can know whether v and T are connected or not by providing the remaining stream (and running the streaming algorithm) in the following way. Connect v and S, u and $S \cup T$. Now note that if there is no edge between v and T, then an easy case-by-case study shows that there are no two vertices in the whole graph G sharing $(1 - \epsilon)c + 1$ common neighbors. If, on the other hand, v connects to all points in T, then clearly u and v share $(1 + \epsilon)c$ common neighbors. Thus if we can distinguish the $mcn(G) \leq (1 - \epsilon)c$ and $mcn(G) \geq (1 + \epsilon)c$ with probability $1 - \delta$, then we can distinguish between the two cases that the edges between v and T all exist and all of them do not exist with success probability $1 - \delta$, which need $\Omega(n^{3/2}\sqrt{c})/2\epsilon c = \Omega(n^{3/2}/\sqrt{c})$ space by Fact 2.

3.3   Independence

Not surprisingly, there are also many problems whose hardness is not due to the extreme case in nature, though the previous lower bounds were proved by considering the extreme cases. We just mention one simple example here to end this section. The problem is to estimate the distance between two vertices on a graph: For two fixed vertices u, v on a graph G which is given in the adjacency stream model, (ϵ, δ)-approximate $d(u, v)$, the distance between u and v on the graph G. It is not hard to see that distinguishing between $d(u, v) \leq 3$ and $d(u, v) \geq n/2$ needs $\Omega(n)$ bits of memory. Actually, consider the graph consisting of two parts. One is a $n/2$-long path connecting u and v, and another part contains $n/4 - 1$ disjoint edges $(u_1, v_1), ..., (u_{n/4-1}, v_{n/4-1})$. We first stream in these two parts, but each edge (u_i, v_i) may or may not exist. Then to know whether a particular edge (u_i, v_i) exists or not, we connect (u, u_i) and (v, v_i). If

(u_i, v_i) exists, then the $d(u, v) = 3$; otherwise it is $n/2$. Thus we need $n/4 - 1$ bits of memory to distinguish these two cases. Note that in [10], a $2t + 1$ spanner is constructed using $O(tn^{1+1/t} \log^2 n)$ space thus the graph distance problem can be approximated up to a factor of $2t + 1$ by using the same amount of space, which implies that the $\Omega(n)$ lower bound is almost optimal for large constant approximation.

4   Discussions

Previous research on streaming algorithms mainly focused on designing space-efficient algorithms for important tasks; usually, log or even constant space complexity is desired. There may be more problems that, though may be very important in practical applications, did not get well studied theoretically simply because a high lower bound can be easily shown (by considering extreme inputs). This paper studies some problems which are hard for general case but easy if the computed quantity is within some range that we care about and/or the practical data are actually in.

Clearly, the same question can be asked for general algorithms. And within the domain of streaming algorithms, the problems studied in this paper happen to be those on graphs, but we believe that there are many more other problems having the same interesting phenomena.

Acknowledgement

The author thanks Sanjeev Arora, Moses Charikar, Yaoyun Shi and Martin Strauss for listening to the results and giving valuable comments.

References

[1] Alon, N., Matias, Y., Szegedy, M.: The space complexity of approximating the frequency moments. Journal of Computer and System Sciences 58(1), 137–147 (1999)

[2] Bar-Yossef, Z., Kumar, R., Sivakumar, D.: Reductions in streaming algorithms, with an application to counting triangles in graphs. In: Proceedings of the thirteenth annual ACM-SIAM symposium on Discrete algorithms (SODA), pp. 623–632. ACM Press, New York (2002)

[3] Bar-Yossef, Z., Jayram, T., Kumar, R., Sivakumar, D.: Information statistics approach to data stream and communication complexity. In: Proceedings of the 43rd Annual IEEE Symposium on Foundations of Computer Science, pp. 209–218. IEEE Computer Society Press, Los Alamitos (2002)

[4] Buchsbaum, A., Giancarlo, R., Westbrook, J.: On finding common neighborhoods in massive graphs. Theoretical Computer Science 299, 707–718 (2003)

[5] Buriol, L., Frahling, G., Leonardi, S., Marchetti-Spaccamela, A., Sohler, C.: Counting Triangles in Data Streams. In: Proceedings of the twenty-fifth ACM SIGMOD-SIGACT-SIGART symposium on Principles of database systems, pp. 253–262. ACM Press, New York (2006)

[6] Chakrabarti, A., Khot, S., Sun, X.: Near-optimal lower bounds on the multi-party communication coplexity of set disjointness. In: Proceedings of the 18th IEEE Conference on Computational Complexity, pp. 107–117. IEEE Computer Society Press, Los Alamitos (2003)

[7] Coppersmith, D., Kumar, R.: An improved data stream algorithm for frequency moments. In: Proceedings of the 15th ACM-SIAM Symposium on Discrete Algorithms (SODA), pp. 151–156. ACM Press, New York (2004)

[8] Derenyi, I., Palla, G., Vicsek, T.: Clique percolation in random networks. Physical Review Letters 94, 160–202 (2005)

[9] Feigenbaum, J., Kannan, S., McGregor, A., Suri, S., Zhang, J.: On graph problems in a semi-streaming model. In: Díaz, J., Karhumäki, J., Lepistö, A., Sannella, D. (eds.) ICALP 2004. LNCS, vol. 3142, pp. 531–543. Springer, Heidelberg (2004)

[10] Feigenbaum, J., Kannan, S., McGregor, A., Suri, S., Zhang, J.: Graph Distances in the Streaming Model: The Value of Space. In: Proceedings of the 16th Symposium on Discrete Algorithms (SODA), pp. 745–754 (2005)

[11] Feigenbaum, J., Kannan, S., Strauss, M., Viswanathan, M.: An approximate L1 difference algorithm for massive data streams. In: Proceedings of the 40th Annual Symposium on Foundations of Computer Science, pp. 501–511 (1999)

[12] Furedi, Z.: New asymptotics for bipartite Turan numbers. Journal of Combinatorial Theory, Series A 75, 141–144 (1996)

[13] Jowhari, H., Ghodsi, M.: New Streaming Algorithms for Counting Triangles in Graphs. In: Proceedings of the Eleventh International Computing and Combinatorics Conference, pp. 710–716 (2005)

[14] Indyk, P., Woodruff, D.: Optimal approximations of the frequency moments of data streams. In: Proceedings of the 37th ACM Symposium on Theory of Computing, pp. 202–208. ACM Press, New York (2005)

[15] Kovari, T., Sos, V.T., Turan, P.: On a problem of K. Zarankiewicz. Colloq. Math, vol. 3, pp. 50–57 (1954)

[16] Muthukrishnan, S.: Data Streams: Algorithms and Applications. Roundations and Trends in Theoretical Computer Science, vol. 1(2), pp. 117–236 (2005)

[17] Saks, M., Sun, X.: Space lower bounds for distance approximation in the data stream model. In: Proceedings on 34th Annual ACM Symposium on Theory of Computing, pp. 360–369. ACM Press, New York (2002)

[18] Shi, X., Adamic, L., Strauss, M.: Networks of strong ties. Physica A 378(1), 33–47 (2007)

A Randomized Approximation Algorithm for Parameterized 3-D Matching Counting Problem[*]

Yunlong Liu[1,2], Jianer Chen[1,3], and Jianxin Wang[1]

[1] College of Information Science and Engineering, Central South University,
Changsha 410083, P.R. China
[2] School of Further Education, Hunan Normal University,
Changsha 410012, P.R. China
[3] Department of Computer Science Texas A&M University,
College Station, TX 77843, USA
hnsdlyl@163.com, chen@cs.tamu.edu, jxwang@mail.csu.edu.cn

Abstract. The computational complexity of counting the number of matchings of size k in a given triple set remains open, and it is conjectured that the problem is infeasible. In this paper, we present a fixed parameter tractable randomized approximation scheme (FPTRAS) for the problem. More precisely, we develop a randomized algorithm that, on given positive real numbers ϵ and δ, and a given set S of n triples and an integer k, produces a number h in time $O(5.48^{3k} n^2 \ln(2/\delta)/\epsilon^2)$ such that

$$prob[(1 - \epsilon)h_0 \leq h \leq (1 + \epsilon)h_0] \geq 1 - \delta$$

where h_0 is the total number of matchings of size k in the triple set S. Our algorithm is based on the recent improved color-coding techniques and the Monte-Carlo self-adjusting coverage algorithm developed by Karp and Luby.

1 Introduction

Counting, like decision, function and enumeration, is a kind of fundamental computation in classical complexity theory, and is an important branch in theoretical computer science. Apart from being mathematically interesting, counting is closely related to important practical applications, such as data processing, reliability analyzing, artificial intelligence, and statistical physics [13].

Counting graph matchings is a central problem in computer science and has received much attention since the seminal work of Valiant, who proved that counting the perfect matchings in a bipartite graph is $\sharp P$-complete [14]. Valiant also proved that counting perfect matchings in a general graph is also $\sharp P$-complete [15]. Vadhan [13] proved that the problems of counting matchings and maximal matchings are all $\sharp P$-complete even restricted to planar bipartite graphs whose

[*] This work is supported by the National Natural Science Foundation of China (60433020) and the Program for New Century Excellent Talents in University (NCET-05-0683)

degree is bounded by a small constant or to k-regular graphs for a constant k. Recently, Flum and Grohe set up a framework for a parameterized complexity theory of counting problems and especially formulated the problem of counting parameterized graph matchings, i.e., the p-♯MATCHING problem. [6]. The computational complexity of p-♯MATCHING remains open and is conjectured to be intractable (i.e., ♯W[1]-hard) [6]. Despite the fact that all the above mentioned counting problems are intractable, many effective approaches have been proposed. Especially, there have been considerable advances in recent years in the design of efficient approximation algorithms for counting graph matchings (see, e.g., [3,5,11] for a survey) or counting parameterized graph matchings [2].

In the current paper, we are focused on a generalization of the bipartite graph matching problem, the 3-D MATCHING problem. In particular, we will study the complexity of counting parameterized 3-D matchings. We start with some definitions related to the problem.

Let A_1, A_2, A_3 be three pairwise disjoint finite symbol sets. A *triple* in $A_1 \times A_2 \times A_3$ is given as (a_1, a_2, a_3), where $a_i \in A_i$ for $i = 1, 2, 3$. A *matching* M in a triple set S is a subset of triples in S such that no two triples in M share a common symbol. A matching is a k-*matching* if it contains exactly k triples.

Definition 1. (Parameterized) 3-D MATCHING: Given a pair (S, k), where S is a triple set and k is an integer, either construct a k-matching in S or report that no such a matching exists.

In its general form, 3-D MATCHING is one of the six basic NP-complete problems according to Garey and Johnson [8]. In recent years, 3-D MATCHING has become a focus in the study of parameterized algorithms. A series of improved algorithms has been developed for the problem [7,9,12]. Currently the best parameterized algorithm for 3-D MATCHING is due to Liu *et al.* [12], and has its running time bounded by $O(2.77^{3k} n^{O(1)})$.

Our focus in the current paper is to count the number of k-matchings in a given triple set, defined as follows.

Definition 2. (Parameterized counting) 3-D MATCHING : Given a pair (S, k), where S is a triple set and k is an integer, count the number of distinct k-matchings in S. Let p-♯3D MATCHING denote this problem.

The p-♯3D MATCHING problem is a generalization of the problem of counting parameterized matchings in bipartite graphs, and the latter has been conjectured to be infeasible [6]. Therefore, p-♯3D MATCHING has no known efficient algorithms. This fact makes approximation algorithms for the problem interesting and meaningful.

2   Preliminaries

Fix three finite symbol sets A_1, A_2, A_3. For a triple $\rho = (a_1, a_2, a_3)$ in $A_1 \times A_2 \times A_3$, denote by $\text{Val}(\rho)$ the set $\{a_1, a_1, a_3\}$, and let $\text{Val}^i(\rho) = \{a_i\}$ for $1 =$

1, 2, 3. For a collection S of triples, define $\text{Val}(S) = \bigcup_{\rho \in S} \text{Val}(\rho)$, and $\text{Val}^i(S) = \bigcup_{\rho \in S} \text{Val}^i(\rho)$, for $1 = 1, 2, 3$. The symbols in $\text{Val}^i(S)$ will be called the *i-th dimension symbols* in S.

Our algorithms take the advantages of the recently developed improved algorithms for the technique of *color-coding*. For this, we briefly review the necessary terminologies and results.

Definition 3. Let F be a finite set. A *k-coloring* of F is a function mapping F to the set $\{1, 2, \ldots, k\}$. A subset W of F is colored *properly* by a coloring f if no two elements in W are colored with the same color under f.

Definition 4. A collection $\mathcal{C}$ of k-colorings of a finite set F is a *k-color coding scheme* if for every subset W of k elements in F, there is a k-coloring in $\mathcal{C}$ that colors W properly. The *size* of the k-color coding scheme $\mathcal{C}$ is the number of k-colorings in $\mathcal{C}$.

The concept of color-coding was introduced by Alon, Yuster, and Zwick [1], who proved that for a set of n elements there exists a k-color coding scheme whose size is bounded by $O(2^{O(k)}n)$. Progress has been made recently. Chen *et al.* [4] presented a construction that gives a k-color coding scheme of size $O(6.4^k n)$ for a set of n elements, and developed an algorithm of running time $O(6.4^k n)$ that generates such a k-color coding scheme.

Theorem 1. ([4]) For any finite set F of n elements and any integer k, $n \geq k$, there is a k-color coding scheme $\mathcal{C}$ of size $O(6.4^k n)$ for the set F. Moreover, such a k-color coding scheme can be constructed in time $O(6.4^k n)$.

Before proceeding to details, we give a brief description of the ideas of our algorithms. Let (S, k) be an input instance of the p-$\sharp$3D MATCHING problem, where S is a set of n triples. Let H be the set of all k-matchings in S. Our objective is to compute the value $|H|$.

Let F be the set of symbols that are in either 2nd-dimension or 3rd-dimension in S. We construct a $(2k)$-color coding scheme $\mathcal{C} = \{f_1, \ldots, f_m\}$ for the set F such that the size m of $\mathcal{C}$ is bounded by $O(6.4^{2k} n)$. For a $(2k)$-coloring f_i in $\mathcal{C}$ and a matching M in S, we say that M is *properly colored* by f_i if all symbols in the 2nd and the 3rd dimensions in M are colored with distinct colors under the coloring f_i.

For $1 \leq i \leq m$, let H_i be the subset of H such that H_i consists all k-matchings in S that are colored properly by the $(2k)$-matching f_i. Since $\mathcal{C}$ is a $(2k)$-color coding scheme for the set F, for every k-matching M in S, there is a $(2k)$-coloring f_i that properly colors M. Therefore, we have $H = \bigcup_{i=1}^m H_i$.

This gives the standard UNION OF SETS problem studied by Karp, Luby, and Madras [10]: given a collection of m sets $H_1, H_2, \ldots, H_m$, compute the value $|H|$, where $H = \bigcup_{i=1}^m H_i$. This problem has been thoroughly studied in [10]. In particular, the following theorem is proved[1].

---

[1] The theorem stated here is more detailed than the one presented in [10]. It is straightforward to follow the original proof in [10] to derive this version of the theorem. Therefore, the proof of the theorem is omitted.

Theorem 2. (Karp-Luby [10]) *Let $H_1, \ldots, H_m$ be m sets such that*
  (1) *for each i, the value $|H_i|$ can be computed in time t_1;*
  (2) *for each i, in time t_2 we can randomly pick an element in H_i with a*
     probability $1/|H_i|$; and
  (3) *for each i and for any $v \in H$, we can decide if $v \in H_i$ in time t_3.*
  Then we can construct a randomized algorithm that for two given positive real
numbers ϵ and δ, and for given m sets satisfying the conditions (1)-(3), generates
a number h in time $O(mt_1 + mt_2t_3 \ln(2/\delta)/\epsilon^2)$ such that

$$prob[(1 - \epsilon)|H| \le h \le (1 + \epsilon)|H|] \ge 1 - \delta,$$

where $H = \bigcup_{i=1}^m H_i$.

Therefore, to develop an effective randomized approximation algorithm for
the p-♯3D MATCHING problem, we only need to show
  (1) how to count the number $|H_i|$ of k-matchings in S that are properly colored
by the $(2k)$-coloring f_i, for each i;
  (2) how to randomly pick, with uniform probability, a k-matching that is
properly colored by the $(2k)$-coloring f_i, for each i; and
  (3) how to determine if a k-matching is properly colored by the $(2k)$-coloring
f_i, for each i.
  In the rest of this paper, we present algorithms that solve the above three
problems. These algorithms combined with Theorem 2 give an algorithm that
effectively computes an approximation of the number of k-matchings in a given
triple set S.

3   Dealing with Properly Colored k-Matchings

As given in the previous section, for the given input instance (S, k) of the p-♯3D
MATCHING problem, let f_i be a $(2k)$-coloring of the symbols in the 2nd and the
3rd dimensions in S, and let H_i be the set of all k-matchings in S that are
properly colored by f_i.

3.1   Computing the Value $|H_i|$

We present an algorithm CM1(S, k, f_i) to count the distinct k-matchings in S
that are properly colored by the $(2k)$-coloring f_i, i.e., CM1(S, k, f_i) computes
the value $|H_i|$. Figure 1 gives the algorithm in detail, where C in the pair (C, N)
denotes a color set, and N denotes the number of k-matchings whose 2nd and 3rd
dimension symbols colored exactly by the color set C, and $cl(\text{Val}^i(\rho))$ denotes
the color of the symbol in column-i in the triple ρ, where $i = 2, 3$.

Theorem 3. *The algorithm CM1(S, k, f_i) runs in time $O(2^{2k}kn)$ and returns
exactly the number of k-matchings in S that are properly colored by f_i.*

Proof. For each i, $1 \le i \le r$, let S_i be the set of triples in S' whose symbols
in column-1 are among $\{x_1, x_2, \cdots, x_i\}$. By induction on i, we first prove that

---

Algorithm CM1(S, k, f_i)

Input: a $(2k)$-coloring f_i of the symbols in the 2nd and 3rd dimensions in the triple set S

Output: The number of k-matchings in S that are properly colored by f_i;

1 remove any triples in S in which any two symbols have the same color ;
2 let the set of remaining triples be S' ;
3 let the symbols in $\mathrm{Val}^1(S')$ be $x_1, x_2, \ldots, x_r$;
4 $\mathcal{Q}_{old} = \{(\emptyset, 0)\}$; $\mathcal{Q}_{new} = \{(\emptyset, 0)\}$;
5 for $i = 1$ to r do
5.1   for each pair (C, N) in $\mathcal{Q}_{old}$ do
5.2   for each $\rho \in S'$ with $\mathrm{Val}^1(\rho) = x_i$ do
5.3   if $C \cap \{cl(\mathrm{Val}^2(\rho)) \cup cl(\mathrm{Val}^3(\rho))\} = \emptyset$
5.4   then $\{C' = C \cup \{cl(\mathrm{Val}^2(\rho)) \cup cl(\mathrm{Val}^3(\rho))\}$;
5.5   if $N = 0$ then $N' = N + 1$ else $N' = N$;
5.6   if the number of colors in C' is no more than $2k$
5.7   then if there exists another pair (C'', N'') in $\mathcal{Q}_{new}$ in which
                C'' is exactly the same color set as C'
5.8   then replace N'' in the pair (C'', N'')
                with $N'' = N'' + N'$;
5.9   else add (C', N') to $\mathcal{Q}_{new}$ directly ; $\}$
5.10 $\mathcal{Q}_{old} = \mathcal{Q}_{new}$;
6 if there exists (C, N) in $\mathcal{Q}_{old}$ in which C contains $2k$ colors then return N
   else return 0 .

Fig. 1. An algorithm computing $|H_i|$

for all $h \leq k$, if S_i has t h-matchings properly colored by the same color set C containing $2h$ distinct colors, then after the i-th execution of the for-loop in step 5 of the algorithm, the super-collection $\mathcal{Q}_{old}$ must contain a pair (C, t).

The initial case $i = 0$ is trivial since $\mathcal{Q}_{old} = \{(\emptyset, 0)\}$.

Now consider a general value $i \geq 1$. Suppose that S_i has t h-matchings ($h \leq k$) properly colored by the same color set C. Let T be the set of triples whose symbol in the 1st dimension is x_i. Moreover, suppose that among the t h-matchings in S_i, exactly u of them are not in the set T, and $t - u$ of them are in T. By induction, the triple set $S_{i-1} = S_i - T$ has exactly u h-matchings properly colored by the color set C. Moreover, there are exactly $t - u$ $(h - 1)$-matchings in S_{i-1} properly colored by $2(h - 1)$ colors that can be combined with the two colors in a triple in T to form the color set C. Let these sets of $2(h - 1)$ colors be $C_1, \ldots, C_g$. By the inductive hypothesis, after the $(i - 1)$-st execution of the for-loop in step 5, the super-collection $\mathcal{Q}_{old}$ must contain a pair (C, u) and the pairs $(C_1, N_1), \ldots, (C_g, N_g)$, where N_j denotes the number of $(h - 1)$-matchings colored properly by the color set C_j, for $1 \leq j \leq g$. It is easy to see that $N_1 + \cdots + N_g = t - u$. Therefore, during the i-th execution of the for-loop in step 5, each of the pairs $(C_1, N_1), \ldots, (C_g, N_g)$ will respectively be combined with the colors of a corresponding triple in T to form the color set C. At the same time, the corresponding N_j is added to N in (C, N). After the

i-th execution of the for-loop in step 5, the value of N in (C, N) will become $N = u + (N_1 + \cdots + N_g) = u + (t - u) = t$. Therefore, the pair (C, t) is included in $\mathcal{Q}_{old}$ by step 5.10, and thus $\mathcal{Q}_{old}$ contains the pair (C, t).

There are two special cases: (1)$u = 0$, it means that each h-matching that uses exactly the same color set C contains one triple in T, and it also means that $S_i - T$ does not have an h-matching that uses exactly the color set C. After the $(i - 1)$-st execution of the for-loop in step 5, the super-collection $\mathcal{Q}_{new}$ contains no pair (C, u), and the pair (C, N_1) produced during the i-th execution of the for-loop in step 5 will directly be added to $\mathcal{Q}_{new}$ by step 5.9. (2) $u = t$, it means that each h-matching that uses exactly the color set C does not contain a triple in T, and it also means that $S_i - T$ has t h-matchings which use exactly the same color set C. By the inductive hypothesis, after the $(i - 1)$-st execution of the for-loop in step 5, the pair (C, t) will have already existed in $\mathcal{Q}_{old}$.

Now let $i = r$, it can be concluded that for any $h \leq k$, if the collection S_r (i.e the original collection S) contains w h-matchings properly colored by the same color set C, then after the r-th execution of the for-loop in step 5, the super-collection $\mathcal{Q}_{old}$ must contain the pair (C, w). In particular, if the triple set S contains t k-matchings properly colored by $2k$ colors (i.e., properly colored by the $(2k)$-coloring f_i), then at the end of the algorithm, the super-collection $\mathcal{Q}_{old}$ contains a pair (C, t), where C is exactly the entire color set used by f_i.

The time complexity of the algorithm $CM1(S, k, f_i)$ can be analyzed as the following. For each $0 \leq h \leq k$ and for each color set C containing $2h$ different colors, the super-collection $\mathcal{Q}_{old}$ keeps at most one pair (C, N). Since there are $\binom{2k}{2h}$ different subsets of $2h$ colors over a total of $2k$ colors, the total number of pairs (C, N) in $\mathcal{Q}_{old}$ is bounded by $\sum_{h=0}^{k} \binom{2k}{2h} \leq 2^{2k}$. For each i, $1 \leq i \leq k$, we examine each pair (C, N) in $\mathcal{Q}_{old}$ in step 5.3 and check if we can construct a larger (C, N) by adding the colors of triple ρ to C. This can be done for each pair (C, N) in time $O(k)$. Step 5.7 can be done in time $O(k)$ by constructing an array of size 2^{2k} to record the status of all color sets. Although in step 5 there are double for-loops (i.e step 5 and step 5.2), the total time of these double for-loops is $O(n)$. Thus, step 5 takes time $O(2^{2k}kn)$. Since step 5 is the dominating step of the algorithm, the algorithm $CM1(S, k, f_i)$ runs in time $O(2^{2k}kn)$.   □

3.2   Random Sampling in the Set H_i

For a $(2k)$-coloring f_i in our $(2k)$-color coding scheme $\mathcal{C}$, let H_i be the set of all k-matchings in S that are colored properly by f_i. In this section, we present a sampling algorithm, which randomly picks a k-matching in H_i with a probability $1/|H_i|$. Figure 2 gives the algorithm in detail, where C in the pair (C, N, M) denotes a color set, N denotes the number of matchings colored exactly by the color set C, and M denotes one matching randomly chosen from these matchings colored by C. In step 5.8, $p(M = M')$ denotes the probability of matching M' being set to M.

Theorem 4. *The algorithm $CM2(S, k, f_i)$ runs in time $O(2^{2k}kn)$, and returns a k-matching properly colored by f_i with a probability $1/|H_i|$.*

Algorithm CM2(S, k, f_i)
Input: f_i is a $(2k)$-coloring on the symbols in the 2nd and 3rd dimensions in S
Output: A random k-matching in S that is properly colored by f_i;

1 remove any triples in S in which any two symbols have the same color ;
2 let the set of remaining triples be S' ;
3 let the symbols in $\text{Val}^1(S')$ be $x_1, x_2, \ldots, x_r$;
4 $\mathcal{Q}_{old} = \{(\emptyset, 0, \emptyset)\}$; $\mathcal{Q}_{new} = \{(\emptyset, 0, \emptyset)\}$;
5 **for** $i = 1$ to r **do**
5.1   **for** each group (C, N, M) in $\mathcal{Q}_{old}$ **do**
5.2   **for** each $\rho \in S'$ with $\text{Val}^1(\rho) = x_i$ **do**
5.3   **if** $C \cap \{cl(\text{Val}^2(\rho)) \cup cl(\text{Val}^3(\rho))\} = \emptyset$
5.4   **then** $\{C' = C \cup \{cl(\text{Val}^2(\rho)) \cup cl(\text{Val}^3(\rho))\}$; $M' = M \cup \{\rho\}$;
5.5   **if** $N = 0$ **then** $N' = N + 1$ **else** $N' = N$;
5.6   **if** the number of triples in M' is no more than k
5.7   **then if** there exists another group (C'', N'', M'') in $\mathcal{Q}_{new}$
                        in which C'' is exactly the same color set as C'
5.8   **then**\{ replace M'' in (C'', N'', M'') with
                           $M = \text{random}(M', M'')$ such that $p(M = M') =$
                           $N'/(N' + N'')$ and $p(M = M'') = N''/(N' + N'')$;
5.9   replace N'' in (C'', N'', M'') with $N'' = N' + N''$;\}
5.10   **else** add (C', N', M') to $\mathcal{Q}_{new}$ directly ; \}
5.11 $\mathcal{Q}_{old} = \mathcal{Q}_{new}$;
6 **if** there exists (C, N, M) in $\mathcal{Q}_{old}$ where C contains $2k$ colors **then** return M
else return "no such a matching exists".

Fig. 2. An algorithm for randomly choosing a properly colored k-matching from S

Proof. For each i, let S_i be the set of triples in S' whose symbols in column-1 are among $\{x_1, x_2, \ldots, x_i\}$. We first prove by induction on i that for all $h \leq k$, if S_i has t h-matchings properly colored by the same color set C which contains $2h$ distinct colors, then the super-collection $\mathcal{Q}_{old}$ must contain a group (C, t, M) after the i-th execution of the for-loop in step 5 of the algorithm, where M is an h-matching randomly chosen from the previous t h-matchings with probability $1/t$. In other words, any h-matching colored by C in S_i can be randomly set to M with the same probability.

Since the algorithm CM2(S, k, f_i) is similar to the algorithm CM1(S, k, f_i), we will only concentrate on the difference, that is, the proof that any h-matching colored by C in S_i can be randomly set to M with the same probability.

Suppose that S_i has t h-matchings ($h \leq k$) properly colored by the same color set C and that T is the set of triples whose first symbol is x_i. Suppose that among the t h-matchings, u of them are not in T, and $t - u$ of them are in T. Then $S_{i-1} = S_i - T$ has u h-matchings properly colored by the same color set C. By the induction, after the $(i - 1)$-st execution of the for-loop in step 5, the super-collection $\mathcal{Q}_{old}$ contains a 3-tuple (C, u, M_0), where M_0 is an h-matching in S_{i-1} randomly picked with a probability $1/u$. Moreover, S_{i-1} has $t - u$ $(h-1)$-matchings properly colored by $2(h-1)$ colors that can be combined with the two colors in a triple in T to form the color set C. Let these sets of $2(h-1)$ colors be $C_1, \ldots, C_g$. Let the number of $(h-1)$-matchings in S_{i-1} that use the color set C_j be N_j, $1 \leq j \leq g$. Obviously, $N_1 + \cdots + N_g = t - u$. Then by induction, after the $(i-1)$-st execution of the for-loop in step 5, for each j, $1 \leq j \leq g$, the super-collection Q_{old} contains a 3-tuple (C_j, N_j, M_j), where N_j is total number of $(h-1)$-matchings in S_{i-1} that are properly colored by the set set C_j, and M_j is an $(h-1)$-matching in S_{i-1} properly colored by the color set C_j and randomly picked with a probability $1/N_j$. Furthermore, let ρ_j, $1 \leq j \leq g$, be the triple in T, whose colors when combined with C_j make the color set C.

During the i-th execution of the for-loop in step 5, according to the order assumed, the group (C, u, M) in Q_{old} will firstly meet with triples in T through running g times of the for-loop in step 5.2, but it will not be modified since the condition in step 5.3 is not satisfied according to the previous assumption. While in the next for-loop in step 5.2, the group (C_1, N_1, M_1) in Q_{old} will be combined with at least one triple (let it be ρ_1) in T. As a result, the probability for h-matching M_0 occurring in $(C, u + N_1, M)$ is $(1/u) * (u/(u+N_1)) = 1/(u+N_1)$, and the probability for h-matching $M_1 \cup \{\rho_1\}$ occurring in $(C, u + N_1, M)$ is $(1/N_1) * (N_1/(u+N_1)) = 1/(u+N_1)$.

After the group (C_2, N_2, M_2) in Q_{old} is combined with one triple (let it be ρ_2) in T through the for-loop in step 5.2, the probability for h-matchings M_0, $M_1 \cup \{\rho_1\}$ occurring in $(C, u+N_1+N_2, M)$ are all $(1/(u+N_1))*((u+N_1)/(u+N_1+N_2)) = 1/(u+N_1+N_2)$, and the probability for h-matching $M_2 \cup \{\rho_2\}$ occurring in $(C, u + N_1 + N_2, M)$ is $(1/N_2) * (N_2/(u + N_1 + N_2)) = 1/(u + N_1 + N_2), \ldots$. Finally, after the group (C_g, N_g, M_g) in Q_{old} is combined with one triple (let it be ρ_g) in T through the for-loop in step 5.2, the probability for h-matching M_0, $M_1 \cup \{\rho_1\}$, $M_2 \cup \{\rho_2\}$, $\ldots$, $M_{g-1} \cup \{\rho_{g-1}\}$ occurring in $(C, u + N_1 + N_2 + \cdots + N_{g-1}+N_g, M)$ are all $(1/(u+N_1+\cdots+N_{g-1}))*((u+N_1+\cdots+N_{g-1}))/((u+N_1+N_2+\cdots+N_{g-1}+N_g)) = 1/(u+N_1+N_2+\cdots+N_{g-1}+N_g) = 1/(u+t-u) = 1/t$, and the probability for h-matching $M_g \cup \{\rho_g\}$ occurring in $(C, u + N_1 + N_2 + \cdots + N_{g-1} + N_g, M)$ is $(1/N_g) * (N_g/(u + N_1 + N_2 + \cdots + N_{g-1} + N_g)) = 1/(u + N_1 + N_2 + \cdots + N_{g-1} + N_g) = 1/(u + t - u) = 1/t$. Furthermore, M_0, $M_1 \cup \{\rho_1\}$, $M_2 \cup \{\rho_2\}$, $\ldots$, $M_g \cup \{\rho_g\}$ are exactly h-matchings colored by color set C and they are arbitrary. Therefore, after the i-th execution of the for-loop in step 5, the super-collection Q_{old} must contain a group (C, t, M), in which M is an h-matching randomly chosen from the previous t h-matchings with probability $1/t$. If these groups in Q_{old} are in other orders, the proof is similar.

Now let $i = r$, it can be concluded that for any $h \leq k$, if the collection S_r (i.e. the original collection S') contains w h-matchings properly colored by the same color set C, then at end of the algorithm, the super-collection Q_{old} must contain the group (C, N, M), in which $N = w$ and M is a k-matching randomly chosen from the previous w k-matchings with probability $1/w$.

The main difference between algorithm $CM2(S, k, f_i)$ and $CM1(S, k, f_i)$ lies in step 5.8. This step can be done by firstly randomly choosing an integer b between 1 and $N' + N''$. If $b > N'$, the matching M'' is set to M in group (C, N, M). Otherwise, the matching M' is set to M in group (C, N, M). Obviously, this step can be done in time $O(1)$ and the time complexity of the rest steps is similar to that of $CM1(S, k, f_i)$. In consequence, the time complexity of the algorithm $CM2(S, k, f_i)$ is $O(2^{2k}kn)$. $\qquad\square$

3.3 On the Membership of the Set H_i

In this last subsection, we discuss how to decide if a k-matching is in the set H_i. In other words, we need to decide if a given k-matching M is properly colored by the $(2k)$-coloring f_i. This is trivial: we only need to go through the symbols in the 2nd and 3rd dimensions in M and check if there are two symbols colored with the same color. This can be easily done in time $O(n)$.

Theorem 5. *For any $(2k)$-coloring f_i in our $(2k)$-color coding scheme $\mathcal{C}$, let H_i be the set of all k-matchings in S that are properly colored by f_i. We can decide for any k-matching M in S if $M \in H_i$ in time $O(kn)$.*

4 Conclusions

Putting all the discussions together, we obtain a randomized approximation algorithm for the p-$\sharp$3D MATCHING problem.

Theorem 6. *There is a randomized approximation algorithm that solves the p-$\sharp$3D MATCHING problem in the following sense: for two given positive real numbers ϵ and δ, and a given instance (S, k) of the p-$\sharp$3D MATCHING problem, where S is a set of n triples and k is an integer, the algorithm generates a number h in time $O(5.48^{3k}n^2 \ln(2/\delta)/\epsilon^2)$ such that*

$$prob[(1 - \epsilon)h_0 \le h \le (1 + \epsilon)h_0] \ge 1 - \delta$$

where h_0 is the total number of k-matchings in the triple set S.

Proof. The algorithm proceeds as follows. For the given instance (S, k) of p-$\sharp$3D MATCHING, let F be the set of the symbols that are in either the 2nd or the 3rd dimension in S. Obviously, $|F| \le 2n$. The algorithm then, using Theorem 1, constructs a $(2k)$-color coding scheme $\mathcal{C}$ for the set F in time $O(6.4^{2k}n)$ such that the size of $\mathcal{C}$ is also bounded by $O(6.4^{2k}n)$. Let $\mathcal{C} = \{f_1, \ldots, f_m\}$, where each f_i is a $(2k)$-coloring of the set F and $m = O(6.4^{2k}n)$. Now for each i, $1 \le i \le m$, we define H_i to be the set of all k-matchings that are properly colored by the $(2k)$-coloring f_i. By Theorem 3, for each i, we can compute the number $|H_i|$ of k-matchings that are properly colored by f_i in time $t_1 = O(2^{2k}kn)$. By Theorem 4, for each i, we can randomly pick in time $t_2 = O(2^{2k}kn)$ a k-matching in H_i with a probability $1/|H_i|$. Finally, by Theorem 5, for any k-matching M in S and for any i, we can decide if M is in H_i, i.e., if M is properly colored by f_i in time $t_3 = O(kn)$. Combining all these with Theorem 2, we obtain an algorithm that for any given positive real numbers ϵ and δ, generates a number h in time

$O(6.4^{2k}n + mt_1 + mt_2t_3 \ln(2/\delta)/\epsilon^2)$ (the first term is for the construction of the $(2k)$-color coding scheme $\mathcal{C}$) such that

$$prob[(1 - \epsilon)h_0 \le h \le (1 + \epsilon)h_0] \ge 1 - \delta$$

where $h_0 = |\bigcup_{i=1}^{m} H_i|$ is the total number of k-matchings in the triple set S. By replacing t_1 by $O(2^{2k}kn)$, t_2 by $O(2^{2k}kn)$, and t_3 by $O(kn)$, we conclude that the running time of the algorithm is $O(5.48^{3k}n^2 \ln(2/\delta)/\epsilon^2)$.   □

According to the literature [2,10], a randomized algorithm Φ_Q for a counting problem Q is a *fully polynomial time randomized approximation scheme* (FPRAS) if for any instance x of Q, and any positive real numbers ϵ and δ, the algorithm Φ_Q runs in time polynomial in $|x|$, $1/\epsilon$, and $\log(1/\delta)$, and produces a number h such that

$$prob[(1 - \epsilon)h_0 \le h \le (1 + \epsilon)h_0] \ge 1 - \delta,$$

where h_0 is the solution to the instance x. There have been some very interesting results in the line of research in this direction. For example, although the problem of counting the number of satisfying assignments for a DNF formula is $\sharp P$-complete [15], the problem has a very nice FPRAS [10].

Arvind and Raman [2] generalized the concept of FPRAS and proposed the concept of *fixed parameter tractable randomized approximation scheme* (FP-TRAS). A parameterized counting problem Q has an FPTRAS if for any instance (x, k) of Q, and any positive real numbers ϵ and δ, the algorithm Φ_Q runs in time $f(k)g(|x|, \epsilon, \delta)$, where f is a fixed recursive function and $g(|x|, \epsilon, \delta)$ is a polynomial of $|x|$, ϵ, and $\log(1/\delta)$, and produces a number h such that

$$prob[(1 - \epsilon)h_0 \le h \le (1 + \epsilon)h_0] \ge 1 - \delta,$$

where h_0 is the solution to the instance (x, k). Arvind and Raman [2] have shown that there are a number of interesting problems that have FPTRAS. In terms of this terminology, the algorithm presented in the current paper is an FPTRAS for the problem p-$\sharp$3D MATCHING, which adds another FPTRAS problem to the literature.

References

1. Alon, N., Yuster, R., Zwick, U.: Color-coding. Journal of the ACM 42, 844–856 (1995)
2. Arvind, V., Raman, V.: Approximation algorithms for some parameterized counting problems. In: Bose, P., Morin, P. (eds.) ISAAC 2002. LNCS, vol. 2518, pp. 453–464. Springer, Heidelberg (2002)
3. Bayati, M., Gamarnik, D., Katz, D., Nair, C., Tetali, P.: Simple deterministic approximation algorithms for counting matchings. In: Proc. 39th Symp. on Theory of Computation (STOC 07) (to appear)
4. Chen, J., Lu, S., Sze, S.-H., Zhang, F.: Improved algorithms for path, matching, and packing problems. In: Proc. 18th Annual ACM-SIAM Symp. on Discrete Algorithms (SODA 07), pp. 298–307. ACM Press, New York (2007)

5. Chien, S.: A determinant-based algorithm for counting perferct matchings in a general graph. In: Proceedings of the 15th Annual ACM-SIAM Symposium on Discrete Algorithms (SODA 04), pp. 728–735. ACM Press, New York (2004)
6. Flum, J., Grohe, M.: The parameterized complexity of counting problems. SIAM J. Comput. 33(4), 892–922 (2004)
7. Fellows, M.R., Knauer, C., Nishimura, N., Ragde, P., Rosamond, F., Stege, U., Thilikos, D., Whitesides, S.: Faster Fixed-parameter tractable algorithms for matching and packing problems. In: Albers, S., Radzik, T. (eds.) ESA 2004. LNCS, vol. 3221, pp. 311–322. Springer, Heidelberg (2004)
8. Garey, M.R., Johnson, D.S.: Computers and Intractability: A Guide to the Theory of NP-Completeness. W. H. Freeman (1979)
9. Koutis, I.: A faster parameterized algorithm for set packing. Information processing letters 94, 7–9 (2005)
10. Karp, R., Luby, M., Madras, N.: Monte-Carlo Approximation Algorithms for Enumerartion Problems. Journal of Algorithms 10, 429–448 (1989)
11. Sankowski, P.: Alternative algorithms for counting all matchings in graph. In: Alt, H., Habib, M. (eds.) STACS 2003. LNCS, vol. 2607, pp. 427–438. Springer, Heidelberg (2003)
12. Liu, Y., Lu, S., Chen, J., Sze, S.-H.: Greedy localization and color-coding: improved matching and packing algorithms. In: Bodlaender, H.L., Langston, M.A. (eds.) IWPEC 2006. LNCS, vol. 4169, pp. 84–95. Springer, Heidelberg (2006)
13. Vadhan, S.P.: The complexity of counting in sparse, regular, and planar graphs. SIAM J. Comput. 31(22), 398–427 (2002)
14. Valiant, L.: The complexity of enumeration and reliability problems. SIAM J. Comput. 8(3), 410–421 (1979)
15. Valiant, L.: The complexity of computing the permanent. Theoretical Computer Science (8), 189–201 (1979)

Optimal Offline Extraction
of Irredundant Motif Bases
(Extended Abstract)

Alberto Apostolico* and Claudia Tagliacollo**

Georgia Institute of Technology & Università di Padova

Abstract. The problem of extracting a basis of irredundant motifs from a sequence is considered. In previous work such bases were built incrementally for all suffixes of the input string s in $O(n^3)$, where n is the length of s. Faster, non-incremental algorithms have been based on the landmark approach to string searching due to Fischer and Paterson, and exhibit respective time bounds of $O(n^2 \log n \log |\Sigma|)$ and $O(|\Sigma| n^2 \log^2 n \log \log n)$, with Σ denoting the alphabet. The algorithm by Fischer and Paterson makes crucial use of the FFT, which is impractical with long sequences.

The algorithm presented in the present paper does not need to resort to the FFT and yet is asymptotically faster than previously available ones. Specifically, an off-line algorithm is presented taking time $O(|\Sigma| n^2)$, which is optimal for finite Σ.

Keywords and Phrases: Design and Analysis of Algorithms, Pattern Matching, Motif Discovery, Irredundant Motif, Basis.

1   Introduction

The extraction from a sequence or sequence ensemble of recurrent patterns consisting of intermixed sequences of solid characters and wildcards finds multiple applications in domains ranging from text processing to computational molecular biology (see, e.g., [11]). Like with most other problems of pattern discovery, the process is often beset by the number of candidates to be considered, which in this particular case can grow exponentially with the input size. Beginning with [8,10], notions of pattern maximality or saturation have been formulated that prove capable of alleviating this problem. This is achieved by the algebraic-flavored notion of a basis, a compact subset of the set of all patterns the elements

---

* Corresponding author. Dipartimento di Ingegneria dell' Informazione, Università di Padova, Padova, Italy *and* College of Computing, Georgia Institute of Technology, 801 Atlantic Drive, Atlanta, GA 30318, USA. `axa@dei.unipd.it` Work Supported in part by the Italian Ministry of University and Research under the Bi-National Project FIRB RBIN04BYZ7, and by the Research Program of Georgia Tech.
** Work performed in part while visiting the College of Computing of the Georgia Institute of Technology.

of which can account, by suitable combination, for any other pattern in the set. The elements of the basis are called irredundant motifs, and a few algorithms have been produced to this date for the extraction of a basis from a sequence. In [2] bases are built incrementally for all suffixes of the input string s in $O(n^3)$, where n is the length of s. Faster algorithms, based on the landmark string searching algorithm by Fischer and Paterson [5], are given in [9] and [7], with respective time bounds of $O(n^2 \log n \log |\Sigma|)$ and $O(|\Sigma| n^2 \log^2 n \log \log n)$. The algorithm in [5] is based on the FFT, which is admittedly impractical for long sequences.

In this work, we design algorithms that do not make use of the FFT and yet are asymptotically faster than previously available ones. Specifically, we present here an off-line algorithm taking time $O(|\Sigma| n^2)$. In a companion paper we also present an incremental algorithm taking time $O(|\Sigma| n^2 \log n)$. The explicit description of a basis requires $\Omega(n^2)$ worst-case time and space, whence the offline algorithm described in the present paper is optimal for finite alphabets.

The paper is organized as follows. Basic definitions and properties are recaptured in the next section. Following that, we describe a basic tool used by our algorithms, which consists of a speed-up in the computation of the occurrence lists of all patterns needed in the computation of a basis. The application of these constructs to the algorithms are then discussed. The paper is self-contained and notation largely conforms to the one adopted in [3,2].

2 Preliminaries

Let Σ be a finite alphabet of *solid* characters, and let '$\bullet$' $\notin \Sigma$ denote a don't-care character, that is, a wildcard matching any of the characters in $\Sigma \cup \{\bullet\}$. A *pattern* is a string over $\Sigma \cup \{\bullet\}$ containing at least one *solid* character. We use σ to denote a generic character from Σ. For characters σ_1 and σ_2, we write $\sigma_1 \preceq \sigma_2$ if and only if σ_1 is a don't care or $\sigma_1 = \sigma_2$.

Given two patterns p_1 and p_2 with $|p_1| \leq |p_2|$, $p_1 \preceq p_2$ holds if $p_1[j] \preceq p_2[j]$, $1 \leq j \leq |p_1|$. We also say in this case that p_1 is a *sub-pattern* of p_2, and that p_2 *implies* or *extends* p_1. If, moreover, the first characters of p_1 and p_2 are matching solid characters, then p_1 is also called a *prefix* of p_2. For example, let $p_1 = ab \bullet \bullet e$, $p_2 = ak \bullet \bullet e$ and $p_3 = abc \bullet e \bullet g$. Then $p_1 \preceq p_3$, and $p_2 \not\preceq p_3$. Note that the $\preceq$ relation is transitive. The following operators are further introduced.

Definition 1. ($\oplus$) *Let* $\sigma_1, \sigma_2 \in \Sigma \cup \bullet$.

$$\sigma_1 \oplus \sigma_2 = \begin{cases} \sigma_1, & \text{if } \sigma_1 = \sigma_2 \\ \bullet, & \text{if } \sigma_1 \neq \sigma_2 \end{cases}$$

Definition 2. (Extended $\oplus$) *Given patterns p_1 and p_2, $p_1 \oplus p_2 = p_1[i] \oplus p_2[i]$, $\forall 1 \leq i \leq \min\{|p_1|, |p_2|\}$.*

Definition 3. (Consensus, Meet) *Given the patterns p_1, p_2, the consensus of p_1 and p_2 is the pattern $p = p_1 \oplus p_2$. Deleting all leading and trailing don't cares from p yields the meet of p_1 and p_2, denoted by $[p_1 \oplus p_2]$.*

For instance, $aac{\bullet}tgcta \oplus caact{\bullet}cat = {\bullet}a{\bullet}{\bullet}t{\bullet}c{\bullet}{\bullet}$, and $[aac{\bullet}tgcta \oplus caact{\bullet}cat]$ $= a{\bullet}{\bullet}t{\bullet}c$. Note that a meet may be the empty word. Let now $s = s_1 s_2 ... s_n$ be a sequence of n over Σ. We use suf_i to denote the suffix $s_i s_{i+1} ... s_n$ of s.

Definition 4. (Autocorrelation) *A pattern p is an* autocorrelation *of s if p is the meet of s and one of its suffixes, i.e., if $p = [s \oplus suf_i]$ for some $1 < i \leq n$.*

For instance, the autocorrelations of $s = acacacacabaaba$ are: $\overline{m}_1 = s \oplus suf_2 = s \oplus suf_{11} = s \oplus suf_{14} = a$, $\overline{m}_2 = s \oplus suf_3 = acacaca{\bullet}a{\bullet}{\bullet}a$, $\overline{m}_3 = s \oplus suf_4 = aba$, $\overline{m}_4 = s \oplus suf_5 = acaca{\bullet}a$, $\overline{m}_5 = s \oplus suf_6 = s \oplus suf_9 = s \oplus suf_8 = s \oplus suf_{10} = s \oplus suf_{12} = a{\bullet}a$, $\overline{m}_6 = s \oplus suf_7 = aca{\bullet}a$.

Definition 5. (Motif) *For a sequence s and positive integer k, $k \leq |s|$, a k-motif of s is a pair $(m, \mathcal{L}_m)$, where m is a pattern such that $|m| \geq 1$ and $m[1]$, $m[|m|]$ are solid characters, and $\mathcal{L}_m = (l_1, l_2, \ldots, l_q)$ with $q \geq k$ is the exhaustive list of the starting position of all occurrences of m in s.*

Note that both components concur to this definition: two distinct location lists correspond to two distinct motifs even if the pattern component is the same and, conversely, motifs that have different location lists are considered to be distinct. In the following, we will denote motifs by their pattern component alone, when this causes no confusion. Consider $s = abcdabcd$. Using the definition of motifs, the different 2-motifs are as follows: $m_1 = ab$ with $\mathcal{L}_{m_1} = \{1, 5\}$, $m_2 = bc$ with $\mathcal{L}_{m_2} = \{2, 6\}$, $m_3 = cd$ with $\mathcal{L}_{m_3} = \{3, 7\}$, $m_4 = abc$ with $\mathcal{L}_{m_4} = \{1, 5\}$, $m_5 = bcd$ with $\mathcal{L}_{m_5} = \{2, 6\}$ and $m_6 = abcd$ with $\mathcal{L}_{m_6} = \{1, 5\}$.

Given a motif m, a *sub-motif* of m is any motif m' that may be obtained from m by *(i)* changing one or more solid characters into don't care, *(ii)* eliminating all resulting don't cares that precede the first remaining solid character or follow the last one, and finally *(iii)* updating $\mathcal{L}_m$ in order to produce the (possibly, augmented) list $\mathcal{L}_{m'}$. We also say that m is a *condensation* for any of its sub-motifs.

We are interested in motifs for which any condensation would disrupt the list of occurrences. A motif with this property has been called *maximal* or *saturated*. In intuitive terms, a motif m is maximal or saturated if we cannot make it more specific while retaining the cardinality of the list $\mathcal{L}_m$ of its occurrences in s. More formally, in a saturated motif m no don't care of m can be replaced by a solid character that appears in all the locations in $\mathcal{L}_m$, nor can m be expanded by a pattern prefix or suffix without affecting the cardinality of $\mathcal{L}_m$.

A motif $(m, \mathcal{L}_m)$ is redundant if m and its location list $\mathcal{L}_m$ can be deduced from the other motifs *without* knowing the input string s. Trivially, every unsaturated motif is redundant. As it turns out, however, saturated motifs may be redundant, too. More formally:

Definition 6. *A saturated motif $(m, \mathcal{L}_m)$, is* redundant *if there exist saturated motifs $(m_i, \mathcal{L}_{m_i})$ $1 \leq i \leq t$, such that*

$$\mathcal{L}_m = (\mathcal{L}_{m_1} + d_1) \cup (\mathcal{L}_{m_2} + d_2) \cup \ldots \cup (\mathcal{L}_{m_p} + d_t)$$

with $0 \leq d_j < |m_j|$.

Here and in the following, $(\mathcal{L}+d)$ is used to denote the list that is obtained by adding a uniform offset d to every element of $\mathcal{L}$. For instance, the saturated motif $m_1 = a \bullet a$ is redundant in $s = acacacacabaaba$, since $\mathcal{L}_{m_1} = \{1, 3, 5, 7, 9, 12\} = (\mathcal{L}_{m_2}) \cup (\mathcal{L}_{m_3}) \cup (\mathcal{L}_{m_4} + 1)$ where $m_2 = acac$, $m_3 = aba$ and $m_4 = ca \bullet a$.

Saturated motifs enjoy some special properties.

Property 1. Let $(m_1, \mathcal{L}_{m_1})$ and $(m_2, \mathcal{L}_{m_2})$ be saturated motifs. Then,

$$m_1 = m_2 \quad \Leftrightarrow \quad \mathcal{L}_{m_1} = \mathcal{L}_{m_2}.$$

We also know that, given a generic pattern m, it is always possible to determine its occurrence list in any sequence s. With a saturated motif m, however, it is possible in addition to retrieve the structure of m from the sole list $\mathcal{L}_m$ in s, simply by taking:

$$m = \left[\bigoplus_{i \in \mathcal{L}_m} suf_i \right].$$

We also have:

Property 2. Let $(m_1, \mathcal{L}_{m_1}), (m_2, \mathcal{L}_{m_2})$ be motifs of s. Then,

$$m_1 \preceq m_2 \Leftrightarrow \mathcal{L}_{m_2} \subseteq \mathcal{L}_{m_1}.$$

Similarly:

Property 3. Let $(m, \mathcal{L}_m)$ be a saturated motif of s. Then $\forall L \subseteq \mathcal{L}_m$ we have

$$m \preceq \left[\bigoplus_{k \in L} suf_k \right].$$

Let now $suf_i(m)$ denote the ith suffix of m.

Definition 7. (Coverage) *The occurrence at j of m_1 is covered by m_2 if $m_1 \preceq suf_i(m_2), j \in \mathcal{L}_{m_2} + i - 1$ for some $suf_i(m_2)$.*

For instance, $\overline{m}_6 = aca \bullet a$ with $\mathcal{L}_{\overline{m}_6} = \{1, 3, 5, 7\}$ is covered at position 5 by $\overline{m}_2 = acacaca \bullet a \bullet \bullet a$, $\mathcal{L}_{\overline{m}_2} = \{1, 3\}$. In fact, let m' be ith suffix of $\overline{m}_3$ with $i = 5$, that is, $m' = aca \bullet a \bullet \bullet \bullet a$. Then $5 \in \mathcal{L}_{\overline{m}_2} + 4$ and $\overline{m}_6 \prec m'$, which together lead to conclude that $\overline{m}_6$ is covered at 5 by $\overline{m}_2$. An alternate definition of the notion of coverage can be based solely on occurrence lists:

Definition 8. (Coverage) *The occurrence at j of m_1 is covered by m_2 if $j \in \mathcal{L}_{m_2} + i \subseteq \mathcal{L}_{m_1}$ for some i.*

In terms of our running example, we have: $5 \in \mathcal{L}_{\overline{m}_2} + 4$ and $\mathcal{L}_{\overline{m}_2} + 4 = \{5, 7\} \subset \mathcal{L}_{\overline{m}_6} = \{1, 3, 5, 7\}$.

A maximal motif that is not redundant is called an *irredundant motif*. Hence a saturated motif $(m, \mathcal{L}_m)$ is irredundant if the components of the pair $(m, \mathcal{L}_m)$ cannot be deduced by the union of a number of other saturated motifs.

We use $\mathcal{B}_i$ to denote the set of irredundant motifs in suf_i. Set $\mathcal{B}_i$ is called the *basis* for the motifs of suf_i. In particular, $\mathcal{B}$ is used to denote the basis of s, which coincides with $\mathcal{B}_1$.

Definition 9. (Basis) *Given a sequence s on an alphabet Σ, let $\mathcal{M}$ be the set of all saturated motifs on s. A set of saturated motifs $\mathcal{B}$ is called a* basis *of $\mathcal{M}$ iff the following hold: (1) for each $m \in \mathcal{B}$, m is irredundant with respect to $\mathcal{B} - \{m\}$, and, (2) let $\mathbf{G}(\mathcal{X})$ be the set of all the redundant maximal motifs generated by the set of motifs $\mathcal{X}$, then $\mathcal{M} = \mathbf{G}(\mathcal{B})$.*

In general, $|\mathcal{M}| = \Omega(2^n)$. Luckily, however, it has been established that the basis of 2-motifs has size linear in $|s|$. As will be recaptured later in the discussion, a simple proof of this fact rests on the circumstance, that all motifs in the basis are autocorrelations of the string s. Before getting to that, we discuss a crucial block of our construction, which is the efficient computation of the lists of occurrences of all meets between suffixes of s. From now on and for the remainder of this paper, treatment will be restricted to 2-motifs.

3   Searching for Pattern Occurrences

String searching, which is the problem of finding all occurrences of an assigned string into a larger text string, is one of the most battered problems of algorithmics. Among the variants of the problem, a prominent role is held by searching for *approximate* occurrences of a solid string (see, e.g., [1,6]), as well as searching for patterns with don't care of the kind considered here. A classical, $O(n \log m \log |\Sigma|)$ time solution based on the FFT was provided in 1974 in a seminal paper by Fischer and Paterson [5] that exploited the convolutory substrate of the problem. More recently, the complexity of that approach was further reduced to $O(n \log n)$ by Cole and Hariharan [4].

All the existing approaches to the extraction of bases of irredundant motifs must solve the problem of finding the occurrences of a special family of patterns with don't cares, namely, the autocorrelations of the input string s or of suffixes thereof. The incremental approach in [2] proceeds by computing those lists for consecutively increasing suffixes of each autocorrelation. This produces the basis associated with each one of the suffixes of s, at the overall cost of $O(n^3)$ time. The approaches of [9,7] compute the occurrences of autocorrelations off-line with Fischer and Paterson [5], hence at an overall cost of $O(n^2 \log n \log |\Sigma|)$. The FFT-based approach does not make use of the fact that the strings being sought are autocorrelations of the input string s. The incremental approach uses this to derive the list of occurrences of consecutive suffixes of the same autocorrelation as consecutive refinements of previously computed lists. However, none of the approaches available takes advantage of the fact, that the patterns of which the occurrences are sought *come all from the set of autocorrelations of the same string*. The analysis and exploitation of such a relationship constitutes a core contribution of the present paper. In a nutshell, if $m = suf_i \oplus suf_j$ has an occurrence at some position k in s, then this induces stringent relationships among the number of don't cares in each of the three patterns $m = suf_i \oplus suf_j, m' = suf_i \oplus suf_k$ and $m'' = suf_j \oplus suf_k$. The specific structure of these relationships depends on whether the alphabet is binary or larger. We examine the case of a binary alphabet first.

3.1 Binary Alphabet

Let $m = [suf_i \oplus suf_j]$ and assume an occurrence of m at k. We establish here a relationship among the number of don't cares that are found respectively in m and in the prefixes of length $|m|$ of $suf_i \oplus suf_k$ and $suf_j \oplus suf_k$, that is, $[suf_i \oplus suf_k]$ and $[suf_j \oplus suf_k]$. Such a relationship will enable us to decide whether $k \in \mathcal{L}_m$ based solely on the knowledge of those three numbers of don't cares. We use d_x to denote the number of don't cares in x and $pref_i(x)$ to denote the prefix of x of length i.

Lemma 1. *Let* $m = [suf_i \oplus suf_j]$, $m' = pref_{|m|}(suf_i \oplus suf_k)$ *and* $m'' = pref_{|m|}(suf_j \oplus suf_k)$.

$$k \in \mathcal{L}_m \;\Leftrightarrow\; d_m = d_{m'} + d_{m''}.$$

Proof. We show first that if m has an occurrence at k this implies the claim. Under such hypotheses, we have $i, j, k \in \mathcal{L}_m$, whence, by Property 3, also $m \preceq m' = pref_{|m|}(suf_i \oplus suf_k)$. Similarly, it must be $m \preceq m''$. Considering then homologous positions in m, m' and m'', the following holds:

- If $m[l] = \sigma$ then, from $m \preceq m'$ and $m \preceq, m''$, we get $m'[l] = m''[l] = \sigma$.
- If $m[l] = \bullet \Leftrightarrow suf_i[l] \neq suf_j[l]$, and one of the following two cases is possible:
  1. $m'[l] = \sigma \Leftrightarrow suf_i[l] = suf_k[l] \Leftrightarrow suf_j[l] \neq suf_k[l] \Leftrightarrow m''[l] = \bullet$.
  2. $m'[l] = \bullet \Leftrightarrow suf_i[l] \neq suf_k[l] \Leftrightarrow suf_j[l] = suf_k[l] \Leftrightarrow m''[l] = \sigma$.
  The last one is summarized by $m[l] = \bullet \Leftrightarrow m'[l] \neq m''[l]$. Note that, since m' and m'' both result from a meet of suf_k with some other suffix of s, then $m'[l] \neq m''[l]$ implies $m'[l] = \bullet$ or $m''[l] = \bullet$.

Thus, in correspondence with every don't care of m, only one of the patterns m' and m'' will have a don't care. Since every solid character of m must also appear in homologous positions in both m' and m'', we have that the total number of don't cares in m' and m'' equals the don't cares in m.

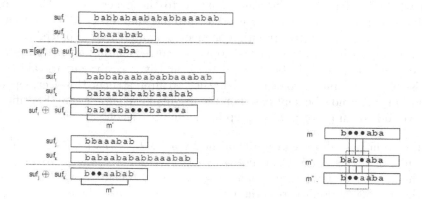

Fig. 1. Illustrating Lemma 1

To prove the converse, we show that if k is not an occurrence of m then this infringes the claimed relationship. Assume then $k \notin \mathcal{L}_m$. Hence, $\exists l$ such that $m[l] = \sigma$ and $suf_k[l] \neq \sigma$. Since $m[l] = \sigma$, it must be $suf_i[l] = \sigma$ and $suf_j[l] = \sigma$, whence $m'[l] = \bullet = m''[l]$. Upon re-examining the distribution of don't cares in m' and m'' with respect to m, we have the following cases:

- $m[l] = \sigma$. This splits into:
  1. $m'[l] = \sigma \Leftrightarrow m''[l] = \sigma$.
  2. $m'[l] = \bullet \Leftrightarrow m''[l] = \bullet$.
- $m[l] = \bullet$. There is no change with respect to the first part of the proof.

We see thus that the difference with respect to the assumption $k \in \mathcal{L}_m$ is posed by some solid characters in m that become don't care in m' and m''. Every don't care in m is balanced by corresponding don't cares in m' and m''. However, we must now add to the equation a positive contribution that amounts to twice the number of positions of suf_k that cause a *mismatch* with m. In other words, when $k \notin \mathcal{L}_m$ we have $d_m < d_{m'} + d_{m''}$, hence $d_m \neq d_{m'} + d_{m''}$. $\square$

Once the number of don't cares in every suffix of each autocorrelation of the binary string s has been tabulated, Lemma 1 makes it possible to decide in constant time whether or not a meet occurs at any given position of s. Tallying don't cares in every suffix of a pattern is trivially accomplished in time linear in the pattern length. Once this is done, the number of don't cares in any substring of that pattern is retrieved by a straightforward subtraction, in constant time. In conclusion, the constant time implementation of occurrence queries based on Lemma 1 requires only a trivial $O(n^2)$ preprocessing of s.

3.2 Alphabets with More Than 2 Symbols

The criterion offered by Lemma 1 can be generalized to a larger alphabet Σ by first generating, from s, $|\Sigma|$ binary instances of the problem, then handling them separately and in analogy to Lemma 1, and finally combining the results in a whole. To handle the instance relative to the generic $\sigma \in \Sigma$, a binary version $\tilde{s}$ of s is built by changing every σ into a '1' and every $\sigma' \neq \sigma$ into a '0'. Upon computing any autocorrelation or meet of $\tilde{s}$, all 0's are replaced by don't cares, yielding what will be referred to as the corresponding *binary projection*. The overall operation will be denoted by $\otimes$ and it is equivalent to taking the bitwise binary product or logical 'and' of the two input strings. However, separate accounting must now be kept based on the origin of each don't care. Specifically, relative to a given pattern m we keep individual tracks of:

1. the number of don't cares originating by a 1×0, with '1' in $\tilde{s}$ and '0' in a suffix of $\tilde{s}$; this number is denoted by d_m^{10};
2. the number of don't cares originating by a 0×1, denoted by d_m^{01};
3. the number of don't cares originating by 0×0, denoted by q_m.

If now $m = pref_{|m|}(suf_i(\tilde{s}) \otimes suf_j(\tilde{s}))$ is one of these binary patterns, we will check its occurrence at k in analogy with Lemma 1 by comparing the don't cares in $m, m' = pref_{|m|}(suf_i(\tilde{s}) \otimes suf_k(\tilde{s}))$, and $m'' = pref_{|m|}(suf_j(\tilde{s}) \oplus suf_k(\tilde{s}))$. Only, this time we will need to distinguish among the three possible origins of the don't cares.

The following relationships among homologous positions of m, m' and m'' are readily checked.

- for any don't care of the type 00 in m, there is one of the type 00 or 01 both in m' and m''.
- for any don't care of the type 10 in m, either:
  - m' has a don't care 10 $\Leftrightarrow$ m'' has a don't care 00.
  - m' has a solid character $\Leftrightarrow$ m'' has a don't care 01.
- for any don't care 01 the situation is dual, and we have one of the following:
  - m'' has a don't care 10 $\Leftrightarrow$ m' has a don't care of type 00.
  - m'' has a solid character $\Leftrightarrow$ m' has a don't care 01.
- every 1 in m has homologous occurrences both in m' and m''.

Lemma 2. *Let* $m = pref_{|m|}(suf_i(\tilde{s}) \otimes suf_j(\tilde{s}))$, $m' = pref_{|m|}(suf_i(\tilde{s}) \otimes suf_k(\tilde{s}))$, $m'' = pref_{|m|}(suf_j(\tilde{s}) \otimes suf_k(\tilde{s}))$, *and set*

$$t = q_m - \frac{(q_{m'} + q_{m''})}{2}.$$

Then,

$$k \in \mathcal{L}_m \Leftrightarrow d_m^{10} - d_{m'}^{10} = d_{m''}^{01} - t.$$

Proof. Observe first that to every 00 don't care of m there corresponds in m' and m'' a pair of don't cares of the same or of 01 type, the latter being counted precisely by the parameter t. Moreover, by the structure of $\otimes$, a 00 don't care in m cannot be covered by a solid character in m' or m''. In other words, the value of t does not depend on whether or not k is an occurrence of m.

Assume now $k \in \mathcal{L}_m$. Then, $d_m^{10} - d_{m'}^{10}$ represents the number of don't cares of m that are covered by a solid character in m'. The only such don't cares of m are of the 10 type, and the only case in which a 10 don't care may originate in m' is by having a corresponding 10 don't care in m. To any don't care covered by m', there corresponds a 01 don't care in m''. However, m'' will contain in general additional don't cares of 01 type, which correspond to the configuration, described earlier, where a 00 don't care of m corresponds a 01 don't care both in m' and m''. This yields the term t in the equality.

For the second part of the proof, assume that the equality holds and yet $k \notin \mathcal{L}_m$. Imagine that, starting with an occurrence of m, one injects mismatches (that is, solid 1 characters, that are transformed into as many 0's in m' e m''). Every new mismatch causes a unit increase in $d_{m'}^{10}$, whereas d_m^{10} and $d_{m''}^{01}$ are not affected. Consequently, $t = d_{m'}^{10} + d_{m''}^{01} - d_m^{10}$ undergoes a unit increase. But this is impossible, by the invariance of t. □

Lemma 3. *Let* $\Sigma = \{\sigma_1, \sigma_2, ..., \sigma_{|\Sigma|}\}$ *and* $m_1, m_2, ..., m_{|\Sigma|}$ *the binary projections of a given meet* $m = [suf_i \oplus suf_j]$.

$$k \in \mathcal{L}_m \Leftrightarrow k \in \mathcal{L}_{m_1} \cap \mathcal{L}_{m_2} \cap ... \cap \mathcal{L}_{|\Sigma|}$$

up to a suitable shift for each list.

Proof. Since the only solid characters of m_i are precisely the occurrences of σ_i in s, we have $m_i \preceq m, \forall i = 1, ..., |\Sigma|$. Then, $\mathcal{L}_m \subseteq \mathcal{L}_{m_i}, \forall i$, hence $\mathcal{L}_m \subseteq \mathcal{L}_{m_1} \cap \mathcal{L}_{m_2} \cap ... \cap \mathcal{L}_{m_{|\Sigma|}}$. Assume an occurrence $l \in \mathcal{L}_{m_1} \cap \mathcal{L}_{m_2} \cap ... \cap \mathcal{L}_{m_{|\Sigma|}}$ but $l \notin \mathcal{L}_m$. This implies the existence of a position in suf_l that causes a mismatch with a solid character, say, σ_k, of m. Consider then the projection involving σ_k. From $l \notin \mathcal{L}_m$ we get $l \notin \mathcal{L}_{m_k}$, hence $l \notin \mathcal{L}_{m_1} \cap \mathcal{L}_{m_2} \cap ... \cap \mathcal{L}_{m_{|\Sigma|}}$. □

Theorem 1. *Let* s *be a string of* n *characters over an alphabet* Σ, *and* m *the meet of any two suffixes of* s. *Following an* $O(|\Sigma|n^2)$ *time preprocessing of* s, *it is possible to decide for any assigned position* k *whether or not* k *is an occurrence of* m *in time* $O(|\Sigma|)$.

Proof. By the preceding properties and discussion. □

4   An $O(|\Sigma|n^2)$ Off-Line Basis Computation

We are now ready to develop in full the criterion upon which our optimal algorithm is built. Recall that in order for a motif to be irredundant it must have at least one occurrence that cannot be deduced from occurrences of other motifs. In [2], such an occurrence is called *maximal* and the motif is correspondingly said to be *exposed* at the corresponding position. Clearly, every motif with a maximal occurrence is saturated. However, not every saturated motif has a maximal occurrence. In fact, the set of irredundant motifs is precisely the subset of saturated motifs with a maximal occurrence. The following known definitions and properties (see, e.g., [2,9]) systematize these notions and a few more important facts.

Definition 10. (Maximal occurrence) *Let* $(m, \mathcal{L}_m)$ *be a motif of* s *and* $j \in \mathcal{L}_m$. *Position* j *is a* maximal occurrence *for* m *if for no* $d' \geq 0$ *and* $(m', \mathcal{L}_{m'})$ *we have* $\mathcal{L}_{m'} \subseteq (\mathcal{L}_m - d')$ *with* $(j - d') \in \mathcal{L}_{m'}$.

For a given $m \in \mathcal{B}$, let $\mathcal{L}_m^{max}$ denote the list of maximal occurrences of m. The following lemma contributes an alternative definition of irredundancy, whereby the problem of identifying the basis for s translates into that of identifying the motifs with at least one maximal occurrence in s.

Lemma 4. $m \in \mathcal{B} \Leftrightarrow |\mathcal{L}_m^{max}| > 0$.

Proof. By the definition of irredundancy. □

Lemma 5. *If $m \in \mathcal{B}$, then*

$$j \in \mathcal{L}_m^{max} \Leftrightarrow [s \oplus suf_{(max\{j,k\}-min\{j,k\})}] = m, \forall k \in \mathcal{L}_m.$$

Proof. W.l.o.g., assume $k > j$ and that $\exists k \in \mathcal{L}_m$, $k > j : w = [s \oplus suf_{k-j}] \neq m$. Let $w = vm'$ with $m' = [suf_j \oplus suf_k]$. Since $j, k \in \mathcal{L}_m$, it must be $m \preceq m'$, hence $m \prec m'$ from the hypothesis. Then, the occurrence at j is covered by m' (or by the motifs that cover m'). Therefore, it must be $m = m' = [suf_j \oplus suf_k], \forall k \in \mathcal{L}_m$ and then $w = vm' = vm$. Likewise, if $w \neq m$, then the occurrence at j is covered by w and thus cannot be maximal. It remains to be shown that the converse is also true, i.e., that

$$[s \oplus suf_{(max\{j,k\}-min\{j,k\})}] = m, \forall k \in \mathcal{L}_m$$

implies that j is a maximal occurrence for m. But it follows from the hypothesis that the occurrence at j cannot be covered by any motif with a list $\mathcal{L} \subset \mathcal{L}_m$, whence this occurrence is maximal. □

Lemma 5 gives a handle to check whether a position i is a maximal occurrence for an assigned motif $(m, \mathcal{L}_m)$. For this, it suffices to check that $[suf_i \oplus suf_k] = m, \forall k \in \mathcal{L}_m$. We note, for future record, that this holds in particular for i equal to the smallest index in $\mathcal{L}_m$.

Theorem 2. *Every irredundant motif is the meet of s and one of its suffixes.*

Proof. Let $m \in \mathcal{B}$. By Lemma 4, we derive that $m \in \mathcal{B} \Leftrightarrow m$ must be exposed at some position of s. Let j be this position. It follows from Lemma 5 that for m to be exposed at j it must be $m = [s \oplus suf_{(max\{j,k\}-min\{j,k\})}], \forall k \in \mathcal{L}_m$. □

We have thus:

Theorem 3. *The number of irredundant motifs in a string of n characters is $O(n)$.*

Proof. Immediate. □

Lemma 6. $\sum_{m \in \mathcal{B}} |\mathcal{L}_m| < 2n$.

Proof. Let m be an irredundant motif in s, with a maximal occurrence at j. >From Lemma 5, it follows that

$$\forall k \in \mathcal{L}_m, \quad m = [s \oplus suf_{(max\{j,k\}-min\{j,k\})}].$$

We charge each term of $\mathcal{L}_m$ other than j to a different shift $(max\{j,k\} - min\{j,k\})$ between s and one of its suffixes, and we charge the occurrence at j to s itself. By the maximality of m, each one of the possible shifts in $\{1, 2, ..., n-1\}$ is only charged once, while s gets charged at most $n - 1$ times overall. □

Lemma 6 shows that the implicit specification of the basis takes linear space. This provides a crucial pillar for an algorithm based entirely on the management of occurrence lists, as is described next.

In the off-line basis computation the only patterns at play are the autocorrelations of the input string s, since the only candidate irredundant motifs consist of the corresponding meets. The main stages of the computation are summarized as follows:

1 compute the occurrence lists of the autocorrelations of s;
2 store these autocorrelations in a trie ;
3 identify the members of the basis by visiting the nodes of the trie.

In order to take advantage of Lemma 1 and its extensions, Stage 1 requires a straightforward preprocessing that consists of computing, for each $m = [s \oplus suf_i]$ the number of don't cares that are present in every suffix of m partitioned, for large alphabets, among the implied binary projections. Once this information is available, that Lemma and its extensions will support the compilation of the occurrence lists of all autocorrelations of s.

Once this is done, the collection of all autocorrelations of s are stored in a trie each node of which stores some additional information, as follows. Let us say that suf_k takes part in a autocorrelation m of s if $m = [suf_k \oplus suf_l]$ for some l. Let v be a node of the trie and denote by $index[k]$ the number of times that suf_k takes part in a autocorrelation of s that terminates at v.

Definition 11. The index of v is $I(v) = max_{1 \le k \le |s|}\{index[k]\}$.

We assume that it is possible to know, for every node of the trie, both $I(v)$ and the suffix of s that produces it. Then, the elements of the basis are identified by comparing, for every terminal node of an autocorrelation, $I(v)$ and $|\mathcal{L}_m|$. In fact, we know that if m has a maximal occurrence at i, then $m = [suf_i \oplus suf_j], \forall j \in \mathcal{L}_m$. Therefore, if $I(v) = |\mathcal{L}_m|$ then the suffix yielding $I(v)$ is a maximal occurrence for m.

Theorem 4. The basis of irredundant 2-motifs in a string s of n characters can be computed in time $(O|\Sigma|n^2)$.

Proof. The dominant term in the computation is finding the occurrence lists for the autocorrelations, which was found to take time $O(|\Sigma|n^2)$. Building the trie takes $O(n^2)$ and the computation of $I(v)$ for each v can be carried out during this construction at no extra cost. The last step only calls for a constant check at each node, which charges $O(n)$ overall. □

Since the explicit description of the $O(n)$ motifs of the basis takes space $O(n^2)$, then this algorithm is optimal for any alphabet of constant size. As it turns out, the computation of the basis may be further simplified when $|\Sigma| = 2$. This rests on some additional properties that will be described only in the full version of the paper.

5 Concluding Remarks

Several issues are still open, notable among them, the existence of an optimal algorithm for general alphabets and of an optimal incremental algorithm for alphabets of constant or unbounded size.

References

1. Apostolico, A., Galil, Z.: Pattern matching algorithms. Oxford University Press, New York (1997)
2. Apostolico, A., Parida, L.: ncremental paradigms of motif discovery. Journal of Computational Biology 11(1), 15–25 (2004)
3. Apostolico, A.: Pattern discovery and the algorithmics of surprise. Artificial Intelligence and Heuristic Methods for Bioinformatics, pp. 111–127 (2003)
4. Cole, R., Hariharan, R.: Verifying candidate matches in sparse and wildcard matching. In: STOC '02. Proceedings of the thiry-fourth annual ACM symposium on Theory of computing, pp. 592–601 (2002)
5. Fischer, M.J., Paterson, M.S.: String matching and other products. In: Karp, R. (ed.) Proceedings of the SIAM-AMS Complexity of Computation, Providence, R.I. American Mathematical Society, pp. 113–125 (1974)
6. Navarro, G.: A guided tour to approximate string matching. ACM Computing Surveys 33(1), 31–88 (2001)
7. Pelfrêne, J., Abdeddaïm, S., Alexandre, J.: Extracting approximate patterns. Journal of Discrete Algorithms 3(2-4), 293–320 (2005)
8. Parida, L.: Algorithmic Techniques in Computational Genomics. PhD thesis, Department of Computer Science, New York University (1998)
9. Pisanti, N., Crochemore, M., Grossi, R., Sagot, M.-F.: Bases of motifs for generating repeated patterns with wild cards. IEEE/ACM Trans. Comput. Biol. Bioinformatics 2(1), 40–50 (2005)
10. Parida, L., Rigoutsos, I., Floratos, A., Platt, D., Gao, Y.: Pattern discovery on character sets and real-valued data: linear bound on irredundant motifs and an efficient polynomial time algorithm. In: Symposium on Discrete Algorithms, pp. 297–308 (2000)
11. Wang, J.T.L., Shapiro, B.A., Shasha, D.E.: Pattern Discovery in Biomolecular Data: Tools, Techniques and Applications. Oxford University Press, Oxford (1999)

Linear Algorithm for Broadcasting in Unicyclic Graphs

(Extended Abstract)

Hovhannes Harutyunyan and Edward Maraachlian

Concordia University, Department of Computer Science and Software Engineering,
Montreal, QC. H4G 1M8, Canada
haruty@cs.concordia.ca, e_maraac@cs.concordia.ca

Abstract. *Broadcasting* is an information dissemination problem in a connected network, in which one node, called the *originator*, disseminates a message to all other nodes by placing a series of calls along the communication lines of the network. Once informed, the nodes aid the originator in distributing the message. Finding the minimum broadcast time of a vertex in an arbitrary graph is NP-complete. The problem is solved polynomially only for trees. It is proved that the complexity of the problem of determining the minimum broadcast time of any vertex in an arbitrary tree $T = (V, E)$ is $\Theta|V|$. In this paper we present an algorithm that determines the broadcast time of any originator in an arbitrary unicyclic graph $G = (V, E)$ in $O(|V|)$ time. This, combined with the obvious lower bound, gives a $\Theta(|V|)$ solution for the problem of broadcasting in unicyclic graphs. As a byproduct, we also find a broadcast center of the unicyclic graph (a vertex in G with the minimum broadcast time).

1   Introduction

Computer networks have become essential in several aspects of modern society. The performance of information dissemination in networks often determines their overall efficiency. One of the fundamental information dissemination problems is broadcasting. Broadcasting is a process in which a single message is sent from one member of a network to all other members. Inefficient broadcasting could degrade the performance of a network seriously. Therefore, it is of a major interest to improve the performance of a network by using efficient broadcasting algorithms.

Broadcasting is an information dissemination problem in a connected network, in which one node, called the *originator*, must distribute a message to all other nodes by placing a series of calls along the communication lines of the network. Once informed, the informed nodes aid the originator in distributing the message. This is assumed to take place in discrete time units. The broadcasting is to be completed as quickly as possible, subject to the following constrains:

- Each call involves only one informed node and one of its uninformed neighbors.
- Each call requires one unit of time.

G. Lin (Ed.): COCOON 2007, LNCS 4598, pp. 372–382, 2007.

– A node can participate in only one call per unit of time.
– In one unit of time, many calls can be performed in parallel.

A *broadcast scheme* of an originator u is a set of calls that completes the broadcasting in the network originated at vertex u.

Formally, any network can be modeled as a connected graph $G = (V, E)$, where V is the set of vertices (or nodes) and E is the set of edges (or communication lines) between the vertices in graph G.

Given a connected graph $G = (V, E)$ and a message originator, vertex u, the *broadcast time* of vertex u, $b(u, G)$ or $b(u)$, is the minimum number of time units required to complete broadcasting from the vertex u. Note that for any vertex u in a connected graph G on n vertices, $b(u) \geq \lceil \log n \rceil$ since during each time unit the number of informed vertices can at most be doubled. The broadcast time $b(G)$ of the graph G is defined as $max\{b(u)|u \in V\}$.

Determination of $b(u, G)$ or $b(u)$ for a vertex u in an arbitrary graph G is NP-complete [17]. The proof of NP-completeness is presented in [20]. Therefore, many papers have presented approximation algorithms to determine the broadcast time of any vertex in G (see [2], [3], [8], [9], [11], [12], [13], [14], [18], [19], [21]). Some of these papers give theoretical bounds on the broadcast time. Given $G = (V, E)$ and the originator u, the heuristic in [18] returns broadcast algorithm with broadcast time at most $b(u, G) + O(\sqrt{|V|})$. The best theoretical upper bound is presented in [9]. Their approximation algorithm generates a broadcast algorithm with broadcast time $O(\frac{log(|V|)}{loglog(|V|)})b(G)$. The heuristics [3] and [16] are the best existing heuristics for broadcasting in practice. Their performance is almost the same for commonly used interconnection networks. However, the broadcast algorithm from [16] outperforms the broadcast algorithm from [3] in three graph models from a network simulator ns-2 (see[1,2,7,22]). Also, its time complexity is $O(|E|)$, while the complexity of the algorithm from [3] is $O(|V|^2 \cdot |E|)$.

Since the problem is NP-complete in general, another direction for research is to design polynomial algorithms that determines the broadcast time of any vertex for a class of graphs. To the best of our knowledge, the only known result in this direction is the linear algorithm (called BROADCAST) designed in [20] which determines the broadcast time of a given vertex in any tree. They also find a broadcast scheme of a given tree $T = (V, E)$ in linear time $O(|V|)$.

In this paper we present an algorithm that determines the broadcast time of any originator in an arbitrary unicyclic graph $G = (V, E)$. The algorithm is linear, $O(|V|)$, for any originator. As a byproduct, we also find a broadcast center (a vertex in G with minimum broadcast time) of the unicyclic graph.

The paper is organized as follows. The next section will present some auxiliary results that will be necessary to understand the algorithm and prove its correctness and linearity. Section 3 presents the actual algorithm with a proof of correctness and a complexity analysis. The final section is a conclusion.

2   Definitions and Auxiliary Results

A unicyclic graph (Fig. 1) is a connected graph with only one cycle. Basically it is a tree with only one extra edge. It can also be seen as a cycle where every vertex on the cycle is the root of a tree. Denote the vertices of the cycle C_k by $r_1, r_2, \cdots, r_k$ and the tree rooted at r_i by T_i, where $1 \le i \le k$. We will use the following definitions and results from [20].

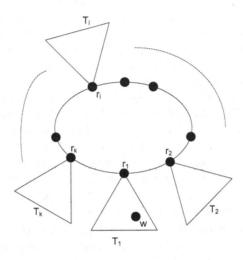

Fig. 1. A unicyclic graph where the vertices r_i, belonging to the cycle C_k, are the roots of the trees T_i for $1 \le i \le k$

Definition 1 ([20]). *The minimum broadcast time, $b(BC, G)$, of the graph $G = (V, E)$ is defined to be the minimum of the broadcast times of all the vertices. $b(BC, G) = min_{u \in V}\{b(u, G)\}$.*

Definition 2 ([20]). *The broadcast center of the graph G, $BC(G)$, is defined to be the set of all vertices whose broadcast time is equal to the minimum broadcast time of the graph, $BC(G) = \{u | b(u, G) = b(BC, G)\}$.*

Theorem 1 ([20]). *Let $v \notin BC(T)$ be a vertex in a tree T such that the shortest distance from v to a vertex $x \in BC(T)$ is k. Then $b(v, T) = k + b(BC, T)$.*

Corollary 1 ([20]). *For any tree T, $BC(T)$ consists of a star with at least two vertices.*

The unicyclic graph can be converted into a tree by cutting one of the edges of the cycle $C_k = r_1, r_2, ..., r_k, r_1$. A simple algorithm to determine the broadcast time of a vertex w in an arbitrary unicyclic graph $G = (V, E)$ would be the following:

SIMPLEBROADCASTALGORITHM(w, G):

1. Extract from G the cycle C_k and the trees T_i for $1 \leq i \leq k$ which are rooted at a vertex on C_k.
2. Cut edge (r_i, r_{i+1}) from the cycle C_k, for $i = 1, 2, ..., k$. Denote the resulting tree by G_i.
3. Apply BROADCAST(w, G_i) for $i = 1, 2, ..., k$ from [20] and choose the tree G_i with the minimum broadcast time $b(w, G_i)$.

The complexity of step 1 of the algorithm is $O(n)$, where $|V| = n$. The complexity of steps 2 and 3 are $O(k)$ and $O(kn)$ respectively. Thus, the total complexity of the above algorithm will be $O(kn)$, which is $O(n^2)$ in the worst case. However, $\Omega(n)$ is an obvious lower bound. In this paper we will show $\Theta(n)$ bound by describing a linear algorithm that determines the broadcast time of any vertex w in an arbitrary unicyclic graph.

Notation 1 *If u and v are two vertices in graph G then (u, v) represents the edge between them, and $d(u, v)$ represents the distance between them.*

Definition 3. *Given trees $T_1 = (V_1, E_1)$, $T_2 = (V_2, E_2)$, $\cdots$, $T_i = (V_i, E_i)$ with roots r_1, r_2, $\cdots$, r_i respectively, the tree $T_{1,2,\cdots,i} = (V, E) = T_1 \oplus T_2 \oplus \cdots \oplus T_i$ is a tree where $V = V_1 \cup V_2 \cup \cdots \cup V_i$ and $E = E_1 \cup E_2 \cup \cdots \cup E_i \cup \{(r_1, r_2), (r_2, r_3), \cdots, (r_{i-1}, r_i)\}$.*

In other words, the trees T_i are connected by adding the edges (r_1, r_2), (r_2, r_3), $\cdots$, (r_{i-1}, r_i).

2.1   The Broadcast Center of the Sum of Two Trees

In this section we will describe how to find a broadcast center and calculate the minimum broadcast time of the sum of two trees.

Lemma 1. *In any tree T, rooted at r, there exists a unique vertex $u \in BC(T)$, called the special broadcast center denoted as $u = SBC(T)$, such that the path joining u and r does not contain any other vertex v such that $v \in BC(T)$.*

Proof. First we will show the existence of vertex u. Let v be a vertex such that $v \in BC(T)$ and let $P = v, v_1, v_2, \cdots, v_k, r$ be the unique path from v to r. Due to Corollary 1 only v_1 can be in $BC(T)$. If $v_1 \in BC(T)$, then it is the required vertex, and $v_1, v_2, \cdots, v_k, r$ is the path from $SBC(T) = v_1$ to the root of T, r. If $v_1 \notin BC(T)$ then v is the required vertex and $P = v, v_1, v_2, \cdots, v_k, r$ is the unique path from $SBC(T) = v$ to the root of T, r. The uniqueness of $u = SBC(T)$ immediately follows from the fact that T is a tree and no cycles are allowed in a tree.

From Lemma 1 it follows that if T_1 and T_2 are two trees with $u_1 = SBC(T_1)$ and $u_2 = SBC(T_2)$, then the path that joins u_1 and u_2 does not contain any other vertex v (different than u_1 and u_2) such that $v \in (BC(T_1) \cup BC(T_2))$.

In the remaining part of this section it is assumed that there are two trees T_1 and T_2 with roots r_1 and r_2 respectively, $T = T_1 \oplus T_2$, $u_1 = SBC(T_1)$, and $u_2 = SBC(T_2)$.

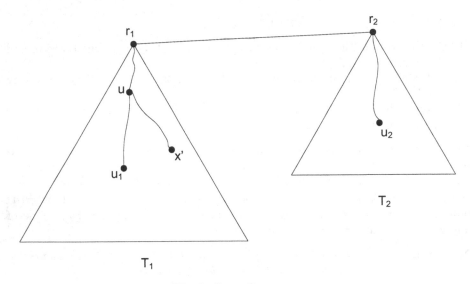

Fig. 2. Sum of two trees

Theorem 2. *There exists a vertex u such that $u \in BC(T_1 \oplus T_2)$ and u is on the path joining $u_1 = SBC(T_1)$ and $u_2 = SBC(T_2)$.*

Proof. We will prove this theorem by contradiction. Assume that there exists a vertex x' (Fig. 2) not on the path from u_1 to u_2 and such that $b(x', T) < b(u, T)$ for all vertices u on the path joining u_1 and u_2. Without loss of generality assume that x' is in T_1. Because T is a tree, there exists a unique path P that joins x' to r_1. Two cases may arise:

Case 1: The path P intersects the path from u_1 to r_1 at a vertex other than u_1. Let $u \in P$ be this intersection vertex. We denote by $t_1(x)$ the minimum time that is needed to inform all the vertices of T_1 starting at the originator x. Therefore, $b(u, T_1) = t_1(u) = d(u, u_1) + b(BC, T_1)$ and $t_1(x') = b(x', T_1) = d(x', u) + (d(u, u_1) + b(BC, T_1)) = d(x', u) + t_1(u)$. Similarly, the minimum time to inform all the vertices in T_2 starting at any originator x will be denoted by $t_2(x)$. We have $t_2(u) = d(u, r_1) + 1 + d(r_2, u_2) + b(BC, T_2) = d(u, u_2) + b(BC, T_2)$ and $t_2(x') = d(x', u) + d(u, u_2) + b(BC, T_2)$. Having calculated $t_1(x)$ and $t_2(x)$ we can calculate $b(x, T)$ since $b(x, T) = t_1(x) + 1$ if $t_1(x) = t_2(x)$ and $b(x, T) = max\{t_1(x), t_2(x)\}$ otherwise. Since, $t_1(x') > t_1(u)$ and $t_2(x') > t_1(u)$ we conclude that $b(x', T) > b(u, T)$ which contradicts the assumption that $b(x', T) < b(u, T)$.

Case 2: The path P, joining x' and r_1, does not intersect the path joining u_1 and r_1. In this case the path P will merge with the path joining u_1 and r_1 at vertex u_1. Using arguments similar to the preceding case we can show that there there is no vertex x' such that $b(x', T) < b(u_1, T)$. The details are omitted.[1]

---

[1] Refer to Fig. 3 for an example illustrating Theorems 2 and 3.

Let u be a vertex such that $u \in BC(T_1 \oplus T_2)$. Theorem 2 confirms the existence of such a vertex on the path joining u_1 and u_2. The position of u can be found as described in the following theorem.

Theorem 3. *Let* $A = b(BC, T_1) - b(BC, T_2)$ *and* $B = d(r_1, u_1) + d(r_2, u_2) + 1$. *Three cases may arise:*

If $B - A < 0$, *then* $d(u, u_1) = 0$, *i.e.* u *coincides with* u_1.

If $A + B < 0$, *then* $d(u, u_1) = B$, *i.e.* u *coincides with* u_2.

If $B - A \geq 0$ *and* $A + B \geq 0$, *then* $d(u, u_1) = \lfloor \frac{B-A}{2} \rfloor$ *or* $d(u, u_1) = \lceil \frac{B-A}{2} \rceil$. *Both positions of* u *have equal broadcast times in the tree* T.

Proof. If $B - A < 0$, then we have

$$b(BC, T_1) > b(BC, T_2) + d(u_1, r_1) + 1 + d(u_2, r_2). \tag{1}$$

We will prove that $u = u_1 \in BC(T)$ by contradiction. Assume that there is another vertex u' on the path joining u_1 and u_2 such that $b(u', T) < b(u, T)$. Because of Theorem 2 we do not have to consider a vertex not belonging to the path joining u_1 and u_2. Using the definitions of the functions $t_1(x)$ and $t_2(x)$ from above we get: $t_1(u') = d(u', u_1) + b(BC, T_1)$, $t_2(u') = d(u', r_1) + 1 + d(u_2, r_2) + b(BC, T_2)$, $t_1(u) = b(BC, T_1)$, and $t_2(u) = d(u_1, r_1) + 1 + d(u_2, r_2) + b(BC, T_2)$. Using the condition in equation 1 we get that $t_1(u) = b(BC, T_1) > t_2(u) = d(u_1, r_1) + 1 + d(u_2, r_2) + b(BC, T_2)$ Therefore, $b(u, T) = t_1(u)$. Similarly, $t_1(u') = d(u', u_1) + b(BC, T_1) > d(u', u_1) + b(BC, T_2) + d(u_1, r_1) + 1 + d(u_2, r_2)$ which implies that $t_1(u') > d(u', u_1) + b(BC, T_2) + d(u_1, u') + d(u', r_1) + 1 + d(u_2, r_2) = 2d(u', u_1) + b(BC, T_2) + d(u', r_1) + 1 + d(u_2, r_2) = 2d(u', u_1) + t_2(u')$ which implies that $t_1(u') > t_2(u')$ and hence $b(u', T) = t_1(u')$. But we have $t_1(u') = d(u', u_1) + b(BC, T_1)$ and $t_1(u) = b(BC, T_1)$ which implies that $t_1(u') > t_1(u)$. Therefore we conclude that $b(u', T) > b(u, T)$ which contradicts the assumption.

The case $A + B < 0$ can be proved similarly.

Note that the two conditions $A + B < 0$ and $B - A < 0$ are mutually exclusive. If either one of them is satisfied the other will not be satisfied. The only remaining case is when both of them are not satisfied i.e. $A + B \geq 0$ and $B - A \geq 0$. Assume the case where $B - A$ is odd, the case if it is even can be dealt with similarly. Without loss of generality, assume that there exists a vertex $u' \in T_1$, on the path joining u_1 and u_2, such that $b(u', T) < b(u, T)$. Two cases may arise:

Case 1: $d(u', u_1) < d(u, u_1) = \lfloor \frac{B-A}{2} \rfloor$. Since $B - A$ is assumed to be odd, $d(u, u_1) = \frac{B-A-1}{2}$. Calculating $t_1(u) = d(u, u_1) + b(BC, T_1)$ and $t_2(u) = d(u, r_1) + 1 + d(u_2, r_2) + b(BC, T_2)$ we deduce that $t_2(u) = (d(u_1, r_1) - d(u, u_1)) + 1 + d(u_2, r_2) + b(BC, T_2)$. Substituting the value of $d(u, u_1)$, and $b(BC, T_2) = b(BC, T_1) - A$ we get: $t_2(u) = [d(u_1, r_1) + 1 + d(u_2, r_2)] + b(BC, T_2) - d(u, u_1) = B + (b(BC, T_1) - A) - \frac{B-A-1}{2} = \frac{B-A+1}{2} + b(BC, T_1) = t_1(u) + 1$. Therefore, $b(u, T) = t_2(u)$. Now consider the vertex u', $t_1(u') = d(u', u_1) + b(BC, T_1)$ and $t_2(u') = d(u', u) + d(u, r_1) + 1 + d(u_2, r_2) + b(BC, T_2)$. Since $d(u', u_1) < d(u, u_1)$, we get $t_1(u') < t_1(u)$. Moreover, $t_2(u') = d(u', u) + t_2(u) > t_1(u')$ since $t_2(u) > t_1(u) > t_1(u')$. Finally we arrive at: $b(u', T) = t_2(u') > b(u, T)$ which contradicts the assumption.

Case 2: $d(u', u_1) > d(u, u_1)$. As it was done in the previous case, we can deduce that $t_2(u) = t_1(u) + 1$ and $b(u, T) = t_2(u)$. Now consider the vertex u'. Calculating $t_1(u')$ and $t_2(u')$ we get: $t_1(u') = d(u', u) + d(u, u_1) + b(BC, T_1)$ and $t_2(u') = d(u', r_1) + 1 + d(r_2, u_2) + b(BC, T_2)$. Using that $d(u, r_1) = d(u, u') + d(u', r_1)$ we get: $t_2(u') = [d(u, r_1) - d(u, u')] + 1 + d(r_2, u_2) + b(BC, T_2)$. Hence, $t_2(u') = t_2(u) - d(u, u') = t_1(u) + 1 - d(u, u')$. On the other hand, $t_1(u') = t_1(u) + d(u, u')$. Since $d(u', u_1) > d(u, u_1)$, we conclude that $d(u, u') \geq 1$. Subtracting $t_2(u')$ from $t_1(u')$ we get: $t_1(u') - t_2(u') = t_1(u) + d(u, u') - [t_1(u) + 1 - d(u, u')]$. Therefore, $t_1(u') - t_2(u') = 2d(u, u') - 1$. Using $d(u, u') \geq 1$, we get that $t_1(u') > t_2(u') + 1$. Hence, we conclude that $b(u', T) = t_1(u')$. So, $b(u', T) = t_1(u') = t_2(u') + 2d(u, u') - 1$. Furthermore, using $t_2(u') = t_2(u) - d(u, u')$ we get $b(u', T) = t_2(u) + d(u, u') - 1 \geq t_2(u)$, which implies that $b(u', T) \geq b(u, T)$ which is a contradiction.

3   The UNICYCLICBROADCAST Algorithm

The algorithm, $UNICYCLICBROADCAST(w, G)$, calculates the broadcast time, $b(w, G)$, of a given vertex w in any unicyclic graph G. For convenience we will assume that w belongs to tree T_1.

3.1   Description of the Algorithm

INPUT: A unicyclic graph G on n vertices and the broadcast originator w.
OUTPUT: Broadcast time of the originator w in G, $b(w, G)$, and a broadcast scheme.

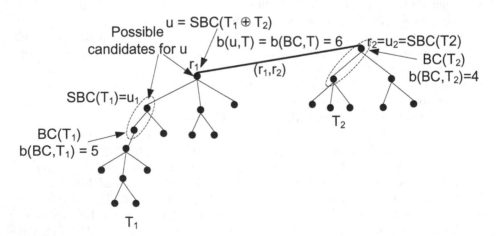

Fig. 3. An example of the sum of two trees. It shows the vertices in the broadcast center and the special broadcast center of T_1, T_2, and $T_1 \oplus T_2$. The minimum broadcast time of the three trees are calculated too.

UNICYCLICBROADCAST(w, G):

1. Extract from G the cycle C_k, consisting of the vertices $\{r_1, r_2, \cdots, r_k\}$, and the trees T_i rooted at the vertices r_i for $1 \leq i \leq k$.
2. For all trees T_i where $1 \leq i \leq k$ calculate and save the positions of $u_i = SBC(T_i)$ relative to r_i, as well as $b(BC, T_i)$.
3. Calculate and save the distance $d(w, u_1)$ and the path joining w and u_1.
4. Construct the trees $T_{1,2,\cdots,i}$, where $2 \leq i \leq k$, and $T_{k,k-1,\cdots,i}$, where $1 \leq i \leq k - 1$. For each tree T, compute and store the position of the $SBC(T)$ and $b(BC, T)$.
5. Construct the spanning trees $T_{j,j+1,\cdots,k,1,2,\cdots,j-1}$, where $1 \leq j \leq k$. For each tree T compute and store $SBC(T)$, $b(BC, T)$, $d(w, SBC(T))$, and $b(w, T)$.
6. Out of the trees generated in the previous step, choose the spanning tree T with the minimum value of $b(w, T)$.
7. Run BROADCAST [20] to find a broadcast scheme for the originator w.

The algorithm first preprocesses the unicyclic graph and calculates the cycle C_k consisting of the vertices $\{r_1, r_2, \cdots, r_k\}$ and the trees T_i rooted at r_i, where $1 \leq i \leq k$. In step 2 the path joining w to u_1 and $d(w, u_1)$ are calculated and saved. This information will be needed to calculate the broadcast time of w. Steps 3 and 4 construct several trees and calculate results that will be used in constructing the k spanning trees of the graph G. More specifically the trees, T_i, $T_{i,i+1,\cdots,k}$ for $1 \leq i \leq k - 1$, and $T_{1,2,\cdots,i}$ for $2 \leq i \leq k$ are constructed. For each spanning tree T the position of the $SBC(T)$ and $b(BC, T)$ are calculated and stored. These results will be useful to calculate the broadcast centers and the minimum broadcast times of the spanning trees. At the end of step 4 only one spanning tree will be constructed. In step 5, the algorithm builds the remaining $k - 1$ spanning trees of the unicyclic graph. It also calculates the broadcast time of w for each one of them. Note that for each spanning tree T, $b(w, T) = d(w, SBC(T)) + b(BC, T)$ where the distance $d(w, SBC(T))$ can be easily calculated by using $d(w, u_1)$ and position of $SBC(T)$ calculated in steps 2 and 4 respectively. The spanning tree that has the minimum broadcast time for w is the required result. Finally in order to obtain the optimal broadcast scheme for the originator w in the unicyclic graph G, the BROADCAST algorithm of [20] is run on the spanning tree T that had the minimum value of $b(w, T)$.

In step 5, $k - 1$ spanning trees are constructed each in a constant time. The construction of each tree can be done easily by observing that the spanning tree $T_{i,i+1,\cdots,k,1,2,\cdots,i-1}$ is the sum of the two trees $T_{i,i+1,\cdots,k}$ and $T_{1,2,\cdots,i-1}$ rooted at r_k and r_1 respectively. These two trees and all the information pertinent to the calculation of the $b(w, T_{i,i+1,\cdots,k,1,2,\cdots,i-1})$ were calculated in step 4.

3.2 Proof of Correctness and Complexity Analysis

Theorem 4. *UNICYCLICBROADCAST(w, G) generates a broadcast scheme for originator w, and finds $b(w, G)$.*

Proof. Let C_k, consisting of the vertices $\{r_1, r_2, \cdots, r_k\}$, represent the cycle in the unicyclic graph G. One of the spanning trees of G is obtained by removing the edge (r_k, r_1). The resulting tree is $T_{1,2,\cdots,k}$. The minimum broadcast time of $T_{1,2,\cdots,k}$ is calculated iteratively by tree summations $T_{1,\cdots,i-1} \oplus T_i$ rooted at r_{i-1} and r_i respectively, where $2 \leq i \leq k$. The remaining minimum spanning trees, $T_{j,j+1,\cdots,k,1,2,\cdots,j-1}$ where $2 \leq j \leq k$, are calculated by performing the summations $T_{j,\cdots,k} \oplus T_{1,\cdots,j-1}$. Theorems 2 and 3 guarantee that a broadcast center and the minimum broadcast time of all the trees $T_{j,j+1,\cdots,k,1,2,\cdots,j-1}$, where $1 \leq j \leq k$, are calculated correctly. Since, the trees $T_{j,j+1,\cdots,k,1,2,\cdots,j-1}$, where $1 \leq j \leq k$, are the all possible spanning trees of the unicyclic graph G, the algorithm correctly finds the spanning tree of the unicyclic graph G that has the minimum broadcast time of all the spanning trees. We will prove the correctness of the algorithm by contradiction. Assume that there exists a broadcast tree T' and a broadcast scheme in G that performs broadcasting in time t such that $t = d(w, BC(T')) + b(BC, T') < b(w, G)$. T' should be one of the trees $T_{j,j+1,\cdots,k,1,2,\cdots,j-1}$, where $1 \leq j \leq k$. But the algorithm correctly calculated the minimum broadcast time of all the spanning trees and chose the spanning tree T with the minimum value of $d(w, BC(T)) + b(BC, T)$. Therefore the existence of T' creates a contradiction.

Theorem 5. *The complexity of the UNICYCLICBROADCAST running on a unicyclic graph $G = (V, E)$ is $O(|V|)$.*

Proof. Steps 1 and 2 of the algorithm can be accomplished by a depth first search in $O(|V|)$ time. In Step 3, BROADCAST [20] is applied on the trees T_i for $1 \leq i \leq k$. The complexity of this step is $O(|V_1| + |V_2| + \cdots + |V_k|) = O(|V|)$. Step 4 is of complexity $O(k)$ since there are k sums to be done, and the sum of two trees T_1 and T_2, $T_1 \oplus T_2$, can be done in a constant time. Moreover, every time a tree T is constructed as $T = T_1 \oplus T_2$, calculating the distance between the root of T and $SBC(T)$, and $b(BC, T)$ can be done in a constant time using the positions of $SBC(T_1)$ and $SBC(T_2)$, and the broadcast times $b(BC, T_1)$ and $b(BC, T_2)$. Step 5 involves the calculation of $k - 1$ spanning trees. Each spanning tree is constructed by summing two trees, hence the complexity of this step is again $O(k)$. Step 6 chooses the spanning tree with the least minimum broadcast time hence its complexity is $O(k)$. Adding all the complexities we get that the complexity of the algorithm is $O(|V| + k)$. However, $k \leq |V|$, so this proves that the complexity of the algorithm UNICYCLICBRDCST is $O(|V|)$.

4   Conclusion

In this paper we presented an algorithm that determines the broadcast time of any vertex w in an arbitrary unicyclic graph $G = (V, E)$ in linear time, $O(|V|)$. Therefore, the complexity of the problem of determining the broadcast time of any vertex in an arbitrary unicyclic graph is $\Theta(|V|)$. The algorithm can find the minimum broadcast time of G as well as the spanning tree of G which is the broadcast tree corresponding to the minimum broadcast time. As a byproduct, the algorithm also finds a broadcast center of the unicyclic graph.

References

1. Aiello, W., Chung, F., Lu, L.: Random evolution in massive graphs. In: FOCS'01. Proceedings of the 42nd Annual IEEE Symposium on Foundations of Computer Science, pp. 510–519 (2001)
2. Bar-Noy, A., Guha, S., Naor, J., Schieber, B.: Multicasting in Heterogeneous Networks. In: STOC'98. Proc. of ACM Symp. on Theory of Computing (1998)
3. Beier, R., Sibeyn, J.F.: A powerful heuristic for telephone gossiping. In: SIROCCO'00. Proc. of the 7th International Colloquium on Structural Information & Communication Complexity, L'Aquila, Italy, pp. 17–36 (2000)
4. Bermond, J.-C., Fraigniaud, P., Peters, J.: Antepenultimate broadcasting. Networks 26, 125–137 (1995)
5. Bermond, J.-C., Hell, P., Liestman, A.L., Peters, J.G.: Sparse broadcast graphs. Discrete Appl. Math. 36, 97–130 (1992)
6. Dinneen, M.J., Fellows, M.R., Faber, V.: Algebraic constructions of efficient broadcast networks. In: Mattson, H.F., Rao, T.R.N., Mora, T. (eds.) Applied Algebra, Algebraic Algorithms and Error-Correcting Codes. LNCS, vol. 539, pp. 152–158. Springer, Heidelberg (1991)
7. Doar, M.B.: A better model for generating test networks. In: IEEE GLOBECOM'96, London, IEEE Computer Society Press, Los Alamitos (1996)
8. Elkin, M., Kortsarz, G.: A combinatorial logarithmic approximation algorithm for the directed telephone broadcast problem. In: STOC'02. Proc. of ACM Symp. on Theory of Computing, pp. 438–447 (2002)
9. Elkin, M., Kortsarz, G.: Sublogarithmic approximation for telephone multicast: path out of jungle. In: SODA'03. Proc. of Symposium on Discrete Algorithms, Baltimore, Maryland, pp. 76–85 (2003)
10. Farley, A.M., Hedetniemi, S.T., Proskurowski, A., Mitchell, S.: Minimum broadcast graphs. Discrete Math. 25, 189–193 (1979)
11. Feige, U., Peleg, D., Raghavan, P., Upfal, E.: Randomized broadcast in networks. In: SIGAL'90. Proc. of International Symposium on Algorithms, pp. 128–137 (1990)
12. Fraigniaud, P., Vial, S.: Approximation algorithms for broadcasting and gossiping. J. Parallel and Distrib. Comput. 43(1), 47–55 (1997)
13. Fraigniaud, P., Vial, S.: Heuristic Algorithms for Personalized Communication Problems in Point-to-Point Networks. In: SIROCCO'97. Proc. of the 4th Colloquium on Structural Information and Communication Complexity, pp. 240–252 (1997)
14. Fraigniaud, P., Vial, S.: Comparison of Heuristics for One-to-All and All-to-All Communication in Partial Meshes. Parallel Processing Letters 9(1), 9–20 (1999)
15. Harutyunyan, H.A., Liestman, A.L.: More broadcast graphs. Discrete Math. 98, 81–102 (1999)
16. Harutyunyan, H.A., Shao, B.: An Efficient Heuristic for Broadcasting in Networks. Journal of Parallel and Distributed Computing (to appear)
17. Johnson, D., Garey, M.: Computers and Intractability: A Guide to the Theory of NP-Completeness. Freeman, San Francisco, CA (1979)
18. Kortsarz, G., Peleg, D.: Approximation algorithms for minimum time broadcast. SIAM J. Discrete Math. 8, 401–427 (1995)

19. Ravi, R.: Rapid Rumor Ramification: Approximating the minimum broadcast time. In: FOCS'94. Proc. of 35th Symposium on Foundation of Computer Science, pp. 202–213 (1994)
20. Slater, P.J., Cockayne, E.J., Hedetniemi, S.T.: Information dissemination in trees. SIAM J.Comput. 10(4), 692–701 (1981)
21. Scheuerman, P., Wu, G.: Heuristic Algorithms for Broadcasting in Point-to-Point Computer Network. IEEE Transactions on Computers C-33(9), 804–811 (1984)
22. Zegura, E.W., Calvert, K., Bhattacharjee, S.: How to model an internetwork. In: INFOCOM'96. Proc. The IEEE Conf. on Computer Communications, San Francisco, CA, IEEE Computer Society Press, Los Alamitos (1996)

An Improved Algorithm for Online Unit Clustering

Hamid Zarrabi-Zadeh and Timothy M. Chan

School of Computer Science, University of Waterloo
Waterloo, Ontario, Canada, N2L 3G1
{hzarrabi,tmchan}@uwaterloo.ca

Abstract. We revisit the *online unit clustering* problem in one dimension which we recently introduced at WAOA'06: given a sequence of n points on the line, the objective is to partition the points into a minimum number of subsets, each enclosable by a unit interval. We present a new randomized online algorithm that achieves expected competitive ratio $11/6$ against oblivious adversaries, improving the previous ratio of $15/8$. This immediately leads to improved upper bounds for the problem in two and higher dimensions as well.

1 Introduction

At WAOA'06 [1], we began investigating an online problem we call *unit clustering*, which is extremely simple to state but turns out to be nontrivial surprisingly:

Given a sequence of n points on the real line, assign points to clusters so that each cluster is enclosable by a unit interval, with the objective of minimizing the number of clusters used.

In the offline setting, variations of this problem frequently appear as textbook exercises and can be solved in $O(n \log n)$ time by a simple greedy algorithm (e.g., see [3]). The problem is equivalent to finding the minimum number of points that stab a given collection of unit intervals (i.e., clique partitioning in unit interval graphs, or coloring unit co-interval graphs), and to finding the maximum number of disjoint intervals in a given collection (i.e., maximum independent set in unit interval graphs). It is the one-dimensional analog of an often-studied and important geometric clustering problem—covering a set of points in d dimensions using a minimum number of unit disks (for example, under the Euclidean or L_∞ metric) [5,6,8,11,12]. This geometric problem has applications in facility location, map labeling, image processing, and other areas.

Online versions of clustering and facility location problems are natural to consider because of practical considerations and have been extensively studied in the literature [2,4,10]. Here, input points are given one by one as a sequence over time, and each point should be assigned to a cluster upon its arrival. The main constraint is that clustering decisions are irrevocable: once formed, clusters cannot be removed or broken up.

G. Lin (Ed.): COCOON 2007, LNCS 4598, pp. 383–393, 2007.

For our one-dimensional problem, it is easy to come up with an algorithm with competitive ratio 2; for example, we can use a naïve grid strategy: build a uniform unit grid and simply place each arriving point in the cluster corresponding to the point's grid cell (for the analysis, just observe that every unit interval intersects at most 2 cells). Alternatively, we can use the most obvious greedy strategy: for each given point, open a new cluster only if the point does not "fit" in any existing cluster; this strategy too has competitive ratio 2.

In the previous paper [1], we have shown that it is possible to obtain an online algorithm with expected competitive ratio strictly less than 2 using randomization; specifically, the ratio obtained is at most $15/8 = 1.875$. This result is a pleasant surprise, considering that ratio 2 is known to be tight (among both deterministic and randomized algorithms) for the related *online unit covering* problem [2,1] where the position of each enclosing unit interval is specified upon its creation, and this position cannot be changed later. Ratio 2 is also known to be tight among deterministic algorithms for the problem of online coloring of (arbitrary rather than unit) co-interval graphs [7,9].

In this paper, we improve our previous result further and obtain a randomized online algorithm for one-dimensional unit clustering with expected competitive ratio at most $11/6 \approx 1.8333$. Automatically, this implies improved online algorithms for geometric unit clustering under the L_∞ metric, with ratio $11/3$ in 2D, for example.

The new algorithm is based on the approach from the previous paper but incorporates several additional ideas. A key difference in the design of the algorithm is to make more uses of randomization (the previous algorithm requires only 2 random bits). The previous algorithm is based on a clever grid approach where windows are formed from pairs of adjacent grid cells, and clusters crossing two adjacent windows are "discouraged"; in the new algorithm, crossings of adjacent windows are discouraged to a "lesser" extent, as controlled by randomization. This calls for other subtle changes in the algorithm, as well as a lengthier case analysis that needs further technical innovations.

2   The New Randomized Algorithm

In this section, we present the new randomized algorithm for the online unit clustering problem in one dimension. The competitive ratio of the algorithm is not necessarily less than 2, but will become less than 2 when combined with the naïve grid strategy as described in Section 5. Our new algorithm is based in part on our previous randomized algorithm [1], although we will keep the presentation self-contained. A key difference is to add an extra level of randomization.

Consider a uniform unit grid on the line, where each grid cell is a half-closed interval of the form $[i, i+1)$. To achieve competitive ratio better than 2, we have to allow clusters to cross grid cells occasionally (for example, just consider the input sequence $\langle \frac{1}{2}, \frac{3}{2}, \frac{5}{2}, \ldots \rangle$, where the naïve grid strategy would require twice as many clusters as the optimum). As in the previous algorithm, we accomplish this by forming *windows* over the line each consisting of two grid cells and permit